T0338553

CAMBRIDGE STUDIES IN ADVANCED MATHEMATICS 211

POLYTOPES AND GRAPHS

This book introduces convex polytopes and their graphs, alongside the results and methodologies required to study them. It guides the reader from the basics to current research, presenting many open problems to facilitate the transition. The book includes results not previously found in other books, such as: the edge connectivity and linkedness of graphs of polytopes; the characterisation of their cycle space; the Minkowski decomposition of polytopes from the perspective of geometric graphs; Lei Xue's recent lower bound theorem on the number of faces of polytopes with a small number of vertices; and Gil Kalai's rigidity proof of the lower bound theorem for simplicial polytopes. This accessible introduction covers prerequisites from linear algebra, graph theory, and polytope theory. Each chapter concludes with exercises of varying difficulty, designed to help the reader engage with new concepts. These features make the book ideal for students and researchers new to the field.

Guillermo Pineda Villavicencio is an Associate Professor in Computer Science and Mathematics at Deakin University, Australia, and a Fellow of AdvanceHE. He conducts research on graph theory and discrete geometry, the construction and analysis of large networks, and applications of mathematics to health informatics. He is an Accredited Member of the Australian Mathematical Society and served on its Council from 2018 to 2022. He is also a Life Member of the Combinatorial Mathematics Society of Australasia.

CAMBRIDGE STUDIES IN ADVANCED MATHEMATICS

All the titles listed below can be obtained from good booksellers or from Cambridge University Press. For a complete series listing, visit www.cambridge.org/mathematics.

Already Published

171 J. Gough & J. Kupsch *Quantum Fields and Processes*
172 T. Ceccherini-Silberstein, F. Scarabotti & F. Tolli *Discrete Harmonic Analysis*
173 P. Garrett *Modern Analysis of Automorphic Forms by Example, I*
174 P. Garrett *Modern Analysis of Automorphic Forms by Example, II*
175 G. Navarro *Character Theory and the McKay Conjecture*
176 P. Fleig, H. P. A. Gustafsson, A. Kleinschmidt & D. Persson *Eisenstein Series and Automorphic Representations*
177 E. Peterson *Formal Geometry and Bordism Operators*
178 A. Ogus *Lectures on Logarithmic Algebraic Geometry*
179 N. Nikolski *Hardy Spaces*
180 D.-C. Cisinski *Higher Categories and Homotopical Algebra*
181 A. Agrachev, D. Barilari & U. Boscain *A Comprehensive Introduction to Sub-Riemannian Geometry*
182 N. Nikolski *Toeplitz Matrices and Operators*
183 A. Yekutieli *Derived Categories*
184 C. Demeter *Fourier Restriction, Decoupling and Applications*
185 D. Barnes & C. Roitzheim *Foundations of Stable Homotopy Theory*
186 V. Vasyunin & A. Volberg *The Bellman Function Technique in Harmonic Analysis*
187 M. Geck & G. Malle *The Character Theory of Finite Groups of Lie Type*
188 B. Richter *Category Theory for Homotopy Theory*
189 R. Willett & G. Yu *Higher Index Theory*
190 A. Bobrowski *Generators of Markov Chains*
191 D. Cao, S. Peng & S. Yan *Singularly Perturbed Methods for Nonlinear Elliptic Problems*
192 E. Kowalski *An Introduction to Probabilistic Number Theory*
193 V. Gorin *Lectures on Random Lozenge Tilings*
194 E. Riehl & D. Verity *Elements of ∞-Category Theory*
195 H. Krause *Homological Theory of Representations*
196 F. Durand & D. Perrin *Dimension Groups and Dynamical Systems*
197 A. Sheffer *Polynomial Methods and Incidence Theory*
198 T. Dobson, A. Malnič & D. Marušič *Symmetry in Graphs*
199 K. S. Kedlaya *p-adic Differential Equations*
200 R. L. Frank, A. Laptev & T. Weidl *Schrödinger Operators:Eigenvalues and Lieb–Thirring Inequalities*
201 J. van Neerven *Functional Analysis*
202 A. Schmeding *An Introduction to Infinite-Dimensional Differential Geometry*
203 F. Cabello Sánchez & J. M. F. Castillo *Homological Methods in Banach Space Theory*
204 G. P. Paternain, M. Salo & G. Uhlmann *Geometric Inverse Problems*
205 V. Platonov, A. Rapinchuk & I. Rapinchuk *Algebraic Groups and Number Theory, I (2nd Edition)*
206 D. Huybrechts *The Geometry of Cubic Hypersurfaces*
207 F. Maggi *Optimal Mass Transport on Euclidean Spaces*
208 R. P. Stanley *Enumerative Combinatorics, II (2nd edition)*
209 M. Kawakita *Complex Algebraic Threefolds*
210 D. Anderson & W. Fulton *Equivariant Cohomology in Algebraic Geometry*

Polytopes and Graphs

GUILLERMO PINEDA VILLAVICENCIO
Deakin University, Australia

Shaftesbury Road, Cambridge CB2 8EA, United Kingdom

One Liberty Plaza, 20th Floor, New York, NY 10006, USA

477 Williamstown Road, Port Melbourne, VIC 3207, Australia

314–321, 3rd Floor, Plot 3, Splendor Forum, Jasola District Centre, New Delhi – 110025, India

103 Penang Road, #05–06/07, Visioncrest Commercial, Singapore 238467

Cambridge University Press is part of Cambridge University Press & Assessment, a department of the University of Cambridge.
We share the University's mission to contribute to society through the pursuit of education, learning and research at the highest international levels of excellence.

www.cambridge.org
Information on this title: www.cambridge.org/9781009257817
DOI: 10.1017/9781009257794

First published 2024

A catalogue record for this publication is available from the British Library

A Cataloging-in-Publication data record for this book is available from the Library of Congress

ISBN 978-1-009-25781-7 Hardback

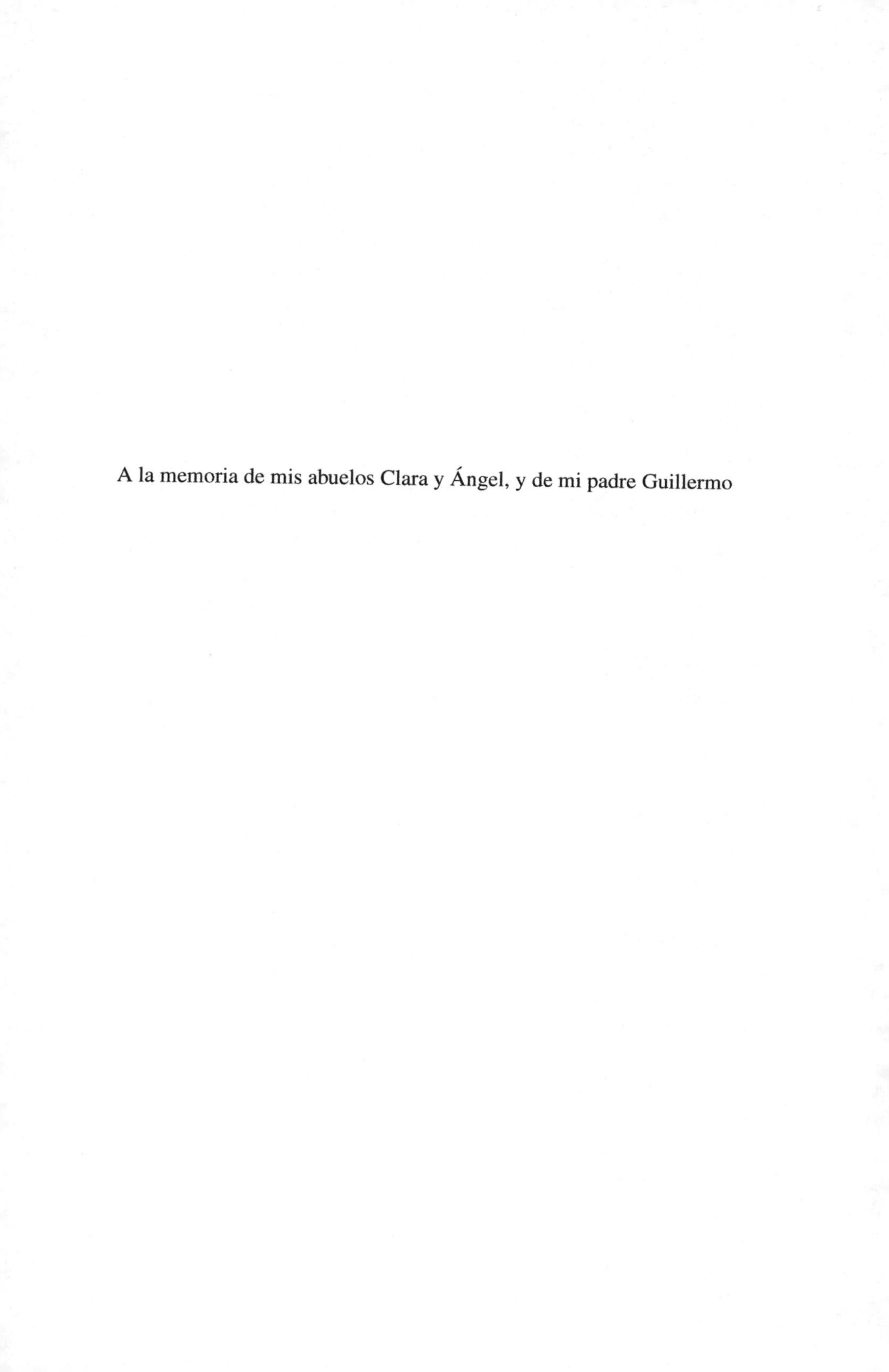

A la memoria de mis abuelos Clara y Ángel, y de mi padre Guillermo

Contents

Preface

This book introduces convex polytopes and graphs the arise from them, alongside the ideas, results, and methodologies from graph theory, convexity, and polytope theory necessary to study these objects. We consider only convex polytopes, and so henceforth we drop the adjective 'convex'. The book grew out of several seminars I gave to colleagues and students at Federation University Australia between 2015 and 2018. During the seminars, it became apparent that research in polytopes and their graphs required an interdisciplinary approach; we needed a solid background on all these three branches of mathematics. This book aspires to give that background, with a focus on active research topics.

The study of polytopes and their graphs has drawn on and contributed to many branches of mathematics such as graph theory, convexity, optimisation, and polytope theory. A case in point is optimisation. Polytopes and, more generally, polyhedra are connected to optimisation via solution spaces of problems in linear programming and via algorithms for these problems such as the simplex method of George B. Dantzig in 1947; see Dantzig (1963). Much beautiful mathematics has been generated through this connection, as featured in the works of Victor Klee, Gil Kalai, and Francisco Santos related to the 1957 conjecture of Warren M. Hirsch and the subsequent Polynomial Hirsch Conjecture. Hirsch originally stated his conjecture in a letter to George B. Dantzig (1963, sec. 7.4). Both conjectures concern lower bounds for the number of steps needed in an application of the simplex method to the graph of the polyhedron in question.

Treatments of polytopes and their graphs have appeared in the literature before, mostly in books on discrete geometry or polytope theory (Brøndsted, 1983; Gruber, 2007; Grünbaum, 2003; Ziegler, 1995). The most comprehensive treatment is that of Grünbaum (2003) followed by that of Ziegler (1995). We have benefitted greatly from these sources, as well as from books on graph theory such as those of Bondy and Murty (2008) and Diestel (2017), and from books on convexity such as that of Webster (1994). Consequently, for many results in the book, especially in Chapters 1 to 3 and the appendices, when we believe that an accessible proof is available elsewhere, we reference it in a footnote in the text.

To promote research in graphs of polytopes, we often compare, contrast, and contextualise results in graph theory. The main idea is to discern to what degree graph-theoretical properties and theorems from graph theory are meaningful and important when restricted to graphs of polytopes. We have highlighted open questions and unexplored areas that seem attractive to us, without trying to be encyclopaedic. Open questions at the end of a chapter are marked with an asterisk and are all compiled in Appendix A.

Each chapter also includes a postscript, highlighting bibliographical notes and further reading. The book ends with a detailed index and a separate index of symbols.

Prerequisites

The book is aimed at undergraduate and graduate students, as well as any researcher interested in the fascinating world of graphs of polytopes. It assumes elementary knowledge of convexity, linear algebra, graph theory, and polytope theory. But the first two chapters and the appendices review and explain the relevant basic concepts from these branches.

How to Use the Book

The amount of material in this book exceeds a realistic plan for a one-semester or two-semester course on convexity, polytope theory, or graphs of polytopes. Chapter 1 reviews the basic concepts on convex sets, while Chapter 2 offers foundational material on polytopes and polyhedra in general. Appendix C is there to provide a basic background on graphs. A course based on the book could include as much or as little material as necessary from these two chapters and the appendix. After that, several study paths are possible, as the figure illustrates.

A one-semester or two-semester course in polytope theory may include material from Chapters 1 to 3 to provide a background on polytopes and enable a proof of Steinitz's theorem. After studying Steinitz's theorem, one may move to Chapter 4, Chapter 7, or Chapter 8. Courses on graphs of polytopes may include Chapters 1 to 4 and any of Chapters 5, 6, and 7. Another course on

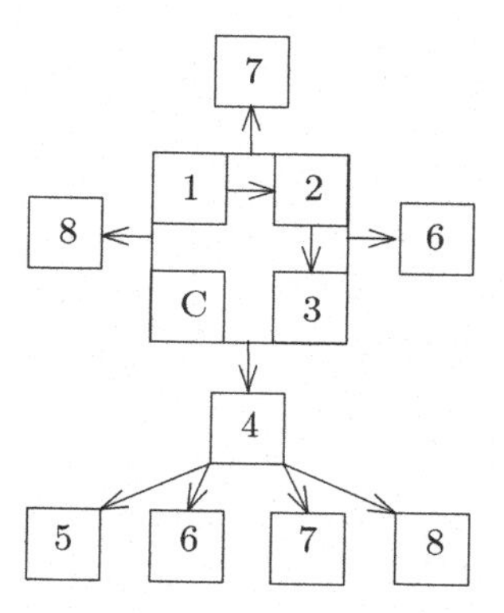

graphs of polytopes may consist of Chapters 1, 2, 3, and 7. Chapters 5 to 8 are mostly independent and can be read in an arbitrary order.

Acknowledgements

I would like to express my gratitude to several people who have assisted me in this project: those who have read little or much of the drafts, those who have provided helpful comments, and those who have given much spiritual support. Many thanks to Michael Joswig and Eran Nevo for inspiring discussions on the topics of the book. Special thanks also to David Yost and Reinier Díaz Millán for reading many chapters and offering invaluable feedback, and to Julien Ugon for helping to present my own proofs, as well as for the many related (and unrelated) conversations. Finally, heartfelt thanks to my family: Clara Aurora Moreno, Antonio (Ángel) Villavicencio, María de los Ángeles Villavicencio, Guillermo Pineda Calás, Clari Merci Pineda, Ramón Lester Turiño, Sharon Ryba-Kahn, and Fleur Pitman.

Feedback

I would appreciate your feedback. Feel free to reach me by email at work@guillermo.com.au.

1

Introduction

Convex polytopes, or simply polytopes, are geometric objects in some space $\mathbb{R}^d$; in fact, they are bounded intersections of finitely many closed halfspaces in $\mathbb{R}^d$. The space $\mathbb{R}^d$ can be regarded as a linear space or an affine space, and its linear or affine subspaces can be described by linear or affine equations. We introduce the basic concepts and results from linear algebra that allow the description and analysis of these subspaces. In particular, we describe the embedding of an affine space into a larger linear space, which often results in a clearer perspective of the initial space. In Chapter 2, we will embed an affine space into a projective space and so projective spaces are introduced in preparation for that embedding. Spaces are fundamental concepts in linear algebra, as are the maps that preserve their inherent structure; Section 1.4 revisits these maps. The principle of duality will surface often in this book, starting with an overview of dual spaces in Section 1.5 and dual sets in Section 1.11.

A polytope can alternatively be described as the convex hull of a finite set of points in $\mathbb{R}^d$ and so it is a convex set. Convex sets are therefore introduced, as well as the basic theorems of Carathéodory and Radon. Section 1.7 revisits topological properties of convex sets, with an emphasis on relative notions as these are based on a more natural setting, the affine hull of the set. We then review the separation and support of convex sets by hyperplanes. A convex set is formed by fitting together other polytopes of smaller dimensions, its faces; Section 1.9 discusses them. Finally, the chapter studies convex cones and lineality spaces of convex sets in $\mathbb{R}^d$; these sets are closely connected to the structure of unbounded convex sets.

1.1 Subspaces

From the outset, we want to make clear that all the spaces considered in this book are of finite dimension and 'concrete'; their underlying fields of scalars are 'concrete': they are either the set $\mathbb{Q}$ of rational numbers, the set $\mathbb{R}$ of real numbers, or in rare occasions the set $\mathbb{C}$ of complex numbers.

The set of all d-tuples $(x_1,\dots,x_d)$ with entries in $\mathbb{R}$ defines the *d-dimensional real linear space* $\mathbb{R}^d$ when endowed with vector addition and scalar multiplication. The elements of $\mathbb{R}^d$, called *vectors*, are always represented as column vectors, written in bold as $\boldsymbol{x},\boldsymbol{y},\boldsymbol{z}$. When the vector coordinates are required, we will then write $\boldsymbol{x}=(x_1,\dots,x_d)^t$ for typographical reasons, where X^t denotes the transpose of a matrix X. The all-one vector and the all-zero vector in $\mathbb{R}^d$ are denoted by $\mathbf{1}_d$ and $\mathbf{0}_d$, respectively; when there is no opportunity for confusion we write simply $\mathbf{0}$ and $\mathbf{1}$.

A *linear subspace* L of $\mathbb{R}^d$ is a nonempty subset that contains the linear combination of any two of its vectors; that is, a set of the form

$$L := \{\alpha_1\boldsymbol{l}_1+\alpha_2\boldsymbol{l}_2 \mid \alpha_1,\alpha_2\in\mathbb{R} \text{ and } \boldsymbol{l}_1,\boldsymbol{l}_2\in L\}; \qquad (1.1.1)$$

here $x := y$ and $y =: x$ define x as an object equal to y. The expression $\alpha_1\boldsymbol{l}_1+\alpha_2\boldsymbol{l}_2$ is the *linear combination* of the vectors $\boldsymbol{l}_1$ and $\boldsymbol{l}_2$.

A point $\boldsymbol{x}$ in $\mathbb{R}^d$ lies in a line ℓ through distinct points $\boldsymbol{a}_1,\boldsymbol{a}_2\in\mathbb{R}^d$ if $\boldsymbol{x}$ can be expressed as $\alpha_1\boldsymbol{a}_1+\alpha_2\boldsymbol{a}_2$ for some scalars $\alpha_1,\alpha_2\in\mathbb{R}$ satisfying $\alpha_1+\alpha_2=1$. The line ℓ can be defined as

$$\ell := \{\alpha_1\boldsymbol{a}_1+\alpha_2\boldsymbol{a}_2 \mid \alpha_1,\alpha_2\in\mathbb{R},\ \alpha_1+\alpha_2=1\}.$$

In the particular case that the line ℓ passes through the origin, ℓ can be defined as $\ell := \{\alpha\boldsymbol{a} \mid \alpha\in\mathbb{R},\ \boldsymbol{a}\in\ell\}$.

Example 1.1.2 (Linear subspaces) The following are examples of linear spaces.

(i) A line in $\mathbb{R}^d$ through the origin is a linear subspace of $\mathbb{R}^d$.
(ii) The set $\mathbb{R}^d$ is a linear subspace of $\mathbb{R}^d$.
(iii) The smallest linear space is $\{\mathbf{0}\}$.
(iv) The set of solutions of a system of homogeneous linear equations in d unknowns is a linear subspace of $\mathbb{R}^d$:

$$L := \left\{ \boldsymbol{x} = \begin{pmatrix} x_1 \\ \vdots \\ x_d \end{pmatrix} \;\middle|\; \begin{array}{ccc} \alpha_{1,1}x_1+\cdots+\alpha_{1,d}x_d & = & 0 \\ & \vdots & \\ \alpha_{n,1}x_1+\cdots+\alpha_{n,d}x_d & = & 0 \end{array} \right\}.$$

A subset $L\subseteq\mathbb{R}^d$ is a linear subspace of $\mathbb{R}^d$ if and only if it is the solution set of a system of homogeneous linear equations in d unknowns (Problem 1.12.2).
(v) The set $\mathbb{R}^{d\times n}$ of $d\times n$ matrices with entries in $\mathbb{R}$ is a linear space with the usual matrix addition and scalar multiplication.

If we want to emphasise the affine properties of $\mathbb{R}^d$, we refer to it as the *d-dimensional real affine space* $\mathbb{A}^d$ and call its elements *points*. As with vectors, our points are always column vectors, written in bold.

A subset A of $\mathbb{R}^d$ is an *affine subspace* if it contains the line between any two of its elements:

$$A = \{\alpha_1 \boldsymbol{a}_1 + \alpha_2 \boldsymbol{a}_2 \mid \alpha_1, \alpha_2 \in \mathbb{R}, \alpha_1 + \alpha_2 = 1, \text{ and } \boldsymbol{a}_1, \boldsymbol{a}_2 \in A\}. \tag{1.1.3}$$

The expression $\alpha_1 \boldsymbol{a}_1 + \alpha_2 \boldsymbol{a}_2$, where $\alpha_1, \alpha_2 \in \mathbb{R}$, $\alpha_1 + \alpha_2 = 1$, and $\boldsymbol{a}_1, \boldsymbol{a}_2 \in A$ is called the *affine combination* of the points $\boldsymbol{a}_1$ and $\boldsymbol{a}_2$.

Example 1.1.4 (Affine subspaces) The following are examples of affine spaces.

(i) A line in $\mathbb{R}^d$ is an affine subspace of $\mathbb{R}^d$.
(ii) The set $\mathbb{R}^d$ is an affine subspace of $\mathbb{R}^d$.
(iii) The smallest affine space is $\varnothing$.
(iv) A set of solutions of a system of nonhomogeneous linear equations in d unknowns is an affine subspace of $\mathbb{R}^d$:

$$A := \left\{ \boldsymbol{x} = \begin{pmatrix} x_1 \\ \vdots \\ x_d \end{pmatrix} \middle| \begin{array}{rcl} \alpha_{1,1}x_1 + \cdots + \alpha_{1,d}x_d & = & b_1 \\ & \vdots & \\ \alpha_{n,1}x_1 + \cdots + \alpha_{n,d}x_d & = & b_d \end{array} \right\}.$$

A subset $A \subseteq \mathbb{R}^d$ is an affine subspace of $\mathbb{R}^d$ if and only if it is the solution set of a system of nonhomogeneous linear equations in d unknowns (Problem 1.12.3).

Expressions (1.1.1) and (1.1.3) suggest a close relation between linear subspaces and nonempty affine subpaces. A nonempty affine subspace A is a *translate* $\boldsymbol{a}_0 + L$ of a linear subspace L by a point $\boldsymbol{a}_0 \in A$. Fixing any point $\boldsymbol{a}_0$ of A and translating any point of A by $-\boldsymbol{a}_0$ gives the linear space $L := \{\boldsymbol{l} \mid \boldsymbol{a}_0 + \boldsymbol{l} \in A\}$. In this setting, the point $\boldsymbol{a}_0$ plays the role of the origin of the subspace. We call the subspace L the *direction* of A and denote it by $\overrightarrow{A}$. Two nonempty affine subspaces are *parallel* if they have the same direction. Often we present a nonempty affine subspace A as the pair $(A, \overrightarrow{A})$ to keep track of its direction.

Theorem 1.1.5[1] *The nonempty affine spaces in $\mathbb{R}^d$ are precisely the translates of linear subspaces of $\mathbb{R}^d$. Furthermore, the direction of each nonempty affine subspace is unique.*

[1] A proof is available in Webster (1994, thm. 1.2.1).

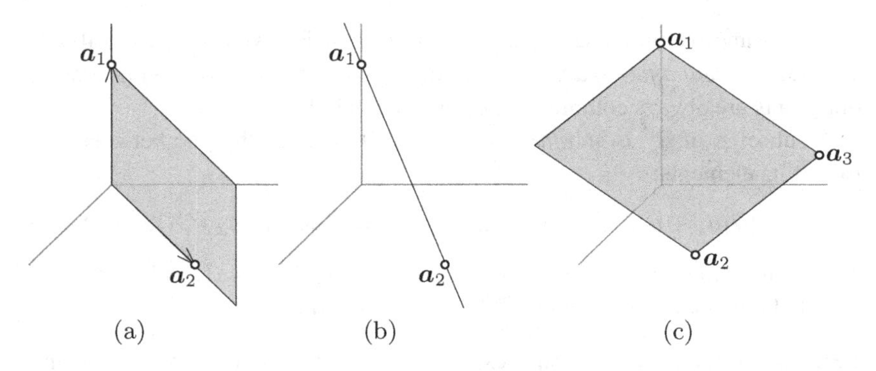

Figure 1.1.1 Affine and linear hulls in $\mathbb{R}^3$. (a) The linear hull of the points $\boldsymbol{a}_1$ and $\boldsymbol{a}_2$, which is a plane in $\mathbb{R}^3$. (b) The affine hull of the points $\boldsymbol{a}_1$ and $\boldsymbol{a}_2$, which is a line in $\mathbb{R}^3$. (c) The affine hull of the points $\boldsymbol{a}_1$, $\boldsymbol{a}_2$, and $\boldsymbol{a}_3$, which is a plane in $\mathbb{R}^3$.

It is obvious that any linear subspace admits an affine structure where the linear space itself acts as its direction. In view of this, it may not always be possible to make a clear distinction between points and vectors of a linear subspace.

Given a subset X of $\mathbb{R}^d$, it is often convenient to describe the smallest linear or affine subspace containing it. These new sets are the *linear hull* and *affine hull* of the set X, respectively. The linear hull is also called the *linear span* and, similarly, the affine hull is also called the *affine span*. The hulls of a set X can be described as the set of combinations of *finitely* many elements of X. Notationally,

$$\operatorname{lin} X := \{\alpha_1 \boldsymbol{a}_1 + \cdots + \alpha_n \boldsymbol{a}_n \mid \boldsymbol{a}_i \in X, \text{ and } \alpha_i \in \mathbb{R}\}.$$

$$\operatorname{aff} X := \left\{ \alpha_1 \boldsymbol{a}_1 + \cdots + \alpha_n \boldsymbol{a}_n \,\middle|\, n \geqslant 1, \boldsymbol{a}_i \in X, \alpha_i \in \mathbb{R}, \text{ and } \sum_{i=1}^{n} \alpha_i = 1 \right\}.$$

Note that $\operatorname{lin} \varnothing = \{\boldsymbol{0}\} = \operatorname{aff}(\varnothing \cup \{\boldsymbol{0}\})$, while $\operatorname{aff} \varnothing = \varnothing$. Figure 1.1.1 depicts examples of affine and linear hulls.

Dependence, Bases, and Dimension

Let $\boldsymbol{x}, \boldsymbol{y}$ be vectors or points of $\mathbb{R}^d$. We say that

$$(x_1, \ldots, x_d)^t = \boldsymbol{x} \leqslant \boldsymbol{y} = (y_1, \ldots, y_d)^t$$

if $x_i \leqslant y_i$ for every $1 \leqslant i \leqslant d$. Similarly, we define the relations $\boldsymbol{x} \geqslant \boldsymbol{y}$, $\boldsymbol{x} > \boldsymbol{y}$, $\boldsymbol{x} < \boldsymbol{y}$, and $\boldsymbol{x} = \boldsymbol{y}$. In contrast, the relation $\boldsymbol{x} \neq \boldsymbol{y}$ means that $x_i \neq y_i$, for some $1 \leqslant i \leqslant d$.

A set $L = \{\boldsymbol{l}_1, \ldots, \boldsymbol{l}_n\}$ of vectors in $\mathbb{R}^d$ is said to be *linearly dependent* if and only if there exist distinct scalars $\alpha_1, \ldots, \alpha_n$, not all zero, such that

$$\alpha_1 \boldsymbol{l}_1 + \cdots + \alpha_n \boldsymbol{l}_n = \boldsymbol{0}_d.$$

If this equation is satisfied only when $\alpha_1 = \cdots = \alpha_n = 0$, the set is said to be *linearly independent*. A set $A = \{\boldsymbol{a}_1, \ldots, \boldsymbol{a}_n\}$ of points in $\mathbb{R}^d$ is said to be *affinely dependent* if and only if there exist distinct scalars $\alpha_1, \ldots, \alpha_n$, not all zero, such that

$$\alpha_1 \boldsymbol{a}_1 + \cdots + \alpha_n \boldsymbol{a}_n = \boldsymbol{0}_d \text{ and } \alpha_1 + \cdots + \alpha_n = 0.$$

If these equations are simultaneously satisfied only when $\alpha_1 = \cdots = \alpha_n = 0$, the set is said to be *affinely independent*.

A *linear basis* of a linear subspace L is a linearly independent set whose linear span is L, while an *affine basis* of an affine subspace A is an affinely independent set whose affine span is A. If an affine space A is written as $\boldsymbol{a}_0 + \vec{A}$, with $\boldsymbol{a}_0 \in A$, and the set $\{\boldsymbol{l}_1, \ldots, \boldsymbol{l}_n\}$ is a linear basis of $\vec{A}$, then the set $\{\boldsymbol{a}_0, \boldsymbol{a}_0 + \boldsymbol{l}_1, \ldots, \boldsymbol{a}_0 + \boldsymbol{l}_n\}$ is an affine basis of A. Each point of a subspace can be expressed uniquely as an affine or linear combination of the elements of a basis of the subspace.

The *dimension* of a linear space L, denoted $\dim L$, is defined as the number of elements of any of its bases, while the *dimension* of a nonempty affine subspace is defined as the dimension of its direction. This definition ensures that when an affine space is a linear subspace, its affine dimension coincides with its linear dimension. The dimension of the empty affine subspace is -1.

The *dot product* $\cdot$ of two vectors $\boldsymbol{x} = (x_1, \ldots, x_d)^t$ and $\boldsymbol{y} = (y_1, \ldots, y_d)^t$ is defined as

$$\boldsymbol{x} \cdot \boldsymbol{y} := x_1 y_1 + \cdots + x_d y_d.$$

If the linear space $\mathbb{R}^d$ is equipped with a dot product, then it becomes a *Euclidean space*, the *d-dimensional Euclidean space*.

Example 1.1.6 recalls the definitions of five important subspaces.

Example 1.1.6 (Five important linear subspaces) Let M be an $n \times d$ matrix. The following subspaces can be defined from M.

(i) The (right) *nullspace* $\operatorname{null} M$ of M is the set of solutions of the homogeneous system $M\boldsymbol{x} = \boldsymbol{0}$. Notationally,

$$\operatorname{null} M := \left\{ \boldsymbol{x} \in \mathbb{R}^d \middle| \; M\boldsymbol{x} = \boldsymbol{0} \right\}. \tag{1.1.6.1}$$

(ii) The *left nullspace* of M is the nullspace of the transpose of M.

(iii) The *column space* $\operatorname{col} M$ of M is the subspace of $\mathbb{R}^n$ spanned by the columns of M, seen as vectors in $\mathbb{R}^n$. Notationally,

$$\operatorname{col} M := \{\boldsymbol{b} \in \mathbb{R}^n \mid \text{The system } M\boldsymbol{x} = \boldsymbol{b} \text{ is solvable}\}. \tag{1.1.6.2}$$

(iv) The *row space* $\operatorname{row} M$ of M is the subspace of $\mathbb{R}^d$ spanned by the rows of M, seen as vectors in $\mathbb{R}^d$. Notationally,

$$\operatorname{row} M := \left\{\boldsymbol{b} \in \mathbb{R}^d \,\middle|\, \text{The system } M^t\boldsymbol{x} = \boldsymbol{b} \text{ is solvable}\right\}. \tag{1.1.6.3}$$

The dimension of $\operatorname{row} M$ equals the dimension of $\operatorname{col} M$, and either is called the *rank* of M.

(v) We say that two vectors $\boldsymbol{x}$ and $\boldsymbol{y}$ are *orthogonal* if $\boldsymbol{x} \cdot \boldsymbol{y} = 0$. Define the *orthogonal complement* of any linear subspace L as the set of vectors orthogonal to every vector in L. Notationally,

$$L^{\perp} := \left\{\boldsymbol{x} \in \mathbb{R}^d \,\middle|\, \boldsymbol{x} \cdot \boldsymbol{l} = 0 \text{ for every } \boldsymbol{l} \in L\right\}. \tag{1.1.6.4}$$

The nullspace of M coincides with the orthogonal complement of its row space, whereas the left nullspace of M coincides with the orthogonal complement of its column space. These five subspaces are closely related, as Problems 1.12.4 and 1.12.5 attest.

Two linear subspaces L and L' of $\mathbb{R}^d$ are *orthogonal* if every vector of L is orthogonal to every vector of L'. A linear space and its orthogonal complement are clearly orthogonal. Two affine spaces are *orthogonal* if their corresponding directions are orthogonal.

We revisit the link between linear and affine spaces, and solutions to linear equations.

Example 1.1.7 (Subspaces and linear equations) In this example, given a linear subspace L and an affine subspace A of $\mathbb{R}^d$, we describe them as solutions of linear equations.

(i) (Linear case) Let $\{\boldsymbol{l}_1, \ldots, \boldsymbol{l}_n\}$ be a set of vectors in $\mathbb{R}^d$ that form a basis of $L^{\perp}$. Letting M be the $n \times d$ matrix with the vectors $\boldsymbol{l}_1, \ldots, \boldsymbol{l}_n$ as rows, we find that L is the set of solutions of the homogeneous system $M\boldsymbol{x} = \boldsymbol{0}$. Notationally,

$$M := \begin{pmatrix} \boldsymbol{l}_1^t \\ \vdots \\ \boldsymbol{l}_n^t \end{pmatrix} \text{ and } L := \left\{\boldsymbol{x} \in \mathbb{R}^d \,\middle|\, M\boldsymbol{x} = \boldsymbol{0}\right\}.$$

(ii) (Affine case) Pick any point $\boldsymbol{a}_0 \in A$ and write $A = \boldsymbol{a}_0 + \vec{A}$ as the translation of its direction by $\boldsymbol{a}_0$. Represent $\vec{A}$ as the set of solutions of a homogeneous system $M\boldsymbol{x} = \boldsymbol{0}$ for some $n \times d$ matrix M, with $n \leqslant d$. Then

$$A = \left\{ \boldsymbol{x} \in \mathbb{R}^d \,\middle|\, M\boldsymbol{x} = \boldsymbol{b} \text{ with } \boldsymbol{b} = M\boldsymbol{a}_0 \right\}.$$

Examples 1.1.4, 1.1.6, and 1.1.7 contain all the ingredients to derive expressions for the dimensions of affine and linear spaces when given as solutions of systems of linear equations.

Proposition 1.1.8 *Let L and A be a linear subspace and an affine subspace of $\mathbb{R}^d$, respectively, given as solutions of systems of linear equations*

$$L = \left\{ \boldsymbol{x} \in \mathbb{R}^d \,\middle|\, M\boldsymbol{x} = \boldsymbol{0} \right\} \text{ and } A = \left\{ \boldsymbol{x} \in \mathbb{R}^d \,\middle|\, M\boldsymbol{x} = \boldsymbol{b} \right\},$$

for some $M \in \mathbb{R}^{n \times d}$ and $\boldsymbol{b} \in \mathbb{R}^n$. Then L is the direction of A and $\dim L = \dim A = d - \operatorname{rank} M$.

Interval Notation

The interval

$$[x, y] = \{z \in \mathbb{R} \mid x \leqslant z \leqslant y\}$$

is the set of real numbers between x and y. Similarly, we define the open interval (x, y) and the halfopen intervals $(x, y]$ and $[x, y)$ of real numbers from x to y. We will often consider intervals on the set $\mathbb{Z}$ of integers. The interval

$$[x...y] = \{z \in \mathbb{Z} \mid x \leqslant z \leqslant y\}$$

is the set of integers between x and y. Similarly, we define the open interval $(x...y)$ and the halfopen intervals $(x...y]$ and $[x \ldots y)$ of integers from x to y.

Sets in General Position

Another demand that will be imposed regularly on a subset of $\mathbb{R}^d$ is that of being in 'general position'. The precise meaning will depend on the nature of the underlying space, but the guiding principle is that the number of elements from the set in any hyperplane does not exceed the number of elements in any basis of the hyperplane. We say that a set of at least $d + 1$ points in $\mathbb{R}^d$ is in *general position* if and only if no $d + 1$ points of the set lie in an (affine) hyperplane; in other words, if and only if every subset of at most $d+1$ points is affinely independent. Similarly, we say that a set of at least d vectors in $\mathbb{R}^d$ is in *general position* if and only if no d vectors of the set lie in a linear hyperplane: if and only if every subset of at most d vectors is linearly independent.

An affine subspace in $\mathbb{R}^d$ of dimension $d-1$ is an *affine hyperplane* or just a *hyperplane*; this is geometrically defined as

$$H_d(\boldsymbol{a},\alpha) := \left\{ \boldsymbol{x} \in \mathbb{R}^d \,\middle|\, \boldsymbol{a} \cdot \boldsymbol{x} = \alpha,\, \alpha \in \mathbb{R},\, \boldsymbol{a} \in \mathbb{R}^d \right\},$$

where $\boldsymbol{a}$ is a nonzero vector called the *normal* of the hyperplane. If $\alpha = 0$, the hyperplane is a linear subspace of dimension $d-1$ that we call a *linear hyperplane*.

1.2 Embedding Affine Spaces into Linear Spaces

In Section 1.1 we defined a nonempty r-dimensional affine subspace A of $\mathbb{A}^d$ as a set of the form $\boldsymbol{a}_0 + \vec{A}$ for some arbitrary element of $\boldsymbol{a}_0 \in A$ and a unique r-dimensional linear subspace $\vec{A}$. Here, we present A in another way, as a subspace of $\mathbb{R}^{d+1}$.

Every nonlinear hyperplane of $\mathbb{R}^{d+1}$ can be thought of as $\mathbb{A}^d$. Consider the concrete nonlinear hyperplane

$$H_A := \left\{ (x_1, \ldots, x_{d+1})^t \in \mathbb{R}^{d+1} \,\middle|\, x_{d+1} = 1 \right\}$$

and an arbitrary point $\boldsymbol{a}_0 = (x_1, \ldots, x_d, 1)^t$ in H_A. The selection of $\boldsymbol{a}_0$ identifies the hyperplane H_A with its direction

$$\vec{H}_A := \left\{ (x_1, \ldots, x_{d+1})^t \in \mathbb{R}^{d+1} \,\middle|\, x_{d+1} = 0 \right\}.$$

To every point $\boldsymbol{a} \in H_A$ there corresponds the vector $\boldsymbol{a}_0 + \boldsymbol{l}$ in $\mathbb{R}^{d+1}$ for some vector $\boldsymbol{l} \in \vec{H}_A$; that is, for an arbitrary $\boldsymbol{a}_0 \in H_A$ there is a bijection σ from H_A to the set of vectors in $\mathbb{R}^{d+1}$ with $x_{d+1} = 1$, which is defined as

$$\sigma(\boldsymbol{a}) = \boldsymbol{a}_0 + \boldsymbol{l}.$$

This embedding of $\mathbb{A}^d$ and its direction into $\mathbb{R}^{d+1}$ is called a *homogenisation* of $\mathbb{A}^d$. We may also say that $\mathbb{R}^{d+1}$ is a homogenisation of $\mathbb{A}^d$. See Fig. 1.2.1.

An *isomorphism* between linear spaces is a bijection that preserves vector addition and scalar multiplication; isomorphic spaces can be regarded as the *same* space, with the difference residing only in the nature or labelling of the vectors. Every d-dimensional linear or affine space over $\mathbb{R}$ is isomorphic to $\mathbb{R}^d$.

Starting from an affine subspace A and its direction $\vec{A}$, we can construct a linear subspace $\widehat{A}$ that contains both A and $\vec{A}$ and is unique up to isomorphism. As a consequence, we talk about the homogenisation $\widehat{A}$ of A. For one construction of the linear subspace $\widehat{A}$, check, for instance, Gallier (2011, ch. 4).

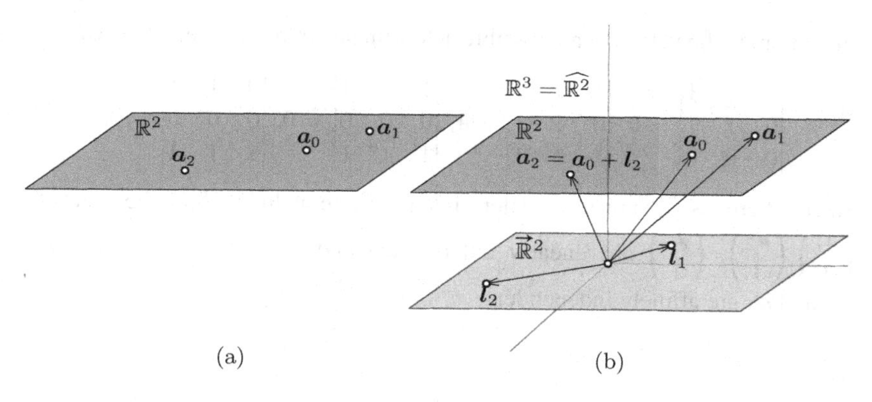

Figure 1.2.1 Homogenisation of the affine space $\mathbb{R}^2$. (a) The affine space $\mathbb{R}^2$. (b) The affine space $\mathbb{R}^2$ and its direction $\vec{\mathbb{R}}^2$ as subspaces of $\mathbb{R}^3$.

In the homogenisation of $\mathbb{A}^d$, a k-dimensional affine subspace A of $\mathbb{A}^d$ corresponds to a $(k+1)$-dimensional linear subspace $\widehat{A}$ of $\mathbb{R}^{d+1}$ spanned by an arbitrary point $\boldsymbol{a}_0$ in A and k linearly independent vectors $\boldsymbol{l}_1, \ldots, \boldsymbol{l}_k$ spanning the direction $\vec{A}$ of A in $\vec{H}_A$; that is, $A = H_A \cap \widehat{A}$ and $\vec{A} = \vec{H}_A \cap \widehat{A}$. Equivalently, we can see that $\widehat{A}$ is spanned by the $k+1$ linearly independent vectors $\boldsymbol{a}_0, \boldsymbol{a}_0 + \boldsymbol{l}_1, \ldots, \boldsymbol{a}_0 + \boldsymbol{l}_k$ and A is spanned by the $k+1$ affinely independent points $\sigma^{-1}(\boldsymbol{a}_0), \sigma^{-1}(\boldsymbol{a}_0 + \boldsymbol{l}_1), \ldots, \sigma^{-1}(\boldsymbol{a}_0 + \boldsymbol{l}_k)$. Affine properties of H_A then reduce to linear properties of $\mathbb{R}^{d+1}$. For instance, a set $\{\boldsymbol{a}_1, \ldots, \boldsymbol{a}_{k+1}\}$ of points in H_A is an affine basis of A if and only if the corresponding set $\{\sigma(\boldsymbol{a}_1), \ldots, \sigma(\boldsymbol{a}_{k+1})\}$ of vectors in $\mathbb{R}^{d+1}$ is a linear basis of $\widehat{A}$. Example 1.2.1 uses this embedding to give a concrete test for affine independence.

Example 1.2.1 (Criterion for checking affine independence) Let a set of affine points $\boldsymbol{a}_1, \ldots, \boldsymbol{a}_n$ in $\mathbb{A}^d$ be given, with $n \leqslant d+1$. Form the n vectors

$$\begin{pmatrix} \boldsymbol{a}_1 \\ 1 \end{pmatrix}, \ldots, \begin{pmatrix} \boldsymbol{a}_n \\ 1 \end{pmatrix}$$

and test their linear independence.

Concretely, verify that the points $\boldsymbol{a}_1 = (1,0,0)^t$, $\boldsymbol{a}_2 = (2,1,0)^t$, and $\boldsymbol{a}_3 = (1,1,0)^t$ are linearly independent. Form the matrix

$$M := \begin{pmatrix} 1 & 2 & 1 \\ 0 & 1 & 1 \\ 0 & 0 & 0 \\ 1 & 1 & 1 \end{pmatrix}$$

and compute the values of all possible determinants of order three, if necessary.

$$\begin{vmatrix} 1 & 2 & 1 \\ 0 & 1 & 1 \\ 0 & 0 & 0 \end{vmatrix} = 0, \begin{vmatrix} 1 & 2 & 1 \\ 0 & 1 & 1 \\ 1 & 1 & 1 \end{vmatrix} = 1, \begin{vmatrix} 1 & 2 & 1 \\ 0 & 0 & 0 \\ 1 & 1 & 1 \end{vmatrix} = 0, \begin{vmatrix} 0 & 1 & 1 \\ 0 & 0 & 0 \\ 1 & 1 & 1 \end{vmatrix} = 0.$$

Since there is a nonzero determinant, we conclude that the vectors $\begin{pmatrix} \boldsymbol{a}_1 \\ 1 \end{pmatrix}, \begin{pmatrix} \boldsymbol{a}_2 \\ 1 \end{pmatrix}, \begin{pmatrix} \boldsymbol{a}_3 \\ 1 \end{pmatrix}$ are linearly independent and, thus, that the points $\boldsymbol{a}_1$, $\boldsymbol{a}_2$, and $\boldsymbol{a}_3$ are affinely independent.

1.3 Projective Spaces

Our discussion of projective spaces is mostly utilitarian and not an aim in itself; in this book the objects live in affine spaces and the proofs mostly begin and end there. Our treatment will be analytic, based on linear algebra. For a synthetic treatment consult Hodge and Pedoe (1994, ch. VI).

The definition of a *d-dimensional projective space* $\mathbb{P}(\mathbb{R}^{d+1})$ over the linear space $\mathbb{R}^{d+1}$ is simple: it is the set of lines in $\mathbb{R}^{d+1}$ that pass through the origin. Each such line is a *projective point*. Projective points are our zero-dimensional projective subspaces. We prefer the notation $\mathbb{P}^d$ to $\mathbb{P}(\mathbb{R}^{d+1})$ when there is no risk of ambiguity. See Fig. 1.3.1(a) for a depiction of $\mathbb{P}(\mathbb{R}^3)$.

Since a line through the origin has the form $\alpha \boldsymbol{x}$ for some nonzero vector $\boldsymbol{x} \in \mathbb{R}^{d+1}$ and every scalar $\alpha \in \mathbb{R}$, we can define an equivalence relation $\sim$ on the nonzero vectors in $\mathbb{R}^{d+1}$ by relating two vectors $\boldsymbol{x}$ and $\boldsymbol{y}$ if they lie in the

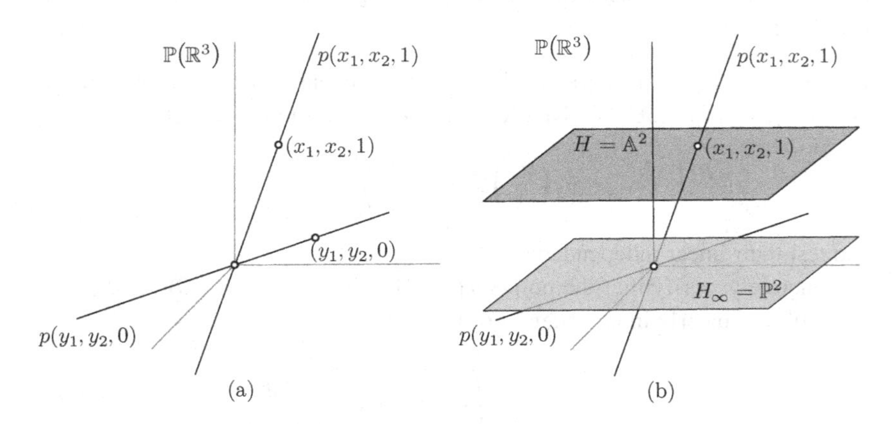

Figure 1.3.1 The projective space $\mathbb{P}(\mathbb{R}^3)$. (a) The affine space $\mathbb{R}^2$. (b) The projective space $\mathbb{P}(\mathbb{R}^3)$ as $H \cup H_\infty = \mathbb{A}^2 \cup \mathbb{P}(\mathbb{R}^2)$.

same line through the origin; that is, if there exists a scalar α such that $\boldsymbol{x} = \alpha \boldsymbol{y}$. In this way, we get a map p that assigns to each nonzero vector $\boldsymbol{x} \in \mathbb{R}^{d+1}$ its equivalence class, or equivalently, the line $\ell_{\boldsymbol{x}} = \alpha \boldsymbol{x}$ that it spans. Notationally,

$$p(\boldsymbol{x}) = \ell_{\boldsymbol{x}}. \tag{1.3.1}$$

This equivalence relation on the vectors in $\mathbb{R}^{d+1}$ also implies that *uniqueness* in projective spaces must be understood as *uniqueness up to a scalar multiplication.*

Constructing projective subspaces from subspaces of $\mathbb{R}^{d+1}$ requires little effort. To every k-dimensional linear subspace L of $\mathbb{R}^{d+1}$ there corresponds a $(k-1)$-dimensional projective space $\mathbb{P}(L)$ defined as the set of lines through the origin that are spanned by the nonzero vectors in L. In other words,

$$p(L \backslash \{\mathbf{0}\}) = \mathbb{P}(L).$$

If $L = \{\mathbf{0}\}$, then $\mathbb{P}(L) = \varnothing$ and $\dim \mathbb{P}(L) = -1$. A one-dimensional projective subspace is a *projective line*, a two-dimensional projective subspace is a *projective plane*, and a $(d-1)$-dimensional projective subspace is a *projective hyperplane*.

An initial advantage of having the underlying linear space $\mathbb{R}^{d+1}$ in $\mathbb{P}(\mathbb{R}^{d+1})$ is that all its projective properties can be verified by linear properties of $\mathbb{R}^{d+1}$. We mention a couple of examples.

(P1) Every two projective lines intersect at a projective point. Every two linear planes (think of the ones defining the projective lines) intersect at a unique line through the origin.

(P2) Every two projective points determine a unique projective line. Every two lines through the origin (think of the ones defining the projective points) determine a unique linear plane.

Homogeneous Coordinates

Another advantage of the underlying linear space $\mathbb{R}^{d+1}$ in $\mathbb{P}(\mathbb{R}^{d+1})$ is the access to its linear bases. Consider the *standard basis* of $\mathbb{R}^{d+1}$, namely

$$\boldsymbol{e}_1 = (1, 0, \ldots, 0)^t, \ldots, \boldsymbol{e}_{d+1} = (0, \ldots, 0, 1)^t.$$

Via the standard basis, we have a bijection that maps the projective point $p(\boldsymbol{x})$ onto the set of coordinates of nonzero vectors of the form $\alpha \boldsymbol{x}$ where α is a scalar. This defines an equivalence class on all the nonzero vectors in $\mathbb{R}^{d+1}$. The equivalence class containing the coordinates $(\alpha_1, \ldots, \alpha_{d+1})$ of the vector $\boldsymbol{x}$ is denoted by $(\alpha_1 : \cdots : \alpha_{d+1})$ and defines the *homogeneous coordinates* of $p(\boldsymbol{x})$. That is,

$$(\beta_1, \dots, \beta_{d+1}) \in (\alpha_1 : \cdots : \alpha_{d+1})$$

if and only if

$$(\alpha_1, \dots, \alpha_{d+1})^t = \lambda(\beta_1, \dots, \beta_{d+1})^t, \text{ for some nonzero } \lambda \in \mathbb{R}.$$

1.4 Maps

Now we revisit the maps that preserve the inherent character of affine, linear, or projective spaces.

A *linear map* is a function φ between two linear subspaces X and Y of $\mathbb{R}^d$ that satisfies

$$\varphi(\alpha_1 \boldsymbol{l}_1 + \alpha_2 \boldsymbol{l}_2) = \alpha_1 \varphi(\boldsymbol{l}_1) + \alpha_2 \varphi(\boldsymbol{l}_2),$$

for all $\alpha_1, \alpha_2 \in \mathbb{R}$ and $\boldsymbol{l}_1, \boldsymbol{l}_2 \in X$. Similarly, an *affine map* is a function ϱ between two affine spaces A and B of $\mathbb{R}^d$ that preserves affine combinations. In other words, it is a function $\varrho \colon A \to B$ that satisfies

$$\varrho(\alpha_1 \boldsymbol{a}_1 + \alpha_2 \boldsymbol{a}_2) = \alpha_1 \varrho(\boldsymbol{a}_1) + \alpha_2 \varrho(\boldsymbol{a}_2),$$

for all $\alpha_1, \alpha_2 \in \mathbb{R}$ with $\alpha_1 + \alpha_2 = 1$ and for all $\boldsymbol{a}_1, \boldsymbol{a}_2 \in A$. As a consequence, every linear map is an affine map.

From the definitions it follows that an affine map ϱ is linear if and only if $\varrho(\boldsymbol{0}) = \boldsymbol{0}$. It also follows that linear maps send linear subspaces into linear subspaces and that affine maps do the same for affine subspaces. In the same vein, injective linear maps send linear subspaces of dimension r into linear subspaces of dimension r, and so do injective affine maps with affine subspaces. See Problem 1.12.6.

Bijective maps between two spaces that respect the structure of the spaces are called *isomorphisms*. Maps from a space to itself are called *transformations*, and bijective transformations are called *automorphisms*.

Projections

Orthogonal projections surface with some regularity in this book; we discuss them henceforth. Let L be a linear subspace of $\mathbb{R}^d$. Then every vector of $\mathbb{R}^d$ can be written uniquely as the sum of a vector in L and a vector in $L^\perp$ (Problem 1.12.4). Notationally, $\mathbb{R}^d = L + L^\perp$. The *orthogonal projection* π_L of any vector $\boldsymbol{x}$ of $\mathbb{R}^d$ onto L is the unique vector $\boldsymbol{l} \in L$ with the property that $\boldsymbol{x} - \boldsymbol{l} \in L^\perp$. See Fig. 1.4.1(a).

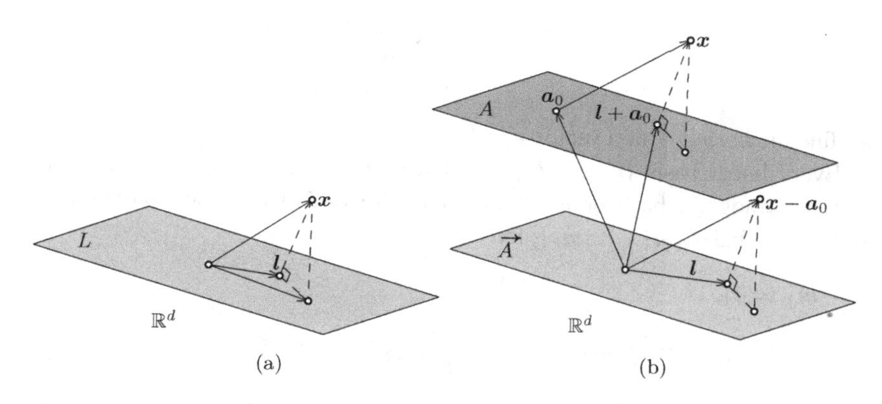

Figure 1.4.1 Orthogonal projections. (a) An orthogonal projection of a vector $\boldsymbol{x}$ onto a linear subspace L of $\mathbb{R}^d$. (b) An orthogonal projection of a vector $\boldsymbol{x}$ onto an affine subspace A of $\mathbb{R}^d$.

The projection $\pi_L(\boldsymbol{x})$ is the point in L closest to x. We provide three expressions for π_L, depending on the presentation of L.

Suppose that $\{\boldsymbol{l}_1, \ldots, \boldsymbol{l}_r\}$ is a basis of L and write $M := (\boldsymbol{l}_1 \cdots \boldsymbol{l}_r)$. Then M is a $d \times r$ matrix whose columns are these basis vectors. The projection $\pi_L(\boldsymbol{x})$ is a point in L and so it can be written as $M\boldsymbol{y}$ for some vector $\boldsymbol{y} \in \mathbb{R}^r$. The vector $\boldsymbol{x} - \pi_L(\boldsymbol{x})$ is in $L^\perp$, which yields that $\boldsymbol{x} - \pi_L(\boldsymbol{x}) \in (\operatorname{col} M)^\perp$ or, equivalently, that $\boldsymbol{x} - \pi_L(\boldsymbol{x}) \in \operatorname{null} M^t$, as $(\operatorname{col} M)^\perp = \operatorname{null} M^t$ (Problem 1.12.5). Putting these elements together we get that

$$M^t(\boldsymbol{x} - \pi_L(\boldsymbol{x})) = \boldsymbol{0}$$
$$M^t(\boldsymbol{x} - M\boldsymbol{y}) = \boldsymbol{0}.$$

In addition, it can be shown that M^tM is *nonsingular* – it has an inverse; see Problem 1.12.9. Hence, an expression for π_L is given by Equation (1.4.1) (Problem 1.12.9).

$$\pi_L(\boldsymbol{x}) = \left(M\left(M^tM\right)^{-1}M^t\right)\boldsymbol{x}. \tag{1.4.1}$$

We provide another expression for π_L, now in terms of orthogonal vectors. A set $X := \{\boldsymbol{l}_1, \ldots, \boldsymbol{l}_r\}$ of vectors of $\mathbb{R}^d$ is *orthogonal* if $\boldsymbol{l}_i \cdot \boldsymbol{l}_j = 0$ whenever $i \neq j$. It follows that X is linearly independent. The set X is an *orthogonal basis* of L if it is both a linear basis of L and an orthogonal set. It is an *orthonormal basis* if it is an orthogonal basis consisting of unit vectors. A *unit vector* is a vector with norm one. Suppose $\{\boldsymbol{l}_1, \ldots, \boldsymbol{l}_r\}$ is an orthogonal basis of L. Then

$$\pi_L(\boldsymbol{x}) = \sum_{i=1}^{r} \frac{\boldsymbol{x} \cdot \boldsymbol{l}_i}{\|\boldsymbol{l}_i\|^2} \boldsymbol{l}_i. \tag{1.4.2}$$

Define the *norm* $\|\cdot\|$ of a vector $\boldsymbol{x}$ in $\mathbb{R}^d$ as $\sqrt{\boldsymbol{x} \cdot \boldsymbol{x}}$.

Every linear basis $\{\boldsymbol{l}_1, \ldots, \boldsymbol{l}_r\}$ of a linear subspace L can be transformed into an orthogonal basis or an orthonormal basis of L if we wish. To obtain an orthogonal basis $\{\boldsymbol{m}_1, \ldots, \boldsymbol{m}_r\}$, let

$$\begin{aligned} \boldsymbol{m}_1 &:= \boldsymbol{l}_1, \\ \boldsymbol{m}_i &:= \boldsymbol{l}_i - \frac{\boldsymbol{l}_i \cdot \boldsymbol{m}_1}{\|\boldsymbol{m}_1\|^2}\boldsymbol{m}_1 - \cdots - \frac{\boldsymbol{l}_i \cdot \boldsymbol{m}_{i-1}}{\|\boldsymbol{m}_{i-1}\|^2}\boldsymbol{m}_{i-1}, \text{ for each } i \in [2\ldots r]. \end{aligned}$$

This is the *Gram–Schmidt orthogonalisation process.* If we are after an orthonormal basis, we divide each vector in $\{\boldsymbol{m}_1, \ldots, \boldsymbol{m}_r\}$ by its norm. Thus the set

$$\left\{ \frac{\boldsymbol{m}_1}{\|\boldsymbol{m}_1\|}, \ldots, \frac{\boldsymbol{m}_r}{\|\boldsymbol{m}_r\|} \right\}$$

is an orthonormal basis of L.

The final expression for π_L is obtained when a linear subspace L is given as a set of solutions of a homogeneous system:

$$L := \left\{ \boldsymbol{x} \in \mathbb{R}^d \middle| \, N\boldsymbol{x} = \boldsymbol{0} \right\}.$$

In this case, we get that

$$\pi_L(\boldsymbol{x}) = \boldsymbol{x} - \left(N^t \left(NN^t\right)^{-1}\right)(N\boldsymbol{x}). \tag{1.4.3}$$

In particular, if L is a linear hyperplane defined as $\left\{\boldsymbol{x} \in \mathbb{R}^d \middle| \, \boldsymbol{a} \cdot \boldsymbol{x} = 0\right\}$ then

$$\pi_L(\boldsymbol{x}) = \boldsymbol{x} - \frac{\boldsymbol{a} \cdot \boldsymbol{x}}{\|\boldsymbol{a}\|^2}\boldsymbol{a}. \tag{1.4.4}$$

Once we have defined projections onto linear spaces it is not difficult to extend the process to affine spaces via their directions. Let $(A, \vec{A})$ be an affine space and let $\boldsymbol{a}_0$ be a fixed point of A. Then the orthogonal projection π_A of any vector $\boldsymbol{x} \in \mathbb{R}^d$ onto A is defined as

$$\pi_A(\boldsymbol{x}) := \boldsymbol{a}_0 + \pi_{\vec{A}}(\boldsymbol{x} - \boldsymbol{a}_0). \tag{1.4.5}$$

See Fig. 1.4.1(b).

In analogy to the linear case, we obtain expressions for π_A depending on the presentation of A. If an affine subspace A is given as the set spanned by an affine basis $\{\boldsymbol{a}_0, \boldsymbol{a}_1, \ldots, \boldsymbol{a}_r\}$ then its direction is spanned by the linear basis

$\{\boldsymbol{a}_1 - \boldsymbol{a}_0, \dots, \boldsymbol{a}_r - \boldsymbol{a}_0\}$. Writing $M := (\boldsymbol{a}_1 - \boldsymbol{a}_0 \cdots \boldsymbol{a}_r - \boldsymbol{a}_0)$, an application of (1.4.1) gives that

$$\pi_{\vec{A}}(\boldsymbol{x}) = \left(M\left(M^t M\right)^{-1} M^t\right)\boldsymbol{x},$$

wherefrom it follows that

$$\pi_A(\boldsymbol{x}) = \boldsymbol{a}_0 + \left(M\left(M^t M\right)^{-1} M^t\right)(\boldsymbol{x} - \boldsymbol{a}_0). \tag{1.4.6}$$

If, instead, an affine subspace A is given as a set of solutions of a nonhomogeneous system

$$A := \left\{\boldsymbol{x} \in \mathbb{R}^d \,\middle|\, N\boldsymbol{x} = N\boldsymbol{a}_0,\ \boldsymbol{a}_0 \in A\right\},$$

then

$$\pi_A(\boldsymbol{x}) = \boldsymbol{x} - \left(N^t\left(NN^t\right)^{-1}\right)(N\boldsymbol{x} - N\boldsymbol{a}_0). \tag{1.4.7}$$

In particular, if $A = \{\boldsymbol{x} \in \mathbb{R}^d \mid \boldsymbol{a}\cdot\boldsymbol{x} = \boldsymbol{a}\cdot\boldsymbol{a}_0,\ \boldsymbol{a}_0 \in A\}$ then

$$\pi_A(\boldsymbol{x}) = \boldsymbol{x} - \frac{\boldsymbol{a}\cdot\boldsymbol{x} - \boldsymbol{a}\cdot\boldsymbol{a}_0}{\|\boldsymbol{a}\|^2}\boldsymbol{a}. \tag{1.4.8}$$

Projective Maps

We define maps between projective spaces through linear maps between the underlying linear spaces. Recall that the map p assigns to each nonzero vector $\boldsymbol{x}$ in $\mathbb{R}^{d+1}$ the line $\alpha\boldsymbol{x}$ (see (1.3.1)). Let φ be a linear map between linear spaces X and Y. Since $\varphi(\alpha\boldsymbol{x}) = \alpha\varphi(\boldsymbol{x})$ for every nonzero vector $\boldsymbol{x}$ and scalar α, the map φ assigns lines through the origin to lines through the origin, provided $\boldsymbol{x} \notin \ker\varphi$. Here, $\ker\varphi$ denotes the *kernel* of φ, the subspace of X consisting of the vectors $\boldsymbol{x}$ for which $\varphi(\boldsymbol{x}) = \boldsymbol{0}$. It then follows that if φ is injective, namely $\ker\varphi = \{\boldsymbol{0}\}$, then it defines a projective map $p(\varphi)\colon \mathbb{P}(X) \to \mathbb{P}(Y)$ given by

$$p(\boldsymbol{x}) \mapsto p(\varphi(\boldsymbol{x})).$$

If φ is not injective, it defines a projective map from $\mathbb{P}(X)\backslash\mathbb{P}(\ker\varphi)$ to $\mathbb{P}(Y)$. Hence we hereafter restrict ourselves to injective linear maps. If a linear φ induces the map $p(\varphi)$, so does the map $\alpha\varphi$ for any nonzero α, and thus the linear map φ is determined up to multiplication by a nonzero scalar.

If there is a bijective projective map, a *projective isomorphism*, between two projective spaces, we say the spaces are (projectively) *isomorphic*. As before, projective maps from a space to itself are called *projective transformations* and bijective transformations are called *projective automorphisms*.

We next define a projective basis. We feel that a *projective basis* in $\mathbb{P}(\mathbb{R}^{d+1})$ is best defined as a set of points $\{p(\boldsymbol{x}_1), \dots, p(\boldsymbol{x}_n)\}$ for which, given any projective basis $\{p(\boldsymbol{y}_1), \ldots, p(\boldsymbol{y}_n)\}$ of $\mathbb{P}(\mathbb{R}^{d+1})$, there is a unique projective automorphism that takes $p(\boldsymbol{x}_i)$ to $p(\boldsymbol{y}_i)$ for all $i \in [1 \ldots n]$. This follows the general scheme that a *basis* in a linear or affine space Z ought to be a set B_1 of elements for which, given any basis B_2 in Z, there is a unique automorphism mapping B_1 onto B_2. This definition, while unusual, could have been equally used for linear and affine bases.

We say that a set of projective points $p(\boldsymbol{x}_1), \ldots, p(\boldsymbol{x}_n)$ is *projectively dependent* in $\mathbb{P}^d$ if the vectors $\boldsymbol{x}_1, \ldots, \boldsymbol{x}_n$ are linearly dependent in $\mathbb{R}^{d+1}$; otherwise we say that the projective points are *projectively independent* in $\mathbb{P}^d$. In the same vein, we say that a set of projective points $p(\boldsymbol{x}_1), \ldots, p(\boldsymbol{x}_n)$ is in *general position* in $\mathbb{P}^d$ if the vectors $\boldsymbol{x}_1, \ldots, \boldsymbol{x}_n$ are in general position in $\mathbb{R}^{d+1}$; that is, if every subset of at most $d+1$ vectors is linearly independent.

Given the close connection between the notions of linear independence in $\mathbb{R}^{d+1}$ and projective independence in $\mathbb{P}(\mathbb{R}^{d+1})$, we may be tempted to say that a set $p(X) := \{p(\boldsymbol{x}_1), \ldots, p(\boldsymbol{x}_n)\}$ of points in $\mathbb{P}(\mathbb{R}^{d+1})$ is a basis if the set $X = \{\boldsymbol{x}_1, \ldots, \boldsymbol{x}_n\}$ is a basis in $\mathbb{R}^{d+1}$. But that would not satisfy our running definition, since no $d+1$ projective points determine a basis of $\mathbb{R}^{d+1}$, not even up to scalar multiplication; for this uniqueness, we need $d+2$ projective points.

Proposition 1.4.9[2] *Let $(\boldsymbol{x}_1, \ldots, \boldsymbol{x}_{d+1})$ and $(\boldsymbol{y}_1, \ldots, \boldsymbol{y}_{d+1})$ be bases of $\mathbb{R}^{d+1}$ such that $p(\boldsymbol{x}_i) = p(\boldsymbol{y}_i)$ for $i \in [1 \ldots d+1]$ and $p(\boldsymbol{x}_1 + \cdots + \boldsymbol{x}_{d+1}) = p(\boldsymbol{y}_1 + \cdots + \boldsymbol{y}_{d+1})$. Then there is a nonzero scalar α such that $\boldsymbol{x}_i = \alpha \boldsymbol{y}_i$, for each $i \in [1 \ldots d+1]$.*

In view of Proposition 1.4.9, we say that a *projective basis* in $\mathbb{P}(\mathbb{R}^{d+1})$ is any set $\{p(\boldsymbol{x}_1), \ldots, p(\boldsymbol{x}_{d+1}), p(\boldsymbol{x}_{d+2})\}$ of $d+2$ projective points in general position. This definition is compatible with our initial definition as the following theorem attests.

Theorem 1.4.10 (Fundamental theorem of projective maps)[3] *Let $\mathbb{P}(\mathbb{R}^r)$ and $\mathbb{P}(\mathbb{R}^s)$ be projective spaces with corresponding projective bases $(\boldsymbol{p}_1, \ldots, \boldsymbol{p}_{r+1})$ and $(\boldsymbol{q}_1, \ldots, \boldsymbol{q}_{s+1})$. Then there exists a unique projective map that sends $\boldsymbol{p}_i$ to $\boldsymbol{q}_i$, for each $i \in [1 \ldots r+1]$. In the case $r = s$, the map is an isomorphism.*

Given an ordered linear basis $(\boldsymbol{x}_1, \ldots, \boldsymbol{x}_{d+1})$ of $\mathbb{R}^{d+1}$, it is customary to take $\boldsymbol{x}_{d+2} = \boldsymbol{x}_1 + \cdots + \boldsymbol{x}_{d+1}$ to form the corresponding projective basis

[2] A proof is available in Berger (2009, lem. 4.4.2).
[3] A proof is available in Berger (2009, lem. 4.5.10).

$\{p(\boldsymbol{x}_1), \ldots, p(\boldsymbol{x}_{d+1}), p(\boldsymbol{x}_{d+2})\}$ of $\mathbb{P}(\mathbb{R}^{d+1})$. The projective basis arising from the standard basis is called the *projective standard basis*:

$$(1 : 0 : \cdots : 0), \ldots, (0 : 0 : \cdots : 1), (1 : 1 : \cdots : 1).$$

1.5 Dual Spaces

This section offers the first instance of the powerful principle of duality in this book; throughout the book, we will come in contact with many other incarnations of this principle. *Duality* roughly involves a pair of objects X and X^*, an involution $X \mapsto X^*$, and a correspondence, often order-reversing, between subsets or properties of X and X^* that translates results on X into results on X^*.

The set of all linear maps between two linear spaces is itself a linear space, where vector addition is defined as

$$(\varphi_1 + \varphi_2)(\boldsymbol{l}) = \varphi_1(\boldsymbol{l}) + \varphi_2(\boldsymbol{l})$$

and scalar multiplication as

$$(\alpha\varphi)(\boldsymbol{l}) = \alpha\varphi(\boldsymbol{l})$$

(Problem 1.12.6). In particular, the set of linear maps from $\mathbb{R}^d$ to $\mathbb{R}$ forms a linear space, called the *dual space* of $\mathbb{R}^d$ and denoted by $(\mathbb{R}^d)^*$.

Elements of a dual space are called *linear functionals*. Common examples of linear functionals include the *zero functional* on $\mathbb{R}^d$, which assigns zero to every vector in $\mathbb{R}^d$, and maps that define affine hyperplanes in $\mathbb{R}^d$, namely $\boldsymbol{x} \mapsto \boldsymbol{a} \cdot \boldsymbol{x}$ for every $\boldsymbol{x} \in \mathbb{R}^d$ and some $\boldsymbol{a} \in \mathbb{R}^d$.

The dual space of $\mathbb{R}^d$ is a linear space of dimension d. If $B = (\boldsymbol{l}_1, \ldots, \boldsymbol{l}_d)$ is an ordered basis of $\mathbb{R}^d$, there exists a uniquely determined basis $\varphi_1, \ldots, \varphi_d$ of $(\mathbb{R}^d)^*$ that is called the *dual basis* of B and satisfies

$$\varphi_i(\boldsymbol{l}_j) = \begin{cases} 1, & \text{if } i = j; \\ 0, & \text{otherwise.} \end{cases}$$

The dual basis of the standard basis of $\mathbb{R}^d$ can then be defined as the basis $\varphi_1, \ldots, \varphi_d$ given by $\varphi_i(x_1, \ldots, x_d) = x_i$.

There is an isomorphism between the space and its dual space. To each linear functional φ of $\mathbb{R}^d$ there corresponds a unique vector $\boldsymbol{a}$ of $\mathbb{R}^d$ such that $\varphi(\boldsymbol{x}) = \boldsymbol{a} \cdot \boldsymbol{x}$ for every $\boldsymbol{x} \in \mathbb{R}^d$. If φ is the zero functional then let $\boldsymbol{a} = \boldsymbol{0}$; otherwise, let $\boldsymbol{a} = \varphi(\boldsymbol{x}_0)\boldsymbol{x}_0$ for any nonzero unit vector $\boldsymbol{x}_0$ in $(\ker\varphi)^\perp$, which is not empty.

How does the dual space $(\mathbb{R}^d)^{**}$ of $(\mathbb{R}^d)^*$, the *double dual space* of $\mathbb{R}^d$ look? Is it what you would expect? An element of $(\mathbb{R}^d)^{**}$ is a linear functional

ϱ that sends a linear functional $\varphi\colon \mathbb{R}^d \to \mathbb{R}$ to an element in $\mathbb{R}$. There is also an isomorphism σ between $\mathbb{R}^d$ and $(\mathbb{R}^d)^{**}$. A common way to define σ is to first define a linear functional $\varrho_{x_0}\colon (\mathbb{R}^d)^* \to \mathbb{R}$ as $\varrho_{x_0}(\varphi) = \varphi(x_0)$ for a fixed $x_0 \in \mathbb{R}^d$ and every $\varphi \in (\mathbb{R}^d)^*$. Then define $\sigma(x) = \varrho_x$ for every $x \in \mathbb{R}^d$. For finite-dimensional spaces it is customary to identify $\mathbb{R}^d$ with $(\mathbb{R}^d)^{**}$ via σ, and we do so in this book.

1.6 Convex Sets

If on the equation of a line through two points $\boldsymbol{a}_1$ and $\boldsymbol{a}_2$ of $\mathbb{R}^d$, namely $\alpha_1\boldsymbol{a}_1 + \alpha_2\boldsymbol{a}_2$ with $\alpha_1 + \alpha_2 = 1$, we impose the additional condition of $\alpha_1, \alpha_2 \geqslant 0$, then we obtain the segment $[\boldsymbol{a}_1, \boldsymbol{a}_2]$ joining the points $\boldsymbol{a}_1$ and $\boldsymbol{a}_2$. Segments give rise to *convex sets*, sets that contain the segment between any of two of its points. That is, X is a *convex set* if

$$X = \{\alpha_1\boldsymbol{a}_1 + \alpha_2\boldsymbol{a}_2 |\ \alpha_1, \alpha_2 \geqslant 0,\ \alpha_1 + \alpha_2 = 1, \text{ and } \boldsymbol{a}_1, \boldsymbol{a}_2 \in X\}.$$

The expression $\alpha_1\boldsymbol{a}_1 + \alpha_2\boldsymbol{a}_2$ where $\alpha_1, \alpha_2 \geqslant 0$, $\alpha_1 + \alpha_2 = 1$, and $\boldsymbol{a}_1, \boldsymbol{a}_2 \in X$ is the *convex combination* of the points $\boldsymbol{a}_1$ and $\boldsymbol{a}_2$. The sets $\mathbb{R}^d$ and $\varnothing$ are convex sets, and so is any singleton in $\mathbb{R}^d$. Figure 1.6.1 shows examples of convex and nonconvex sets in $\mathbb{R}^3$.

In analogy to the affine and linear cases, given a set X in $\mathbb{R}^d$, the smallest convex set containing X is the *convex hull* of X and is denoted by $\operatorname{conv} X$. The convex hull of a set can be described as the set of all convex combinations of *finitely* many elements of X, namely

$$\operatorname{conv} X = \left\{ \alpha_1\boldsymbol{a}_1 + \cdots + \alpha_n\boldsymbol{a}_n \,\middle|\, \boldsymbol{a}_i \in X,\ \alpha_i \geqslant 0, \text{ and } \sum_{i=1}^{n} \alpha_i = 1 \right\}.$$

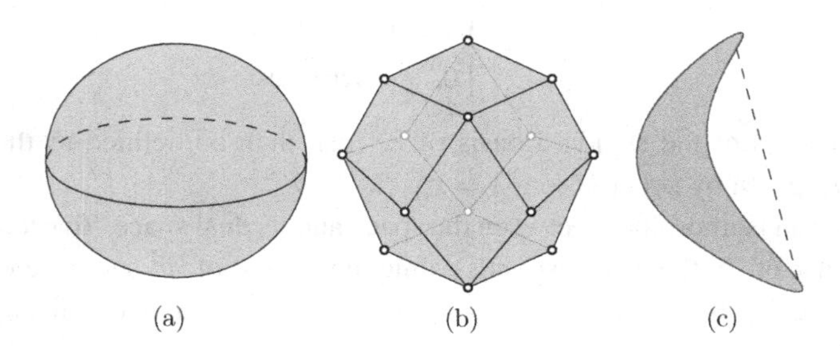

Figure 1.6.1 Convex and nonconvex sets. (a) A 3-dimensional ball, a convex set. (b) The rhombic dodecahedron, a convex 3-polytope. (c) A nonconvex set.

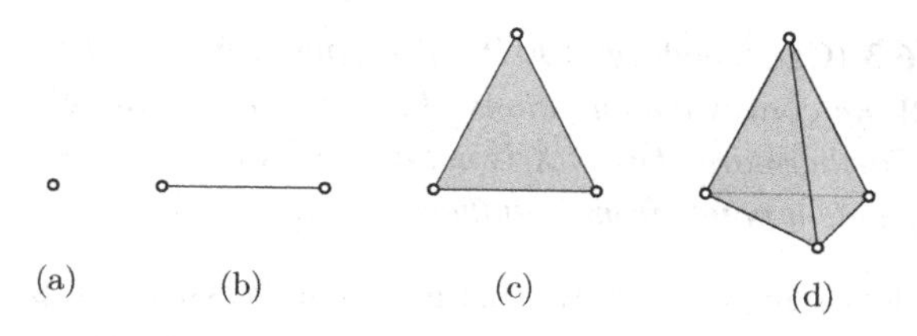

Figure 1.6.2 Simplices in $\mathbb{R}^3$.

Hyperplanes and the halfspaces in $\mathbb{R}^d$ bounded by a hyperplane are examples of convex sets. Each hyperplane $H_d(\boldsymbol{a},\alpha)$ in $\mathbb{R}^d$ with normal vector $\boldsymbol{a}$ and constant α determines or bounds four halfspaces in $\mathbb{R}^d$: the open halfspaces $H_d^+(\boldsymbol{a},\alpha)$ and $H_d^-(\boldsymbol{a},\alpha)$, and the closed halfspaces $H_d^+[\boldsymbol{a},\alpha]$ and $H_d^-[\boldsymbol{a},\alpha]$. In formulas we have that

$$\begin{aligned} H_d^+(\boldsymbol{a},\alpha) &= \left\{\boldsymbol{x} \in \mathbb{R}^d : \boldsymbol{a}\cdot\boldsymbol{x} > \alpha\right\}, \; H_d^-(\boldsymbol{a},\alpha) = \left\{\boldsymbol{x} \in \mathbb{R}^d : \boldsymbol{a}\cdot\boldsymbol{x} < \alpha\right\}, \\ H_d^+[\boldsymbol{a},\alpha] &= \left\{\boldsymbol{x} \in \mathbb{R}^d : \boldsymbol{a}\cdot\boldsymbol{x} \geqslant \alpha\right\}, \; H_d^-[\boldsymbol{a},\alpha] = \left\{\boldsymbol{x} \in \mathbb{R}^d : \boldsymbol{a}\cdot\boldsymbol{x} \leqslant \alpha\right\}. \end{aligned} \tag{1.6.1}$$

Define the Euclidean *distance* between any two vectors $\boldsymbol{x}$ and $\boldsymbol{y}$ in $\mathbb{R}^d$ as $\|\boldsymbol{x}-\boldsymbol{y}\|$. The distance function turns $\mathbb{R}^d$ into a metric space and enables more examples of convex sets such as the d-dimensional *open ball*, or simply the *open d-ball* $B_d(\boldsymbol{a},r)$, and the d-dimensional *closed ball*, or simply the *closed d-ball* $B_d[\boldsymbol{a},r]$, both with centre $\boldsymbol{a}$ and radius r:

$$\begin{aligned} B_d(\boldsymbol{a},r) &= \left\{\boldsymbol{x} \in \mathbb{R}^d \,\middle|\, \|\boldsymbol{x}-\boldsymbol{a}\| < r\right\}, \\ B_d[\boldsymbol{a},r] &= \left\{\boldsymbol{x} \in \mathbb{R}^d \,\middle|\, \|\boldsymbol{x}-\boldsymbol{a}\| \leqslant r\right\}. \end{aligned} \tag{1.6.2}$$

In all the formulas above, if there is no place for confusion we drop the subindex d. The ball $B_d[\boldsymbol{0},1]$ is referred to as the *closed unit d-ball*, while the ball $B_d(\boldsymbol{0},1)$ is referred to as the *open unit d-ball*.

Convex polytopes are other examples of convex sets. A *convex polytope* is the convex hull of a finite set of points in $\mathbb{R}^d$ (see Fig. 1.6.1(b)). In this book we speak only of convex polytopes. Hence we drop the adjective ‘convex’ hereafter. *Simplices* are an important class of polytopes; they are convex hulls of affinely independent points in $\mathbb{R}^d$. Figure 1.6.2 depicts all the simplices in $\mathbb{R}^3$.

It turns out that to get the convex hull of a set X, we do not require the finite convex combinations of all the points of X but rather the finite combinations of all collections of affinely independent points of X. Carathéordory’s theorem elucidates these remarks.

Theorem 1.6.3 (Carathéodory, 1907)[4] *The convex hull of a set X in $\mathbb{R}^d$ is formed by* all *the convex combinations of at most $d+1$ affinely independent points of X. Furthermore, if* conv X *is not a simplex then no fixed collection of affinely independent points from X suffices to span* conv X.

Carathéordory's theorem shows that a simplex possesses a proper *convex basis*: a fixed subset of affinely independent points that uniquely generate each element of the simplex. Since the set $\{\boldsymbol{x}_1, \ldots, \boldsymbol{x}_r\}$ of a simplex $T := \operatorname{conv}\{\boldsymbol{x}_1, \ldots, \boldsymbol{x}_r\}$ is affinely independent, every point $\boldsymbol{x}$ of T has a unique representation as an affine combination of $\{\boldsymbol{x}_1, \ldots, \boldsymbol{x}_r\}$ and, in particular, as a convex combination of $\{\boldsymbol{x}_1, \ldots, \boldsymbol{x}_r\}$.

Remark 1.6.4 If, for a set $X \subseteq \mathbb{R}^d$, conv X is not a simplex, then the notion of 'convex basis' is not available for (at least) two reasons:

(i) no fixed, finite collection of affinely independent points from X would suffice to generate conv X (by Carathéordory's theorem), and
(ii) some points in conv X have no unique representation as the convex combination of a fixed subset of affinely independent points from X.

Let X be a d-dimensional *sphere*, or simply *d-sphere* $S_d(\boldsymbol{a}, r)$, namely a set of the form $\{\boldsymbol{x} \in \mathbb{R}^{d+1} \mid \|\boldsymbol{x} - \boldsymbol{a}\| = r\}$. The convex hull of X is the closed ball $B_d[\boldsymbol{a}, r]$. The ball $B_d[\boldsymbol{a}, r]$ exemplifies Remark 1.6.4(ii) in the case of X being an infinite set. Every point in $B_d[\boldsymbol{a}, r]$ is a convex combination of at most two points from X, every point in the open ball admits more than one representation, and no finite subset of $S_d(\boldsymbol{a}, r)$ spans $B_d[\boldsymbol{a}, r]$. The justification for (ii) in the case of X being a finite set is given by Radon's theorem.

We will be mostly dealing with the sphere $S_d(\mathbf{0}, 1)$, which is referred to as the *unit d-sphere* and is denoted by $\mathbb{S}^d$.

Theorem 1.6.5 (Radon, 1921)[5] *Let X be a finite set of affinely dependent points in $\mathbb{R}^d$. Then the set X can be partitioned into subsets X_+ and X_- whose convex hulls intersect. Furthermore,* conv X_+ *and* conv X_- *can be assumed to be simplices.*

A *Radon partition* $\{X_+, X_-\}$ of a set $X \subseteq \mathbb{R}^d$ is a partition of X such that $\operatorname{conv} X_+ \cap \operatorname{conv} X_- \neq \emptyset$. A *Radon point* is a point in $\operatorname{conv} X_+ \cap \operatorname{conv} X_-$, a point that admits more than one representation as a convex hull of points of X. Figure 1.6.3 shows Radon partitions of four points in $\mathbb{R}^2$.

4 A proof is available in Webster (1994, thm. 2.2.4).
5 A proof is available in Webster (1994, thm. 2.2.5).

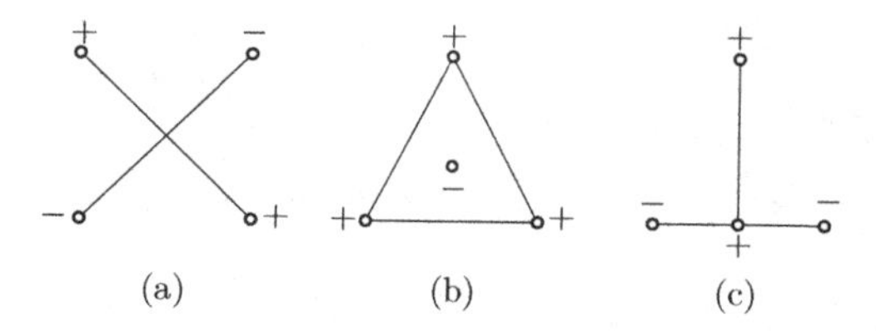

Figure 1.6.3 Radon partitions of four points in $\mathbb{R}^2$. Elements of the sets X_+ and X_- in Radon's theorem (1.6.5) are labelled with a + and − sign, respectively. Then $\operatorname{conv} X_+ \cap \operatorname{conv} X_- \neq \varnothing$. (a) Two line segments intersecting at a point; the convex hull of the four points is a quadrilateral. (b) A triangle and a point contained in the triangle; the convex hull of the four points is a triangle. (c) Two line segments intersecting at a point; the convex hull of the four points is a triangle.

An implication of the lack of a 'basis' for a convex set is that its *dimension* is defined as the dimension of its affine hull. It follows that $\dim \varnothing = -1$, since $\operatorname{aff} \varnothing = \varnothing$. As it should be, the dimension of a convex set is an invariant that does not dependent on the space in which the set is embedded.

Convexity is preserved by a number of operations; the final theorem of this section gathers some of these. Before stating the theorem, we require some basic definitions.

Definition 1.6.6 The *Minkowski sum* or *sum* $X + Y$ of two sets X and Y is the set

$$X + Y := \{x + y \mid x \in X, y \in Y\}.$$

The sets X and Y are the *summands* of $X + Y$.

The *scalar multiple* αX of a set X is the set

$$\alpha X := \{\alpha \boldsymbol{x} \mid \boldsymbol{x} \in X, \alpha \in \mathbb{R}\}.$$

Minkowski sums and scalar multiples of sets are illustrated in Fig. 1.6.4.

Let $\varphi \colon X \to Y$ be a function and let $B \subseteq Y$. The *preimage* $\varphi^{-1}(B)$ of B is the subset $\{x \in X \mid \varphi(x) \in B\}$ of X. The preimage of a function is well defined even if the function is not a bijection. In the case that φ is a bijection, $\varphi^{-1}(B)$ coincides with the image of B under the inverse function φ^{-1} of φ.

Theorem 1.6.7[6] *The following operations in $\mathbb{R}^d$ all return convex sets.*

(i) *The intersection of an arbitrary family of convex sets.*
(ii) *The Minkowski sum of convex sets.*

[6] A proof is available in Webster (1994, sec. 2.1).

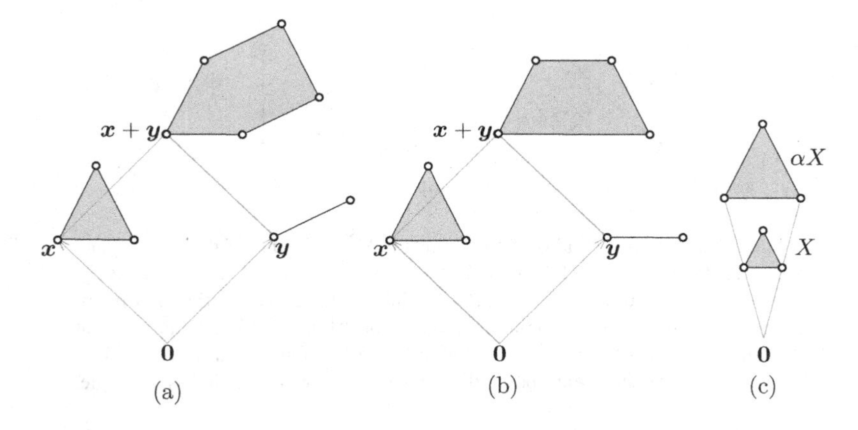

Figure 1.6.4 Minkowski sums and scalar multiple of sets. (a)–(b) Minkowski sums of a triangle and a segment. (c) Scalar multiple of a triangle.

(iii) *The scalar multiple of a convex set.*
(iv) *The image of a convex set under an affine map.*
(v) *The preimage of a convex set under an affine map.*

A function is *convexity-preserving* if it maps convex sets to convex sets. Every affine map is a convexity-preserving function (Theorem 1.6.7(iv)), but not every convexity-preserving function is an affine map (Problem 1.12.13). It is, however, true that every injective convexity-preserving function $\varphi\colon \mathbb{R}^r \to \mathbb{R}^s$ with $r \geqslant 2$ is an affine map (Meyer and Kay, 1973).

1.7 Interior, Boundary, and Closure

This section considers the topological notions of interior, boundary, and closure in the context of convex sets.

The *interior* of a set X in $\mathbb{R}^d$ is defined as the set of all points of X that are centres of some open d-dimensional ball that lies in X. Notationally,

$$\operatorname{int} X := \left\{ \boldsymbol{x} \in \mathbb{R}^d \,\middle|\, \text{there exists } r > 0 \text{ such that } B_d(\boldsymbol{x}, r) \subseteq X \right\}. \tag{1.7.1}$$

The points in the interior of X are its *interior points*.

An immediate consequence of the above definition is that a set X in $\mathbb{R}^d$ of dimension less than d has an empty interior. We benefit from studying the interior of X relative to the smallest affine space containing X; this is the *relative interior* of X. Notationally,

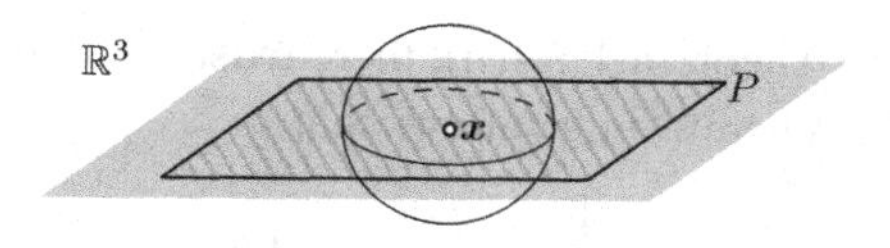

Figure 1.7.1 Difference between the relative interior and the interior of a 2-polytope, or *polygon*, P in $\mathbb{R}^3$. The affine hull of P is a plane; its relative interior is highlighted in a tiling pattern. However, the interior of P is empty: no ball in $\mathbb{R}^3$ centred at a point of P is fully contained in P.

$$\operatorname{rint} X := \left\{ \boldsymbol{x} \in \mathbb{R}^d \,\middle|\, \text{there exists } r > 0 \text{ such that } (B_d(\boldsymbol{x}, r) \cap \operatorname{aff} X) \subseteq X \right\}. \tag{1.7.2}$$

The points in the relative interior of X are its *relative interior points*.

The relative interior of a set is a more natural concept than that of its interior and, as such, we discuss it in more detail. The geometric difference between the relative interior and the interior of a set is captured in Fig. 1.7.1.

For a set $X \subseteq \mathbb{R}^d$, it is obvious that $\operatorname{int} X \subseteq \operatorname{rint} X \subseteq X$ (Problem 1.12.14). In the case of X being a nonempty convex set, it is not straightforward but it is true that $\operatorname{int} X = \operatorname{rint} X$ if and only if $\operatorname{int} X \neq \emptyset$ (Problem 1.12.15(ii)); the necessity of the statement relies on the nontrivial assertion that $\operatorname{rint} X \neq \emptyset$.

Theorem 1.7.3[7] *If X is a nonempty convex set in $\mathbb{R}^d$, then* $\operatorname{rint} X \neq \emptyset$.

We unveil some geometric properties of the relative interior of a convex set.

Theorem 1.7.4[8] *Let X be a convex set in $\mathbb{R}^d$ and let $\boldsymbol{a} \in \mathbb{R}^d$. The following statements hold.*

(i) *If $\boldsymbol{a} \in \operatorname{rint} X$, then the halfopen segment $[\boldsymbol{a}, \boldsymbol{b})$ lies in $\operatorname{rint} X$ for every $\boldsymbol{b} \in X$.*
(ii) *If $\boldsymbol{a} \in \operatorname{rint} X$ then, for each $\boldsymbol{b} \in \operatorname{aff} X$, there exists $\alpha > 1$ such that $\alpha \boldsymbol{a} + (1-\alpha)\boldsymbol{b} \in X$.*
(iii) *If, for each $\boldsymbol{b} \in X$, there exists $\alpha > 1$ such that $\alpha \boldsymbol{a} + (1-\alpha)\boldsymbol{b} \in X$, then $\boldsymbol{a} \in \operatorname{rint} X$.*

As a corollary of Theorem 1.7.4 we get another geometric property of relative interior points.

Corollary 1.7.5 *Let X be a convex set in $\mathbb{R}^d$ and let $\boldsymbol{a} \in \operatorname{rint} X$. Then, for each $\boldsymbol{b} \in \operatorname{aff} X \setminus \{\boldsymbol{a}\}$, there is a point $\boldsymbol{c} \in X \setminus \{\boldsymbol{a}\}$ such that $\boldsymbol{a} \in (\boldsymbol{b}, \boldsymbol{c})$.*

[7] A proof is available in Webster (1994, thm. 2.3.1).
[8] A proof is available in Webster (1994, sec. 2.3).

Proof According to Theorem 1.7.4(ii), there exists an $\alpha > 1$ such that $\boldsymbol{c} := \alpha\boldsymbol{a} + (1-\alpha)\boldsymbol{b} \in X$. Wherefrom it follows that

$$\boldsymbol{a} = \frac{1}{\alpha}\boldsymbol{c} + \left(1 - \frac{1}{\alpha}\right)\boldsymbol{b},$$

and with this equality, the corollary. □

Perhaps the most useful geometric property of the relative interior is the following.

Theorem 1.7.6[9] *Let $X \subseteq \mathbb{R}^d$ be defined as* $\operatorname{conv}\{\boldsymbol{x}_1, \ldots, \boldsymbol{x}_n\}$ *and let $\boldsymbol{a} \in X$. The point $\boldsymbol{a}$ is in* $\operatorname{rint} X$ *if and only if there exist scalars $\alpha_1, \ldots, \alpha_n > 0$ with $\sum_{i=1}^{n} \alpha_i = 1$ such that*

$$\boldsymbol{a} = \sum_{i=1}^{n} \alpha_i \boldsymbol{x}_i.$$

A *closure point* of a set $X \subseteq \mathbb{R}^d$ is a point $\boldsymbol{x}$ in which every open d-dimensional ball with centre at $\boldsymbol{x}$ meets X. The set of closure points is the *closure* of X and is denoted $\operatorname{cl} X$. Notationally,

$$\operatorname{cl} X := \left\{ \boldsymbol{x} \in \mathbb{R}^d \,\middle|\, \text{for every } r > 0,\ B_d(\boldsymbol{x}, r) \cap X \neq \emptyset \right\}.$$

A sequence $\boldsymbol{x}_1, \ldots, \boldsymbol{x}_n, \ldots$ of points in $\mathbb{R}^d$, written as $(\boldsymbol{x}_n)$, is said to *converge* to a point $\boldsymbol{x}$ if $\|\boldsymbol{x}_n - \boldsymbol{x}\|$ tends to zero as n tends to infinity. Sequences provide a characterisation of closure points.

Theorem 1.7.7[10] *Let $X \subseteq \mathbb{R}^d$ be a set in $\mathbb{R}^d$. A point $\boldsymbol{x} \in \mathbb{R}^d$ is a closure point of X if and only if there exists a sequence of points of X converging to $\boldsymbol{x}$.*

A *boundary point* of a convex set $X \subseteq \mathbb{R}^d$ is a point $\boldsymbol{x}$ in which every open ball with centre at $\boldsymbol{x}$ meets both X and $\mathbb{R}^d \backslash X$. The set of boundary points is the *boundary* of X and is denoted $\operatorname{bd} X$. The definition of a boundary point gives the following at once.

Proposition 1.7.8 *Let $X \subseteq \mathbb{R}^d$ be a set and let $\boldsymbol{x} \in \operatorname{bd} X$. The following assertions hold.*

(i) $\operatorname{bd} X = \operatorname{cl} X \backslash \operatorname{int} X$.
(ii) *There exists a sequence of points of X converging to $\boldsymbol{x}$.*
(iii) *There exists a sequence of points of $\mathbb{R}^d \backslash X$ converging to $\boldsymbol{x}$.*

[9] A proof is available in Webster (1994, thm. 2.3.7).
[10] A proof is available in Webster (1994, thm. 1.8.2).

The boundary of a convex set X with respect to aff X defines its relative boundary. Formally, a *relative boundary point* of a convex set $X \subseteq \mathbb{R}^d$ is a point in $\mathrm{cl}\, X \setminus \mathrm{rint}\, X$. The set of boundary points is the *relative boundary* of X and is denoted rbd X. If $X \subseteq \mathbb{R}^d$ is full dimensional then $\mathrm{rbd}\, X = \mathrm{bd}\, X$, while if $\dim X < d$ then $\mathrm{rbd}\, X \subset \mathrm{bd}\, X = \mathrm{cl}\, X$.

Theorem 1.7.9[11] *Let $X \subseteq \mathbb{R}^d$ be a convex set. Then* rint X, int X, *and* cl X *are all convex sets.*

1.8 Separation and Support

Two disjoint nonempty compact convex sets can be 'separated' by a hyperplane (Theorem 1.8.5); this is an intuitive and fundamental result in convexity. Also intuitive and fundamental is the result that a closed convex set is the intersection of (possibly infinitely many) halfspaces (Theorem 1.8.3) that 'support' the set. This section explores this kind of result.

Every nonempty closed set X contains a point $\boldsymbol{x}_0$ closest to a given point $\boldsymbol{a}$ of $\mathbb{R}^d$. If we add convexity then the point $\boldsymbol{x}_0$ is unique and the angle between the vectors $\boldsymbol{a} - \boldsymbol{x}$ and $\boldsymbol{x} - \boldsymbol{x}_0$ is nonacute for every $\boldsymbol{x} \in X$. The details are captured in Theorem 1.8.1.

Theorem 1.8.1[12] *Let X be a nonempty closed convex set in $\mathbb{R}^d$ and let $\boldsymbol{a} \in \mathbb{R}^d$. Then there exists a unique point $\boldsymbol{x}_0 \in X$ that is closest to $\boldsymbol{a}$; notationally,*

$$\|\boldsymbol{a} - \boldsymbol{x}_0\| = \inf\{\|\boldsymbol{a} - \boldsymbol{x}\| \mid \boldsymbol{x} \in X\}.$$

Moreover, the angle between the vectors $\boldsymbol{a} - \boldsymbol{x}_0$ and $\boldsymbol{x} - \boldsymbol{x}_0$ is nonacute for every $\boldsymbol{x} \in X$; notationally,

$$(\boldsymbol{a} - \boldsymbol{x}_0) \cdot (\boldsymbol{x} - \boldsymbol{x}_0) \leqslant 0 \textit{ for every } \boldsymbol{x} \in X.$$

Figure 1.8.1 depicts a geometric description of the theorem.

We proceed with a number of definitions. Let X be a set in $\mathbb{R}^d$. A *supporting halfspace* of X is a closed halfspace in $\mathbb{R}^d$ that contains X and whose bounding hyperplane meets cl X. A *supporting hyperplane* of X in $\mathbb{R}^d$ is a hyperplane H that bounds a supporting halfspace of X. In the case $X \subseteq H$, the supporting hyperplane is said to be *trivial*; otherwise it is said to be *nontrivial*. Notationally, a hyperplane $H(\boldsymbol{a},\alpha)$ is a supporting hyperplane of X if and only if

[11] A proof is available in Webster (1994, thm. 2.3.5).
[12] A proof is available in Webster (1994, thm. 2.4.1).

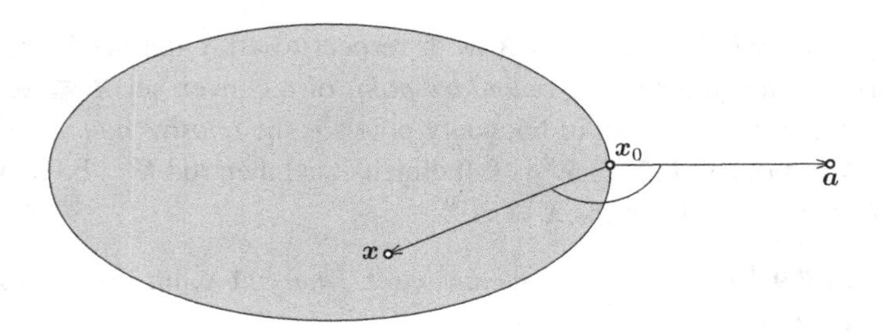

Figure 1.8.1 The point $\boldsymbol{x}_0$ in a closed convex set that is closest to a point $\boldsymbol{a} \in \mathbb{R}^d$; the nonacute angle between $\boldsymbol{a} - \boldsymbol{x}_0$ and $\boldsymbol{x} - \boldsymbol{x}_0$ is also shown.

$$\text{either } \alpha = \sup_{x \in X} \boldsymbol{a} \cdot \boldsymbol{x} \quad \text{or} \quad \alpha = \inf_{x \in X} \boldsymbol{a} \cdot \boldsymbol{x}.$$

And a supporting hyperplane $H(\boldsymbol{a}, \alpha)$ is nontrivial if and only if

$$\text{either } \inf_{x \in X} \boldsymbol{a} \cdot \boldsymbol{x} < \sup_{x \in X} \boldsymbol{a} \cdot \boldsymbol{x} = \alpha \quad \text{or} \quad \sup_{x \in X} \boldsymbol{a} \cdot \boldsymbol{x} > \inf_{x \in X} \boldsymbol{a} \cdot \boldsymbol{x} = \alpha.$$

A supporting hyperplane of X is said to *support* X at the points where it intersects $\operatorname{cl} X$.

Theorem 1.8.1 facilitates the proof of a number of results; the first such result comes next.

Corollary 1.8.2 *Let $X \subseteq \mathbb{R}^d$ be a nonempty closed convex set and let $\boldsymbol{a} \notin X$. Then there exists a hyperplane that does not contain $\boldsymbol{a}$ and that supports X at the point $\boldsymbol{x}_0 \in X$ closest to $\boldsymbol{a}$.*

Proof Thanks to Theorem 1.8.1, we find a point $\boldsymbol{x}_0 \in X$ closest to $\boldsymbol{a}$ such that

$$(\boldsymbol{a} - \boldsymbol{x}_0) \cdot (\boldsymbol{x} - \boldsymbol{x}_0) \leqslant 0 \text{ for every } \boldsymbol{x} \in X$$

or, equivalently, that

$$(\boldsymbol{a} - \boldsymbol{x}_0) \cdot \boldsymbol{x} \leqslant (\boldsymbol{a} - \boldsymbol{x}_0) \cdot \boldsymbol{x}_0 \text{ for every } \boldsymbol{x} \in X.$$

Define $\boldsymbol{b} := \boldsymbol{a} - \boldsymbol{x}_0$ and $\beta := (\boldsymbol{a} - \boldsymbol{x}_0) \cdot \boldsymbol{x}_0$. It immediately follows that $H^-[\boldsymbol{b}, \beta]$ is a supporting halfspace of X and that $H(\boldsymbol{b}, \beta)$ supports X at $\boldsymbol{x}_0$.

Suppose that $\boldsymbol{a} \in H^-[\boldsymbol{b}, \beta]$. Then

$$0 \geqslant (\boldsymbol{a} - \boldsymbol{x}_0) \cdot \boldsymbol{a} - (\boldsymbol{a} - \boldsymbol{x}_0) \cdot \boldsymbol{x}_0 = \|\boldsymbol{a} - \boldsymbol{x}_0\|^2,$$

which implies that $\boldsymbol{a} = \boldsymbol{x}_0$ (Problem 1.12.1(ii)). This is a contradiction because $\boldsymbol{x}_0 \in X$ and $\boldsymbol{a} \notin X$. □

Another consequence of Theorem 1.8.1 is that a closed convex set is the set of solutions of a system of (possibly infinitely many) linear inequalities.

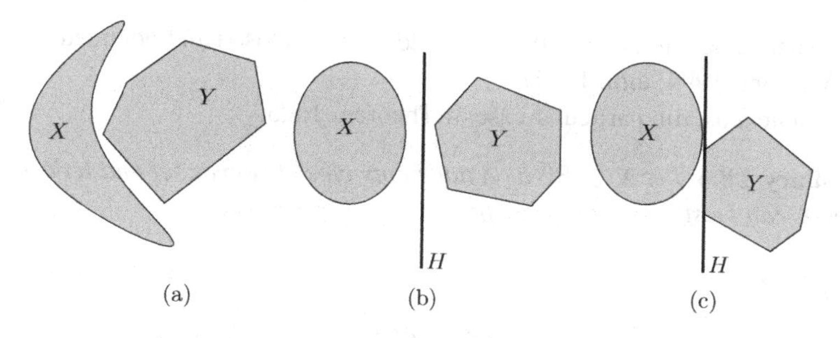

Figure 1.8.2 Examples of the separation of sets $X, Y \subseteq \mathbb{R}^2$. (a) Sets X and Y that cannot be separated. (b) Sets X and Y that are strictly separated by the hyperplane H. (c) Sets X and Y that are separated but not strictly by the hyperplane H.

Theorem 1.8.3[13] *A nonempty closed convex set in $\mathbb{R}^d$ is the intersection of its supporting halfspaces.*

It seems intuitive that through each relative boundary point of a closed convex set passes a nontrivial supporting hyperplane; this intuition is correct.

Theorem 1.8.4[14] *Let X be a nonempty convex set in $\mathbb{R}^d$ and let $\boldsymbol{x}_0 \in \operatorname{bd} X$. Then there exists a hyperplane in $\mathbb{R}^d$ supporting X at $\boldsymbol{x}_0$. In the case $\boldsymbol{x}_0 \in \operatorname{rbd} X$, the hyperplane can be assumed to be nontrivial.*

A hyperplane in $\mathbb{R}^d$ separates the space into two closed halfspaces. This fact gives rise to the important concept of separation of convex sets. Let X and Y be two sets in $\mathbb{R}^d$ and let H be a hyperplane in $\mathbb{R}^d$. The sets X and Y are said to be *separated* by H if X and Y lie in opposite closed halfspaces defined by H. And the sets X and Y are said to be *strictly separated* by H if X and Y lie in opposite open halfspaces defined by H. Figure 1.8.2(b)–(c) exemplifies these new notions.

Not every two disjoint sets in $\mathbb{R}^d$ can be separated by a hyperplane, as Fig. 1.8.2(a) shows. But every two disjoint convex sets can. We, however, content ourselves with an instance of this assertion.

Theorem 1.8.5 (Separation theorem)[15] *Let X and Y be disjoint nonempty convex sets in $\mathbb{R}^d$. Suppose that X is closed and Y is compact. Then X and Y can be strictly separated by a hyperplane in $\mathbb{R}^d$.*

Recall that a set X in $\mathbb{R}^d$ is said to be *compact* if each sequence of its points contains a subsequence that converges to a point of X. Compact sets in $\mathbb{R}^d$

[13] A proof is available in Webster (1994, cor. 2.4.8).
[14] A proof is available in Webster (1994, thm. 2.4.12).
[15] A proof is available in Webster (1994, thm. 2.4.6).

have a neat characterisation: they are precisely the closed and bounded sets in $\mathbb{R}^d$ (Webster, 1994, thm. 1.8.4).

We state a useful, particular case of Theorem 1.8.5.

Corollary 1.8.6 *Let $X \subseteq \mathbb{R}^d$ be a nonempty closed convex set and let $\boldsymbol{a} \notin X$. Then X can be strictly separated from $\boldsymbol{a}$ by a hyperplane.*

1.9 Faces

Convex sets are structured around other convex sets, their faces; this feature makes their mathematics amenable to inductive arguments. A convex subset F of a convex set $X \subseteq \mathbb{R}^d$ is a *face* of X if each time the relative interior of a segment in X meets F then the segment is fully contained in F. By the convexity of F, a segment is in F if its endpoints are. Hence the definition of a face amounts to the following.

$$\text{If } \alpha \boldsymbol{x}_1 + (1-\alpha)\boldsymbol{x}_2 \in F \text{ with } \boldsymbol{x}_1, \boldsymbol{x}_2 \in X \text{ and } \alpha \in (0,1), \text{ then } \boldsymbol{x}_1, \boldsymbol{x}_2 \in F. \tag{1.9.1}$$

It is clear that $\emptyset$ and X itself are faces of X; these are its *improper* faces. Every other face of X is a *proper* face.

A k-dimensional face of a convex set X is referred to as a *k-face*. The set of k-faces of X is denoted by $\mathcal{F}_k(X)$ and the set of all faces of X is denoted by $\mathcal{F}(X)$. The 0-faces are called *extreme points*, and the set of extreme points of X is denoted by $\operatorname{ext} X$.

Proposition 1.9.2 *Let X be a convex set. An extreme point $\boldsymbol{v}$ of X is not in any segment $(\boldsymbol{x}_1, \boldsymbol{x}_2)$ with $\boldsymbol{x}_1, \boldsymbol{x}_2 \in X$ or, equivalently, the set $X \backslash \{\boldsymbol{v}\}$ is again convex.*

Another immediate consequence of (1.9.1) is that 'is a face of' is a transitive relation on the faces of a convex set X.

Proposition 1.9.3 *Let X be a convex set in $\mathbb{R}^d$. If F is a face of X and I is a face of F, then I is a face of X.*

We now characterise faces.

Theorem 1.9.4[16] *Let X be a convex set in $\mathbb{R}^d$ and let F be a convex subset of X. Then F is a face of X if and only if $X \backslash F$ is convex and $F = X \cap \operatorname{aff} F$.*

[16] A proof is available in Webster (1994, thm. 2.6.2).

We list a corollary of Theorem 1.9.4.

Corollary 1.9.5 *If X is a convex set in $\mathbb{R}^d$ and F is a proper face of X, then* $\dim F < \dim X$.

Proof Suppose $\dim F = \dim X$. Then, $\operatorname{aff} F = \operatorname{aff} X$ and, by Theorem 1.9.4, $F = X \cap \operatorname{aff} X = X$. Hence, a proper face F of X must satisfy $\dim F < \dim X$. □

The next theorem reveals five ways in which faces appear.

Theorem 1.9.6 *Let $X \subseteq \mathbb{R}^d$ be a convex set. Then the following assertions hold.*

(i) *The intersection of any nonempty family of faces of X is a face of X.*
(ii) *The intersection of any nonempty family of faces of X can be expressed as the intersection of at most $d + 1$ members of the family.*
(iii) *The intersection of X and any of its supporting hyperplanes is a face of X.*
(iv) *Let $\varphi\colon \mathbb{R}^d \to \mathbb{R}^d$ be an affine transformation. If F is a face of X, then the set $\varphi^{-1}(F)$ is a face of $\varphi^{-1}(X)$.*
(v) *Let $\varphi\colon \mathbb{R}^d \to \mathbb{R}^d$ be an injective affine transformation and let $F \subseteq X$. The set F is a face of X if and only if the set $\varphi(F)$ is a face of $\varphi(X)$.*

Proof Take $\boldsymbol{x}_1, \boldsymbol{x}_2 \in X$ and $\alpha \in (0, 1)$.

(i) Let I be some index set and let $\{F_i \mid i \in I\}$ be a family of faces of X. Let $F := \bigcap_{i\in I} F_i$. Then F is a convex subset of X (Theorem 1.6.7). Suppose that $\alpha\boldsymbol{x}_1 + (1-\alpha)\boldsymbol{x}_2 \in F$. It follows that $\alpha\boldsymbol{x}_1 + (1-\alpha)\boldsymbol{x}_2 \in F_i$ for each $i \in I$. Since F_i is a face, we get that $\boldsymbol{x}_1, \boldsymbol{x}_2 \in F_i$ for each $i \in I$. This confirms that $\boldsymbol{x}_1, \boldsymbol{x}_2 \in F$ and that F is a face of X.

(ii) Let I be some index set; let $\{F_i \mid i \in I\}$ be a family of distinct faces of X and let $F := \bigcap_{i\in I} F_i$. We show that there are elements $i_1, \ldots, i_n$ of I, with $n \leqslant d + 1$, such that

$$F = F_{i_1} \cap \cdots \cap F_{i_n}. \tag{1.9.6.1}$$

Pick any element $i_1 \in I$. If $F = F_{i_1}$ then we are done; otherwise $F \subset F_{i_1}$, $\dim F < \dim F_{i_1}$ by Corollary 1.9.5, and there is another index i_2 of I such that

$$F \subseteq F_{i_1} \cap F_{i_2} \subset F_{i_1}.$$

The selection of i_2 is possible because the intersection of all these faces F_i (with $i \in I$) is F and $\dim F < \dim F_{i_1}$. If the statement is false, then there are indices $i_1, \ldots, i_n$ of I such that $n \geqslant d + 2$, and

$$F \subset F_{i_1} \cap \cdots \cap F_{i_n} \subset \cdots \subset F_{i_1} \cap F_{i_2} \subset F_{i_1}.$$

Additionally, as a consequence of Corollary 1.9.5, we have that $\dim F_{i_1} \leqslant d$ and

$$\dim F < \dim(F_{i_1} \cap F_{i_2} \cap \cdots \cap F_{i_n}) < \cdots < \dim(F_{i_1} \cap F_{i_2}) < \dim F_{i_1}.$$

But from this series of inequalities, it follows that

$$-1 \leqslant \dim F < \dim(F_{i_1} \cap F_{i_2} \cap \cdots \cap F_{i_n}) < \dim F_{i_1} - (n-1) \leqslant d - n + 1,$$

which yields that $-1 < d - n + 1$, a contradiction for $n \geqslant d + 2$. Hence (1.9.6.1) holds and $n \leqslant d + 1$, as desired.

(iii) Let $H := H(\boldsymbol{b}, \beta)$ be a supporting hyperplane of X with $X \subseteq H^-[\boldsymbol{b}, \beta]$. Then $\boldsymbol{b} \cdot \boldsymbol{x} \leqslant \beta$ for every $\boldsymbol{x} \in X$. Let $F := X \cap H$. It follows that F is a convex subset of X. Now suppose that $\alpha\boldsymbol{x}_1 + (1-\alpha)\boldsymbol{x}_2 \in F$. Since $\alpha\boldsymbol{x}_1 + (1-\alpha)\boldsymbol{x}_2 \in H$, we must have that

$$\beta = \boldsymbol{b} \cdot (\alpha\boldsymbol{x}_1 + (1-\alpha)\boldsymbol{x}_2) = \alpha\boldsymbol{b} \cdot \boldsymbol{x}_1 + (1-\alpha)\boldsymbol{b} \cdot \boldsymbol{x}_2 \leqslant \alpha\beta + (1-\alpha)\beta = \beta.$$

Hence, $\boldsymbol{b} \cdot \boldsymbol{x}_1 = \boldsymbol{b} \cdot \boldsymbol{x}_2 = \beta$ since $\alpha \in (0, 1)$. This proves that $\boldsymbol{x}_1, \boldsymbol{x}_2 \in X \cap H$ and that F is a face of X.

(iv) Take $\boldsymbol{y}_1, \boldsymbol{y}_2 \in \varphi^{-1}(X)$. Then $\varphi(\boldsymbol{y}_1), \varphi(\boldsymbol{y}_2) \in X$. From F being a convex set of X, it follows that $\varphi^{-1}(F)$ is a convex subset of $\varphi^{-1}(X)$ (Theorem 1.6.7(v)). Suppose that

$$\alpha\boldsymbol{y}_1 + (1-\alpha)\boldsymbol{y}_2 \in \varphi^{-1}(F).$$

Then $\varphi(\alpha\boldsymbol{y}_1 + (1-\alpha)\boldsymbol{y}_2) \in F$. Since φ is an affine transformation, we have that

$$\varphi(\alpha\boldsymbol{y}_1 + (1-\alpha)\boldsymbol{y}_2) = \alpha\varphi(\boldsymbol{y}_1) + (1-\alpha)\varphi(\boldsymbol{y}_2),$$

and since F is a face of X we have that $\varphi(\boldsymbol{y}_1), \varphi(\boldsymbol{y}_2) \in F$. Hence $\boldsymbol{y}_1, \boldsymbol{y}_2 \in \varphi^{-1}(F)$.

(v) The proof goes along the lines of that for (iv) and is left to the reader. □

The condition of φ being injective in Theorem 1.9.6(v) cannot be removed. We can find many examples of convex sets X, faces F of X, and noninjective affine transformations φ for which $\varphi(F)$ is not a face of $\varphi(X)$. Consider, for instance, the two-dimensional unit ball $X := B[\boldsymbol{0}, 1]$ and the orthogonal projection φ of $\mathbb{R}^2$ onto the x-axis. Then $\varphi(X) = [-\boldsymbol{1}, \boldsymbol{1}]$. It is now clear that, for any extreme point $\boldsymbol{x}$ of X other than $\pm\boldsymbol{1}$, $\varphi(\boldsymbol{x})$ is not a face of $[-\boldsymbol{1}, \boldsymbol{1}]$.

A face of a convex set X is *exposed* if it is either an improper face of X or is of the form $X \cap H$ where H is a supporting hyperplane of X (Theorem 1.9.6(iii)); Figure 1.9.1(b) gives examples of exposed faces. Suppose

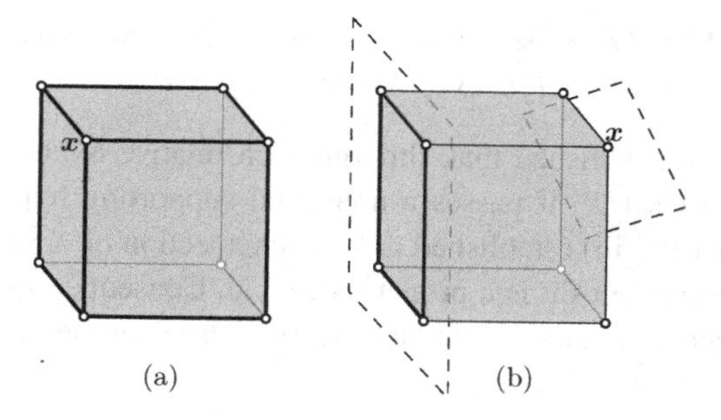

Figure 1.9.1 Faces of convex sets. (a) An extreme point $\boldsymbol{x}$ as the intersection of three 2-faces. (b) A 2-face and an extreme point $\boldsymbol{x}$ as exposed faces.

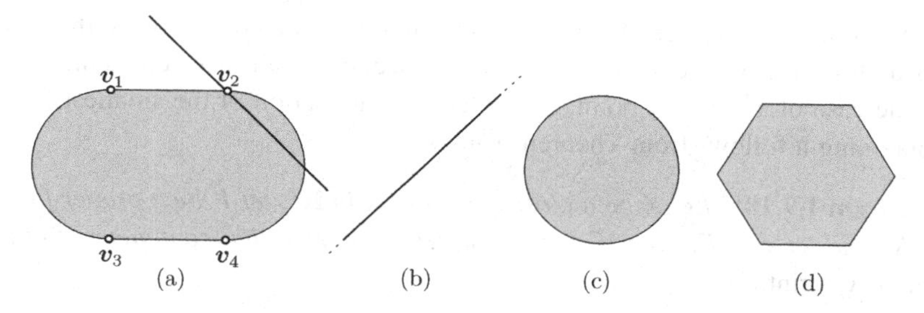

Figure 1.9.2 Faces in closed convex sets. (a) A two-dimensional closed convex set X with faces that are not exposed. The set X is the convex hull of two closed discs; its boundary consists of two closed segments $[\boldsymbol{v}_1, \boldsymbol{v}_2]$ and $[\boldsymbol{v}_3, \boldsymbol{v}_4]$ and two open halfcircles. The extreme points are the points $\boldsymbol{v}_1, \boldsymbol{v}_2, \boldsymbol{v}_3, \boldsymbol{v}_4$ and the points on each of the two open halfcircles, and the 1-faces are the two aforementioned segments. The extreme points $\boldsymbol{v}_1, \boldsymbol{v}_2, \boldsymbol{v}_3, \boldsymbol{v}_4$ are not exposed: a supporting hyperplane of X passing through precisely one of them must also meet one of the two open halfcircles; every other face of X is exposed. (b) A convex set with no extreme points. (c) A two-dimensional closed convex set with extreme points but no 1-faces. (d) A two-dimensional closed convex set with extreme points and 1-faces.

that $H = \{\boldsymbol{x} \in \mathbb{R}^d \mid \boldsymbol{a} \cdot \boldsymbol{x} = \alpha\}$ and that X lies in the supporting halfspace $H^-[\boldsymbol{a},\alpha]$. From the definition it follows that an exposed face $F := X \cap H$ maximises the linear functional $\boldsymbol{a} \cdot \boldsymbol{x}$ over X:

$$\alpha = \max\{\boldsymbol{a} \cdot \boldsymbol{x} \mid \boldsymbol{x} \in X\} \text{ and } F = \{\boldsymbol{x} \in X \mid \boldsymbol{a} \cdot \boldsymbol{x} = \alpha\}. \tag{1.9.7}$$

This face is often denoted by $F(X,\boldsymbol{a})$. An inequality $\boldsymbol{c} \cdot \boldsymbol{x} \leqslant \gamma$ is *valid* for a set in $\mathbb{R}^d$ if it is satisfied for every point $\boldsymbol{x}$ in the set. The inequality $\boldsymbol{a} \cdot \boldsymbol{x} \leqslant \alpha$ that defines $H^-[\boldsymbol{a},\alpha]$ is valid for X.

Not every face of a convex set X is exposed, as Fig. 1.9.2(a) shows. However, every intersection of exposed faces of X is exposed.

Proposition 1.9.8[17] *Let X be a convex set in $\mathbb{R}^d$. Then the intersection of any nonempty family of exposed faces of X is also exposed.*

Theorem 1.8.4 established that, through each relative boundary point of a nonempty convex set X, it passes a nontrivial supporting hyperplane of X, while Theorem 1.9.6(iii) established that the intersection of X and a nontrivial supporting hyperplane of it is a proper face of X. Consequently, some proper faces of X meet its relative boundary. More is true: proper faces of X are contained in the relative boundary of X.

Theorem 1.9.9[18] *If X is a closed convex set in $\mathbb{R}^d$ and F is a proper face of X, then $F \subseteq \operatorname{rbd} X$.*

According to Theorem 1.9.6(i), for any subset Y of a convex set X there is a smallest face of X containing Y, the intersection of all faces containing Y. In the case of Y being a point, a simple characterisation of the smallest face containing it follows from Theorem 1.9.9.

Theorem 1.9.10[19] *Let X be a closed convex set in $\mathbb{R}^d$, let F be a proper face of X, and let $\boldsymbol{x} \in F$. Then F is the smallest face of X that contains $\boldsymbol{x}$ if and only if $\boldsymbol{x} \in \operatorname{rint} F$.*

Figure 1.9.2(b) depicts a closed convex set in $\mathbb{R}^2$ with no extreme points. But every bounded, closed, and convex set in $\mathbb{R}^d$ has extreme points and, moreover, it is spanned by them. This is perhaps the most important result on the facial structure of compact convex sets.

Theorem 1.9.11 (Minkowski–Krein–Milman's theorem)[20] *Let X be a compact convex set in $\mathbb{R}^d$. Then X is the convex hull of its extreme points. Notationally,*

$$X = \operatorname{conv}(\operatorname{ext} X).$$

A simple consequence of Minkowski–Krein–Milman's theorem (1.9.11) is that every compact convex set has extreme points, but this does not extend to higher dimensional faces. Figure 1.9.2(c) depicts a compact convex set with no 1-faces. In fact, a compact convex set in $\mathbb{R}^d$ can have a very diverse face-dimension pattern. The *face-dimension pattern* of a compact convex set X is an increasing sequence $(d_1, \ldots, d_n)$ of positive integers that encode all possible positive dimensions of faces in X. We have all the possible patterns in $\mathbb{R}^2$:

[17] A proof is available in Webster (1994, thm. 2.6.17).
[18] A proof is available in Brøndsted (1983, thm. 5.3).
[19] A proof is available in Brøndsted (1983, thm. 5.6).
[20] A proof is available in Brøndsted (1983, thm. 5.10).

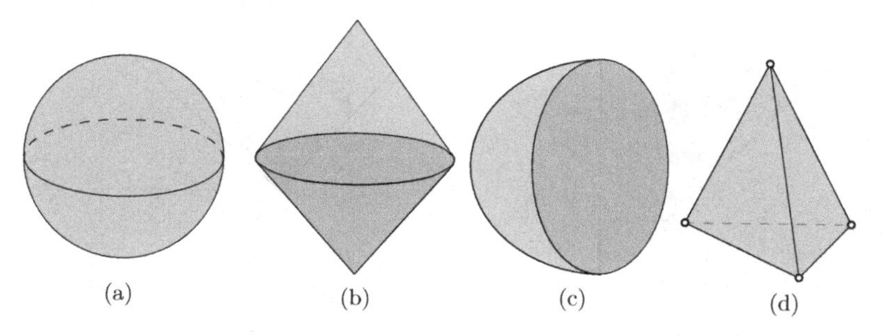

Figure 1.9.3 All the possible face-dimension patterns for three-dimensional compact convex sets in $\mathbb{R}^3$. (a) The three-dimensional unit ball, which has pattern (3). (b) A set obtained as the convex hull of a circle on a plane and two points at opposite sides of the plane; the set has pattern $(1,3)$. (c) A set obtained by intersecting a closed halfspace with the three-dimensional, unit ball; the set has pattern $(2,3)$. (d) A 3-simplex, which has pattern $(1,2,3)$.

the pattern (), exemplified by a singleton; the pattern (1), exemplified by a line segment; the pattern (2), exemplified by Fig. 1.9.2(c); and the pattern $(1,2)$, exemplified by Fig. 1.9.2(d). This gives a total of 2^2 patterns in $\mathbb{R}^2$. Figure 1.9.3 shows examples with all the possible face-dimension patterns in three-dimensional compact convex sets: (3), $(1,3)$, $(2,3)$, and $(1,2,3)$; these examples together with the lower dimensional examples give a total of 2^3 patterns in $\mathbb{R}^3$.

A result of Roshchina et al. (2018) states that for any finite, increasing sequence of positive integers, there exists a compact convex set in $\mathbb{R}^d$ that has extreme points and faces with dimensions only from this prescribed sequence; in other words, for any of the 2^d possible face-dimension patterns, there is a compact convex set in $\mathbb{R}^d$ exhibiting that pattern.

1.10 Cones and Lineality Spaces

If on the line through the points $\boldsymbol{a}_1$ and $\boldsymbol{a}_1 + \boldsymbol{a}_2$, namely the set

$$\left\{ \boldsymbol{a}_1 + \alpha \boldsymbol{a}_2 \,\middle|\, \boldsymbol{a}_1 \in \mathbb{R}^d, \boldsymbol{0} \neq \boldsymbol{a}_2 \in \mathbb{R}^d, \text{ and every } \alpha \in \mathbb{R} \right\},$$

we impose the condition of $\alpha \geqslant 0$, we arrive at the definition of a *ray*.

A subset X of $\mathbb{R}^d$ is a convex *cone* if it is convex and it contains the ray passing through any of its points and the origin; that is, X is a set of the form

$$\{\alpha_1 \boldsymbol{a}_1 + \alpha_2 \boldsymbol{a}_2 | \; \alpha_1, \alpha_2 \geqslant 0 \text{ and } \boldsymbol{a}_1, \boldsymbol{a}_2 \in X\}. \tag{1.10.1}$$

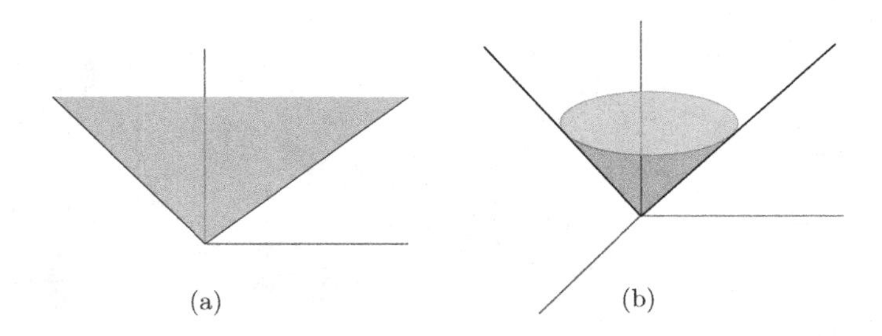

Figure 1.10.1 Parts of cones. (a) Part of a cone in $\mathbb{R}^2$. (b) Part of a cone in $\mathbb{R}^3$.

The expression $\alpha_1 \boldsymbol{a}_1 + \alpha_2 \boldsymbol{a}_2$ where $\alpha_1, \alpha_2 \geqslant 0$ is the *positive combination* or *conical combination* of the points $\boldsymbol{a}_1$ and $\boldsymbol{a}_2$. We consider only convex cones, so we will drop the adjective convex hereafter, unless we want to reinforce the convexity of the cone. Figure 1.10.1 depicts examples of cones in $\mathbb{R}^3$.

The *positive hull* or *conical hull* of a set $X \subseteq \mathbb{R}^d$, denoted cone X, is the smallest cone containing X. It is also defined as the set of all positive combinations of finitely many points of X. That is,

$$\operatorname{cone} X = \{\alpha_1 \boldsymbol{a}_1 + \cdots + \alpha_n \boldsymbol{a}_n |\ \boldsymbol{a}_i \in X \text{ and } \alpha_i \geqslant 0\}.$$

In this case, we say that cone X is *generated* by X and call the elements of X *generators*. And if X is finite, then we call the cone a *V-cone*.

A set $X := \{\boldsymbol{a}_1, \ldots, \boldsymbol{a}_n\}$ is *positively dependent* if some $\boldsymbol{a}_i$ is a positive combination of the others; otherwise it is *positively independent*. The set X *positively spans* a linear subspace L if any vector of the subpace can be expressed as a positive combination of elements of X. And the set X is a *positive basis* of L if it is positively independent and positively spans the subspace. The next characterisation of positively spanning sets is well known.

Theorem 1.10.2 (Davis, 1954, thm. 3.6) *Let $X := \{\boldsymbol{a}_1, \ldots, \boldsymbol{a}_n\}$ with each $\boldsymbol{a}_i \neq \boldsymbol{0}$ such that X linearly spans $\mathbb{R}^d$. Then the following statements are equivalent.*

(i) *The set X positively spans $\mathbb{R}^d$.*
(ii) *For every $i \in [1 \ldots n]$, the point $-\boldsymbol{a}_i \in \operatorname{cone}(X \setminus \{\boldsymbol{a}_i\})$.*
(iii) $\boldsymbol{0} = \sum_{i=1}^{n} \alpha_i \boldsymbol{a}_i$, *with $\alpha_i > 0$ for $i \in [1 \ldots n]$.*

It will often be useful to call translations of convex cones *affine convex cones*, following the same analogy between linear spaces and affine spaces. An *affine convex cone* A in $\mathbb{R}^d$ is a set of the form $\boldsymbol{a}_0 + C$ where $\boldsymbol{a}_0 \in \mathbb{R}^d$ and C is a convex cone in $\mathbb{R}^d$, the translation of C by $\boldsymbol{a}_0$. We often call the point

$\boldsymbol{a}_0$ the *apex* of the cone, or we say that the affine convex cone is *based at* $\boldsymbol{a}_0$. Every (standard) convex cone is an affine convex cone with $\mathbf{0}$ as the apex. In this way, we have that hyperplanes and halfspaces are affine convex cones.

Lineality Spaces and Recession Cones

The dimension of the dual set of a set (Section 1.11) and the structure of an unbounded convex set can be described by the lines and rays that are contained in the set, and specifically by the lineality space and recession cone of the set.

Let X be a nonempty convex set in $\mathbb{R}^d$. The *lineality space* of X is defined as

$$\operatorname{lineal} X := \left\{ \boldsymbol{y} \in \mathbb{R}^d \,\middle|\, \boldsymbol{x} + \alpha \boldsymbol{y} \in X \text{ for all } \boldsymbol{x} \in X \text{ and all } \alpha \in \mathbb{R} \right\}, \tag{1.10.3}$$

and the *recession cone* of X is defined as

$$\operatorname{rec} X := \left\{ \boldsymbol{y} \in \mathbb{R}^d \,\middle|\, \boldsymbol{x} + \alpha \boldsymbol{y} \in X \text{ for all } \boldsymbol{x} \in X \text{ and all } \alpha \geqslant 0 \right\}. \tag{1.10.4}$$

From the definitions, it easily follows that

$$\operatorname{lineal} X = \operatorname{rec} X \cap (-\operatorname{rec} X) \tag{1.10.5}$$

and that the set $\operatorname{lineal} X$ is a linear subspace of $\mathbb{R}^d$.

A convex set X is *pointed* if it contains no lines, which together with the definition of $\operatorname{lineal} X$ implies the following.

Proposition 1.10.6 *A nonempty convex set X in $\mathbb{R}^d$ is pointed if and only if* $\operatorname{lineal} X = \{\mathbf{0}\}$.

Another basic property of lineality spaces is stated below.

Proposition 1.10.7[21] *If X is a nonempty closed convex set in $\mathbb{R}^d$, then*

$$X = \operatorname{lineal} X + \left(X \cap (\operatorname{lineal} X)^{\perp} \right),$$

where $X \cap (\operatorname{lineal} X)^{\perp}$ is a pointed closed convex set.

According to Proposition 1.10.7, every point $\boldsymbol{x}$ of a closed convex set X can be uniquely written as a sum of a point in $\operatorname{lineal} X$ and one in a pointed closed convex set. This decomposition often makes it possible to focus on pointed closed convex sets.

Definitions (1.10.3) and (1.10.4) also ensure, albeit not at once, that the recession cone is a convex cone and that if X is a cone, then $\operatorname{rec} X = X$ (Problem 1.12.17).

[21] A proof is available in Webster (1994, thm. 2.5.8).

1.11 Dual Sets

This section explores another manifestation of duality.

Definition 1.11.1 (Dual set) With each set $X \subseteq \mathbb{R}^d$, we associate the set $X^* \subseteq \mathbb{R}^d$ defined as

$$X^* := \left\{ y \in \mathbb{R}^d \,\middle|\, y \cdot x \leqslant 1 \text{ for every } x \in X \right\}.$$

The set X^* is said to be the *dual* of X.

From this definition, it is plain that the dual of $\varnothing$ and $\{\mathbf{0}\}$ is $\mathbb{R}^d$ and that the dual of $\mathbb{R}^d$ is $\{\mathbf{0}\}$. Moreover, it follows that the set X^* can also be expressed as

$$X^* = \bigcap_{x \in X} \{x\}^* = \bigcap_{x \in X} H^-[x, 1]. \tag{1.11.2}$$

The geometric relation between $x \in X$ and $\{x\}^* = H^-[x, 1]$ is depicted in Fig. 1.11.1. The next proposition is also immediate from the definition of dual sets.

Proposition 1.11.3 *Let X be a subset in $\mathbb{R}^d$. Then X^* is a closed convex set in $\mathbb{R}^d$ that contains the origin.*

Example 1.11.4 We find the dual of a closed ball. Let $r > 0$. Each nonzero point $y := (y_1, \ldots, y_d)^t$ of $B_d^*[\mathbf{0}_d, r]$ satisfies $y \cdot x \leqslant 1$ for every point $x \in B_d[\mathbf{0}_d, r]$ and, in particular, for the point $x_y := r y / \|y\|$ of $B_d[\mathbf{0}_d, r]$. Then

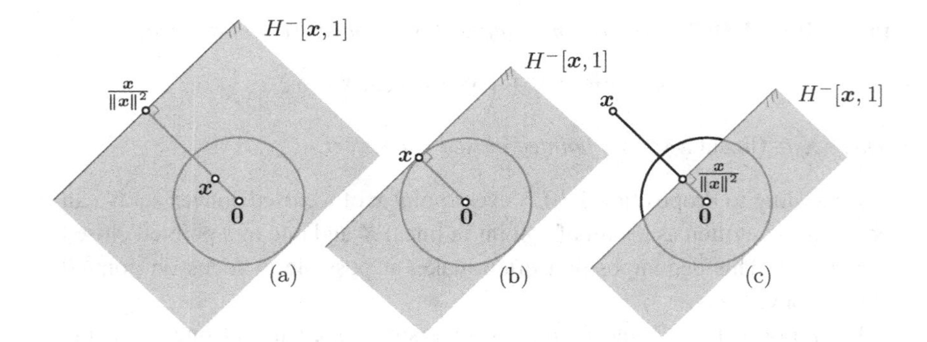

Figure 1.11.1 Geometric relation between x and $H^-[x, 1]$ in $\mathbb{R}^2$. The halfspace $H^-[x, 1]$ is highlighted in grey. The hyperplane bounding $H^-[x, 1]$ passes through $x / \|x\|^2$. Also, the unit ball is depicted to better see the relative Euclidean distance between x and $\mathbf{0}$. (a) $\|x\| < 1$. (b) $\|x\| = 1$. (c) $\|x\| > 1$.

$$\boldsymbol{y} \cdot \boldsymbol{x}_y = \boldsymbol{y} \cdot \frac{r\boldsymbol{y}}{\|\boldsymbol{y}\|} = r\|\boldsymbol{y}\| \leqslant 1.$$

Hence

$$B_d^*[\mathbf{0}_d, r] \subseteq \left\{ \boldsymbol{z} \in \mathbb{R}^d \,\middle|\, \|\boldsymbol{z}\| \leqslant 1/r \right\}.$$

Take $\boldsymbol{z} \in \mathbb{R}^d$ such that $\|\boldsymbol{z}\| \leqslant 1/r$. Then, for every point $\boldsymbol{x}$ in $B_d[\mathbf{0}_d, r]$, the Cauchy–Schwarz inequality (Problem 1.12.1(iii)) ensures that

$$\boldsymbol{z} \cdot \boldsymbol{x} \leqslant |\boldsymbol{z} \cdot \boldsymbol{x}| \leqslant \|\boldsymbol{z}\| \|\boldsymbol{x}\| \leqslant \frac{1}{r} r = 1;$$

here $|x|$ denotes the absolute value of the real number x. Hence, $\boldsymbol{z} \in B_d^*[\mathbf{0}_d, r]$ and

$$B_d^*[\mathbf{0}_d, r] = \left\{ \boldsymbol{z} \in \mathbb{R}^d \,\middle|\, \|\boldsymbol{z}\| \leqslant 1/r \right\} = B_d[\mathbf{0}_d, 1/r]. \qquad (1.11.4.1)$$

This concludes the example.

Example 1.11.4 illustrates another consequence of Definition 1.11.1, the order-reversing inclusion between subsets of X and subsets of X^*:

$$Y \subseteq X \implies X^* \subseteq Y^*. \qquad (1.11.5)$$

For any set $X \subseteq \mathbb{R}^d$, the dual set X^{**} of X^* is well defined. Additionally, if $\boldsymbol{x} \in X$ then, for every $\boldsymbol{y} \in X^*$, we have that $\boldsymbol{x} \cdot \boldsymbol{y} \leqslant 1$ by Definition 1.11.1. Hence $\boldsymbol{x} \in X^{**}$ and

$$X \subseteq X^{**}. \qquad (1.11.6)$$

Combining (1.11.2), (1.11.5), and (1.11.6), we get that the set $X^{**} \subseteq \mathbb{R}^d$ is a closed convex set containing X and the origin. Thus, X^{**} contains the smallest closed convex set in $\mathbb{R}^d$ that contains X and the origin, namely $\mathrm{cl}(\mathrm{conv}(X \cup \{\mathbf{0}\}))$. But more is true: $X^{**} = \mathrm{cl}(\mathrm{conv}(X \cup \{\mathbf{0}\}))$ (Webster, 1994, thm. 2.8.3). The next theorem follows from this discussion.

Theorem 1.11.7[22] *If $X \subseteq \mathbb{R}^d$ is a closed convex set that contains the origin, then $X^{**} = X$.*

Because of Theorem 1.11.7, we are mostly interested in dual sets of closed convex sets that contains the origin.

While taking duals is an involution in the class of closed convex sets in $\mathbb{R}^d$ that contain the origin (Theorem 1.11.7), compactness is not necessarily preserved by this involution. Compactness is, however, preserved in a special subclass.

[22] A proof is available in Webster (1994, thm. 2.8.3).

Theorem 1.11.8[23] *Let $X \subseteq \mathbb{R}^d$ be a closed convex set that contains the origin. The set X^* is bounded if and only if the set X contains the origin in its interior, and vice versa.*

We state a consequence of Theorems 1.11.7 and 1.11.8.

Corollary 1.11.9 *If $X \subseteq \mathbb{R}^d$ is a compact convex set that contains the origin in its interior, then so is X^*. In addition, we have that $X^{**} = X$.*

Linear Subspaces and Cones

In the case of linear subspaces and convex cones, Definition 1.11.1 for the dual set can be sharpened.

Let L be a linear subspace of $\mathbb{R}^d$. We show that if $\boldsymbol{y} \in L^*$, then $\boldsymbol{y} \cdot \boldsymbol{x} = 0$ for every $\boldsymbol{x} \in L$. Suppose otherwise: $\boldsymbol{y} \cdot \boldsymbol{x} \neq 0$ for some $\boldsymbol{x} \in L$. Because L is a linear space, $\alpha\boldsymbol{x} \in L$ for each $\alpha \in \mathbb{R}$. In the case $\boldsymbol{y} \cdot \boldsymbol{x} > 0$, we choose $\alpha > 0$ sufficiently large, and in the case $\boldsymbol{y} \cdot \boldsymbol{x} < 0$, we choose $\alpha < 0$ with $|\alpha|$ sufficiently large. In either case, our chosen α would cause $\boldsymbol{y} \cdot (\alpha\boldsymbol{x}) > 1$. This contradiction validates our claim. Hence,

$$L^* = \left\{ \boldsymbol{y} \in \mathbb{R}^d \,\middle|\, \boldsymbol{y} \cdot \boldsymbol{x} = 0 \text{ for every } \boldsymbol{x} \in L \right\} = L^{\perp}. \tag{1.11.10}$$

In other words, the dual of a linear subspace coincides with its orthogonal complement (Example 1.1.6). Therefore, we will use (1.11.10) as the definition of the *dual linear subspace* L^* of L. It is instructive to compare this discussion on dual linear subspaces with our discussion on dual spaces in Section 1.5.

Let C be a convex cone. The analysis in the previous paragraph also proves that if $\boldsymbol{y} \in C^*$, then $\boldsymbol{y} \cdot \boldsymbol{x} \leqslant 0$ for every $\boldsymbol{x} \in C$. Suppose otherwise: $\boldsymbol{y} \cdot \boldsymbol{x} > 0$ for some $\boldsymbol{x} \in C$. Then, since $\alpha\boldsymbol{x} \in C$ for any $\alpha > 0$, choosing α sufficiently large would cause $\boldsymbol{y} \cdot (\alpha\boldsymbol{x}) > 1$. This contradiction validates our claim. Hence,

$$C^* = \left\{ \boldsymbol{y} \in \mathbb{R}^d \,\middle|\, \boldsymbol{y} \cdot \boldsymbol{x} \leqslant 0 \text{ for every } \boldsymbol{x} \in C \right\} = \bigcap_{\boldsymbol{x} \in C} H^-[\boldsymbol{x}, 0]. \tag{1.11.11}$$

It is clear that $\alpha_1\boldsymbol{y}_1 + \alpha_2\boldsymbol{y}_2 \in C^*$ for every $\boldsymbol{y}_1, \boldsymbol{y}_2 \in C^*$ and every $\alpha_1, \alpha_2 \geqslant 0$. Hence, C^* is a closed convex cone by (1.10.1) and (1.11.11). As a consequence, we will use (1.11.11) as the definition of the *dual cone* C^* of C. Note that our dual cone is sometimes called the polar cone of C; see, for instance, Lauritzen (2013, sec. 3.4)

[23] A proof is available in Webster (1994, thm. 2.8.4).

1.12 Problems

1.12.1 Let x, y, z be three vectors in $\mathbb{R}^d$ and let $\alpha \in \mathbb{R}$. Prove the following properties of the dot product.

(i) $x \cdot y = y \cdot x$.
(ii) $x \cdot x = \|x\|^2 \geqslant 0$, and $x \cdot x = 0$ if and only if $x = \mathbf{0}$.
(iii) $|x \cdot y| \leqslant \|x\| \|y\|$, and equality holds if and only if one vector is a scalar multiple of the other (Cauchy–Schwarz inequality).
(iv) $\|x + y\| \leqslant \|x\| + \|y\|$ (triangle inequality).
(v) $|\,\|x\| - \|y\|\,| \leqslant \|x - y\|$.
(vi) $(\alpha x) \cdot y = \alpha(x \cdot y)$.
(vii) $x \cdot (y + z) = x \cdot y + x \cdot z$.

1.12.2 Prove that a set of $\mathbb{R}^d$ is a linear subspace if and only if it is the solution of a system of homogenous linear equations.

1.12.3 Prove that a set of $\mathbb{A}^d$ is an affine subspace if and only if it is the solution of a system of linear equations.

1.12.4 Let L be a linear subspace of $\mathbb{R}^d$. Prove the following assertions.

(i) $(L^{\perp})^{\perp} = L$.
(ii) $L^{\perp} \cap L = \{\mathbf{0}\}$.
(iii) $\dim L^{\perp} = d - \dim L$.
(iv) Every vector of $\mathbb{R}^d$ can be written uniquely as the sum of a vector in L and a vector in $L^{\perp}$.

1.12.5 Let $M \in \mathbb{R}^{n \times d}$. Prove the following assertions.

(i) $\operatorname{null} M = (\operatorname{row} M)^{\perp}$.
(ii) $\operatorname{null} M^t = (\operatorname{col} M)^{\perp}$.
(iii) $\dim(\operatorname{null} M) + \operatorname{rank} M = d$ (nullity–rank theorem).

1.12.6 Prove the following assertions related to linear and affine maps.

(i) Not every affine map is a linear map.
(ii) An affine map ϱ is linear if and only if $\varrho(\mathbf{0}) = \mathbf{0}$.
(iii) Any affine map can be obtained as a translation of some unique linear map.
(iv) Linear maps send linear subspaces into linear subspaces, and affine maps do the same for affine subspaces.

(v) Injective linear maps send linear subspaces of dimension r into linear subspaces of dimension r, and so do injective affine maps with affine subspaces.
(vi) The set of linear automorphisms form a group, called the *general linear group*, under composition of functions.
(vii) The set of affine automorphisms form a group, called the *affine group*, under composition of functions.
(viii) The set of all linear maps between two linear spaces is itself a linear space.

1.12.7 Define the *image* of a linear map $\varphi\colon X \to Y$ as the linear subspace of Y consisting of the images of X under φ, and define the *rank* of φ as the dimension of its image. Prove that the rank of φ coincides with the rank of any matrix representing it.

1.12.8 (Continuity of linear and affine maps) This exercise explores continuity of linear and affine maps.

A map $\varphi\colon X \to \mathbb{R}^s$ defined on a nonempty set $X \subseteq \mathbb{R}^r$ is said to satisfy a *Lipschitz condition* on X if there exists a real number α such that, for all $\boldsymbol{x}_1, \boldsymbol{x}_2 \in X$, we have that

$$\|\varphi(\boldsymbol{x}_1) - \varphi(\boldsymbol{x}_2)\| \leqslant \alpha \|\boldsymbol{x}_1 - \boldsymbol{x}_2\|.$$

Prove the following.

(i) If a map $\varphi\colon X \to \mathbb{R}^s$ satisfies the Lipschitz condition on a subset $X \subseteq \mathbb{R}^r$, then φ is continuous on X.
(ii) Every linear functional $\varphi(\boldsymbol{x}) := \boldsymbol{a} \cdot \boldsymbol{x}$ for some $\boldsymbol{a} \in \mathbb{R}^d$ is continuous on $\mathbb{R}^d$.
(iii) Every linear map $\varphi\colon \mathbb{R}^r \to \mathbb{R}^s$ is continuous on $\mathbb{R}^r$.
(iv) Every affine map $\varphi\colon \mathbb{R}^r \to \mathbb{R}^s$ is continuous on $\mathbb{R}^r$.

1.12.9 Suppose that $B := \{\boldsymbol{l}_1, \ldots, \boldsymbol{l}_r\}$ is a basis of a linear subspace L of $\mathbb{R}^d$ and let $\boldsymbol{x}$ be any vector of $\mathbb{R}^d$. Write $M := (\boldsymbol{l}_1 \cdots \boldsymbol{l}_r)$. Then M is a $d \times r$ matrix whose columns are these basis vectors. Prove the following.

(i) (1.4.1): $\pi_L(\boldsymbol{x}) := (M(M^t M)^{-1} M^t)\boldsymbol{x}$.
(ii) (1.4.2): if B is an orthogonal basis of L, then $\pi_L(\boldsymbol{x}) := \sum_{i=1}^{r} \frac{\boldsymbol{x} \cdot \boldsymbol{l}_i}{\|\boldsymbol{l}_i\|^2} \boldsymbol{l}_i$.
(iii) The orthogonal projection is a linear transformation.
(iv) $M^t M$ is nonsingular.

1.12.10 Let $L := \left\{x \in \mathbb{R}^d \middle| Nx = \mathbf{0}\right\}$ be a linear subspace of $\mathbb{R}^d$ and let x be any vector of $\mathbb{R}^d$. Prove (1.4.3): $\pi_L(x) = x - (N^t(NN^t)^{-1})(Nx)$.

1.12.11 Let A be an affine subspace of $\mathbb{R}^d$ and let $a_0 \in A$. Suppose that $B := \{a_0, a_1, \ldots, a_r\}$ is a basis of A of $\mathbb{R}^d$ and let x be any vector of $\mathbb{R}^d$. Prove (1.4.6): write $M := (a_1 - a_0 \cdots a_r - a_0)$; then

$$\pi_A(x) = a_0 + \left(M \left(M^t M\right)^{-1} M^t\right)(x - a_0).$$

1.12.12 Let $A := \left\{x \in \mathbb{R}^d \middle| Nx = Na_0,\ a_0 \in A\right\}$ be an affine subspace of $\mathbb{R}^d$ and let x be any vector of $\mathbb{R}^d$. Prove (1.4.7), namely

$$\pi_A(x) = x - (N^t(NN^t)^{-1})(Nx - Na_0).$$

1.12.13 Find convexity-preserving functions that are not affine.

1.12.14 Let $X, Y \subseteq \mathbb{R}^d$ be sets. Prove the following.

(i) $\operatorname{int} X \subseteq \operatorname{rint} X \subseteq X$.
(ii) If $X \subseteq Y$ and $\operatorname{aff} X = \operatorname{aff} Y$, then $\operatorname{rint} X \subseteq \operatorname{rint} Y$.
(iii) $\operatorname{rint}(\operatorname{rint} X) = \operatorname{rint} X$.
(iv) If $\operatorname{rint} X \subseteq Y \subseteq X$ then $\operatorname{rint} X = \operatorname{rint} Y$.
(v) $\operatorname{rint}(X + x) = \operatorname{rint} X + x$ for every $x \in \mathbb{R}^d$.

1.12.15 Let $X \subseteq \mathbb{R}^d$ be a convex set. Prove that $\operatorname{int} X = \operatorname{rint} X$ if and only if $\operatorname{int} X \neq \varnothing$.

1.12.16 (Supporting function) Let $X \subseteq \mathbb{R}^d$ be a nonempty convex set. For every $y \in \mathbb{R}^d$, define the *supporting function* h of X as

$$h(X, y) := \sup\{x \cdot y \mid x \in X\}. \tag{1.12.0.1}$$

It follows that, if $h(X, a) < \infty$ and $a \neq \mathbf{0}_d$, then the hyperplane

$$\left\{z \in \mathbb{R}^d \middle| a \cdot z = h(X, a)\right\}$$

is a supporting hyperplane of X with normal vector a.

Prove that, for any nonempty convex sets $X \subseteq \mathbb{R}^d$ and $Y \subseteq \mathbb{R}^d$, the function h satisfies the following.

(i) $X = \{x \in \mathbb{R}^d \mid x \cdot a \leqslant h(X, a) \text{ for all } a \in \mathbb{R}^d\}$.
(ii) $h(X, \alpha x) = \alpha h(X, x)$ for all $\alpha \geqslant 0$ and all $x \in \mathbb{R}^d$.
(iii) $h(\alpha X, x) = \alpha h(X, x)$ for all $\alpha \geqslant 0$ and all $x \in \mathbb{R}^d$.
(iv) $h(X, x + y) \leqslant h(X, x) + h(X, y)$ for all $x, y \in \mathbb{R}^d$.
(v) $h(X, x) \leqslant h(Y, x)$ if and only if $X \subseteq Y$ for all $x \in \mathbb{R}^d$.

(vi) $h(X+Y,\boldsymbol{x}) = h(X,\boldsymbol{x}) + h(Y,\boldsymbol{x})$.
(vii) If, in addition, X and Y are closed sets that satisfy $h(X,\boldsymbol{x}) = h(Y,\boldsymbol{x})$ for all $\boldsymbol{x} \in \mathbb{R}^d$, then $X = Y$.

1.12.17 Let X be a nonempty closed convex set set in $\mathbb{R}^d$. Prove the following.

(i) lineal X is a linear subspace of $\mathbb{R}^d$.
(ii) rec X is a convex cone.
(iii) If X is a convex cone, then rec $X = X$.
(iv) lineal $\left(X \cap (\text{lineal } X)^{\perp}\right) = \{\mathbf{0}\}$.

1.13 Postscript

The information related to linear subspaces and linear maps (Sections 1.1 and 1.4) can be found in most linear algebra books; for instance, in Shifrin and Adams (2011). For the material on affine subspaces, while fairly standard, one may need look outside linear algebra books; for instance in Webster (1994, chap. 1) or Lauritzen (2013, chap. 2). The presentation in Section 1.2 on the embedding of affine spaces into linear spaces follows that of Berger (2009, ch. 3) and Gallier (2011, ch. 4). The material on projective spaces and projective maps can be found in Berger (2009, ch. 4), Gallier (2011, ch. 5), and Reid and Szendroi (2005, ch. 5); if there is a need for a concise review of results in projective geometry that we did not cover, albeit with no proofs, we recommend Fortuna et al. (2016, ch. 1). The section on dual spaces is based on Halmos (1974, secs. 13–20,67–69).

The presentation of basic convexity in Sections 1.6 to 1.11 is standard and can be found elsewhere, for instance, in Webster (1994); Soltan (2015); Lauritzen (2013); Brøndsted (1983).

Carathéodory's theorem (1.6.3), Radon's theorem (1.6.5), and a theorem of Helly (1923) all appeared in the first half of the 20th century (Carathéodory, 1907; Radon, 1921). Since then they have sparked a great deal of interest in intersection and covering patterns of convex sets. Equally influential was the second wave of such theorems that appeared in the second half of the 20th century, including the colourful Carathéodory theorem (Bárány, 1982) and Tveberg's theorem (Tverberg, 1966). All these results are covered with care in Bárány (2021).

Radon's theorem (1.6.5) can be restated in terms of affine maps: for every $n \geqslant d+1$ and every affine map φ from the $(n-1)$-simplex to $\mathbb{R}^d$, there exists a pair of disjoint faces of the simplex whose φ-images intersect. The equivalence between the statement in Theorem 1.6.5 and this affine formulation stems from noticing that every set $X \subseteq \mathbb{R}^d$ with cardinality n determines a unique affine map φ that takes the extreme points of the $(n-1)$-simplex to the elements of X and, in this way, each face of the simplex is mapped to the convex hull of the images of its extreme points. A topological version of Radon's theorem, due to Bajmóczy and Bárány (1979), replaces the adjective 'affine' with the adjective 'continuous', and thus relaxes the condition on φ. These topics and their topological versions are also presented in Bárány (2021).

The main concepts related to separation and support originated in Minkowski (1896), including the separation theorem (1.8.5). The converse of Theorem 1.8.1 is also true: if $X \subseteq \mathbb{R}^d$ is a nonempty set in which, to each point $\boldsymbol{a}$ in $\mathbb{R}^d$, there is a unique point in X closest to $\boldsymbol{a}$, then X must be closed and convex; this was shown independently by Bunt (1934) and Motzkin (1935).

A more general version of Minkowski–Krein–Milman's theorem (1.9.11) appeared in Krein and Milman (1940). Theorem 1.9.11 is often called Minkowski's theorem because Minkowski (1911, pp. 131–229) proved the finite-dimensional version that we presented.

2

Polytopes

Convex polytopes can be equivalently defined as bounded intersections of finitely many halfspaces in some $\mathbb{R}^d$, and as convex hulls of finitely many points in $\mathbb{R}^d$. A halfspace is defined by a linear inequality, and each nonempty closed convex set in $\mathbb{R}^d$ is the set of solutions of a system of possibly infinitely many linear inequalities. If we have a finite number of inequalities, then the set is a *polyhedron*. Polyhedra are therefore generalisations of polytopes and polyhedral cones. Many assertions in this chapter, for instance the facial structure of polytopes, are derived from analogous assertions about polyhedra.

In this chapter, we learn how to preprocess objects via projective transformations to simplify problem-solving. We then discuss common examples of polytopes such as pyramids, prisms, simple polytopes, and simplicial polytopes. Section 2.10 considers a construction method for polytopes that inductively adds a vertex at each step. For visualising low-dimensional polytopes, we study Schlegel diagrams, a special type of polytopal complex. We also examine key results in polytope theory such as the Euler–Poincaré–Schläfli equation, the 1971 theorem of Bruggesser and Mani on the existence of shellings (orderings of the facets of a polytope with very useful properties), and the Dehn–Sommerville equations for simplicial polytopes. The chapter ends with Gale transforms, a useful device to study polytopes with a small number of vertices.

2.1 Polyhedra

Polyhedra are convex sets that generalise polytopes and polyhedral cones; the latter are defined later in this section. Polyhedra come in two formats: H-polyhedra and V-polyhedra.

An *H-polyhedron P* is the set of solutions of a system of finitely many linear inequalities. Notationally,

$$P = \left\{ \begin{pmatrix} x_1 \\ \vdots \\ x_d \end{pmatrix} \in \mathbb{R}^d \;\middle|\; \begin{matrix} \alpha_{1,1}x_1 + \cdots + \alpha_{1,d}x_d & \leqslant & b_1 \\ & \vdots & \\ \alpha_{n,1}x_1 + \cdots + \alpha_{n,d}x_d & \leqslant & b_n \end{matrix} \right\}, \text{ or alternatively,}$$

$$P = \left\{ \boldsymbol{x} \in \mathbb{R}^d \;\middle|\; M\boldsymbol{x} \leqslant \boldsymbol{b}, \text{ with } M = \begin{pmatrix} \alpha_{1,1} & \cdots & \alpha_{1,d} \\ & \vdots & \\ \alpha_{n,1} & \cdots & \alpha_{n,d} \end{pmatrix}, \boldsymbol{b} = \begin{pmatrix} b_1 \\ \vdots \\ b_n \end{pmatrix} \right\}$$
$$=: P(M, \boldsymbol{b}).$$

We always assume that no two inequalities in the system are the same. For the polyhedron P, we say that an inequality is *redundant* if its elimination does not alter P; otherwise it is *irredundant*. An *H-description* of the polyhedron is a definition of it as an H-polyhedron. When defining H-descriptions, we favour *irredundant* ones, which include only irredundant inequalities; otherwise the H-description is *redundant*.

The definition of an H-polyhedron yields the following at once.

Proposition 2.1.1 *An H-polyhedron in $\mathbb{R}^d$ is a closed convex set.*

Proof Each closed halfspace in $\mathbb{R}^d$ is a closed convex set (see Section 1.6). Besides, the intersection of an arbitrary family of convex sets is a convex set (Theorem 1.6.7), and the intersection of an arbitrary family of closed sets is a closed set. Since an H-polyhedron is the intersection of closed halfspaces, the result follows. □

We next characterise the nonempty H-polyhedra that are cones. Let $\mathbb{R}^{n\times d}$ denote the linear space of $n \times d$ matrices with entries in $\mathbb{R}$.

Proposition 2.1.2 *Let $M \in \mathbb{R}^{n\times d}$. A nonempty H-polyhedron $P(M, \boldsymbol{b})$ in $\mathbb{R}^d$ is a cone if and only if $\boldsymbol{b} = \boldsymbol{0}_n$. It is pointed if and only if* $\operatorname{rank} M = d$.

Proof Suppose $P := P(M, \boldsymbol{0}_n)$; we show that P is a cone. Take $\boldsymbol{a}_1, \boldsymbol{a}_2 \in P$ and $\alpha_1, \alpha_2 \geqslant 0$. From $M\boldsymbol{a}_1 \leqslant \boldsymbol{0}_n$, $M\boldsymbol{a}_2 \leqslant \boldsymbol{0}_n$, and $\alpha_1, \alpha_2 \geqslant 0$, it follows that

$$M(\alpha_1\boldsymbol{a}_1 + \alpha_2\boldsymbol{a}_2) = \alpha_1 M\boldsymbol{a}_1 + \alpha_2 M\boldsymbol{a}_2 \leqslant \boldsymbol{0}_n.$$

Hence $\alpha_1\boldsymbol{a}_1 + \alpha_2\boldsymbol{a}_2 \in P$, implying that P is a cone.

Suppose that $P := P(M, \boldsymbol{b})$ is a cone and $X := \{\boldsymbol{x} \in \mathbb{R}^d \mid M\boldsymbol{x} \leqslant \boldsymbol{0}_n\}$; we show that $P = X$. The point $\boldsymbol{0}_d$ is in P, and so $\boldsymbol{b} \geqslant \boldsymbol{0}_n$. It then follows that $X \subseteq P$. By way of contradiction, suppose that there exists $\boldsymbol{y} \in P \backslash X$. Then $\boldsymbol{r}_i \cdot \boldsymbol{y} > 0$ for some row vector $\boldsymbol{r}_i^t$ of M. For the corresponding entry b_i of $\boldsymbol{b}$, we find that $\boldsymbol{r}_i \cdot \boldsymbol{y} \leqslant b_i$. Since P is a cone, we have that $\alpha\boldsymbol{y} \in P$ for every

$\alpha \geqslant 0$, and consequently that $\boldsymbol{r}_i \cdot (\alpha \boldsymbol{y}) \leqslant b_i$ for every $\alpha \geqslant 0$. But the term $\alpha \boldsymbol{r}_i \cdot \boldsymbol{y}$ cannot be bounded by b_i for every $\alpha \geqslant 0$. This contradiction shows that such a point $\boldsymbol{y}$ does not exist, which implies that $P = X$.

We now prove the second part of the theorem. Let $P = P(M, \mathbf{0}_n)$. Suppose $\operatorname{rank} M < d$; we show that P contains a line. The columns $\boldsymbol{c}_1, \ldots, \boldsymbol{c}_d$ of M are linearly dependent:

$$a_1 \boldsymbol{c}_1 + \cdots + a_d \boldsymbol{c}_d = \mathbf{0}_n \text{ for some } \boldsymbol{a} = (a_1, \ldots, a_d)^t \in \mathbb{R}^d \text{ with } \boldsymbol{a} \neq \mathbf{0}_d.$$

Therefore, $M\boldsymbol{a} = \mathbf{0}_n$. This implies that the line $\{\alpha \boldsymbol{a} \mid \alpha \in \mathbb{R}\}$ is in P. Now suppose that P contains the line $\ell := \{\boldsymbol{a}_1 + \alpha \boldsymbol{a}_2 \mid \alpha \in \mathbb{R}\}$, for some $\boldsymbol{a}_1 \in \mathbb{R}^d$ and $\boldsymbol{a}_2 \neq \mathbf{0}_d$; we show that $M\boldsymbol{a}_2 = \mathbf{0}_n$, which would imply that $\operatorname{rank} M < d$. Suppose that $M\boldsymbol{a}_2 \neq \mathbf{0}_n$. This implies that $\boldsymbol{r}_i \cdot \boldsymbol{a}_2 \neq 0$ for some row $\boldsymbol{r}_i^t$ of M. Then we can find $\beta \in \mathbb{R}$ for which

$$\boldsymbol{r}_i(\boldsymbol{a}_1 + \beta \boldsymbol{a}_2) = \boldsymbol{r}_i \cdot \boldsymbol{a}_1 + \beta \boldsymbol{r}_i \cdot \boldsymbol{a}_2 > 0.$$

This gives that $\ell \nsubseteq P$, a contradiction that settles the part. This completes the proof of the proposition. □

In view of Proposition 2.1.2 we call a set of the form $P(M, \mathbf{0})$ an *H-cone*.

Orthogonally projecting an H-polyhedron onto an affine space produces another H-polyhedron; we present a particular instance of this statement, a geometric interpretation of the so-called *Fourier–Motzkin elimination* (Fourier, 1827; Motzkin, 1936).

Proposition 2.1.3 (Fourier–Motzkin elimination)[1] *Let P be an H-polyhedron in $\mathbb{R}^d$ and let π be the orthogonal projection*

$$(x_1, \ldots, x_d)^t \mapsto (x_1, \ldots, x_{d-1})^t.$$

Then $\pi(P)$ is another H-polyhedron.

We now introduce another type of polyhedra. A *V-polyhedron* P is the sum of a convex hull of finitely many points and a finitely generated cone. Notationally,

$$P = \operatorname{conv} X + \operatorname{cone} Y \text{ for some finite subsets } X, Y \text{ of } \mathbb{R}^d. \tag{2.1.4}$$

This definition implies that V-cones are V-polyhedra (where $X = \{\mathbf{0}_d\}$). A convex cone is *polyhedral* if it is a V-cone or an H-cone.

Definition (2.1.4) also yields that a V-polyhedron is a polytope if and only if $\operatorname{cone} Y = \{\mathbf{0}\}$. We show next that $\operatorname{cone} Y = \{\mathbf{0}\}$ amounts to saying that the V-polyhedron is bounded.

[1] A proof is available in Ziegler (1995, sec. 1.2).

Theorem 2.1.5 *A V-polyhedron is a polytope if and only if it is bounded.*

Proof Let $P := \operatorname{conv} X + \operatorname{cone} Y$ in $\mathbb{R}^d$ for some finite subsets X, Y of $\mathbb{R}^d$. We show that P is bounded if and only if $\operatorname{cone} Y = \{\mathbf{0}\}$.

Suppose that P is bounded. Then $\operatorname{cone} Y = \{\mathbf{0}\}$: if there were a nonzero point $z \in \operatorname{cone} Y$, then $\alpha z \in \operatorname{cone} Y$ for each $\alpha \geqslant 0$, which would violate the boundedness of P.

Suppose that $\operatorname{cone} Y = \{\mathbf{0}\}$. We show that $\operatorname{conv} X$ is bounded. Let $X = \{\boldsymbol{x}_1, \ldots, \boldsymbol{x}_r\}$ and take $z \in \operatorname{conv} X$. Then z can be written as $z = \alpha_1 \boldsymbol{x}_1 + \cdots + \alpha_r \boldsymbol{x}_r$ with $\alpha_i \geqslant 0$ and $\sum_{i=1}^{r} \alpha_i = 1$. It follows that

$$
\begin{aligned}
\|z\| &= \|\alpha_1 \boldsymbol{x}_1 + \cdots + \alpha_r \boldsymbol{x}_r\| \\
&\leqslant \|\alpha_1 \boldsymbol{x}_1\| + \cdots + \|\alpha_r \boldsymbol{x}_r\| \text{ (by the triangle inequality)} \\
&= \alpha_1 \|\boldsymbol{x}_1\| + \cdots + \alpha_r \|\boldsymbol{x}_r\| \text{ (as } \alpha_i \geqslant 0 \text{ for each } i \in [1 \ldots r]) \\
&\leqslant \|\boldsymbol{x}_1\| + \cdots + \|\boldsymbol{x}_r\| \text{ (as } \alpha_i \in [0,1] \text{ for each } i \in [1 \ldots r]).
\end{aligned}
$$

Hence P is bounded. □

Example 2.1.6 (d-cube) We present a *d-dimensional cube* or simply a *d-cube* $Q(d)$ as an H-polyhedron and as a V-polyhedron. Figure 2.1.1 shows cubes in $\mathbb{R}^3$.

Consider the *standard basis* of $\mathbb{R}^d$, namely

$$
\boldsymbol{e}_1 = (1,0,\ldots,0)^t, \ldots, \boldsymbol{e}_d = (0,\ldots,0,1)^t.
$$

Let M be the $2d \times d$ matrix with rows $\boldsymbol{e}_1^t, -\boldsymbol{e}_1^t, \ldots, \boldsymbol{e}_d^t, -\boldsymbol{e}_d^t$. Then

$$
\begin{aligned}
Q(d) &= \left\{ \begin{pmatrix} x_1 \\ \vdots \\ x_d \end{pmatrix} \in \mathbb{R}^d \;\middle|\; \begin{pmatrix} \boldsymbol{e}_1^t \\ -\boldsymbol{e}_1^t \\ \vdots \\ \boldsymbol{e}_d^t \\ -\boldsymbol{e}_d^t \end{pmatrix} \begin{pmatrix} x_1 \\ \vdots \\ x_d \end{pmatrix} \leqslant \begin{pmatrix} 1 \\ \vdots \\ 1 \end{pmatrix} \right\} \\
&= \left\{ \begin{pmatrix} x_1 \\ \vdots \\ x_d \end{pmatrix} \in \mathbb{R}^d \;\middle|\; |x_1| \leqslant 1, \ldots, |x_d| \leqslant 1 \right\}.
\end{aligned}
$$

Let X be the set of 2^d vectors $(\pm 1, \ldots, \pm 1)^t$ in $\mathbb{R}^d$. A d-cube can be alternatively defined as the convex hull of X:

$$
Q(d) = \operatorname{conv} X.
$$

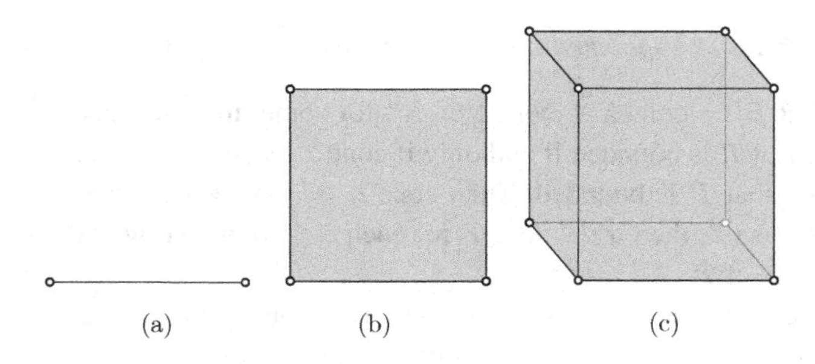

Figure 2.1.1 Cubes in $\mathbb{R}^3$. (a) A 1-cube. (b) A 2-cube. (c) A 3-cube.

Lineality Spaces and Recession Cones

We can readily recover the recession cone, and thus the lineality space, of a polyhedron from its description (see (1.10.5)); these details and the link between the recession cone and the homogenisation cone of a polyhedron ensue.

Definition 2.1.7 (Homogenisation cone) The *homogenisation cone* $\widehat{P}$ of a polyhedron P is a cone in $\mathbb{R}^{d+1}$ whose description is as follows. If $P = P(M,\boldsymbol{b}) \subseteq \mathbb{R}^d$ for some $M \in \mathbb{R}^{n\times d}$ and $\boldsymbol{b} \in \mathbb{R}^n$, then

$$\widehat{P} := \left\{ \begin{pmatrix} \boldsymbol{x} \\ y \end{pmatrix} \in \mathbb{R}^{d+1} \,\middle|\, \boldsymbol{x} \in \mathbb{R}^d,\ y \in \mathbb{R},\ \begin{pmatrix} M & -\boldsymbol{b} \\ \boldsymbol{0}_d^t & -1 \end{pmatrix} \begin{pmatrix} \boldsymbol{x} \\ y \end{pmatrix} \leqslant \boldsymbol{0}_{n+1} \right\}; \quad (2.1.7.1)$$

and if $P = \operatorname{conv} X + \operatorname{cone} Y \subseteq \mathbb{R}^d$ with $X := \{\boldsymbol{a}_1, \dots, \boldsymbol{a}_r\}$ and $Y := \{\boldsymbol{c}_1, \dots, \boldsymbol{c}_s\}$, then

$$\widehat{P} := \operatorname{cone}\left\{ \begin{pmatrix} \boldsymbol{a}_1 \\ 1 \end{pmatrix}, \dots, \begin{pmatrix} \boldsymbol{a}_r \\ 1 \end{pmatrix}, \begin{pmatrix} \boldsymbol{c}_1 \\ 0 \end{pmatrix}, \dots, \begin{pmatrix} \boldsymbol{c}_s \\ 0 \end{pmatrix} \right\}. \quad (2.1.7.2)$$

Figure 2.1.2 sketches the homogenisation cone of a polytope. Compare this terminology with the homogenisation of affine spaces presented in Section 1.2.

Remark 2.1.8 From the definition of a polyhedron $P \subseteq \mathbb{R}^d$ and its homogenisation cone $\widehat{P} \subseteq \mathbb{R}^{d+1}$, it follows that

$$\boldsymbol{x} \in P \text{ if and only if } \begin{pmatrix} \boldsymbol{x} \\ 1 \end{pmatrix} \in \widehat{P}.$$

Theorem 2.1.9 *Let P be a polyhedron in $\mathbb{R}^d$. Then the following hold:*

(i) *If $P = P(M,\boldsymbol{b})$ with $M \in \mathbb{R}^{n\times d}$ and $\boldsymbol{b} \in \mathbb{R}^n$, then*

$$\operatorname{rec} P = \left\{ \boldsymbol{y} \in \mathbb{R}^d \,\middle|\, M\boldsymbol{y} \leqslant \boldsymbol{0}_n \right\} \text{ and } \operatorname{lineal} P = \left\{ \boldsymbol{y} \in \mathbb{R}^d \,\middle|\, M\boldsymbol{y} = \boldsymbol{0}_n \right\}.$$

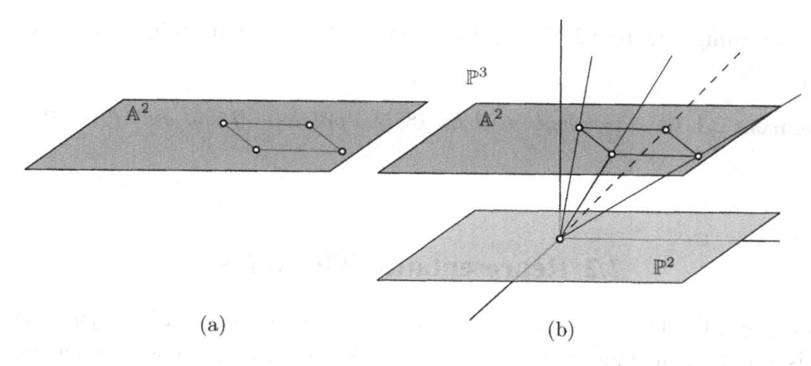

Figure 2.1.2 An affine polytope in $\mathbb{A}^2$ and its homogenisation cone in $\mathbb{P}^3$. (a) A polygon in $\mathbb{A}^2$. (b) The homogenisation cone of the polygon in $\mathbb{P}^3$.

(ii) *If* $P = \operatorname{conv} X + \operatorname{cone} Y$ *for some finite subsets* X, Y *of* $\mathbb{R}^d$, *then* $\operatorname{rec} P = \operatorname{cone} Y$.

(iii) $\widehat{P} = \left\{ \alpha \begin{pmatrix} \boldsymbol{x} \\ 1 \end{pmatrix} \middle| \; \boldsymbol{x} \in P, \alpha > 0 \right\} + \left\{ \begin{pmatrix} \boldsymbol{y} \\ 0 \end{pmatrix} \middle| \; \boldsymbol{y} \in \operatorname{rec} P \right\}$

Proof (i) We prove the equality related to $\operatorname{rec} P$; the one related to lineal P would then follow from (1.10.5).

Suppose that $\boldsymbol{y} \in \operatorname{rec} P$. Then $\boldsymbol{x} + \alpha \boldsymbol{y} \in P$ for each $\boldsymbol{x} \in P$ and each $\alpha \geqslant 0$. In particular, for a fixed $\boldsymbol{x}' \in P$ and every $\alpha \geqslant 0$, we have that $M(\boldsymbol{x}' + \alpha \boldsymbol{y}) \leqslant \boldsymbol{b}$ or, equivalently, that

$$\alpha M \boldsymbol{y} \leqslant \boldsymbol{b} - M \boldsymbol{x}'. \tag{2.1.9.1}$$

Since the right hand side of (2.1.9.1) is a fixed vector, the inequality wouldn't hold for every $\alpha \geqslant 0$ in case that $M\boldsymbol{y} > \boldsymbol{0}$. As a consequence, $M\boldsymbol{y} \leqslant \boldsymbol{0}$, and $\operatorname{rec} P \subseteq \{\boldsymbol{y} \in \mathbb{R}^d \mid M\boldsymbol{y} \leqslant \boldsymbol{0}\}$.

Suppose that $\boldsymbol{y} \in \mathbb{R}^d$ satisfies $M\boldsymbol{y} \leqslant \boldsymbol{0}$. Then $\alpha M \boldsymbol{y} \leqslant \boldsymbol{0}$ for each $\alpha \geqslant 0$. It follows, for every $\boldsymbol{x} \in P$, that $M(\boldsymbol{x} + \boldsymbol{\alpha} \boldsymbol{y}) \leqslant \boldsymbol{b}$; that is, $\boldsymbol{x} + \boldsymbol{\alpha} \boldsymbol{y} \in P$ and $\{\boldsymbol{y} \in \mathbb{R}^d \mid M\boldsymbol{y} \leqslant \boldsymbol{0}\} \subseteq \operatorname{rec} P$. This proves (i).

(ii) Suppose that $\boldsymbol{y} \in \operatorname{rec} P$. Then, for a fixed $\boldsymbol{x}' \in \operatorname{conv} X$ and each $\alpha \geqslant 0$, we have that $\boldsymbol{x}' + \alpha \boldsymbol{y} \in \operatorname{conv} X + \operatorname{cone} Y$. The set $\operatorname{conv} X$ is bounded, and so $\boldsymbol{y} \in \operatorname{cone} Y$. Hence $\operatorname{rec} P \subseteq \operatorname{cone} Y$.

Suppose that $\boldsymbol{y} \in \operatorname{cone} Y$. Then, for each $\alpha \geqslant 0$, we have that $\alpha \boldsymbol{y} \in \operatorname{cone} Y$. It follows that $\boldsymbol{x} + \alpha \boldsymbol{y} \in \operatorname{conv} X + \operatorname{cone} Y$, for each $\alpha \geqslant 0$ and each $\boldsymbol{x} \in P$. Hence $\operatorname{cone} Y \subseteq \operatorname{rec} P$.

(iii) This follows easily from the definitions of a recession cone, given in (1.10.4) and a homogenisation cone (Definition 2.1.7). □

Combining Theorem 2.1.5 and Theorem 2.1.9 gives another characterisation of polytopes.

Theorem 2.1.10 *A polyhedron P in $\mathbb{R}^d$ is a polytope if and only if* $\operatorname{rec} P = \{\mathbf{0}\}$.

2.2 Representation Theorems

Example 2.1.6 describes a d-cube as an H-polyhedron and as a V-polyhedron. This is not a coincidence; H-polyhedra and V-polyhedra are two independent mathematical representations of the same objects.

It has become standard practice to resort to the representation theorem for cones (2.2.1) to prove the representation theorem for polyhedra (2.2.2) and then obtain the representation theorem for polytopes (2.2.4) as a particular case of the one for polyhedra. We follow this approach as well.

Theorem 2.2.1 (Representation theorem for cones)[2] *A subset of $\mathbb{R}^d$ is a V-cone if and only if it is an H-cone.*

Theorem 2.2.2 (Representation theorem for polyhedra) *A subset of $\mathbb{R}^d$ is a V-polyhedron if and only if it is an H-polyhedron.*

Proof In both directions of the proof, given a polyhedron P in $\mathbb{R}^d$, we construct its homogenisation cone $\widehat{P}$ in $\mathbb{R}^{d+1}$ (Definition 2.1.7), which has the property that

$$\boldsymbol{x} \in P \text{ if and only if } \begin{pmatrix} \boldsymbol{x} \\ 1 \end{pmatrix} \in \widehat{P} \qquad (2.2.2.1)$$

(see Remark 2.1.8) and then resort to the representation theorem for cones (2.2.1).

Suppose $P = P(M, \boldsymbol{b})$ is an H-polyhedron in $\mathbb{R}^d$ for some $n \times d$ matrix M and some vector $\boldsymbol{b} \in \mathbb{R}^n$; we represent P as a V-polyhedron. The homogenisation cone $\widehat{P}$ has the form

$$\begin{aligned} \widehat{P} &= \left\{ \begin{pmatrix} \boldsymbol{x} \\ y \end{pmatrix} \in \mathbb{R}^{d+1} \;\middle|\; \boldsymbol{x} \in \mathbb{R}^d,\ y \in \mathbb{R},\ y \geqslant 0,\ M\boldsymbol{x} \leqslant y\boldsymbol{b} \right\} \\ &= \left\{ \begin{pmatrix} \boldsymbol{x} \\ y \end{pmatrix} \in \mathbb{R}^{d+1} \;\middle|\; \boldsymbol{x} \in \mathbb{R}^d,\ y \in \mathbb{R},\ \begin{pmatrix} M & -\boldsymbol{b} \\ \mathbf{0}_d^t & -1 \end{pmatrix} \begin{pmatrix} \boldsymbol{x} \\ y \end{pmatrix} \leqslant \mathbf{0}_{n+1} \right\}. \end{aligned}$$

[2] A proof is available in Ziegler (1995, sec. 1.3).

The representation theorem for cones (2.2.1) ensures that $\widehat{P}$ can be represented as a V-cone in $\mathbb{R}^{d+1}$, say

$$\widehat{P} = \operatorname{cone}\left\{\begin{pmatrix}\boldsymbol{a}_1\\ \alpha_1\end{pmatrix}, \dots, \begin{pmatrix}\boldsymbol{a}_m\\ \alpha_m\end{pmatrix}\right\}. \tag{2.2.2.2}$$

Since $y \geqslant 0$ and the elements of $\widehat{P}$ are positive combinations of the generators of $\widehat{P}$, we may assume that $\alpha_i = 0$ or 1 for each $i \in [1 \dots m]$. Without loss of generality, we further assume that $\alpha_i = 1$ for $i \in [1 \dots r]$ and $\alpha_j = 0$ for $j \in [r+1 \dots m]$. We then partition the set $\{\boldsymbol{a}_1, \dots, \boldsymbol{a}_r, \boldsymbol{a}_{r+1}, \dots, \boldsymbol{a}_m\}$ into subsets X and Y according to the sign of α_i:

$$X = \{\boldsymbol{a}_1, \dots, \boldsymbol{a}_r\} \text{ and } Y = \{\boldsymbol{a}_{r+1}, \dots, \boldsymbol{a}_m\}.$$

By (2.2.2.1) and (2.2.2.2), P can be expressed as

$$P = \operatorname{conv} X + \operatorname{cone} Y,$$

which is a representation of it as a V-polyhedron.

Now suppose that $P = \operatorname{conv} X + \operatorname{cone} Y$ is a V-polyhedron in $\mathbb{R}^d$ where $X := \{\boldsymbol{a}_1, \dots, \boldsymbol{a}_r\}$ and $Y := \{\boldsymbol{c}_1, \dots, \boldsymbol{c}_s\}$. The homogenisation cone of P has the form

$$\widehat{P} = \operatorname{cone}\left\{\begin{pmatrix}\boldsymbol{a}_1\\ 1\end{pmatrix}, \dots, \begin{pmatrix}\boldsymbol{a}_r\\ 1\end{pmatrix}, \begin{pmatrix}\boldsymbol{c}_1\\ 0\end{pmatrix}, \dots, \begin{pmatrix}\boldsymbol{c}_s\\ 0\end{pmatrix}\right\}.$$

By the representation theorem for cones, $\widehat{P}$ can be written as an H-cone $P(N, \boldsymbol{0}_n)$ for some $n \times (d+1)$ matrix N. Let M be the $n \times d$ matrix formed by the first d columns of N and let $-\boldsymbol{b}$ be the last column of N. It follows that

$$\begin{aligned}\widehat{P} &= \left\{\begin{pmatrix}\boldsymbol{x}\\ y\end{pmatrix} \in \mathbb{R}^{d+1} \,\middle|\, \boldsymbol{x} \in \mathbb{R}^d,\ y \in \mathbb{R},\ N\begin{pmatrix}\boldsymbol{x}\\ y\end{pmatrix} \leqslant \boldsymbol{0}_n\right\}\\ &= \left\{\begin{pmatrix}\boldsymbol{x}\\ y\end{pmatrix} \in \mathbb{R}^{d+1} \,\middle|\, \boldsymbol{x} \in \mathbb{R}^d,\ y \in \mathbb{R},\ M\boldsymbol{x} \leqslant y\boldsymbol{b}\right\}.\end{aligned}$$

From (2.2.2.1) we now find that

$$P = \left\{\boldsymbol{x} \in \mathbb{R}^d \,\middle|\, M\boldsymbol{x} \leqslant \boldsymbol{b}\right\},$$

which is a representation of P as an H-polyhedron. This completes the proof of the theorem. □

By the representation theorem for polyhedra (2.2.2), a set is a bounded V-polyhedron if and only if it is a bounded H-polyhedron. Combining this with Theorem 2.1.5, we get at once a characterisation of polytopes.

Theorem 2.2.3 (Polytopes as bounded polyhedra) *A subset of* $\mathbb{R}^d$ *is a polytope if and only if it is a bounded polyhedron.*

Thanks to Theorem 2.2.3, we now have two alternative ways of describing polytopes. An *H-polytope* is a bounded H-polyhedron and a *V-polytope* is a bounded V-polyhedron. Moreover, since polytopes are bounded polyhedra, the representation theorem for polyhedra yields that H-polytopes and V-polytopes are equivalent from a mathematical point of view.

Theorem 2.2.4 (Representation theorem for polytopes) *A subset of* $\mathbb{R}^d$ *is a V-polytope if and only if it is an H-polytope.*

From a computational point of view, a V-polytope is, however, different from an H-polytope. It is trivial to decide whether a given point is in an H-polytope: saying yes if the point satisfies each inequality and no otherwise. It is also trivial to compute the maximum of a linear functional over a V-polytope: evaluate the function at each point in V and return a maximum value. But the standard method to decide whether a point is in a V-polytope is polynomially equivalent to the *basic problem from linear programming* (Fukuda, 2022), the problem of maximising a linear objective function subject to a finite set of linear inequalities. And maximising a linear functional over an H-polytope is essentially the basic problem of linear programming when the intersection of the inequalities is bounded. While linear programming problems, also called *linear programs*, can be solved in polynomial time (Khachiyan, 1979), solving them is certainly not trivial.

2.3 Faces

Convex sets are structured around their faces and these faces can be very heterogeneous: there are convex sets with both exposed and unexposed faces, convex sets with bounded and unbounded faces, and convex sets with a finite number of faces of some dimension and an infinite number of faces of another dimension (Section 1.9). In contrast, the facial structure of polyhedra possesses many attractive properties that are not shared by general convex sets (Theorem 2.3.1). In this section, our focus narrows to explore the faces of polyhedra.

Every proper face of a polyhedron is exposed and is contained in a *facet* of the polyhedron, a face whose dimension is one less than that of the polyhedron. It is also the case that every face of the polyhedron is another polyhedron. We offer the facial structure of polyhedra thereafter.

For a polyhedron $P := \{x \in \mathbb{R}^d \mid Mx \leqslant \mathbf{1}\}$, we say that an inequality is *active* at a subset Y of P if the inequality is satisfied with equality for all points of Y.

Theorem 2.3.1 (Facial structure of polyhedra) *Let $r_1, \ldots, r_n$ be nonzero vectors in $\mathbb{R}^d$, and let P be a d-dimensional polyhedron in $\mathbb{R}^d$ with the H-description*

$$P = \bigcap_{i=1}^{n} \left\{x \in \mathbb{R}^d \mid r_i \cdot x \leqslant 1\right\} = \left\{x \in \mathbb{R}^d \middle| \, Mx \leqslant \mathbf{1}_n\right\},$$

where M is the matrix with the rows $r_1^t, \ldots, r_n^t$. Then the following hold:

(i) *The interior and boundary of P can be expressed as follows:*

$$\operatorname{int} P = \left\{x \in \mathbb{R}^d \middle| \, Mx < \mathbf{1}_n\right\},$$

$$\operatorname{bd} P = \bigcup_{i=1}^{n} \left(P \cap \left\{x \in \mathbb{R}^d \middle| \, r_i \cdot x = 1\right\}\right).$$

(ii) *Every facet of P is exposed, and of the form $P \cap \{x \in \mathbb{R}^d \mid r_j \cdot x = 1\}$ for some $j \in [1 \ldots n]$.*

(iii) *Every set $P \cap \{x \in \mathbb{R}^d \mid r_j \cdot x = 1\}$ is a facet of P if and only if the H-description of P is irredundant.*

(iv) *Every proper face F of P is the intersection of the facets of P that contain it. Thus, if there are t facets containing F and we let $M'x \leqslant \mathbf{1}_t$ be the subsystem of $Mx \leqslant \mathbf{1}_n$ formed by the t inequalities of $Mx \leqslant \mathbf{1}_n$ active at F, then F has the form $\{x \in P \mid M'x = \mathbf{1}_t\}$.*

(v) *Every proper face of P is exposed and polyhedral.*

(vi) *The number of faces of P is finite.*

(vii) *The faces of a face F of P are precisely the faces of P that are contained in F.*

(viii) *For any two proper faces F, K of P, with K not contained in F, there is a facet containing K but not F, and vice versa. In particular, for any two distinct vertices, there is a facet containing one but not the other.*

(ix) *For every proper face F of P, it holds that* $\operatorname{lineal} F = \operatorname{lineal} P$.

Proof Let $H_i := \{x \in \mathbb{R}^d \mid r_i \cdot x = 1\}$ and $H_i^- := \{x \in \mathbb{R}^d \mid r_i \cdot x \leqslant 1\}$, for each $i \in [1 \ldots n]$.

(i) The proof of (i) is simple. The interior of P is the intersection of the interiors of the supporting halfspaces of P (Theorem 1.8.3), which can be found among the halfspaces $H_1^-, \ldots, H_n^-$. Furthermore, the interior of H_i^- is $H_i^- \backslash H_i$. Thus, it follows that

$$\operatorname{int} P = \bigcap_{i=1}^{n} H_i^- \setminus H_i = \left\{ \boldsymbol{x} \in \mathbb{R}^d \,\middle|\, M\boldsymbol{x} < \boldsymbol{1}_n \right\}. \tag{2.3.1.1}$$

Since P is a closed set, the other part follows from (2.3.1.1) and the assertion that $\operatorname{bd} P = P \setminus \operatorname{int} P$ (Proposition 1.7.8).

(ii) Let F be a facet of P. Take $\boldsymbol{z} \in \operatorname{rint} F$. Since $F \subseteq \operatorname{bd} P$ (Theorem 1.9.9), Part (i) gives that $\boldsymbol{z} \in (P \cap H_j)$ for some $j \in [1 \dots n]$. Furthermore, Theorem 1.9.6(iii) yields that $P \cap H_j$ is a proper face of P. According to Theorem 1.9.10, F is the smallest face of P containing $\boldsymbol{z}$ and so $F \subseteq P \cap H_j$. This implies that $F = P \cap H_j$ as $d - 1 = \dim F \leqslant \dim(P \cap H_j) \leqslant d - 1$. Hence F is exposed.

(iii) If the H-description of P is redundant, then there is an index $j \in [1 \dots n]$ such that

$$P = \bigcap_{\substack{i=1 \\ i \neq j}}^{n} H_i^-. \tag{2.3.1.2}$$

We show that $P \cap H_j$ is not a facet of P. Suppose otherwise. Let $\boldsymbol{z} \in \operatorname{rint}(P \cap H_j)$. From (i) and (2.3.1.2) follows the existence of an index $\ell \in [1 \dots n]$ with $\ell \neq j$ such that $\boldsymbol{z} \in (P \cap H_\ell)$. Since $P \cap H_\ell$ is a face of P and since $P \cap H_j$ is a facet and is the smallest face of P containing $\boldsymbol{z}$ (Theorem 1.9.10), we get that $P \cap H_j = P \cap H_\ell$. This implies that $H_j^- = H_\ell^-$, contradicting our running assumption that no two closed halfspaces in the description of a polyhedron are identical. This shows that $P \cap H_j$ is not a facet.

Suppose that the H-description of P is irredundant. Then, for each $i \in [1 \dots n]$, every hyperplane H_i supports P, and so every set $F_i := P \cap H_i$ is a proper face of P by Theorem 1.9.6(iii). Pick $j \in [1 \dots n]$; we show that F_j is a facet of P.

Because the H-description of P is irredundant, there is a point $\boldsymbol{y}_j \in \mathbb{R}^d$ such that

$$\boldsymbol{r}_j \cdot \boldsymbol{y}_j > 1 \text{ and } \boldsymbol{r}_i \cdot \boldsymbol{y}_j \leqslant 1 \text{ for each } i \in [1 \dots n] \text{ with } i \neq j. \tag{2.3.1.3}$$

Now choose a point $\boldsymbol{z} \in \operatorname{int} P$ (which exists by Theorem 1.7.3). Then

$$\boldsymbol{r}_i \cdot \boldsymbol{z} < 1 \text{ for each } i \in [1 \dots n]. \tag{2.3.1.4}$$

Because of (2.3.1.3) and (2.3.1.4), we can find a number $\alpha_j \in (0, 1)$ such that the point $\boldsymbol{u}_j := \alpha_j \boldsymbol{y}_j + (1 - \alpha_j)\boldsymbol{z}$ of the segment $[\boldsymbol{y}_j, \boldsymbol{z}]$ satisfies

$$\boldsymbol{r}_j \cdot \boldsymbol{u}_j = 1 \text{ and } \boldsymbol{r}_i \cdot \boldsymbol{u}_j < 1 \text{ for each } i \in [1 \dots n] \text{ with } i \neq j, \tag{2.3.1.5}$$

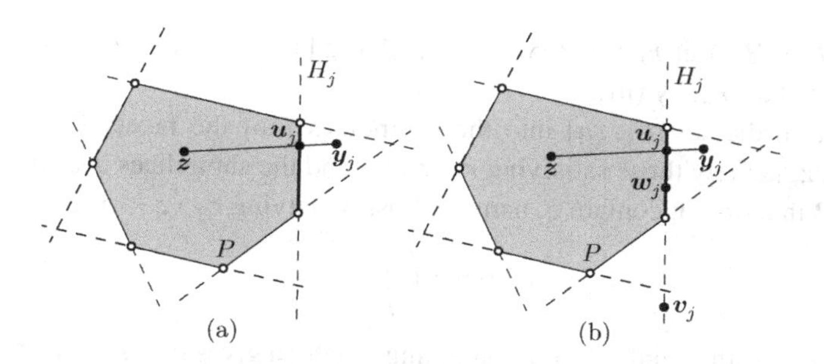

Figure 2.3.1 Auxiliary figure for Theorem 2.3.1. Depicted is a polyhedron P, a hyperplane H_j supporting P, a point z_j in the interior of P, and a point $\boldsymbol{y}_j$ outside P. (a) A point $\boldsymbol{u}_j$ in the segment $[\boldsymbol{y}_j, \boldsymbol{z}]$ satisfying (2.3.1.5). (b) A point $\boldsymbol{v}_j$ lying in a line between the points $\boldsymbol{u}_j$ and $\boldsymbol{w}_j$ of F_j.

namely $\alpha_j = (1 - \boldsymbol{r}_j \cdot \boldsymbol{z})/(\boldsymbol{r}_j \cdot \boldsymbol{y}_j - \boldsymbol{r}_j \cdot \boldsymbol{z})$; see Fig. 2.3.1(a). From Condition (2.3.1.5) it follows that $\boldsymbol{u}_j$ is in F_j. We show that

$$\operatorname{aff} F_j = H_j. \tag{2.3.1.6}$$

Because $F_j = P \cap H_j$, it is clear that $\operatorname{aff} F_j \subseteq H_j$; we prove the other direction. Let $\boldsymbol{v}_j \in H_j$. Choose $\beta_j > 0$ so that

$$\beta_j(\boldsymbol{r}_i \cdot \boldsymbol{v}_j - \boldsymbol{r}_i \cdot \boldsymbol{u}_j) \leqslant 1 - \boldsymbol{r}_i \cdot \boldsymbol{u}_j \text{ for each } i \in [1 \ldots n] \text{ with } i \neq j,$$

which is possible because of Condition (2.3.1.5). The choice of β_j ensures that the point $\boldsymbol{w}_j := \beta_j \boldsymbol{v}_j + (1 - \beta_j)\boldsymbol{u}_j$ satisfies the conditions $\boldsymbol{r}_j \cdot \boldsymbol{w}_j = 1$ and $\boldsymbol{r}_i \cdot \boldsymbol{w}_j \leqslant 1$ for each $i \neq j$:

$$\begin{aligned}
\boldsymbol{r}_j \cdot \boldsymbol{w}_j &= \boldsymbol{r}_j \cdot (\beta_j \boldsymbol{v}_j + (1 - \beta_j)\boldsymbol{u}_j) = \beta_j \boldsymbol{r}_j \cdot \boldsymbol{v}_j + (1 - \beta_j)\boldsymbol{r}_j \cdot \boldsymbol{u}_j \\
&= \beta_j + (1 - \beta_j) = 1, (\text{since } \boldsymbol{v}_j, \boldsymbol{u}_j \in H_j) \\
\boldsymbol{r}_i \cdot \boldsymbol{w}_j &= \boldsymbol{r}_i \cdot (\beta_j \boldsymbol{v}_j + (1 - \beta_j)\boldsymbol{u}_j) = \beta_j \boldsymbol{r}_i \cdot \boldsymbol{v}_j + (1 - \beta_j)\boldsymbol{r}_i \cdot \boldsymbol{u}_j \\
&= \beta_j(\boldsymbol{r}_i \cdot \boldsymbol{v}_j - \boldsymbol{r}_i \cdot \boldsymbol{u}_j) + \boldsymbol{r}_i \cdot \boldsymbol{u}_j \leqslant 1.
\end{aligned}$$

As a consequence, we have that $\boldsymbol{w}_j \in F_j$ and

$$\boldsymbol{v}_j = \frac{1}{\beta_j}\boldsymbol{w}_j + \left(1 - \frac{1}{\beta_j}\right)\boldsymbol{u}_j,$$

and so $\boldsymbol{v}_j$ is in the line between $\boldsymbol{w}_j$ and $\boldsymbol{u}_j$, two points of F_j; see Fig. 2.3.1(b). Hence $\boldsymbol{v}_j \in \operatorname{aff} F_j$. This proves (2.3.1.6) and, with it, that F_j is a facet of P.

(iv) Without loss of generality, we assume that the H-description of P is irredundant. Let F be a proper face of P and let $\boldsymbol{z} \in \operatorname{rint} F$. Since $F \subseteq \operatorname{rbd} P$ (Theorem 1.9.9), part (i) yields that $F \subseteq P \cap H_\ell$ for some $\ell \in [1 \ldots n]$, and so $z \in P \cap H_\ell$. Let $F_r := P \cap H_r$ for each $r \in [1 \dots n]$. Then F_r is a facet for each $r \in [1 \dots n]$, by (iii).

Partition the set $[1 \dots n]$ into the subindices I of the facets F_i of P that contain z, namely those satisfying $\boldsymbol{r}_i \cdot z = 1$ and the subindices J of the facets F_j of P that do not contain z, namely those satisfying $\boldsymbol{r}_j \cdot z < 1$. Let

$$K := \bigcap_{i \in I} F_i.$$

The face F is the smallest face containing z, which gives that $F \subseteq F_i$ for each $i \in I$ and therefore that $F \subseteq K$. Thus K is a face of P that contains F. We show that $z \in \operatorname{rint} K$, which gives us $F = K$.

Because $\boldsymbol{r}_j \cdot z < 1$ for each $j \in J$, we can choose a radius $r > 0$ small enough that the open ball $B(z,r)$ satisfies

$$B(z,r) \subseteq \bigcap_{j \in J} \left\{ \boldsymbol{x} \in \mathbb{R}^d \,\middle|\, \boldsymbol{r}_j \cdot \boldsymbol{x} < 1 \right\}. \tag{2.3.1.7}$$

The definition of K ensures that

$$\operatorname{aff} K \subseteq \bigcap_{i \in I} \left\{ \boldsymbol{x} \in \mathbb{R}^d \,\middle|\, \boldsymbol{r}_i \cdot \boldsymbol{x} = 1 \right\}. \tag{2.3.1.8}$$

By combining (2.3.1.7) and (2.3.1.8) we finally get that

$$B(z,r) \cap \operatorname{aff} K \subseteq K,$$

and therefore we conclude that $z \in \operatorname{rint} K$ (see (1.7.2)). Hence $F = K$ and F is the intersection of the facets that contain it. From (iii), it follows that if $t := |I|$, then

$$F = \{\boldsymbol{x} \in P \mid M'\boldsymbol{x} = \mathbf{1}_t\}$$

for the subsystem $M'\boldsymbol{x} \leqslant \mathbf{1}_t$ of $M\boldsymbol{x} \leqslant \mathbf{1}_n$ formed by the t inequalities of $M\boldsymbol{x} \leqslant \mathbf{1}_n$ active at F.

(v) From (ii) we get that every facet of P is exposed, and from (iv) that every proper face F of P is the intersection of the facets that contain it. That F is exposed now follows from Proposition 1.9.8, which states that the intersection of exposed faces is also exposed.

The assertion of F being polyhedral is an immediate consequence of (iv) as

$$F = \{\boldsymbol{x} \in P \mid M'\boldsymbol{x} = \mathbf{1}'\}$$

for the subsystem $M'\boldsymbol{x} \leqslant \mathbf{1}_t$ of $M\boldsymbol{x} \leqslant \mathbf{1}_n$ formed by the t inequalities of $M\boldsymbol{x} \leqslant \mathbf{1}_n$ active at F.

(vi) This is an easy consequence of (iv). There is a finite number of facets in P and every face of P is the intersection of the facets that contain it.

(vii) Let $K \subseteq F$. If K is a face of P, then the definition of a face (see (1.9.1)) yields that K is a face of the face F. Moreover, since F is a face of P, the transitivity of the relation 'is a face of' on the faces of P (Proposition 1.9.3) ensures that K is a face of P.

(viii) This is a direct consequence of (iv).

(ix) Suppose that $\{\boldsymbol{x} \in \mathbb{R}^d \mid M\boldsymbol{x} \leqslant \mathbf{1}\}$ is an irredundant description of P. Let

$$\left\{\boldsymbol{x} \in \mathbb{R}^d \middle| M'\boldsymbol{x} \leqslant \mathbf{1}'\right\}$$

be the subsystem of $M\boldsymbol{x} \leqslant \mathbf{1}$ comprising all the inequalities of P active at the face F and let

$$\left\{\boldsymbol{x} \in \mathbb{R}^d \middle| M''\boldsymbol{x} \leqslant \mathbf{1}''\right\}$$

be the remaining inequalities of $M\boldsymbol{x} \leqslant \mathbf{1}$. Then according to (iv),

$$F = \left\{\boldsymbol{x} \in \mathbb{R}^d \middle| M'\boldsymbol{x} = \mathbf{1}' \text{ and } M''\boldsymbol{x} \leqslant \mathbf{1}''\right\}. \tag{2.3.1.9}$$

It is now clear from (2.3.1.9) and Theorem 2.1.9 that

$$\begin{aligned} \operatorname{lineal} F &= \left\{\boldsymbol{x} \in \mathbb{R}^d \middle| M'\boldsymbol{x} = \mathbf{0}' \text{ and } M''\boldsymbol{x} = \mathbf{0}''\right\} = \left\{\boldsymbol{x} \in \mathbb{R}^d \middle| M\boldsymbol{x} = \mathbf{0}\right\} \\ &= \operatorname{lineal} P. \end{aligned}$$

This settles the part and, with it, the theorem. □

One of the consequences of Theorem 2.3.1 is that a face F of an H-polyhedron P is the intersection of P with the solution of a system of linear equations. This gives rise to an expression for the dimension of F thanks to Proposition 1.1.8.

Proposition 2.3.2 *Let $M \in \mathbb{R}^{n\times d}$ and let $P := P(M, \mathbf{1}_n)$ be an irredundant H-description of a nonempty d-dimensional polyhedron in $\mathbb{R}^d$. Suppose that a proper face F of P is the solution of the subsystem $M'\boldsymbol{x} \leqslant \mathbf{1}'$ of $M\boldsymbol{x} \leqslant \mathbf{1}_n$ formed by the inequalities of $M\boldsymbol{x} \leqslant \mathbf{1}_n$ active at F:*

$$F = \left\{\boldsymbol{x} \in P \middle| M'\boldsymbol{x} = \mathbf{1}'\right\}.$$

Then

$$\begin{aligned} \operatorname{aff} F &= \left\{\boldsymbol{x} \in \mathbb{R}^d \middle| M'\boldsymbol{x} = \mathbf{1}'\right\}, \\ \dim F &= d - \operatorname{rank} M'. \end{aligned}$$

Proof It suffices to prove the expression for aff F; once this is given, the expression for dim F follows at once from Proposition 1.1.8.

Let $A := \{x \in \mathbb{R}^d \mid M'x = \mathbf{1}'\}$, let $\vec{A} := \{x \in \mathbb{R}^d \mid M'x = \mathbf{0}'\}$ be the direction of A, and let $\{x \in \mathbb{R}^d \mid M''x \leqslant \mathbf{1}''\}$ be the subsystem of $Mx \leqslant \mathbf{1}_n$ comprising all the inequalities of P not active at F. From the definition of F, we have that $F \subseteq A$ and

$$F = \left\{x \in \mathbb{R}^d \middle| \, M'x = \mathbf{1}' \text{ and } M''x \leqslant \mathbf{1}''\right\}. \tag{2.3.2.1}$$

Let $r := d - \operatorname{rank} M'$. Then, according to Proposition 1.1.8, $\dim A = \dim \vec{A} = r$. We find $r + 1$ affinely independent points in F, which will show that aff $F = A$.

We first choose a basis $\boldsymbol{l}_1, \ldots, \boldsymbol{l}_r$ of $\vec{A}$. Then $M'\boldsymbol{l}_i = \mathbf{0}'$ for each $i \in [1 \ldots r]$. Let $z \in \operatorname{rint} F$. Then F is the smallest face containing z (Theorem 1.9.10), which implies that $M''z < \mathbf{1}''$ and $z \notin \vec{A}$. Then for a sufficiently small $\varepsilon > 0$, the points $z, z + \varepsilon\boldsymbol{l}_1, \ldots, z + \varepsilon\boldsymbol{l}_r$ are affinely independent. Additionally, by choosing ε appropriately, we can ensure that all these points are contained within F, with each satisfying (2.3.2.1):

$$\begin{aligned} M'(z + \varepsilon\boldsymbol{l}_i) &= M'z + \varepsilon M'\boldsymbol{l}_i = \mathbf{1}', \\ M''(z + \varepsilon\boldsymbol{l}_i) &= M''z + \varepsilon M''\boldsymbol{l}_i \\ &\leqslant \mathbf{1}'' + \varepsilon M''\boldsymbol{l}_i \leqslant \mathbf{1}''. \end{aligned}$$

Thus $\dim F = r$, concluding that aff $F = A$. □

In the realm of polyhedra, it is customary to call extreme points *vertices* and 1-faces *edges*. We follow the same convention from now on, and for a polyhedron P we denote by $\mathcal{V}(P)$ the set of its vertices and by $v(P)$ the number of elements in $\mathcal{V}(P)$. We also denote by $\mathcal{E}(P)$ the set of edges of P and by $e(P)$ the number of elements in $\mathcal{E}(P)$. The undirected graph formed by the vertices and edges of P, denoted by $G(\mathcal{C})$, is the *graph* of the polyhedron P.

A polyhedron $P(M, \mathbf{1})$ in $\mathbb{R}^d$ is pointed if and only if $\operatorname{rank} M = d$ (Proposition 2.1.2). An alternative characterisation involves the existence of minimal faces. A *minimal face* of a polyhedron P is a proper face that contains no other face of P.

Theorem 2.3.3 (Hoffman and Kruskal 1956) [3]*Let P be a d-dimensional polyhedron in $\mathbb{R}^d$ with the irredundant H-description*

$$P = \left\{x \in \mathbb{R}^d \middle| \, Mx \leqslant \mathbf{1}\right\} \textit{ for some } M \in \mathbb{R}^{n \times d}.$$

[3] A proof is available in Conforti et al. (2014, Thm. 3.3).

Then the following hold:

(i) *A proper face F of P is minimal if and only if*

$$F = \left\{ \boldsymbol{x} \in \mathbb{R}^d \,\middle|\, M'\boldsymbol{x} = \mathbf{1}' \right\} \text{ for some subsystem } M'\boldsymbol{x} \leqslant \mathbf{1}' \text{ of } M\boldsymbol{x} \leqslant \mathbf{1}.$$

That is, $M'\boldsymbol{x} \leqslant \mathbf{1}'$ is formed by the inequalities of $M\boldsymbol{x} \leqslant \mathbf{1}_n$ active at F.

(ii) rank $M' =$ rank M.

(iii) *A minimal face of P is a translate of the lineality space of P.*

A consequence of Theorem 2.3.3 is that a nonempty polyhedron is pointed if and only if its minimal faces are vertices; this is the origin of the term “pointed”: the minimal faces are points. And if the polyhedron has a vertex, then its lineality space must be $\{\mathbf{0}\}$.

Corollary 2.3.4 *A nonempty polyhedron is pointed if and only if it has a vertex.*

A polytope of dimension d is refer to as a *d-polytope*. A *flag* of a d-polytope is a sequence of faces such that each face is a proper face of the next face in the sequence: a sequence

$$F_1 \subset \cdots \subset F_\ell$$

of faces such that $-1 \leqslant \dim F_1 < \cdots < \dim F_\ell \leqslant d$. A flag is *complete* if it includes faces of every dimension from -1 to d.

Theorem 2.3.5 *Let P be a d-polytope in $\mathbb{R}^d$. For every i-face F_i and every j-face F_j of P such that $-1 \leqslant i < j - 1$ and $F_i \subset F_j$, there is a flag in P such that*

$$F_i \subset F_{i+1} \subset \cdots \subset F_{j-1} \subset F_j,$$

and F_ℓ is a facet of $F_{\ell+1}$ for each $\ell \in [i \ldots j-1]$.

Proof First suppose that $-1 < i$, so that $F_i \neq \emptyset$. Then the face F_i is a proper face of F_j by Theorem 2.3.1(v). Since F_j is a polytope there exists a facet F_{j-1} of F_j containing F_i (Theorem 2.3.1). In the case $i = j-2$ we are done. Otherwise, we argue as before, replacing F_j by F_{j-1}, and this argument is repeated $j - i - 2$ times. In this way, we get the desired flag

$$F_i \subset F_{i+1} \subset \cdots \subset F_{j-1} \subset F_j.$$

Now assume that $i = -1$, so that $F_i = \emptyset$. Since F_j is a nonempty polytope, we can find a vertex F_{i+1} in F_j. If $j = 1$ we are home; otherwise we reason as in the previous case for the faces F_{i+1} and F_j, and again obtain the desired flag. □

The proof of Theorem 2.3.5 works for a polyhedron whenever $0 \leqslant i < j-1$. The case $i = -1$ requires the polyhedron to have a vertex, i.e., to be pointed (Corollary 2.3.4), which is not always possible as affine subspaces attest.

A consequence of Theorem 2.3.5 is that if a d-polytope P has a face of dimension k, then P has faces of all dimensions from k to d. Moreover, P contains a vertex (Corollary 2.3.4). We have just established the following.

Corollary 2.3.6 *A d-polytope contains faces of every dimension from* 0 *to* $d-1$.

We denote by f_k the number of k-faces in a d-polytope P. By virtue of Corollary 2.3.6, $f_k(P) \geqslant 1$ for each $k \in [0 \ldots d-1]$ and $f_{-1}(P) = f_d(P) = 1$. The sequence $(f_0, \ldots, f_{d-1})$ is the *f-vector* of P. The f-vector of a polytope plays a central role in the combinatorial theory of polytopes; see, for instance, Chapter 8.

Theorem 1.9.4 characterises faces of convex sets. We refine it next.

Theorem 2.3.7 *Let P be a d-polytope in $\mathbb{R}^d$ with vertex set V and let $W \subseteq V$. Then* $\operatorname{conv} W$ *is a face of P if and only if* $\operatorname{conv}(V \backslash W) \cap \operatorname{aff} W = \emptyset$.

Proof Let $\operatorname{conv} W$ be a face of P. For each vertex $\boldsymbol{u} \in V \backslash W$, the set $P \backslash \{\boldsymbol{u}\}$ is convex (Theorem 1.9.4) and contains W, which yields that $\operatorname{conv} W \subseteq P \backslash \{\boldsymbol{u}\}$. Thus $V \backslash W \subseteq P \backslash \operatorname{conv} W$. Additionally, since $\operatorname{conv} W$ is a face of P we have that $P \backslash \operatorname{conv} W$ is convex (Theorem 1.9.4), which in turn yields that

$$\operatorname{conv}(V \backslash W) \subseteq P \backslash \operatorname{conv} W.$$

Combining this inclusion with $\operatorname{aff} W \cap P = \operatorname{conv} W$ gives that

$$\operatorname{conv}(V \backslash W) \cap \operatorname{aff} W = \emptyset,$$

the necessity of the condition.

Suppose that $V = \{\boldsymbol{v}_1, \ldots, \boldsymbol{v}_n\}$ and $W = \{\boldsymbol{v}_1, \ldots, \boldsymbol{v}_m\}$ with $1 \leqslant m < n$ and that the subset W satisfies $\operatorname{conv}(V \backslash W) \cap \operatorname{aff} W = \emptyset$. If $W = \emptyset$ or V then $\operatorname{conv} W$ is an improper face of P. Suppose otherwise.

Take $\boldsymbol{w} \in \operatorname{conv} W$, and suppose that $\boldsymbol{w} = \alpha \boldsymbol{x} + (1-\alpha)\boldsymbol{y}$ for $\alpha \in [0, 1]$ and $\boldsymbol{x}, \boldsymbol{y} \in P$. Write $\boldsymbol{x}$ and $\boldsymbol{y}$ as a convex combination of V; that is, find scalars $\beta_1, \ldots, \beta_n, \gamma_1, \ldots, \gamma_n \geqslant 0$ such that

$$\sum_{i=1}^{n} \beta_i = \sum_{i=1}^{n} \gamma_i = 1$$

and

$$x = \beta_1 v_1 + \cdots + \beta_n v_n \text{ and } y = \gamma_1 v_1 + \cdots + \gamma_n v_n.$$

Let $\zeta_i := \alpha\beta_i + (1-\alpha)\gamma_i$ for each $i \in [1 \ldots n]$. Then we have that, for each $i \in [1 \ldots n]$, $\zeta_i \geqslant 0$ and $\sum_{i=1}^{n} \zeta_i = 1$. From these expressions of x and y we get that

$$w = \zeta_1 v_1 + \cdots + \zeta_m v_m + \zeta_{m+1} v_{m+1} + \cdots + \zeta_n v_n. \tag{2.3.7.1}$$

Let $\lambda := \zeta_{m+1} + \cdots + \zeta_n$. Suppose that $\lambda > 0$. Then rearranging (2.3.7.1) gives that

$$\frac{1}{\lambda} w - \frac{\zeta_1}{\lambda} v_1 - \cdots - \frac{\zeta_m}{\lambda} v_m = \frac{\zeta_{m+1}}{\lambda} v_{m+1} + \cdots + \frac{\zeta_{m+1}}{\lambda} v_n.$$

The right-hand side is a point in conv $V \backslash W$ (call it z) and the left-hand side expresses z as a point in aff W; note that $1/\lambda - \zeta_1/\lambda - \cdots - \zeta_m/\lambda = 1$. This contradicts the hypothesis $\operatorname{conv}(V \backslash W) \cap \operatorname{aff} W = \varnothing$. Thus $\lambda = 0$, which yields that $\zeta_i = 0$ for each $i \in [m+1 \ldots n]$. The equalities $\zeta_i = 0$ for each $i \in [m+1 \ldots n]$ imply the equalities $\beta_i = 0$ and $\gamma_i = 0$, for each $i \in [m+1 \ldots n]$. Hence x and y are both in conv W, and so conv W is a face of P by (1.9.1). □

Face Lattices

A relation $\leqslant$ on a nonempty set $\mathcal{L}$ is a *partial order* if it is *reflexive*: for every $x \in \mathcal{L}, x \leqslant x$; *antisymmetric*: for every $x, y \in \mathcal{L}$, $x \leqslant y$; and $y \leqslant x$ imply that $x = y$; and *transitive*: for every $x, y, z \in \mathcal{L}$, $x \leqslant y$ and $y \leqslant z$ imply that $x \leqslant z$. A *partially ordered set*, or just *poset*, is a pair $(\mathcal{L}, \leqslant)$ consisting of a nonempty set $\mathcal{L}$ and a partial order $\leqslant$; we write just $\mathcal{L}$ instead of $(\mathcal{L}, \leqslant)$ when the relation is clear from the context. Two elements x and y are said to be *related* or *comparable* if $x \leqslant y$ or $y \leqslant x$; otherwise they are *unrelated* or *incomparable*.

A poset $(\mathcal{L}, \leqslant)$ is *finite* if the set $\mathcal{L}$ is finite. The *Boolean poset* B_n, for some n, is a basic example of a finite poset; it consists of all subsets of a set of n elements, an *n-set* for short, ordered by inclusion.

A poset under which every two elements are comparable is a *linear order*. Any subset of a poset $\mathcal{L}$ is itself a poset, with the partial order induced from $\mathcal{L}$. A linearly ordered subset of $\mathcal{L}$ is a *chain*, whose *length* is the number of elements minus one. An *antichain* is a set of pairwise incomparable elements in $\mathcal{L}$.

Two posets $\mathcal{L}$ and $\mathcal{L}'$ are *isomorphic* if there is an order-preserving bijection σ from $\mathcal{L}$ to $\mathcal{L}'$: for all $x, y \in \mathcal{L}$, $x \leqslant y$ is in $\mathcal{L}$ if and only if $\sigma(x) \leqslant \sigma(y)$ is in $\mathcal{L}'$.

A poset is *bounded* if it contains both a unique maximal element 1 and a unique minimal element 0. It is *graded* if it is bounded and every maximal chain has the same length. A poset is a *lattice* $\mathcal{L}$ if (i) it is bounded, (ii) every pair of elements x and y has a unique minimal upper bound, called the *join* of x and y, and (iii) every pair of elements has a unique maximal lower bound, called the *meet* of x and y. In a graded lattice $\mathcal{L}$, the minimal elements in $\mathcal{L}\setminus\{0\}$ are called *atoms* while the maximal elements in $\mathcal{L}\setminus\{1\}$ are *coatoms*. A graded lattice is *atomic* if every element is a join of atoms and is *coatomic* if every element is a meet of coatoms.

Our interest in posets and lattices stems from the next definition.

Definition 2.3.8 (Face lattice of a polytope) The *face lattice* of a polytope P is the lattice $\mathcal{L}(P)$ of all faces of the polytope, partially ordered by inclusion.

We represent a finite poset $\mathcal{L}$ by a *Hasse diagram*. Each element of $\mathcal{L}$ is represented by a distinct point so that whenever $x \leqslant y$ the point representing x is drawn lower than the point representing y. The face lattice of 3-cube is depicted in Fig. 2.3.2. The empty face is the minimal element and is placed at

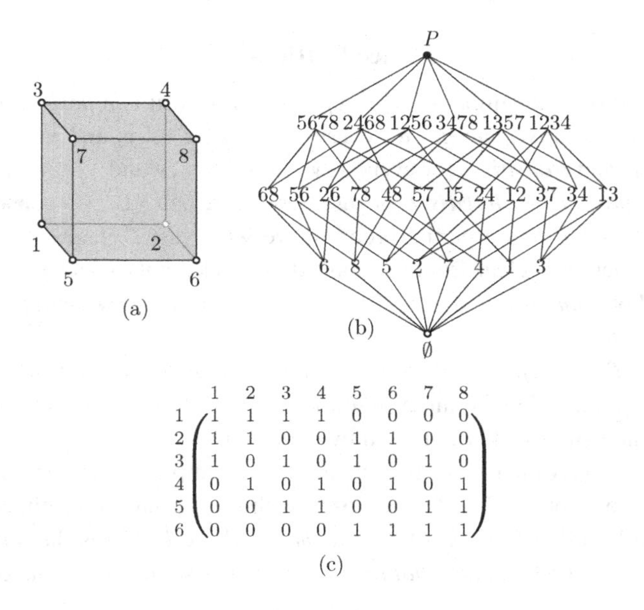

Figure 2.3.2 The face lattice of a 3-cube. (a) A 3-cube with the vertices labelled. The label of each face consists of the vertices contained in it. (b) A Hasse diagram encoding the face lattice of the 3-cube P. (c) A facet-vertex incidence matrix encoding the face lattice of the 3-cube.

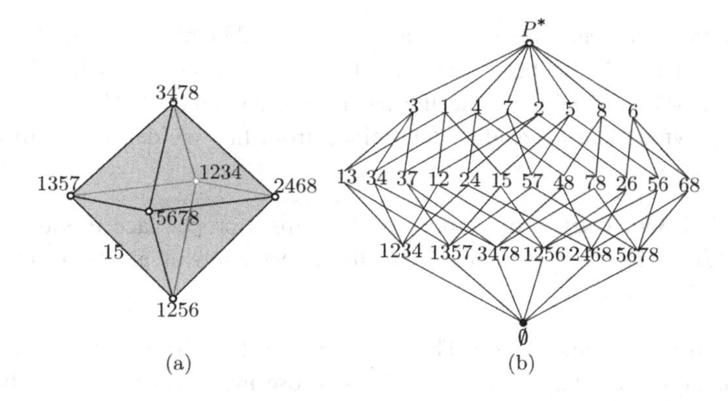

Figure 2.3.3 The face lattice of a 3-crosspolytope. (a) A 3-crosspolytope with the vertices labelled. The label of each face consists of the facets that contain it. For example, the vertex 1256 is contained in the facets $1, 2, 5, 6$, and the edge 15 is contained in the facets $1, 5$ and contains the vertices 1256 and 1357. (b) A Hasse diagram encoding the face lattice of the 3-crosspolytope P^*.

level -1; the vertices are at level 0, the edges at level 1, the 2-faces at level 2, and the polytope at level 3.

Let $(\mathcal{L}, \leqslant)$ be a lattice and let $\mathcal{L}'$ be a nonempty subset of $\mathcal{L}$. Then the partial order $\leqslant$ on $\mathcal{L}$ induces a partial order on $\mathcal{L}'$. The poset $(\mathcal{L}', \leqslant)$ is a *sublattice* of $(\mathcal{L}, \leqslant)$ if, for every two elements x and y of $\mathcal{L}'$, the join and meet of x and y are both in $\mathcal{L}'$. The poset $\mathcal{L}' := \{6, 68, 56, 26, 5678, 2468, 1256, P\}$ in Fig. 2.3.2(b) with the inherited partial order is a sublattice of the face lattice of the 3-cube.

The *opposite poset* $\mathcal{L}^*$ of a poset $\mathcal{L}$ is a poset with the same underlying set $\mathcal{L}$ and relation $\leqslant$, and where $x \leqslant y$ is in $\mathcal{L}^*$ if and only if $y \leqslant x$ is in $\mathcal{L}$. As we will see in Corollary 2.4.11, the opposite of the face lattice of a polytope P is the face lattice of the dual polytope P^* of P. Figure 2.3.3 shows the face lattice $\mathcal{L}(P^*)$ of the 3-crosspolytope P^* as the opposite of the face lattice $\mathcal{L}(P)$ of the 3-cube P; the face lattice $\mathcal{L}(P)$ has been rotated 180° to obtain $\mathcal{L}(P^*)$.

An *antiisomorphism* from a poset $(\mathcal{L}, \leqslant)$ to a poset $(\mathcal{L}', \leqslant)$ is an order-reversing bijection ψ from $\mathcal{L}$ to $\mathcal{L}'$: for all $x, y \in \mathcal{L}$, $x \leqslant y$ is in $\mathcal{L}$ if and only if $\psi(x) \geqslant \psi(y)$. If there is an antiisomorphism between two posets, we say that the posets are *antiisomorphic*. A lattice and its opposite lattice are antiisomorphic.

A *facet-vertex incidence matrix* of a polytope encodes its face lattice in a more efficient way than a Hasse diagram. Each row of the matrix represents a facet, each column a vertex, and the entry (i, j) has a 1 if the facet i contains the vertex j and 0 otherwise. Figure 2.3.2(c) depicts a facet-vertex incidence matrix, where the facet 1234 has label 1, the facet 1256 has label 2, the facet 1357 has label 3, the facet 2468 has label 4, the facet 3478 has label 5, and the facet 5678 has label 6. The other faces of the polytope can be readily determined from this incidence by virtue of Theorem 2.3.1.

Remark 2.3.9 A set X of vertices of P forms a proper face if and only if no vertex in $\mathcal{V}(P)\backslash X$ is contained in the intersection of all the facets that contain X.

We illustrate Remark 2.3.9. The set $X := \{7,8\}$ is a face of the 3-cube; it is contained in the facets 5678 and 3478 whose intersection is precisely 78. However, the set $X' := \{6,7\}$ is not a face; it is contained only in the facet 5678, but there are other vertices in the facet.

Next we gather the main properties of face lattices of polytopes.

Theorem 2.3.10 *Let $\mathcal{L}$ be the face lattice of a polytope.*

(i) *The elements 0 and 1 correspond to the empty face and the polytope, respectively.*
(ii) *The minimal elements in $\mathcal{L}\backslash\{0\}$, the atoms of the lattice, are the vertices of the polytope.*
(iii) *Every face of the polytope is a join of vertices.*
(iv) *The maximal elements in $\mathcal{L}\backslash\{1\}$, the coatoms of the lattice, are the facets of the polytope.*
(v) *Every face of the polytope is the intersection of facets.*
(vi) *The lattice $\mathcal{L}$ is finite, graded, atomic, and coatomic.*

Two polytopes P and P' are *combinatorially isomorphic* if their face lattices are isomorphic. We may also say that the polytopes P and P' are of the same *combinatorial type*. Unless otherwise stated we do not distinguish between combinatorially isomorphic polytopes and thus write $P = P'$.

We often need to embed or 'realise' a polytope or a combinatorial type in some space $\mathbb{R}^d$. A *realisation* of a polytope P with vertices $\boldsymbol{v}_1, \ldots, \boldsymbol{v}_n$ is a polytope $P' := \operatorname{conv}\{\boldsymbol{u}_1, \ldots, \boldsymbol{u}_n\}$ where for each $i \in [i \ldots n]$ $\boldsymbol{u}_i$ is a point in $\mathbb{R}^d$ and the mapping $\boldsymbol{v}_1 \mapsto \boldsymbol{u}_1, \ldots, \boldsymbol{v}_n \mapsto \boldsymbol{u}_n$ is an isomorphism of the face lattices of P and P'. In this way, the polytope P' is an embedding in $\mathbb{R}^d$ of the combinatorial type of P. Researchers are often interested in the set of all realisations of a combinatorial type, which is formalised by the *realisation space of a polytope*. Realisation spaces of polytopes are the central topic of Richter-Gebert (2006).

2.4 Dual Polytopes

In the case of polyhedra, the definition of dual set gives rise to the *dual polyhedron* P^* of a polyhedron P. If the polyhedron happens to be a cone we will use the equivalent definition of the dual cone for the dual polyhedron. There is a recipe to go from a polyhedron that contains the origin to its dual.

Theorem 2.4.1 *The dual of a V-polyhedron in $\mathbb{R}^d$ that contains the origin is an H-polyhedron in $\mathbb{R}^d$ that contains the origin, and vice versa. More precisely, let $X := \{\boldsymbol{x}_1, \ldots, \boldsymbol{x}_r\} \subseteq \mathbb{R}^d$ and $Y := \{\boldsymbol{y}_1, \ldots, \boldsymbol{y}_s\} \subseteq \mathbb{R}^d$, let M be the matrix with rows $\boldsymbol{x}_1^t, \ldots, \boldsymbol{x}_r^t$, and let N be the matrix with rows $\boldsymbol{y}_1^t, \ldots, \boldsymbol{y}_s^t$. Then the following hold:*

(i) *If $P := \operatorname{conv}(X \cup \{\boldsymbol{0}_d\}) + \operatorname{cone} Y$ then $P^* = \{z \in \mathbb{R}^d \mid Mz \leqslant \boldsymbol{1}_r,\ Nz \leqslant \boldsymbol{0}_s\}$.*

(ii) *If instead*

$$P := \left\{ z \in \mathbb{R}^d \middle| \begin{pmatrix} M \\ N \end{pmatrix} z \leqslant \begin{pmatrix} \boldsymbol{1}_r \\ \boldsymbol{0}_s \end{pmatrix} \right\} = \left\{ z \in \mathbb{R}^d \middle| \ Mz \leqslant \boldsymbol{1}_r,\ Nz \leqslant \boldsymbol{0}_s \right\},$$

then $P^* = \operatorname{conv}(X \cup \{\boldsymbol{0}_d\}) + \operatorname{cone} Y$.

Proof (i) Suppose that $P := \operatorname{conv}(X \cup \{\boldsymbol{0}_d\}) + \operatorname{cone} Y$ and that $\boldsymbol{w} \in P^*$. Additionally, let

$$Q := \left\{ z \in \mathbb{R}^d \middle| \ Mz \leqslant \boldsymbol{1}_r,\ Nz \leqslant \boldsymbol{0}_s \right\}.$$

Since $\boldsymbol{x}_1, \ldots, \boldsymbol{x}_r \in P$ we find that $\boldsymbol{w} \cdot \boldsymbol{x}_i \leqslant 1$ for $i \in [1 \ldots r]$ by Definition 1.11.1. And since $\boldsymbol{y}_1, \ldots, \boldsymbol{y}_s \in P$ and $\operatorname{cone} Y \subseteq P$ we find that $\boldsymbol{w} \cdot \boldsymbol{y}_j \leqslant 0$ for $j \in [1 \ldots s]$ by (1.11.11). Hence $\boldsymbol{w} \in Q$ and $P^* \subseteq Q$.

Suppose that $\boldsymbol{w} \in Q$. Take $\boldsymbol{u} \in P$. Then there exist scalars $\alpha_1 \geqslant 0, \ldots, \alpha_r \geqslant 0$ with $\sum_{i=1}^r \alpha_i = 1$ and scalars $\beta_1 \geqslant 0, \ldots, \beta_s \geqslant 0$ for which

$$\begin{aligned}
\boldsymbol{w} \cdot \boldsymbol{u} &= \boldsymbol{w} \cdot (\alpha_1 \boldsymbol{x}_1 + \cdots + \alpha_r \boldsymbol{x}_r + \beta_1 \boldsymbol{y}_1 + \cdots + \beta_s \boldsymbol{y}_s) \\
&= \alpha_1 \boldsymbol{w} \cdot \boldsymbol{x}_1 + \cdots + \alpha_r \boldsymbol{w} \cdot \boldsymbol{x}_r + \beta_1 \boldsymbol{w} \cdot \boldsymbol{y}_1 + \cdots + \beta_s \boldsymbol{w} \cdot \boldsymbol{y}_s \\
&\leqslant \alpha_1 + \cdots \alpha_r + 0 + \cdots + 0 = 1.
\end{aligned}$$

The last inequality follows from the definition of Q. Hence $\boldsymbol{w} \in P^*$ and $Q \subseteq P^*$. As a consequence, $P^* = Q$.

(ii) Suppose that P is given as in (ii) and that $Q := \operatorname{conv}(X \cup \{\boldsymbol{0}\}) + \operatorname{cone} Y$. By (i) we have that $Q^* = P$. And from $Q^{**} = Q$ (Theorem 1.11.7) it follows that $P^* = Q^{**} = Q$, as desired. □

In the particular case of cones, Theorem 2.4.1 reduces to the following.

Theorem 2.4.2 *The dual cone of a V-cone in* $\mathbb{R}^d$ *is an H-cone in* $\mathbb{R}^d$*, and vice versa. More precisely,*

(i) *if* $C := \text{cone}\{\boldsymbol{x}_1, \dots, \boldsymbol{x}_n\}$ *then, letting* $M \in \mathbb{R}^{n\times d}$ *be the matrix with rows* $\boldsymbol{x}_1^t, \dots, \boldsymbol{x}_n^t$*, we have that* $C^* = \{\boldsymbol{y} \in \mathbb{R}^d \mid M\boldsymbol{y} \leqslant \boldsymbol{0}_n\}$*; and*
(ii) *if* $C := P(M, \boldsymbol{0}_n)$ *where* M *is a* $n \times d$ *matrix with rows* $\boldsymbol{x}_1^t, \dots, \boldsymbol{x}_n^t$*, then* $C^* = \text{cone}\{\boldsymbol{x}_1, \dots, \boldsymbol{x}_n\}$.

Polytopes

If a polytope P in $\mathbb{R}^d$ contains the origin in its interior then the dual set P^* of P is a polyhedron by Theorem 2.4.1 and is bounded by Theorem 1.11.8. Thus P^* is a polytope (Theorem 2.2.3), and so we call it the *dual polytope* of P. The bounded case of Theorem 2.4.1 gives a recipe to go from a polytope that contains the origin in its interior to its dual; this is summarised next.

Theorem 2.4.3 *Let* $X := \{\boldsymbol{x}_1, \dots, \boldsymbol{x}_r\} \subseteq \mathbb{R}^d$ *and let* M *be the matrix with rows* $\boldsymbol{x}_1^t, \dots, \boldsymbol{x}_r^t$*. Then the following hold:*

(i) *If* $P := \text{conv}\, X$ *and* P *contains the origin in its interior, then*

$$P^* = \left\{ z \in \mathbb{R}^d \,\middle|\, Mz \leqslant \boldsymbol{1}_r \right\}$$

and P^* *contains the origin in its interior.*
(ii) *If instead* $P := \{z \in \mathbb{R}^d \mid Mz \leqslant \boldsymbol{1}_r\}$ *and* P *contains the origin in its interior then* $P^* = \text{conv}\, X$ *and* P^* *contains the origin in its interior.*
(iii) *Suppose that* $P := \text{conv}\, X$ *contains the origin in its interior. Then* $\mathcal{V}(P) = X$ *if and only if* $\{z \in \mathbb{R}^d \mid Mz \leqslant \boldsymbol{1}_r\}$ *is an irredundant H-description of* P^*.

Proof Parts (i) and (ii) are the bounded case of Theorem 2.4.1. We prove (iii). Since $\boldsymbol{0}_d \in \text{int}\, P$, we have that $\#X \geqslant 2$. We prove the contrapositive of both directions.

Suppose that $\mathcal{V}(P) \subset X$, say $\boldsymbol{x}_\ell \notin \mathcal{V}(P)$, and let

$$P_\ell := \text{conv}(X \backslash \{\boldsymbol{x}_\ell\}) \text{ and } Q_\ell := \{z \in \mathbb{R}^d \mid M_\ell z \leqslant \boldsymbol{1}_{r-1}\},$$

where M_ℓ is obtained from M by removing the row $\boldsymbol{x}_\ell^t$. Because $\boldsymbol{0}_d \in \text{int}\, P$, it follows that $\boldsymbol{0}_d \in \text{int}\, P_\ell$, and so Part (i) yields that $P_\ell^* = Q_\ell$. Besides, Minkowski–Krein–Milman's theorem (1.9.11) ensures that $P = \text{conv}\, \mathcal{V}(P)$, and so $P = P_\ell$. From $P = P_\ell$ it follows that $P^* = P_\ell^*$, which implies that the H-description $\{z \in \mathbb{R}^d \mid Mz \leqslant \boldsymbol{1}_r\}$ of P^* is redundant.

Suppose that the H-description $\{z \in \mathbb{R}^d \mid Mz \leqslant \mathbf{1}_r\}$ of P^* is redundant, say

$$P^* = \left\{z \in \mathbb{R}^d \middle| \, M_\ell z \leqslant \mathbf{1}_{r-1}\right\}$$

where M_ℓ is obtained from M by removing the row $\boldsymbol{x}_\ell^t$. Then $\mathbf{0}_d \in \operatorname{int} P^*$. Let $P_\ell := \operatorname{conv}(X \backslash \{\boldsymbol{x}_\ell\})$. Part (ii) yields that $P^{**} = P_\ell$. The polytope P contains the origin in its interior, which implies that $P = P^{**}$ (Corollary 1.11.9); that is, $P = P_\ell$. Again by Minkowski–Krein–Milman's theorem (1.9.11), we have that $P = P_\ell = \operatorname{conv} \mathcal{V}(P)$, resulting in $\mathcal{V}(P) \subset X$. □

Example 2.4.4 We find the dual of a d-cube $Q(d)$ that is given as an H-polytope (Example 2.1.6) in two different ways: (i) reasoning as in Example 1.11.4 and (ii) following the recipe of Theorem 2.4.3. An H-description of $Q(d)$ is as follows:

$$\begin{aligned} Q(d) &= \left\{(x_1, \ldots, x_d)^t \middle| \, |x_1| \leqslant 1, \ldots, |x_d| \leqslant 1\right\} \\ &= \left\{ z \in \mathbb{R}^d \,\middle|\, \begin{pmatrix} \boldsymbol{e}_1^t \\ -\boldsymbol{e}_1^t \\ \vdots \\ \boldsymbol{e}_d^t \\ -\boldsymbol{e}_d^t \end{pmatrix} z \leqslant \mathbf{1}_{2d} \right\}. \end{aligned} \tag{2.4.4.1}$$

(i) Each point $\boldsymbol{y} := (y_1, \ldots, y_d)^t$ of $Q(d)^*$ satisfies $\boldsymbol{y} \cdot \boldsymbol{x} \leqslant 1$ for every point $\boldsymbol{x} \in Q(d)$ (Definition 1.11.1), and, in particular, for the point $\boldsymbol{x}_y := (\operatorname{sign} y_1, \ldots, \operatorname{sign} y_d)^t$ of $Q(d)$. Here, $\operatorname{sign} y$ denotes the sign function: $\operatorname{sign} y = -1$ if $y < 0$, $\operatorname{sign} y = 0$ if $y = 0$, and $\operatorname{sign} y = 1$ if $y > 0$. Then

$$\boldsymbol{y} \cdot \boldsymbol{x}_y = y_1 \operatorname{sign} y_1 + \cdots + y_d \operatorname{sign} y_d = |y_1| + \cdots + |y_d| \leqslant 1.$$

Hence

$$Q(d)^* \subseteq \left\{(z_1, \ldots, z_d)^t \in \mathbb{R}^d \middle| \, |z_1| + \cdots + |z_d| \leqslant 1\right\}.$$

Take $z \in \mathbb{R}^d$ such that $|z_1| + \cdots + |z_d| \leqslant 1$. Then, for every point $\boldsymbol{x}$ in $Q(d)$, we have that

$$\begin{aligned} z \cdot \boldsymbol{x} = z_1 x_1 + \cdots + z_d x_d &\leqslant |z_1||x_1| + \cdots + |z_d||x_d| \\ &\leqslant |z_1| + \cdots + |z_d| \leqslant 1. \end{aligned}$$

Hence $z \in Q(d)^*$, and

$$Q(d)^* = \left\{ (z_1, \ldots, z_d)^t \in \mathbb{R}^d \,\middle|\, |z_1| + \cdots + |z_d| \leqslant 1 \right\}$$
$$= \left\{ z \in \mathbb{R}^d \,\middle|\, \begin{pmatrix} 1 & 1 & \cdots & 1 & 1 \\ -1 & 1 & \cdots & 1 & 1 \\ & & \cdots & & \\ -1 & -1 & \cdots & -1 & -1 \end{pmatrix} z \leqslant \mathbf{1}_{2^d} \right\}.$$

(ii) Applying the recipe of Theorem 2.4.3 to (2.4.4.1) we get that

$$Q(d)^* = \operatorname{conv}\{\boldsymbol{e}_1, -\boldsymbol{e}_1, \ldots, \boldsymbol{e}_d, -\boldsymbol{e}_d\}.$$

The polytope $Q(d)^*$ is known as a *d-crosspolytope* and is denoted by $I(d)$.

Dimension of the Dual Polytope

The lineality space of a polyhedron is closely linked to the dimension of the dual polyhedron. The next proposition gives the relevant result.

Proposition 2.4.5 *If P is a polyhedron in $\mathbb{R}^d$ that contains the origin, then*

(i) aff P^* *is the orthogonal complement of* lineal P,
(ii) $\dim P^* = d - \dim(\operatorname{lineal} P)$, *and*
(iii) $\dim P = d - \dim(\operatorname{lineal} P^*)$.

Proof (i) Suppose that P is given as the H-polyhedron

$$P = \left\{ z \in \mathbb{R}^d \,\middle|\, \begin{pmatrix} M \\ N \end{pmatrix} z \leqslant \begin{pmatrix} \mathbf{1}_r \\ \mathbf{0}_s \end{pmatrix} \right\} = \left\{ z \in \mathbb{R}^d \,\middle|\, Az \leqslant \begin{pmatrix} \mathbf{1}_r \\ \mathbf{0}_s \end{pmatrix} \right\},$$

where M is the matrix in $\mathbb{R}^{r \times d}$ with rows $\boldsymbol{x}_1^t, \ldots, \boldsymbol{x}_r^t$, N is the matrix in $\mathbb{R}^{s \times d}$ with rows $\boldsymbol{y}_1^t, \ldots, \boldsymbol{y}_s^t$, and A is the matrix in $\mathbb{R}^{(r+s) \times d}$ with rows $\boldsymbol{x}_1^t, \ldots, \boldsymbol{x}_r^t, \boldsymbol{y}_1^t, \ldots, \boldsymbol{y}_s^t$.

By Theorem 2.4.1, the dual P^* of P can be written as

$$P^* = \operatorname{conv}(X \cup \{\mathbf{0}_d\}) + \operatorname{cone} Y,$$

where $X = \{\boldsymbol{x}_1, \ldots, \boldsymbol{x}_r\}$ and $Y = \{\boldsymbol{y}_1, \ldots, \boldsymbol{y}_s\}$. Since $\mathbf{0} \in P^*$, we have that the affine hull of P^* coincides with its linear hull. As a consequence, we further have that aff P^* is linearly spanned by $X \cup Y$ and coincides with the row space of A (see Example 1.1.6).

According to Theorem 2.1.9, lineal $P = \{z \in \mathbb{R}^d \mid Az = \mathbf{0}_{r+s}\}$; that is, lineal P is the nullspace of A. The row space of A is the orthogonal complement of its nullspace by the nullity–rank theorem (Problem 1.12.5), and therefore

$$\operatorname{aff} P^* = (\operatorname{lineal} P)^{\perp},$$

as desired.

(ii) The nullity–rank theorem applied to (i) gives (ii).

(iii) The polyhedron P contains the origin, and so $P^{**} = P$ by Theorem 1.11.7. Moreover, P^* is another polyhedron in $\mathbb{R}^d$ that contains the origin (Theorem 2.4.1). Part (iii) is confirmed by applying (ii) to P^*. □

We remark that Proposition 2.4.5 remains true in the more general case of P being a closed convex set in $\mathbb{R}^d$ that contains the origin. The subsequent corollary of Proposition 2.4.5 follows at once.

Corollary 2.4.6 *If P is a pointed, full-dimensional polyhedron in $\mathbb{R}^d$ that contains the origin, then so is the dual of P.*

Proof By Theorem 2.4.1, the dual P^* of P is a polyhedron that contains the origin.

If P is pointed, then $\operatorname{lineal} P = \{\mathbf{0}\}$ (Section 1.10). By Proposition 2.4.5, this implies that $\dim P^* = d$. Since P is full-dimensional, Proposition 2.4.5 again yields that $\operatorname{lineal} P^* = \{\mathbf{0}\}$, which is equivalent to saying that P^* is pointed. Hence P^* is pointed and full-dimensional.

Finally, the dual polyhedron contains the origin (Proposition 1.11.3), and so the corollary follows. □

The subsequent statement is a consequence of Theorem 2.4.3 and Corollary 2.4.6.

Proposition 2.4.7 *If a d-polytope contains the origin in its interior, then its dual is also a d-polytope that contains the origin in its interior.*

Conjugate Faces

Let P be a polytope that contains the origin in its interior. We next explore the relationship between the faces of P and the faces of the dual polytope P^* of P. For a face F of P, we define the set

$$F^{\vartriangle} := \{\boldsymbol{y} \in P^* \mid \boldsymbol{y} \cdot \boldsymbol{x} = 1 \text{ for every } \boldsymbol{x} \in F\} = \bigcap_{\boldsymbol{x} \in F} \left(P^* \cap H(\boldsymbol{x}, 1)\right). \quad (2.4.8)$$

For exposed faces I and F of P, Definition (2.4.8) gives that

$$\text{if } I \text{ is a face of } F \text{ and } F \text{ is a face of } P \text{ then } I^{\vartriangle} \supseteq F^{\vartriangle}. \quad (2.4.9)$$

The next theorem motivates the definition (2.4.8).

Theorem 2.4.10 *Let $P \subseteq \mathbb{R}^d$ be a d-polytope that contains the origin in its interior. Suppose that F is a proper face of P. Then the following hold:*

(i) *$F^\triangle$ is a proper face of the dual polytope P^*.*
(ii) *A point $\boldsymbol{a}$ is in F if and only if $H(\boldsymbol{a}, 1)$ is a hyperplane supporting P^* and containing $F^\triangle$.*
(iii) *The point $\boldsymbol{a}$ is in* rint F *if and only if $F^\triangle = P^* \cap H(\boldsymbol{a}, 1)$.*
(iv) *$F^{\triangle\triangle} = F$.*
(v) *There exists an antiisomophism ψ from the face lattice $\mathcal{L}(P)$ of P to the face lattice $\mathcal{L}(P^*)$ of P^* that sends each face F of P onto the face $F^\triangle$ of P^*.*

Proof The proofs of (i)–(iv) follow from Brøndsted (1983, thms. 6.6, 6.7). Part (v) is a direct consequence of (i), (iv), and (2.4.9). □

Let P be a d-polytope that contains the origin in its interior. For an exposed face F of P, by virtue of Theorem 2.4.10(i) we say that the face $F^\triangle$ of the dual polytope P^* is the *conjugate face* of F. And by virtue of Theorem 2.4.10(iv), we have that the conjugate face of $F^\triangle$ is F. We often say that F and $F^\triangle$ are conjugate.

A direct consequence of Theorem 2.4.10(v) is the following.

Corollary 2.4.11 *Let $P \subseteq \mathbb{R}^d$ be a d-polytope that contains the origin in its interior and let P^* be the dual polytope of P. Then the face lattice $\mathcal{L}(P^*)$ of P^* is isomorphic to the opposite lattice $\mathcal{L}(P)^*$ of the face lattice $\mathcal{L}(P)$ of P.*

For any d-polytope P in $\mathbb{R}^d$, there is a d-polytope Q in $\mathbb{R}^d$ that contains the origin in its interior and that is combinatorially isomorphic to P; we can obtain Q by translating P or changing the coordinates of P. As a consequence, the face lattice of P is isomorphic to the face lattice of Q and antiisomorphic to the face lattice of Q^*. The existence of Q allows us to define the 'dual polytope' for any polytope, not just for a polytope that contains the origin in its interior. We say that a polytope P^* is the (combinatorial) *dual polytope* of P if the face lattice of P^* in antiisomorphic to the face lattice of P.

We now present a relation between the dimensions of F and $F^\triangle$.

Theorem 2.4.12 *Let $P \subseteq \mathbb{R}^d$ be a d-polytope that contains the origin in its interior, and let P^* be the dual polytope of P. If F and $F^\triangle$ are conjugate faces of P and P^*, respectively, then* $\dim F + \dim F^\triangle = d - 1$.

Proof According to Theorem 2.3.5, every k-face F_k of P is part of a complete flag

$$\emptyset = F_{-1} \subset F_0 \subset \cdots F_k \subset \cdots \subset F_{d-1} \subset F_d = P \qquad (2.4.12.1)$$

of faces of P such that $\dim F_i = i$, for each $i \in [-1 \ldots d]$. By Theorem 2.4.10(v), computing the conjugate of every face in (2.4.12.1) yields a new complete flag

$$\emptyset = F_d^{\vartriangle} \subset F_{d-1}^{\vartriangle} \subset \cdots F_k^{\vartriangle} \subset \cdots \subset F_0^{\vartriangle} \subset F_{-1}^{\vartriangle} = P^*$$

of faces of P^* such that $\dim F_i^{\vartriangle} > \dim F_{i+1}^{\vartriangle}$, for each $i \in [-1 \ldots d-1]$. Since P^* is also a d-polytope (Proposition 2.4.7), we must have that $\dim F_i^{\vartriangle} = \dim F_{i+1}^{\vartriangle} + 1$, for each $i \in [-1 \ldots d-1]$, and that $\dim F_d^{\vartriangle} = -1$. It follows that $\dim F_i + \dim F_i^{\vartriangle} = d - 1$ for each $i \in [-1 \ldots d]$, as desired. □

Some results related to the facial structure of a polytope are easier to prove if duality is invoked. We give four examples.

Theorem 2.4.13 *A d-polytope has at least $d+1$ facets.*

Proof Without loss of generality, suppose that P is a d-polytope in $\mathbb{R}^d$ that contains the origin in its interior. The dual polytope P^* of P is another d-polytope that contains the origin in its interior (Proposition 2.4.7). Moreover, P^* can be expressed as $\operatorname{conv} \mathcal{V}(P^*)$ by Minkowski–Krein–Milman's theorem (1.9.11). Since P^* is d-dimensional, the number of affinely independent points in P^* is $d+1$ and thus $v(P^*) \geqslant d+1$. By Corollary 2.4.11, the number of facets of P is $v(P^*)$, and is at least $d+1$. □

Theorem 2.4.14 *A vertex of a d-polytope P in $\mathbb{R}^d$ is contained in at least d edges of P.*

Proof Without loss of generality, suppose that P contains the origin in its interior. Let $\boldsymbol{v}$ be a vertex of P and let ψ be an antiisomorphism from $\mathcal{L}(P)$ to $\mathcal{L}(P^*)$. From Theorem 2.4.12 it follows that $\dim \psi(\boldsymbol{v}) = d-1$, and from Theorem 2.4.13 it follows that the number of $(d-2)$-faces in $\psi(\boldsymbol{v})$ is at least d, say $R_1, \ldots, R_d$. Hence $\psi^{-1}(R_1), \ldots, \psi^{-1}(R_d)$ are all edges of P containing $\boldsymbol{v}$ (Corollary 2.4.11), concluding the proof of the theorem. □

Theorem 2.4.14 yields a useful inequality between f_0 and f_1 and, by duality, between f_{d-2} and f_{d-1}.

Corollary 2.4.15 *If P is a d-polytope, then*

$$2f_1(P) \geqslant df_0(P), \; 2f_{d-2}(P) \geqslant df_{d-1}(P).$$

Proof Each edge of P contains precisely two vertices and each vertex is incident with at least d edges by Theorem 2.4.14. Hence $2f_1(P) \geqslant df_0(P)$.

By applying this inequality to the dual polytope P^* of P, we get that $2f_1(P^*) \geqslant df_0(P^*)$. Hence $2f_{d-2}(P) \geqslant df_{d-1}(P)$. □

Theorem 2.4.16 *Let $P \subseteq \mathbb{R}^d$ be a d-polytope. For $-1 \leqslant k < h \leqslant d-1$, each k-face F of P is the intersection of at least $h-k+1$ h-faces of P that contain it. In the particular case $k = d-2$, F is the intersection of exactly two facets of P.*

Proof The result is true for $d = 2$ and every $-1 \leqslant k < h \leqslant 1$. Thus, assume that $d \geqslant 3$ and the statement is true for every $(d-1)$-polytope and every pair of numbers k, h satisfying $-1 \leqslant k < h \leqslant d-2$.

Without loss of generality, suppose that P is a d-polytope that contains the origin in its interior. We first prove the result for $h = d-1$ and every $-1 \leqslant k < h$. Let ψ be an antiisomorphism from $\mathcal{L}(P)$ to $\mathcal{L}(P^*)$. Because $\dim F = k$, from Theorem 2.4.12 it follows that $\dim \psi(F) = d-1-k$, which amounts to $\psi(F)$ having at least $d-k$ vertices, say $\boldsymbol{v}_1, \dots, \boldsymbol{v}_{d-k}$; in the case $k = d-2$, $\psi(F)$ is an edge and has exactly two vertices. By Theorem 2.4.10, the faces $\psi^{-1}(\boldsymbol{v}_1), \dots, \psi^{-1}(\boldsymbol{v}_{d-k})$ of P are all facets that contain F. We know from Theorem 2.3.1 that the face F is the intersection of the facets of P containing it. Therefore, there are at least $d-k$ such facets. This settles the case $h = d-1$.

We now pick a facet J of P containing F. By the induction hypothesis, F is the intersection of at least $h-k+1$ h-faces of J that contain it for every $-1 \leqslant k < h \leqslant d-2$. Each face of J is a face of P and so the statement follows for $h \leqslant d-2$ as well. □

2.5 Preprocessing

While most of the proofs in this book live entirely in an affine space, it is sometimes convenient to enlarge the affine space into a real projective space, preprocess our objects, and then return to the affine world with simpler objects.

We digress temporarily to introduce embeddings of affine spaces into projective spaces.

Embedding Affine Spaces into Projective Spaces

We slightly vary the model of $\mathbb{P}(\mathbb{R}^{d+1})$ presented in Section 1.3 so that it now completes a d-dimensional affine space H by adding the points contained in the direction of H. In this new model we keep the close relation between $\mathbb{P}(\mathbb{R}^{d+1})$ and $\mathbb{R}^{d+1}$, which has proven very useful. The main idea has its seeds in the embedding of an affine space into a linear space, as discussed in Section 1.2.

We first embed the d-dimensional affine space $\mathbb{A}^d$ into $\mathbb{R}^{d+1}$ by associating $\mathbb{A}^d$ with the nonlinear hyperplane

$$H := \left\{(x_1, \ldots, x_{d+1})^t \in \mathbb{R}^{d+1} \middle| \, x_{d+1} = 1\right\}.$$

In the embedding of $\mathbb{A}^d$ into $\mathbb{R}^{d+1}$ described in Section 1.2, the linear hyperplane

$$H_\infty := \left\{(x_1, \ldots, x_{d+1})^t \in \mathbb{R}^{d+1} \middle| \, x_{d+1} = 0\right\}$$

parallel to H plays the role of the direction of H. In our new model of $\mathbb{P}^d$, the hyperplane H_∞ will also play an important role.

To every line $p(\boldsymbol{x})$ in $\mathbb{R}^{d+1}$ that is not contained in the linear hyperplane H_∞ we assign the unique point $(\alpha_1, \ldots, \alpha_d, 1)^t$ in the intersection of $p(\boldsymbol{x})$ with H. And to every line $p(\boldsymbol{x})$ in H_∞ we assign the homogeneous coordinates of $p(\boldsymbol{x})$, namely $(\alpha_1 : \cdots : \alpha_d : 0)$; we call the lines $p(\boldsymbol{x})$ in H_∞ *points at infinity*. A point at infinity in H_∞ can be thought of as the asymptotic direction of all lines in H parallel to the point. The hyperplane H_∞ is often called the *hyperplane at infinity*.

The projective points therefore decompose into two types: those represented by an affine point $(\alpha_1, \ldots, \alpha_d, 1)^t$ in H, which can also be seen as a vector of $\mathbb{R}^{d+1}$, and those represented by the lines in H_∞ that pass through the origin, or equivalently, by homogeneous coordinates of the form $(\alpha_1 : \cdots : \alpha_d : 0)$. The hyperplane H_∞ defines a $(d-1)$-dimensional projective subspace of $\mathbb{P}^d$: it is the set of lines through the origin in the linear subspace H_∞. The subsequent decompositions of $\mathbb{P}^d$ follow at once (see Fig. 1.3.1(b)):

$$\mathbb{P}^d = H \cup H_\infty = \mathbb{A}^d \cup \mathbb{P}^{d-1}. \tag{2.5.1}$$

It is instructive to compare Fig. 1.2.1(b) with Fig. 1.3.1(b).

As before, any k-dimensional linear subspace L of $\mathbb{R}^{d+1}$ defines a $(k-1)$-dimensional projective subspace whose projective points are either the affine points in $L \cap H$ or the points at infinity in $L \cap H_\infty$.

In this embedding of $\mathbb{A}^d$ into $\mathbb{P}^d$, the space $\mathbb{P}^d$ is the *projective closure* or *projective completion* of $\mathbb{A}^d$. We can naturally complete an affine subspace A of H. Consider the direction $\overrightarrow{A}$ of A and the homogenisation $\widehat{A}$ of A (see Section 1.2). Then $\widehat{A}$ is a linear subspace of $\mathbb{R}^{d+1}$ that contains both A and $\overrightarrow{A}$. The *projective closure* of A is the projective space $\mathbb{P}(\widehat{A})$; that is, it is the projective space defined as $\mathbb{P}(\overrightarrow{A})$ together with the set $p(A)$ of lines that pass through the origin and are spanned by the points of A. The elements of $\mathbb{P}(\overrightarrow{A})$

are the *points of infinity* of $\mathbb{P}(\widehat{A})$. We obtain a decomposition of $\mathbb{P}(\widehat{A})$ similar to that in (2.5.1):

$$\mathbb{P}(\widehat{A}) = A \cup \mathbb{P}(\overrightarrow{A}).$$

If the affine space A is defined by the system of linear equations

$$\begin{cases} \alpha_{1,1}x_1 + \cdots + \alpha_{1,d}x_d + b_1 &= 0 \\ &\vdots \\ \alpha_{n,1}x_1 + \cdots + \alpha_{n,d}x_d + b_n &= 0, \end{cases}$$

then its closure is defined by the system of homogeneous linear equations

$$\begin{cases} \alpha_{1,1}x_1 + \cdots + \alpha_{1,d}x_d + b_1x_{d+1} &= 0 \\ &\vdots \\ \alpha_{n,1}x_1 + \cdots + \alpha_{n,d}x_d + b_nx_{d+1} &= 0. \end{cases}$$

Scheme for Preprocessing Affine Objects

The idea goes as follows. There is an affine object P (in most instances, a polytope) embedded in a d-dimensional affine space H^e. Projectively complete H^e by adding the hyperplane at infinity H^e_∞. Consider another nonlinear hyperplane H^p that is nonparallel to H^e and denote by H^p_∞ its corresponding hyperplane at infinity. Assume that the object P lies in H^e, in the open halfspace defined by H^p_∞ and containing H^p; following Ziegler (1995, sec. 2.6), if P is positioned as described, we say that the hyperplane H^p is *admissible* for P. In this case, the hyperplane H^p intersects every line passing through the origin and a point of P (Fig. 2.5.1).

With the projective completions of H^e and H^p in place so that the hyperplane H^p is admissible for P, we then describe a projective transformation $\zeta : \mathbb{P}\left(\mathbb{R}^{d+1}\right) \to \mathbb{P}\left(\mathbb{R}^{d+1}\right)$ mapping P onto a 'deformed' object P' in H^p; see Section 1.4 for information on projective maps. The affine space H^p and the object P' are subsequently used instead of H^e and P. Essentially, the affine object P' is geometrically realised by the intersection of H^p and a projective object $p(P)$ consisting of the lines passing through the origin and through a point of P. We require that this projective transformation ζ is *admissible* for P: no point of P lies in H^p_∞.

Let us make our explanation concrete. Take

$$H^e := \left\{ \begin{pmatrix} \boldsymbol{x} \\ x_{d+1} \end{pmatrix} \middle| \; \boldsymbol{x} \in \mathbb{R}^d \text{ and } x_{d+1} = 1 \right\},$$

$$H^e_\infty := \left\{ \begin{pmatrix} \boldsymbol{x} \\ x_{d+1} \end{pmatrix} \middle| \; \boldsymbol{x} \in \mathbb{R}^d \text{ and } x_{d+1} = 0 \right\},$$

$$H^p := \left\{ \begin{pmatrix} \boldsymbol{x} \\ x_{d+1} \end{pmatrix} \middle| \; \boldsymbol{x}, \boldsymbol{a} \in \mathbb{R}^d \text{ and } \boldsymbol{a} \cdot \boldsymbol{x} + a_{d+1} x_{d+1} = 1 \right\},$$

$$H^p_\infty := \left\{ \begin{pmatrix} \boldsymbol{x} \\ x_{d+1} \end{pmatrix} \middle| \; \boldsymbol{x}, \boldsymbol{a} \in \mathbb{R}^d \text{ and } \boldsymbol{a} \cdot \boldsymbol{x} + a_{d+1} x_{d+1} = 0 \right\},$$

so that H^p and H^e are nonparallel. Then the admissibility of the hyperplane H^p for P amounts to saying that

$$\boldsymbol{a} \cdot \boldsymbol{v} + a_{d+1} v_{d+1} > 0$$

for every point $(\boldsymbol{v}, v_{d+1})^t$ of P.

We let the projective transformation ζ be induced by the identity linear map in $\mathbb{R}^{d+1}$ and we associate it with a perspectivity ϱ that goes from $H^e \backslash (H^e \cap H^p_\infty)$ to H^p and is centred at $\mathbf{0}$. In this association, if $\varrho(z_1) = z_2$ then $\zeta(p(z_1)) = p(z_2)$. An affine *perspectivity* centred at $\mathbf{0}$ is a function between affine hyperplanes K and K' that maps a point z of K to a point z' of K' whenever z, z', and $\mathbf{0}$ are collinear; see Fig. 2.5.1(a).

Let the map ϱ fix the points in $H^e \cap H^p$ and map each point in $H^e \backslash (H^p \cup H^p_\infty)$ to the point in H^p lying on the same line through the origin. See Fig. 2.5.1(a). In formulas we get that the map acts as

$$\begin{pmatrix} \boldsymbol{x} \\ 1 \end{pmatrix} \in H^e \mapsto \frac{1}{\boldsymbol{a}\boldsymbol{x} + a_{d+1}} \begin{pmatrix} \boldsymbol{x} \\ 1 \end{pmatrix} \in H^p$$

provided $\boldsymbol{a}\boldsymbol{x} + a_{d+1} \neq 0$.

We extend the map ϱ via the transformation ζ for the remaining points of H^e. Take a point $\boldsymbol{y}$ in $H^e \cap H^p_\infty$ and consider any line ℓ^e in H^e through $\boldsymbol{y}$ that is not contained in $H^e \cap H^p_\infty$. Obtain the line ℓ^p in H^p that is the intersection of H^p and the linear plane in $\mathbb{R}^{d+1}$ spanned by $\mathbf{0}$ and the line ℓ^e. Then ζ maps the point $\boldsymbol{y}$ to the line $p(\boldsymbol{y})$ in H^p_∞, which is the asymptotic direction of all lines in H^p parallel to ℓ^p. Irrespective of the line through $\boldsymbol{y}$ and not contained in $H^e \cap H^p_\infty$ that one chooses, we always obtain a line parallel to ℓ^p, namely a line with asymptotic direction $p(\boldsymbol{y})$. It is customary to say that the map ζ 'sends every object in $H^e \cap H^p_\infty$ to infinity'. See Fig. 2.5.1(b).

The map ζ so defined is clearly projective: some intersecting lines are mapped onto parallel lines. As pointed out by Ziegler (1995, sec. 2.6), it is

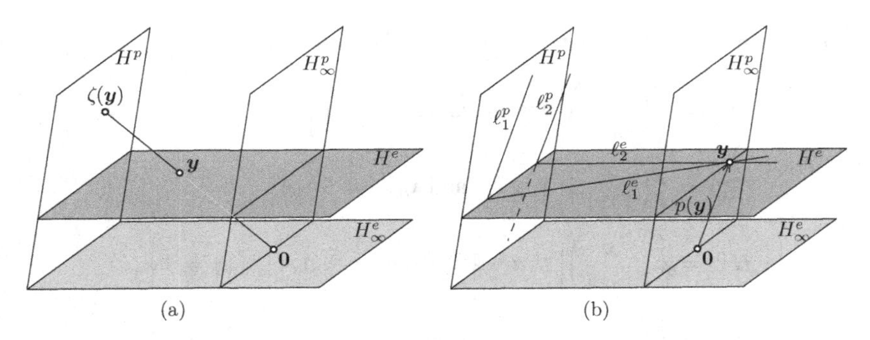

Figure 2.5.1 Mapping points in H^e onto $H^p \cup H^p_\infty$. (a) Mapping of a point in $H^e \backslash (H^p \cup H^p_\infty)$. (b) Mapping of a point in $H^e \cap H^p_\infty$.

hardly ever necessary to produce concrete formulas for the projective map. It suffices to understand how the map treats affine spaces, which is given in Proposition 2.5.2.

Proposition 2.5.2 *Let $\zeta : \mathbb{P}(\mathbb{R}^{d+1}) \rightarrow \mathbb{P}(\mathbb{R}^{d+1})$ be the projective map previously defined. Then ζ takes an affine k-space $A \not\subset H^e \cap H^p_\infty$ onto an affine k-space $\zeta(A) \subseteq H^p$ and takes an affine k-space $A \subset H^e \cap H^p_\infty$ onto a linear $(k+1)$-space $\zeta(A) \subset H^p_\infty$.*

Remark 2.5.3 Most of the time, everything boils down to recognising how the projective transformation treats lines and their intersections; this is summarised in (A)–(D) below, although it is an immediate corollary of Proposition 2.5.2. Figure 2.5.2 depicts the mapping of two lines in H^e onto H^p.

(A) The map ζ carries a line ℓ^e in H^e into a line ℓ^p in H^p, except those contain in $H^e \cap H^p_\infty$.

(B) Let ℓ^e_1 and ℓ^e_2 be two lines of H^e that are not contained in $H^e \cap H^p_\infty$ and intersect outside $H^e \cap H^p_\infty$. Then they are mapped onto intersecting lines ℓ^p_1 and ℓ^p_2 in H^p. If the lines ℓ^e_1 and ℓ^e_2 intersect in the positive open halfspace of H^p_∞, then the lines ℓ^p_1 and ℓ^p_2 intersect in the positive open halfspace of H^e_∞ (Fig. 2.5.2(b)). If in turn the lines ℓ^e_1 and ℓ^e_2 intersect in the negative open halfspace of H^p_∞, then the lines ℓ^p_1 and ℓ^p_2 intersect in the negative open halfspace of H^e_∞ (Fig. 2.5.2(d)).

(C) Let ℓ^e_1 and ℓ^e_2 be two lines of H^e that are not contained in $H^e \cap H^p_\infty$ but intersect inside $H^e \cap H^p_\infty$. Then they are mapped onto parallel lines ℓ^p_1 and ℓ^p_2 in H^p (Fig. 2.5.2(a)).

(D) Let ℓ^e_1 and ℓ^e_2 be two parallel lines of H^e that are not contained in $H^e \cap H^p_\infty$. If they are not parallel to a line in $H^e \cap H^p_\infty$, then they are mapped onto lines ℓ^p_1 and ℓ^p_2 of H^p that intersect at $H^p \cap H^e_\infty$; otherwise they are mapped onto lines parallel to a line in $H^p \cap H^e_\infty$ (Fig. 2.5.2(c)).

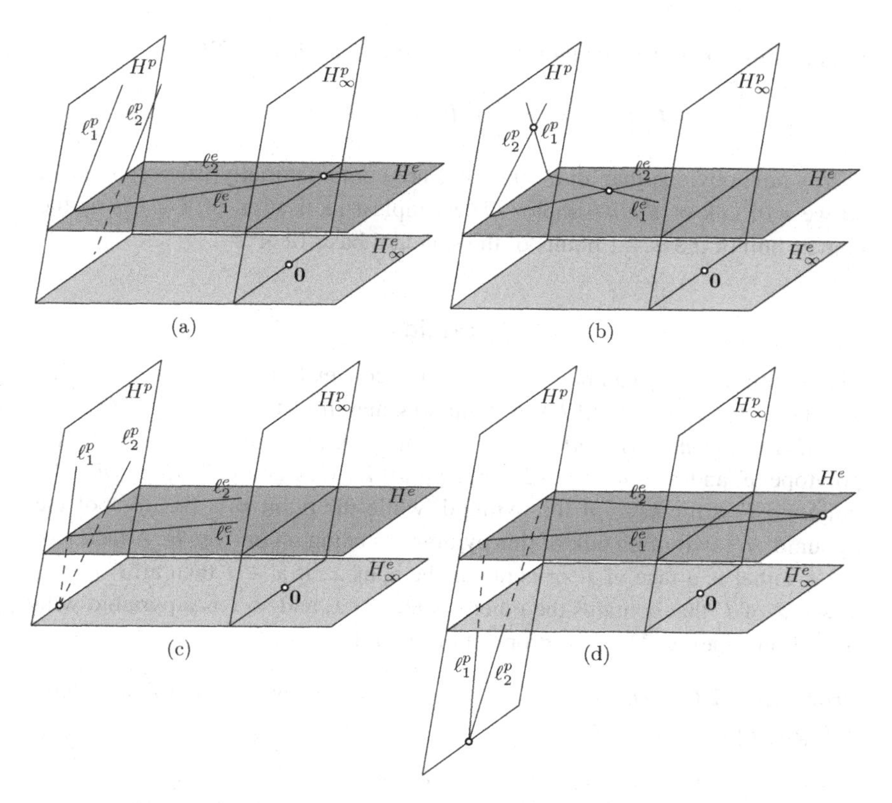

Figure 2.5.2 Mapping of two lines of H^e onto H^p. (a) The lines intersect at $H^e \cap H^p_\infty$. (b) The lines intersect inside the positive open halfspace defined by H^p_∞ and containing H^p. (c) The lines are parallel in H^e. (d) The lines intersect inside the negative open halfspace defined by H^p_∞ and not containing H^p.

2.6 Examples

This section examines particular examples of polytopes, with emphasis on their combinatorial properties.

Simplices

The d-simplex, denoted $T(d)$, is the d-polytope with the smallest number of vertices. As we saw in Chapter 1, it is the convex hull of $d + 1$ affinely independent points in $\mathbb{R}^d$. Figure 1.6.2 shows simplices in $\mathbb{R}^3$. For every $k \in [0 \dots d]$, the k-faces of a d-simplex are simplices of smaller dimension and every $k + 1$ vertices yields a k-face. Thus the f-vector of $T(d)$ can be easily computed.

Proposition 2.6.1 *The number f_k of k-faces of a d-simplex $T(d)$ is*

$$f_k(T(d)) = \binom{d+1}{k+1}, \textit{ for every } k \in [-1 \dots d].$$

It is now obvious that all d-simplices are combinatorially isomorphic, and so we will talk of *the* d-simplex. The simplest realisation of $T(d)$ is as the convex hull of the $d+1$ points of the standard basis of $\mathbb{R}^{d+1}$.

Pyramids

The d-simplex $T(d)$ can be also seen as the convex hull of a facet F of $T(d)$ and the vertex in $T(d)\backslash(\operatorname{aff} F)$. A pyramid generalises this construction.

A *d-dimensional pyramid* or *d-pyramid* in $\mathbb{R}^d$ is the convex hull of a $(d-1)$-polytope F and a point $\boldsymbol{x} \in \mathbb{R}^d$ not on $\operatorname{aff} F$; it is denoted by $\operatorname{pyr} F$. The polytope F is the *base* of the pyramid, while the point $\boldsymbol{x}$ is the *apex* of the pyramid. We will often talk of this pyramid as being on or over F. A face I of $\operatorname{pyr} F$ either is a face of F or contains the apex $\boldsymbol{x}$. If $\boldsymbol{x} \in I$ then $\operatorname{aff} F \cap I$ is a face J of F that contains the other vertices of I, and so I is a pyramid with base J and apex $\boldsymbol{x}$. The next proposition should now be clear.

Proposition 2.6.2 *The number f_k of k-faces of a d-pyramid* $\operatorname{pyr} F$ *with base F is given by*

$$f_k(\operatorname{pyr} F) = f_k(F) + f_{k-1}(F), \textit{ for every } k \in [0 \dots d].$$

The pyramid construction can be generalised. Every d-polytope P is a 0-fold d-pyramid with P as base. And a *1-fold d-pyramid* with base F is simply a d-pyramid over a $(d-1)$-polytope F. If P is a pyramid over a base Q that is itself a $(d-1)$-pyramid over a $(d-2)$-polytope F, then we say that P is *two-fold d-pyramid* over the base F. In general, an *r-fold d-pyramid* P is a pyramid over a base Q that is itself an $(r-1)$-fold $(d-1)$-pyramid and the bases of P and Q coincide. In other words, P is an r-fold d-pyramid over a $(d-r)$-dimensional base F and is denoted by $\operatorname{pyr}_r F$. An inductive application of Proposition 2.6.2 yields the number of faces of $\operatorname{pyr}_r F$.

Proposition 2.6.3 *The number f_k of k-faces of an r-fold d-pyramid* $\operatorname{pyr}_r F$ *with base F is given by*

$$f_k(\operatorname{pyr}_r F) = \sum_{i=0}^{r} \binom{r}{i} f_{k-i}(F), \textit{ for every } k \in [0 \dots d].$$

If the dimension of a d-pyramid or an r-fold d-pyramid is clear from the context or is nonessential, then we simply write pyramid or r-fold pyramid.

Bipyramids

Let F be a $(d-1)$-polytope in $\mathbb{R}^d$ and let $I = [\boldsymbol{x}, \boldsymbol{y}]$ be a line segment in $\mathbb{R}^d$ such that $\operatorname{rint} I \cap \operatorname{rint} F$ is a unique point. Then a *d-bipyramid P* in $\mathbb{R}^d$ is the convex hull of F and I; it is denoted by $\operatorname{bipyr} F$. The polytope F is the *base* of the bipyramid, while the segment I is the *axis* of the bipyramid. We will often talk of a bipyramid *on* or *over* F. A face of $\operatorname{bipyr} F$ is either a proper face of F, a pyramid with a base in F and an apex in $\{\boldsymbol{x}, \boldsymbol{y}\}$, or a vertex in $\{\boldsymbol{x}, \boldsymbol{y}\}$. The next proposition should now be clear.

Proposition 2.6.4 *The number f_k of k-faces of a d-bipyramid* $\operatorname{bipyr} F$ *with base F is given by*

$$f_k(\operatorname{bipyr} F) = \begin{cases} f_k(F) + 2 f_{k-1}(F), & \text{if } k \in [0 \ldots d-2]; \\ 2 f_{d-2}(F), & \text{if } k = d-1. \end{cases}$$

The definition of an r-fold d-bipyramid follows the same idea as that of an r-fold d-pyramid. Every d-polytope P is a 0-fold d-bipyramid with P as base. An *r-fold d-bipyramid* P is a bipyramid over a base Q that is itself an $(r-1)$-fold $(d-1)$-bipyramid, and the bases of P and Q coincide. In other words, P is an r-fold d-bipyramid over a $(d-r)$-dimensional base F and is denoted by $\operatorname{bipyr}_r F$.

If the dimension of a d-bipyramid or an r-fold d-bipyramid is clear from the context or is nonessential, then we simply write bipyramid or r-fold bipyramid.

We met a $(d-1)$-fold d-bipyramid in Chapter 1, the d-crosspolytope $I(d)$. In Chapter 1, the d-crosspolytope appeared as the dual of the d-cube. As a $(d-1)$-fold d-bipyramid, $I(d)$ can be realised as the convex hull of d segments that are pairwise orthogonal and have a common midpoint. It follows that $I(d)$ is a bipyramid over $I(d-1)$, wherefrom we get the number f_k of k-faces. Figure 2.6.1 shows crosspolytopes in $\mathbb{R}^3$.

Proposition 2.6.5 *The number f_k of k-faces of a d-crosspolytope $I(d)$ is given by*

$$f_k(I(d)) = 2^{k+1}\binom{d}{k+1}, \text{ for every } k \in [-1 \ldots d-1].$$

Prisms

Let F be a $(d-1)$-polytope in $\mathbb{R}^d$ and let $I = [\mathbf{0}, \boldsymbol{x}]$ be a line segment in $\mathbb{R}^d$ such that aff I is not parallel to any line in aff F. Then the *d-prism* P with base F and axis I, denoted prism F, is the Minkowski sum $F + I$. This amounts to saying that prism $F = \operatorname{conv}(F \cup (F + \boldsymbol{x}))$. A k-face of prism F is either a k-face of F, a k-face of $F + \boldsymbol{x}$, or the sum of I and some $(k-1)$-face of F. The next proposition should now be clear.

Proposition 2.6.6 *The number f_k of k-faces of a d-prism with base F is given by*

$$f_k(\text{prism } F) = \begin{cases} 2f_k(F), & \textit{if } k = 0; \\ 2f_k(F) + f_{k-1}(F), & \textit{if } k \in [1 \ldots d]. \end{cases}$$

Continuing with the analogy to both pyramids and bipyramids, we define r-fold d-prisms. Every d-polytope P is a 0-fold d-prism with P as base. An *r-fold d-prism* P is a prism over a base Q that is itself an $(r-1)$-fold $(d-1)$-prism and the bases of P and Q coincide. In other words, P is an r-fold d-prism over a $(d-r)$-dimensional base F and is denoted by $\text{prism}_r\, F$.

We have special names for some prisms. A d-prism with a simplex as a base is a *simplicial d-prism*. A $(d-1)$-fold d-prism, which is also a d-fold d-prism, is a *d-parallelotope*; it is the sum of d segments with a common point such that no segment is in the affine hull of the others.

We met a d-parallelotope in Example 2.1.6, the d-cube $Q(d)$. There, we realised $Q(d)$ as the convex hull of 2^d vectors $(\pm 1, \ldots, \pm 1)^t$. Here, we obtain $Q(d)$ as the sum of d segments that are pairwise orthogonal and have equal length. It follows that $Q(d)$ is a prism over $Q(d-1)$, wherefrom we get the number f_k of k-faces. Figure 2.1.1 shows cubes in $\mathbb{R}^3$.

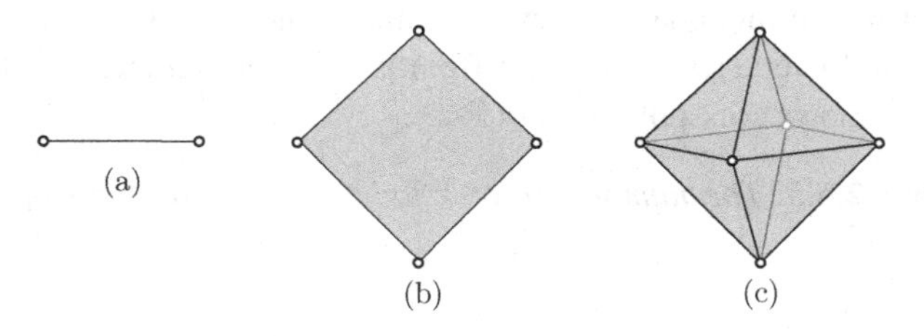

Figure 2.6.1 Crosspolytopes in $\mathbb{R}^3$. (a) A 1-crosspolytope. (b) A 2-crosspolytope. (c) A 3-crosspolytope, usually known as an octahedron.

Proposition 2.6.7 *The number f_k of k-faces of a d-cube $Q(d)$ is given by*

$$f_k(Q(d)) = 2^{d-k}\binom{d}{k}, \text{ for every } k \in [0\ldots d].$$

If the dimension of a d-prism or an r-fold d-prism is clear from the context or is nonessential, then we simply write prism or r-fold prism.

Wedges

Let P be a d-polytope in $\mathbb{R}^d$. Embed $P \times \{0\}$ in the hyperplane

$$H := \left\{ \begin{pmatrix} \boldsymbol{x} \\ x_{d+1} \end{pmatrix} \in \mathbb{R}^{d+1} \,\middle|\, x_{d+1} = 0 \right\}$$

of $\mathbb{R}^{d+1}$, let F be a proper face of P, and let C be the halfcylinder $P \times [0, \infty) \subset \mathbb{R}^{d+1}$. We cut the halfcylinder with a hyperplane H' through $F \times \{0\}$ so that C is partitioned into two parts, one bounded and one unbounded. The *wedge* of P at F is the bounded part; it is denoted by $\mathrm{W}_F(P)$. The sets P and $H' \cap C$ define facets of $\mathrm{W}_F(P)$ that are combinatorially isomorphic to P and intersect at the face $F \times \{0\}$; the facets P and $H' \cap C$ are the *bases* of $\mathrm{W}_F(P)$. See an example in Fig. 2.6.2.

The wedge W over a d-polytope $P \times \{0\} \subseteq \mathbb{R}^{d+1}$ at a face $F \times \{0\}$ of $P \times \{0\}$ is combinatorially isomorphic to a prism Q over $P \times \{0\}$ where the face prism$(F \times \{0\})$ of Q has collapsed into $F \times \{0\}$. Some proper k-faces of W will be wedges defined as the wedge of a $(k-1)$-face $J \times \{0\}$ of $P \times \{0\}$ at a proper face $(F \cap J) \times \{0\}$. Some proper k-faces of W are k-prims over $(k-1)$-faces of $P \times \{0\}$ disjoint from $F \times \{0\}$; these prisms are the *vertical faces* of W. These descriptions together with Proposition 2.6.6 give the following.

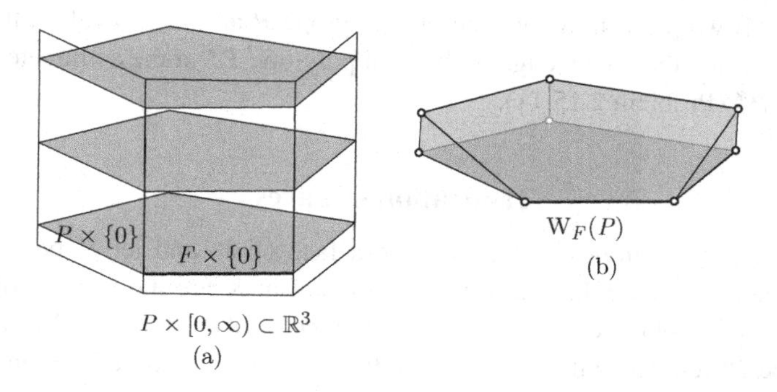

Figure 2.6.2 The wedge of the pentagon P. (a) The cylinder $P \times \mathbb{R}$ in $\mathbb{R}^3$. (b) The wedge over P at a facet of P.

Proposition 2.6.8 *Let $P \subseteq \mathbb{R}^{d+1}$ be a d-polytope and F a proper face of P. A k-face of the wedge W of P at F is either a k-face of one of the bases of W, or the wedge of a $(k-1)$-face J of P at the proper face $F \cap J$, or a vertical k-face.*

Proposition 2.6.9 *The number f_k of k-faces of the wedge W of a d-polytope P at a facet F of P is given by*

$$f_k(W) = \begin{cases} 2f_k(P) - f_k(F), & \text{if } k = 0; \\ 2f_k(P) + f_{k-1}(P) - f_k(F) - f_{k-1}(F), & \text{if } k \in [1 \dots d+1]. \end{cases}$$

Dual Wedges

Let P be a d-polytope in $\mathbb{R}^d$. Embed $P \times \{0\}$ in the hyperplane

$$H := \left\{ \begin{pmatrix} \boldsymbol{x} \\ x_{d+1} \end{pmatrix} \in \mathbb{R}^{d+1} \,\middle|\, x_{d+1} = 0 \right\}$$

of $\mathbb{R}^{d+1}$ and let $\boldsymbol{v}$ be a vertex of P. The *dual wedge* of P at $\boldsymbol{v}$, denoted by $\mathrm{dW}_v(P)$, is the $(d+1)$-polytope

$$\mathrm{dW}_v(P) := \mathrm{conv}\left((P \times \{0\}) \cup (\boldsymbol{v} \times \{-1\}) \cup (\boldsymbol{v} \times \{1\})\right).$$

The facial structure of the dual wedge is plain from its description.

Proposition 2.6.10 *Let $P \times \{0\} \subseteq \mathbb{R}^{d+1}$ be a d-polytope and $\boldsymbol{v}$ a vertex of P. A k-face of the dual wedge of P at $\boldsymbol{v}$ is either a k-face of P not containing $\boldsymbol{v}$, or the dual wedge of a $(k-1)$-face of P at $\boldsymbol{v}$, or a pyramid with apex $\boldsymbol{v} \times \{-1\}$ or $\boldsymbol{v} \times \{1\}$ over a $(k-1)$-face of P not containing $\boldsymbol{v}$.*

As the name indicates, the dual wedge is in some sense the dual operation of a wedge: if we perform the wedge of a polytope P at a facet F of P, then we are performing the dual wedge of the dual polytope P^* at the conjugate vertex of F in P^* (Problem 2.15.11).

Truncation of Faces

Let P be a d-polytope in $\mathbb{R}^d$, let F be a face of P, and let K be a closed halfspace in $\mathbb{R}^d$ such that the vertices of P not in K are the vertices of F. A polytope P' is obtained by *truncating the face* F of P if $P' = P \cap K$. The polytope P' retains all the old facets of P, except F if it was a facet, and gains a new facet.

Truncating faces is a flexible operation. We can see that a simplicial d-prism is obtained from a d-simplex T by truncating a vertex of T.

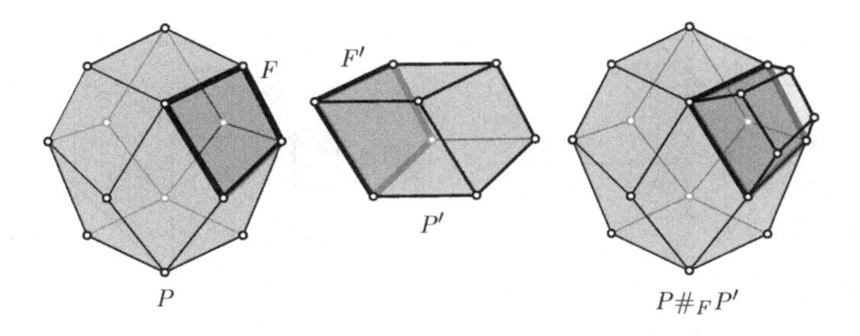

Figure 2.6.3 Connected sum of two polytopes.

Connected Sums

Two polytopes P and P' are *projectively isomorphic* if there is a projective isomorphism ζ permissible for P such that $\zeta(P) = P'$.

Let P and P' be two d-polytopes with a facet F of P projectively isomorphic to a facet F' of P'. The *connected sum* $P\#_F Q$ of P and P' is obtained by 'gluing' P and P' along F and F'; if the facet F is of no importance, we simply write $P\#P'$. Projective transformations on the polytopes P and P' may be required for the connected sum to be convex. A common method is first to assume that P and P' are realised so that $P \cap P' = F = F'$, and then to apply a projective transformation ζ to P' so that ζ fixes F' and $\operatorname{conv}(P \cup \zeta(P'))$ becomes a realisation of $P\#_F P'$; Problem 2.15.7 asks for the details of this transformation. The connected sum of two polytopes is depicted in Fig. 2.6.3. The faces of $P\#_F P'$ are described next.

Proposition 2.6.11 *Let P and P' be two d-polytopes with a facet F of P projectively isomorphic to a facet F' of P'. Then the proper faces of $P\#_F P'$ consist of all the proper faces of P and P', except for the facets F and F'.*

The connected sum of a d-simplex and a d-polytope with a simplex facet is called *stacking*; this sum is always possible (Problem 2.15.9). The *stacked polytopes* are the polytopes obtained from a simplex by successive stacking. The dual operation of stacking is truncating a vertex: if we stack over a facet F of a polytope P, then the conjugate vertex of F in the dual P^* of P gets truncated (Problem 2.15.10). In particular, the dual of a stacked polytope is a *truncated polytope*, a polytope obtained from a simplex by repeatedly truncating vertices.

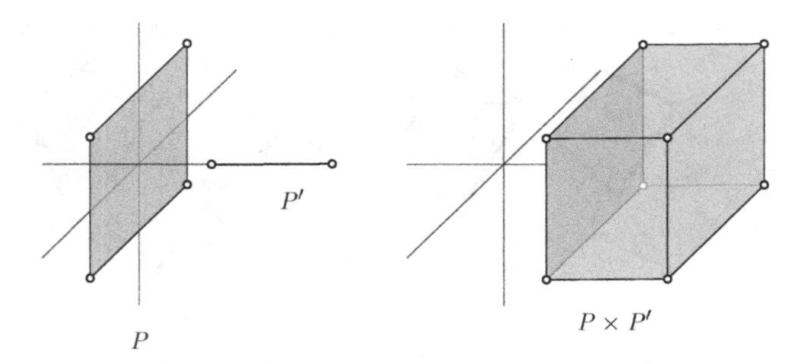

Figure 2.6.4 Cartesian products of two polytopes. The Cartesian product of the 2-cube $P := \text{conv}\,\{(-1,\,-1)^t, (1,\,-1)^t, (-1,1)^t, (1,1)^t\}$ and the segment $P' := \text{conv}\{(1), (2)\}$.

Cartesian Products

The *Cartesian product* $P \times P'$ of a d-polytope $P \subset \mathbb{R}^d$ and a d'-polytope $P' \subset \mathbb{R}^{d'}$ is the Cartesian product of the sets P and P':

$$P \times P' = \left\{ \begin{pmatrix} p \\ p' \end{pmatrix} \in \mathbb{R}^{d+d'} \,\middle|\, p \in P,\, p' \in P \right\}. \tag{2.6.12}$$

The resulting polytope is $(d + d')$-dimensional. An example is depicted in Fig. 2.6.4.

The characterisation of the faces of a Cartesian product is presented in Proposition 2.6.13.

Proposition 2.6.13 *Let $P \subset \mathbb{R}^d$ be a d-polytope and $P' \subset \mathbb{R}^{d'}$ a d'-polytope. The k-faces of the Cartesian product $P \times P'$ are precisely the Cartesian products of an i-face F of P and a j-face F' of P' such that $i + j = k$, for each $k \in [0 \dots d + d']$.*

Free Joins

Let $P \subset \mathbb{R}^{d+d'+1}$ be a d-polytope and $P' \subset \mathbb{R}^{d+d'+1}$ a d'-polytope such that their affine hulls are skew; two affine spaces are *skew* if they do not intersect and no line from one space is parallel to a line from the other. The *free join* $P * P'$ of the polytopes P and P' is the $(d + d' + 1)$-polytope $\text{conv}(P \cup P')$. For a concrete setting, let P be a d-polytope and P' a d'-polytope both of which are embedded in $\mathbb{R}^{d+d'+1}$ as follows:

$$P = \operatorname{conv}\left\{ \begin{pmatrix} \boldsymbol{p} \\ \mathbf{0}_{d'} \\ 0 \end{pmatrix} \in \mathbb{R}^{d+d'+1} \,\middle|\, \boldsymbol{p} \in P \right\},$$

$$P' = \operatorname{conv}\left\{ \begin{pmatrix} \mathbf{0}_{d} \\ \boldsymbol{p}' \\ 1 \end{pmatrix} \in \mathbb{R}^{d+d'+1} \,\middle|\, \boldsymbol{p}' \in P' \right\}.$$

Then

$$P * P' = \operatorname{conv}\left(\left\{ \begin{pmatrix} \boldsymbol{p} \\ \mathbf{0}_{d'} \\ 0 \end{pmatrix} \in \mathbb{R}^{d+d'+1} \,\middle|\, \boldsymbol{p} \in P \right\} \bigcup \left\{ \begin{pmatrix} \mathbf{0}_{d} \\ \boldsymbol{p}' \\ 1 \end{pmatrix} \in \mathbb{R}^{d+d'+1} \,\middle|\, \boldsymbol{p}' \in P' \right\} \right). \tag{2.6.14}$$

A pyramid over a d-polytope P with apex $\boldsymbol{x}$ is the free join of P and the point $\boldsymbol{x}$; in this case, we write $P * \boldsymbol{x}$ rather than $P * \{\boldsymbol{x}\}$.

The characterisation of the faces of a free join is presented in Proposition 2.6.15.

Proposition 2.6.15 *Let $P \subset \mathbb{R}^{d+d'+1}$ be a d-polytope and $P' \subset \mathbb{R}^{d+d'+1}$ a d'-polytope such that* aff P *and* aff P' *are skew. The k-faces of the free join $P * P'$ are precisely the free joins of an i-face F of P and a j-face F' of P' such that $i + j + 1 = k$, for $k \in [0 \dots d + d']$.*

A consequence of Proposition 2.6.15 is a formula for the number of faces of a free join.

Corollary 2.6.16 *Let $P \subset \mathbb{R}^{d+d'+1}$ be a d-polytope and $P' \subset \mathbb{R}^{d+d'+1}$ a d'-polytope such that* aff P *and* aff P' *are skew. The number f_k of k-faces of $P * P'$ is given by*

$$f_k(P * P') = \sum_{i=-1}^{k} f_i(P) f_{k-i-1}(P), \text{ for every } k \in [0 \dots d + d'].$$

Direct Sums

The *direct sum* $P \oplus P'$ of a d-polytope $P \subset \mathbb{R}^d$ and a d'-polytope $P' \subset \mathbb{R}^{d'}$ with the origin in their relative interiors is the $(d + d')$-polytope

$$P \oplus P' = \operatorname{conv}\left(\left\{ \begin{pmatrix} \boldsymbol{p} \\ \mathbf{0}_{d'} \end{pmatrix} \in \mathbb{R}^{d+d'} \,\middle|\, \boldsymbol{p} \in P \right\} \bigcup \left\{ \begin{pmatrix} \mathbf{0}_{d} \\ \boldsymbol{p}' \end{pmatrix} \in \mathbb{R}^{d+d'} \,\middle|\, \boldsymbol{p}' \in P' \right\} \right). \tag{2.6.17}$$

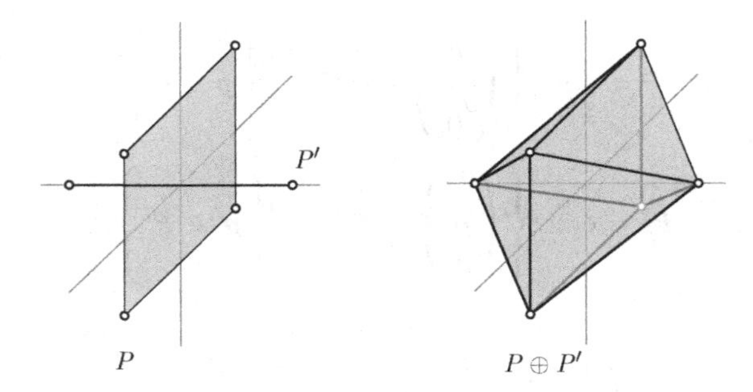

Figure 2.6.5 Direct sum of two polytopes. The direct sum of the 2-cube $P := \operatorname{conv}\{(-1,-1)^t,(1,-1)^t,(-1,1)^t,(1,1)^t\}$ and the segment $P' := \operatorname{conv}\{(-1),(1)\}$.

The resulting polytope lies in $\mathbb{R}^{d+d'}$ and has $f_0(P) + f_0(P')$ vertices and $f_{d-1}(P) \times f_{d'-1}(P')$ facets. An example is depicted in Fig. 2.6.5.

From the definition, it is clear that the direct sum $P \oplus P'$ is a projection of the join $P * P'$. It is not so clear but true that the direct sum $P \oplus P'$ is closely related to the Cartesian product $P \times P'$ by duality.

Proposition 2.6.18 *If $P \subset \mathbb{R}^d$ is a d-polytope and $P' \subset \mathbb{R}^{d'}$ is a d'-polytope such that the origin is in their relative interiors, then*

$$P \oplus P' = (P^* \times (P')^*)^*.$$

2.7 Face Figures

Face figures of a d-polytope P are polytopes whose face lattices are formed by the set of faces F between some i-face F_i and some j-face F_j of P such that $F_i \subseteq F \subseteq F_j$, for $-1 \leqslant i < j \leqslant d$. The most useful of the face figures is the vertex figure, the case when $i = 0$ and $j = d$, and as such, vertex figures are the focus of this section.

Vertex figures exist around each of the vertices of a polytope; they contain information on the facial structure of the polytope and the dual polytope. Let P be a d-polytope in $\mathbb{R}^d$, let $\boldsymbol{v}$ be a vertex of P, and let H be a hyperplane in $\mathbb{R}^d$ that has $\boldsymbol{v}$ on one side of H and the remaining vertices of P on the other side. The *vertex figure* $P/\boldsymbol{v}$ of P at $\boldsymbol{v}$ is the set $P \cap H$; see Fig. 2.7.1.

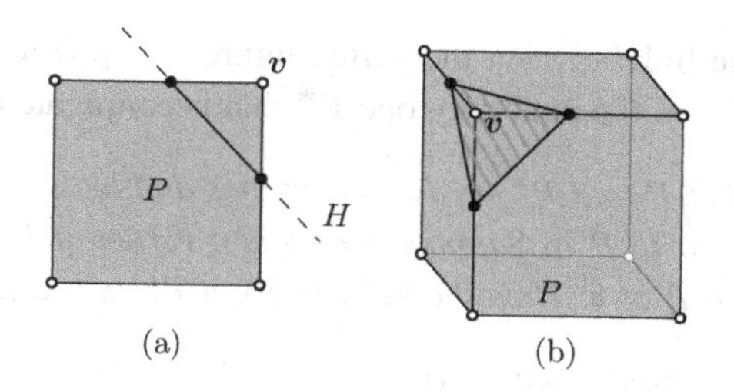

Figure 2.7.1 Vertex figures of polytopes. (a) A segment as the vertex figure of a 2-polytope. (b) A triangle as the vertex figure of a 3-polytope.

Theorem 2.7.1[4] *Let P be a d-polytope in $\mathbb{R}^d$ and let H be a hyperplane in $\mathbb{R}^d$ such that H intersects the interior of P. Then the following hold:*

(i) *The polytope $P' := P \cap H$ is $(d-1)$-dimensional.*
(ii) *If F is a k-face of P, then the set $F' := F \cap H$ is a k'-face of P' with $k' \leqslant k$; in the case of F and F' being proper faces and H not being a supporting hyperplane of F at F', we have that $k' = k - 1$.*
(iii) *If F' is a k'-face of P' but not of P, then there is a unique k-face F of P such that $F' = H \cap F$ and $k' = k - 1$.*

While many choices are possible for a hyperplane that defines a vertex figure Q of a polytope, all of them produce the same face lattice of Q. In other words, the combinatorics of Q is independent of the hyperplane.

Theorem 2.7.2 *Let P be a d-polytope and let $\boldsymbol{v}$ be a vertex of P. Suppose that H is a hyperplane in $\mathbb{R}^d$ such that $H \cap P$ is the vertex figure $P/\boldsymbol{v}$ of P at $\boldsymbol{v}$. Then there is a bijection σ from the k-faces F of P that contain $\boldsymbol{v}$ to the $(k-1)$-faces F' of $P/\boldsymbol{v}$, given by*

$$\sigma(F) = H \cap F =: F',$$
$$\sigma^{-1}(F') = \operatorname{aff}(\{\boldsymbol{v}\} \cup F') =: F.$$

Proof The hyperplane H intersects $\operatorname{int} P$, and so Theorem 2.7.1(i) implies that $P/\boldsymbol{v}$ is a $(d-1)$-polytope. If F is a k-face of P that contains $\boldsymbol{v}$, then H does not support F, which implies that $F' := F \cap H$ is a k'-face of $P/\boldsymbol{v}$ of dimension $\dim F - 1$ (Theorem 2.7.1(ii)). Moreover, for each k'-face F' of $P/\boldsymbol{v}$, we have that F' is not a face of P. Therefore, Theorem 2.7.1(iii) gives the existence of a unique k-face F of P that contains $\boldsymbol{v}$ and satisfies $k' = k - 1$. It is now clear that σ is a bijection (by Theorem 2.7.1). □

[4] A proof is available in Brøndsted (1983, thm. 11.1).

There is a close link between the vertex figure of a polytope P at a vertex $\boldsymbol{v} \in P$ and the facet of the dual polytope P^* that is conjugate to $\boldsymbol{v}$.

Theorem 2.7.3 *Let P and P^* be dual polytopes and let ψ be an antiisomorphism from $\mathcal{L}(P)$ to $\mathcal{L}(P^*)$. Suppose that $\boldsymbol{v}$ is a vertex of P and that $P/\boldsymbol{v}$ is the vertex figure of P at $\boldsymbol{v}$. Then the facet $\psi(\boldsymbol{v})$ of P^* is a dual of $P/\boldsymbol{v}$.*

Proof This is a consequence of Theorem 2.7.2 and the fact that $\mathcal{L}(P)$ is the opposite of $\mathcal{L}(P^*)$ (Corollary 2.4.11).

From Theorem 2.7.2, it follows that the face lattice $\mathcal{L}(P/\boldsymbol{v})$ of $P/\boldsymbol{v}$ is isomorphic to the sublattice $\mathcal{L}_{\boldsymbol{v}}$ of $\mathcal{L}(P)$ formed by the faces of P containing $\boldsymbol{v}$. And, since $\mathcal{L}(P)$ is the opposite of $\mathcal{L}(P^*)$, $\mathcal{L}_{\boldsymbol{v}}$ is antiisomorphic to the sublattice $\mathcal{L}(\psi(\boldsymbol{v}))$ of $\mathcal{L}(P^*)$ corresponding to the facet $\psi(\boldsymbol{v})$ of P^*. Hence $\mathcal{L}(P/\boldsymbol{v})$ is antiisomorphic to $\mathcal{L}(\psi(\boldsymbol{v}))$, as desired. □

Face figures generalise vertex figures; they can be obtained as an iterated vertex figure. Let P be a d-polytope, F_i an i-face of P, and F_j a j-face of P such that $-1 \leqslant i < j \leqslant d$. The set of faces F of P such that $F_i \subseteq F \subseteq F_j$ is a *face figure* F_j/F_i of P. The vertex figure $P/\boldsymbol{v}$ of P at a vertex $\boldsymbol{v}$ is recovered when $F_i = \boldsymbol{v}$ and $F_j = P$. It is useful to consider F_j as a j-polytope and F_i as a face of F_j, as we will see in Chapter 8. In this way, we can extend Theorem 2.7.3 to all face figures. If we consider F_j and its dual polytope F_j^* as j-polytopes and let ψ_j be an antiisomorphism from $\mathcal{L}(F_j)$ to $\mathcal{L}(F_j^*)$, then the face figure F_j/F_i is combinatorially isomorphic to the dual of the face $\psi_j(F_i)$ in F_j^*.

Theorem 2.7.4[5] *Let P be a d-polytope, F_i an i-face of P, and F_j a j-face of P such that $-1 \leqslant i < j \leqslant d$. Then the face figure F_j/F_i is combinatorially isomorphic to a $(j-1-i)$-polytope Q. Additionally, to each face F of P such that $F_i \subseteq F \subseteq F_j$ there corresponds a face in Q of dimension* $\dim(F) - 1 - i$.

This is a good place to introduce some graph-theoretical terminology in order to streamline our future statements. For an edge $e = \operatorname{conv}\{\boldsymbol{x}, \boldsymbol{y}\}$ of a polytope P, we write $e = [\boldsymbol{x}, \boldsymbol{y}]$ or $e = \boldsymbol{x}\boldsymbol{y}$, and we say that the vertices $\boldsymbol{x}$ and $\boldsymbol{y}$ are *adjacent* or *neighbours*, and that the edge e is *incident* with $\boldsymbol{x}$ and $\boldsymbol{y}$. We denote the set of neighbours of a vertex $\boldsymbol{x}$ in P by $\mathcal{N}_P(\boldsymbol{x})$:

$$\mathcal{N}_P(\boldsymbol{x}) = \{\boldsymbol{y} \in \mathcal{V}(P) |\ \boldsymbol{x}\boldsymbol{y} \in \mathcal{E}(P)\}\,; \qquad (2.7.5)$$

we often drop the symbol P if the reference is clear from the context.

[5] A proof is available in Brøndsted (1983, thm. 11.4).

In a polytope, issuing a ray from each edge containing a given vertex produces a cone that contains the polytope.

Theorem 2.7.6 *Let P be a polytope in $\mathbb{R}^d$ and let $\boldsymbol{v} \in \mathcal{V}(P)$. Then the affine convex cone based at $\boldsymbol{v}$ and spanned by the neighbours of $\boldsymbol{v}$ in P contains P. Notationally,*

$$P \subseteq \boldsymbol{v} + \operatorname{cone}\{\boldsymbol{u} - \boldsymbol{v} |\ \boldsymbol{u} \in \mathcal{N}_P(\boldsymbol{v})\}.$$

Proof Let H be a hyperplane in $\mathbb{R}^d$ such that $P \cap H$ defines the vertex figure $P/\boldsymbol{v}$ of P at $\boldsymbol{v}$. Take any other vertex $\boldsymbol{y}$ of P. Then the segment $[\boldsymbol{v}, \boldsymbol{y}]$ intersects H at a point $\boldsymbol{y}'$, a point of $P/\boldsymbol{v}$. It follows that $\boldsymbol{y}$ lies in the ray $\{\boldsymbol{v}+\alpha(\boldsymbol{y}'-\boldsymbol{v}) \mid \alpha \geqslant 0\}$, which in turn implies that

$$P \subseteq \left\{\boldsymbol{v} + \alpha(\boldsymbol{y}' - \boldsymbol{v}) \middle|\ \text{for all } \boldsymbol{y}' \in P/\boldsymbol{v}, \alpha \geqslant 0\right\}. \tag{2.7.6.1}$$

Every point $\boldsymbol{y}' \in P/\boldsymbol{v}$ is in the convex hull of $\mathcal{V}(P/\boldsymbol{v})$ and so

$$\begin{aligned}&\left\{\boldsymbol{v} + \alpha(\boldsymbol{y}' - \boldsymbol{v}) \middle|\ \text{for all } \boldsymbol{y}' \in P/\boldsymbol{v}, \alpha \geqslant 0\right\}\\ &\qquad\subseteq \boldsymbol{v} + \operatorname{cone}\{\boldsymbol{w} - \boldsymbol{v} |\ \text{for all } \boldsymbol{w} \in \mathcal{V}(P/\boldsymbol{v})\}.\end{aligned} \tag{2.7.6.2}$$

Additionally, every vertex of $P/\boldsymbol{v}$ lies in a ray from $\boldsymbol{v}$ to a neighbour of $\boldsymbol{v}$. Thus

$$\boldsymbol{v} + \operatorname{cone}\{\boldsymbol{w} - \boldsymbol{v} |\ \text{for all } \boldsymbol{w} \in \mathcal{V}(P/\boldsymbol{v})\} \subseteq \boldsymbol{v} + \operatorname{cone}\{\boldsymbol{u} - \boldsymbol{v} |\ \boldsymbol{u} \in \mathcal{N}_P(\boldsymbol{v})\}. \tag{2.7.6.3}$$

Combining (2.7.6.1), (2.7.6.2), and (2.7.6.3), we get that

$$P \subseteq \boldsymbol{v} + \operatorname{cone}\{\boldsymbol{u} - \boldsymbol{v} |\ \boldsymbol{u} \in \mathcal{N}_P(\boldsymbol{v})\},$$

the desired conclusion. □

An immediate corollary of Theorem 2.7.6 is the following.

Corollary 2.7.7 *Let P be a d-polytope in $\mathbb{R}^d$ and let $\boldsymbol{v} \in \mathcal{V}(P)$. Then*

$$\operatorname{aff}\left(\{\boldsymbol{v}\} \cup \mathcal{N}_P(\boldsymbol{v})\right) = \mathbb{R}^d.$$

Proof If the the statement were false, then the set $\{\boldsymbol{v}\} \cup \mathcal{N}(\boldsymbol{v})$ would lie in a hyperplane H in $\mathbb{R}^d$. This would in turn imply that the affine convex cone $C := \boldsymbol{v} + \operatorname{cone}\{\boldsymbol{u} - \boldsymbol{v} \mid \boldsymbol{u} \in \mathcal{N}(\boldsymbol{v})\}$ is in H. However, $P \subset C$ (Theorem 2.7.6), which would lead to the contradictory conclusion that a d-polytope lies in H. Hence the corollary follows. □

Another corollary of Theorem 2.7.6 is the following.

Corollary 2.7.8 *Let P be a d-polytope in $\mathbb{R}^d$ and let $\boldsymbol{v} \in \mathcal{V}(P)$. Suppose that H is a hyperplane in $\mathbb{R}^d$ and K is a closed halfspace defined by H. If H contains $\boldsymbol{v}$ and K contains $\mathcal{N}_P(\boldsymbol{v})$, then K is a supporting halfspace of P; that is, $P \subseteq K$.*

Proof The vertices $\{\boldsymbol{v}\} \cup \mathcal{N}(\boldsymbol{v})$ of P all lie in K, and so K, being an affine convex cone itself, contains the affine convex cone $C := \boldsymbol{v} + \operatorname{cone}\{\boldsymbol{u} - \boldsymbol{v} \mid \boldsymbol{u} \in \mathcal{N}(\boldsymbol{v})\}$. By Theorem 2.7.6, $P \subseteq C$. Hence $P \subseteq K$. □

2.8 Simple and Simplicial Polytopes

A polytope is *simplicial* if every facet is a simplex. And a d-polytope is *simple* if every vertex is contained in precisely d facets; otherwise the d-polytope is *nonsimple*. It is clear that a simplex is a simple polytope. A face of a polytope that is itself a simple polytope is a *simple face*; otherwise the face is *nonsimple*. Trivially, every vertex, edge, and 2-polytope is simple. Simplicial and simple polytopes are closely related by duality.

Theorem 2.8.1 *A polytope is simple if and only if its dual polytope is simplicial.*

Proof Let P be a d-polytope, P^* the dual polytope of P, and ψ an antiisomorphism from $\mathcal{L}(P)$ to $\mathcal{L}(P^*)$. Suppose that P is simple. Then every vertex $\boldsymbol{v}$ of P is contained in precisely d facets. Since $\mathcal{L}(P)$ is the opposite of $\mathcal{L}(P^*)$, the facet $\psi(\boldsymbol{v})$ of P^* that is conjugate to $\boldsymbol{v}$ contains precisely d vertices; that is, $\psi(\boldsymbol{v})$ is a simplex (Problem 2.15.5). Every facet of P^* is the conjugate of some vertex of P; hence P^* is simplicial.

Suppose that P^* is simplicial. Then every facet F of P^* is a simplex; it has precisely d vertices. Because $\mathcal{L}(P^*)$ is the opposite of $\mathcal{L}(P)$, the vertex $\psi^{-1}(F)$ of P is contained in precisely d facets. Additionally, every vertex of P is the conjugate of some facet of P^*. Hence P is simple. □

In the same way that we gave special names to the 0-faces, 1-faces, and $(d-1)$-faces of a d-polytope P, we give the name *ridge* to a $(d-2)$-face of P.

We define simple polytopes by the number of facets in which each vertex is contained. An alternative definition could have considered the number of edges.

Theorem 2.8.2 *A d-polytope is simple if and only if each vertex is incident with precisely d edges.*

Proof Let P be a d-polytope, P^* the dual polytope of P, and ψ an antiisomorphism from $\mathcal{L}(P)$ to $\mathcal{L}(P^*)$. A vertex $\boldsymbol{v}$ of P is contained in precisely d edges if and only if the $(d-1)$-face $\psi(\boldsymbol{v})$ of P^* contains precisely d ridges of P^*. And a $(d-1)$-face has precisely d ridges of P^* if and only if it is a $(d-1)$-simplex (Problem 2.15.5). Thus, every vertex of P is incident with precisely d edges if and only if P^* is simplicial, and thus the result now follows from Theorem 2.8.1. □

A d-polytope P is *k-simplicial* if each k-face of P is a simplex; every polytope is 1-simplicial and simplicial polytopes are $(d-1)$-simplicial. The polytope P is said to be *k-simple* if each $(d-1-k)$-face is contained in precisely $k+1$ facets; every polytope is 1-simple and simple polytopes are $(d-1)$-simple. Theorem 2.8.1 ensures that a polytope P is simple if and only if P^* is simplicial. This generalises to k-simplicial and k-simple polytopes: P is k-simplicial if and only if P^* is k-simple (Problem 2.15.8).

According to Theorem 2.4.16, a k-face F of a d-polytope is contained in at least $d-k$ facets of the polytope. This lower bound is met with equality in the case of simple polytopes.

Theorem 2.8.3 *Let P be a simple d-polytope, k a number in* $[0\dots d-1]$, *and* $F_1,\dots,F_{d-k}$ *facets of P. Then*

$$F := \bigcap_{i=1}^{d-k} F_i$$

is either $\emptyset$ or a k-face of P.

Proof Let P^* be the dual polytope of P, and let ψ be an antiisomorphism from $\mathcal{L}(P)$ to $\mathcal{L}(P^*)$. If $F=\emptyset$ there is nothing to prove, so suppose otherwise. The face F is the largest face of P contained in the facets $F_1,\dots,F_{d-k}$, and so the face $\psi(F)$ is the smallest face of P^* containing the vertices $\psi(F_1),\dots,\psi(F_{d-k})$. Since P^* is simplicial, each facet of P^* is a simplex (Theorem 2.8.1), which implies that every face of P^* is simplex. Thus $\psi(F)$ is a $(d-k-1)$-simplex. Theorem 2.4.12 now ensures that F is a k-face of P. □

Theorem 2.8.3 gives that a j-face of a simple d-polytope is contained in exactly $d-j$ facets. We can give the exact number of k-faces containing the j-face.

Theorem 2.8.4 *Let P be a simple d-polytope and let* $0 \leqslant j \leqslant k \leqslant d$. *Then there are precisely*

$$\binom{d-j}{d-k}$$

k-faces of P containing a given j-face of P.

Proof Let J be a j-face of P. If $k = d$ then P is the only d-face containing J. Let P^* be the dual polytope of P and let ψ be an antiisomorphism from $\mathcal{L}(P)$ to $\mathcal{L}(P^*)$. If instead $k < d$, then $\psi(J)$ is a $(d-1-j)$-face of P^* and the number of k-faces of P containing J coincides with the number of $(d-1-k)$-faces of $\psi(J)$. Since $\psi(J)$ is a $(d-1-j)$-simplex, it contains

$$\binom{d-j}{d-k}$$

$(d-1-k)$-faces by Proposition 2.6.1. The proof is now complete. □

The application of duality in the proof of Theorem 2.8.4 also gives that each vertex of a k-face F in a simple polytope is contained in precisely k $(k-1)$-faces of F, which ensures that F is also a simple polytope (by definition).

Theorem 2.8.5 *Every proper face of a simple polytope is another simple polytope.*

Another important property of simple polytopes is that every k-subset of edges incident with a vertex defines a k-face. It is easy to find examples of polytopes that do not satisfy the latter property. For instance, consider a 3-crosspolytope I as a bipyramid over a quadrangle Q. Then no two edges of Q sharing a vertex define a 2-face of I.

Theorem 2.8.6 *Let P be a simple d-polytope and $k \in [0 \ldots d-1]$. Suppose that $\boldsymbol{v}$ is a vertex of P, $\boldsymbol{v}\boldsymbol{v}_1, \ldots, \boldsymbol{v}\boldsymbol{v}_k$ are k edges of P that are incident with $\boldsymbol{v}$, and F is the smallest face of P containing these edges. Then F is a simple k-face of P.*

Proof Let P^* be the dual polytope of P and let ψ be an antiisomorphism from $\mathcal{L}(P)$ to $\mathcal{L}(P^*)$. By duality, the face $\psi(F)$ is the largest face of P^* contained in the $(d-2)$-faces $\psi(\boldsymbol{v}\boldsymbol{v}_1), \ldots, \psi(\boldsymbol{v}\boldsymbol{v}_k)$ of the facet $\psi(\boldsymbol{v})$:

$$\psi(F) = \bigcap_{i=1}^{k} \psi(\boldsymbol{v}\boldsymbol{v}_i).$$

The facet $\psi(\boldsymbol{v})$ is a simplex (by Theorem 2.7.3 or Theorem 2.8.1). From Theorem 2.8.3, it follows that $\psi(F)$ is a $(d-1-k)$-face of $\psi(\boldsymbol{v})$. Since $\psi(F)$ is contained in precisely k $(d-2)$-faces of $\psi(\boldsymbol{v})$, the k edges $\boldsymbol{v}\boldsymbol{v}_1, \ldots, \boldsymbol{v}\boldsymbol{v}_k$ are the only edges of P that are incident with $\boldsymbol{v}$ and are contained in F.

The face F is the conjugate of $\psi(F)$, and so it is a k-face of P. Thus, by Theorem 2.8.5 it is a simple polytope. □

The proof of Theorem 2.8.6 also yields a slightly more general result.

Theorem 2.8.7 *Let P be a d-polytope and $k \in [0 \dots d-1]$. Suppose that $\boldsymbol{v}$ is a vertex of P incident with precisely d edges in P. Further suppose that $\boldsymbol{v}\boldsymbol{v}_1, \dots, \boldsymbol{v}\boldsymbol{v}_k$ are k edges of P that are incident with $\boldsymbol{v}$ and that F is the smallest face of P containing these edges. Then F is a k-face of P and these k edges are the only edges of F incident with $\boldsymbol{v}$.*

By Theorem 2.8.7, vertices in a d-polytope P that are incident with precisely d edges behave as vertices of a simple d-polytope. In view of this, we will say that such a vertex is *simple*; a vertex in P incident with more than d edges is *nonsimple*. Since every vertex is a simple 0-polytope, the expressions 'simple vertex' and 'nonsimple vertex' will refer only to the nature of the vertex in relation to the ambient polytope, and they should cause no confusion.

A d-simplex is both simple and simplicial, which characterises d-simplices for $d \geqslant 3$.

Theorem 2.8.8 *A simple and simplicial polytope is a simplex or a 2-polytope.*

Proof The case of two dimensions is trivial, so let P be a d-polytope that is both simple and simplicial for $d \geqslant 3$. Take a vertex $\boldsymbol{v}_0$ of P. Since P is simple, the vertex $\boldsymbol{v}_0$ is incident with precisely d edges $\boldsymbol{v}_0\boldsymbol{v}_1, \dots, \boldsymbol{v}_0\boldsymbol{v}_d$ (Theorem 2.8.2). Let $X := \{\boldsymbol{v}_0, \boldsymbol{v}_1, \dots, \boldsymbol{v}_d\}$ and let $T := \operatorname{conv} X$. Because P is simplicial, every $d-1$ of these edges defines a simplex facet of P (Theorem 2.8.6). It follows that, for $d \geqslant 3$, every pair of vertices in X are adjacent in P. As a result, every vertex $\boldsymbol{v}_i$ in X is adjacent to precisely d other vertices in X. Hence $T \subseteq P$.

Consider any supporting halfspace K of T and let H be a hyperplane bounding K. Then $T \cap H$ is a proper face of T and thus it contains a vertex $\boldsymbol{v}_\ell$ with $\ell \in [0 \dots d]$. Since H contains $\boldsymbol{v}_\ell$ and K contains all the neighbours of $\boldsymbol{v}_\ell$ in P, Corollary 2.7.8 ensures that K is a supporting halfspace of P. In other words, every supporting halfspace of T is a supporting halfspace of P. A polytope is the intersection of its supporting halfspaces (Theorem 1.8.3). Hence $P \subseteq T$, concluding that $P = T$. □

Similarly, for $d \geqslant 3$, d-cubes are the only simple and cubical polytopes. A polytope is *cubical* if every facet is a cube. A proof for this result follows from Blind and Blind (1998, sec. 7).

Theorem 2.8.9 (Blind and Blind, 1998) *A simple and cubical polytope is a cube or a 2-polytope.*

2.9 Cyclic and Neighbourly Polytopes

Cyclic polytopes are the d-polytopes with the maximum number of k-faces among the d-polytope with n vertices, for each $k \in [0 \ldots d-1]$ (Chapter 8). Because of this, they feature in a number of fundamental results in polytope theory, in particular in the upper bound theorem of McMullen (1970). This section studies them.

The *moment curve* μ_d in $\mathbb{R}^d$ is defined, for $x \in [a,b]$, as

$$\mu_d(x) = (x, x^2, \ldots, x^d)^t. \tag{2.9.1}$$

We next describe some of the properties of the moment curve.

Proposition 2.9.2 (Properties of the moment curve) *The moment curve μ_d in $\mathbb{R}^d$ has the following properties:*

(i) *Every $d+1$ points on μ_d are affinely independent.*
(ii) *If d distinct points on μ_d lie in a hyperplane H of $\mathbb{R}^d$, then the curve at each intersection with H passes from one side of H to the other side.*

Proof Let H be a hyperplane in $\mathbb{R}^d$ defined as

$$H := \left\{ y \in \mathbb{R}^d \middle| \; y \cdot (a_1, \ldots, a_d)^t = -a_0 \right\}.$$

Then, for some $i \in [0 \ldots d]$, we have $a_i \neq 0$. With the numbers $a_0, \ldots, a_d$, we now define a nonzero polynomial $p_H(x)$ of degree at most d:

$$p_H(x) = a_0 + a_1 x + \cdots + a_d x^d.$$

It follows that a point $\mu_d(x_i)$ is in H if and only if x_i is a root of p_H. The polynomial p_H has at most d roots, and so no $d+1$ points on μ_d can lie in H; this shows (i). Suppose that H contains exactly d distinct points of μ_d. Then p_H has d simple roots. In a small neighbourhood of a simple root, the polynomial is either increasing or decreasing, which causes the moment curve to pass from one side of H to the other side; this proves (ii) □

A *cyclic d-polytope* $C(n,d)$ is the convex hull of $n \geqslant d+1$ points $\mu_d(x_1), \ldots, \mu_d(x_n)$ on the moment curve satisfying $x_1 < \cdots < x_n$. Properties of cyclic polytopes are explained from properties of the moment curve.

Proposition 2.9.3 *Let P be a cyclic d-polytope on n vertices. Then*

(i) *P is simplicial; and*
(ii) *every set of k vertices of P, with $2k \leqslant d$, forms a $(k-1)$-face.*

Proof (i) This ensues from Proposition 2.9.2(i).

(ii) Let P be the convex hull of the n points $\mu_d(x_1), \ldots, \mu_d(x_n)$ with $x_1 < \cdots < x_n$. Set $X := \{x_1, \ldots, x_n\}$. Among the elements of X, select a k-subset X_k satisfying $x'_1 < \cdots < x'_k$. With these k numbers, we define a polynomial

$$p_k(x) := (x - x'_1)^2 \cdots (x - x'_k)^2 = a_0 + a_1 x + \cdots + a_{2k} x^{2k}$$

of degree $2k \leqslant d$. And with the coefficients $a_0, \ldots, a_{2k}$, we define a hyperplane H_k in $\mathbb{R}^d$

$$H_k := \left\{ y \in \mathbb{R}^d \,\middle|\, y \cdot (a_1, \ldots, a_{2k}, 0, \ldots, 0)^t = -a_0 \right\}.$$

It follows that all the points $\mu_d(x'_i)$ with $x'_i \in X_k$ are in H_k, and that any other point $\mu_d(x_j)$ with $x_j \in X \backslash X_k$ lies in the same side of H_k, as the expression

$$\begin{aligned} \mu_d(x_j) \cdot (a_1, \ldots, a_{2k}, 0, \ldots, 0)^t &= -a_0 + p_k(x_j) \\ &= -a_0 + (x_j - x'_1)^2 \cdots (x_j - x'_k)^2 \\ &> -a_0 \end{aligned}$$

attests. Hence H_k supports P at conv $\{\mu_d(x'_1), \ldots, \mu_d(x'_k)\}$, which is a $(k-1)$-simplex by (i). This proves (ii). □

Gale (1963) provided a criterion to tell which d-subsets of vertices of a cyclic d-polytope form a facet. The criterion relies on a linear ordering $\leqslant$ on the vertices $\mu_d(x_1), \ldots, \mu_d(x_n)$ of a cyclic d-polytope P given by $\mu_d(x_i) \leqslant \mu_d(x_j)$ if and only if $x_i \leqslant x_j$. Henceforth, we implicitly assume that a cyclic polytope is coupled with this vertex ordering.

Theorem 2.9.4 (Gale's evenness condition) *Let P be a cyclic d-polytope. A d-subset X of $\mathcal{V}(P)$ is the vertex set of a facet of P if and only if, for every two distinct vertices in $\mathcal{V}(P) \backslash X$, the number of elements of X between them is even.*

We put Gale's evenness condition into practice.

Example 2.9.5 Consider a cyclic 3-polytope P on seven vertices and the following sets:

$$\begin{aligned} X_1 &:= \{\mu_3(4), \mu_3(5), \mu_3(6)\}, \\ X_2 &:= \{\mu_3(3), \mu_3(4), \mu_3(6)\}, \\ X_3 &:= \{\mu_3(1), \mu_3(3), \mu_3(4)\}. \end{aligned}$$

See Fig. 2.9.1. Between any two vertices $\mu_3(x_i)$ and $\mu_3(x_j)$ of P outside X_1, there are zero elements of X_1, since $x_i, x_j \in [0 \ldots 3]$; here, we use the linear ordering of the vertices of P. Thus X_1 is the vertex set of a facet of P.

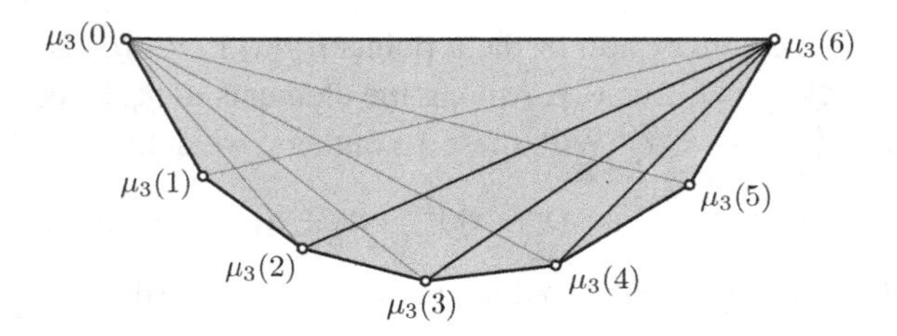

Figure 2.9.1 A cyclic 3-polytope on seven vertices.

Take any two vertices $\mu_3(x_i)$ and $\mu_3(x_j)$ of $P \backslash X_2$. If $x_i, x_j \in [0 \dots 2]$, then there are zero elements of X_2 between $\mu_3(x_i)$ and $\mu_3(x_j)$. So suppose $x_j = 5$. Then $x_i \in [0 \dots 2]$, and the vertices $\mu_3(3)$ and $\mu_3(4)$ of X_2 are between $\mu_3(x_i)$ and $\mu_3(x_j)$. Hence X_2 is also the vertex set of a facet of P.

Take the vertices $\mu_3(0)$ and $\mu_3(2)$ of $P \backslash X_3$. Between $\mu_3(0)$ and $\mu_3(2)$, there is exactly one vertex of X_3, namely $\mu_3(1)$. Hence X_3 is not the vertex set of a facet of P.

Proof of Gale's evenness condition (Theorem 2.9.4) Let H be a hyperplane in $\mathbb{R}^d$ that is spanned by X. The set X determines a facet of P if and only if all the vertices in $\mathcal{V}(P) \backslash X$ lie in the same side of H. Take any two distinct vertices $\mu_d(x_i), \mu_d(x_j) \in \mathcal{V}(P) \backslash X$ with $x_i < x_j$. The moment curve at each intersection with H passes from one side of H to the other side (Proposition 2.9.2(ii)). Therefore, $\mu_d(x_i)$ and $\mu_d(x_j)$ lie in the same side of H if and only if, while traversing the curve from $\mu_d(x_i)$ to $\mu_d(x_j)$, we encounter an even number of vertices of X. □

One consequence of Gale's evenness condition is that every cyclic d-polytope on n vertices has the same facet-vertex incidence matrix (Section 2.3). Thus, every two cyclic d-polytopes on n vertices are combinatorially isomorphic, and we can just talk of the cyclic d-polytope on n vertices, namely $C(n,d)$.

A second consequence of Gale's evenness condition is that we can talk of linear orderings of vertices of $C(n,d)$ satisfying the condition. We say that a linear ordering

$$\boldsymbol{u}_1 <' \cdots <' \boldsymbol{u}_n$$

of the vertices of $C(n,d)$ is *cyclic* if it satisfies Gale's evenness condition: a d-subset X of $\mathcal{V}(C(n,d))$ determines a facet of $C(n,d)$ if and only if, between any two vertices of $\mathcal{V}(C(n,d)) \backslash X$, there is an even number of elements of X in the vertex ordering $<'$. The aforementioned linear ordering $<$ of the vertices of $C(n,d)$ given by

$$\mu_d(x_1) < \cdots < \mu_d(x_n) \text{ whenever } x_1 < \cdots < x_n$$

is cyclic. The following two assertions are now plain.

Proposition 2.9.6 *A simplicial d-polytope P is combinatorially isomorphic to $C(n, d)$ if and only if some ordering*

$$\boldsymbol{u}_1 < \cdots < \boldsymbol{u}_n$$

of the vertices of P is cyclic.

Lemma 2.9.7 *Let d be even. If the ordering $\boldsymbol{u}_1 < \cdots < \boldsymbol{u}_n$ of the vertices of $C(n, d)$ is cyclic, then so is the ordering*

$$\boldsymbol{u}_i < \cdots < \boldsymbol{u}_n < \boldsymbol{u}_1 < \cdots < \boldsymbol{u}_{i-1},$$

for each $i \in [1 \ldots n]$.

Gale's evenness condition explains the vertex figures of cyclic polytopes.

Theorem 2.9.8 (Vertex figures of cyclic polytopes) *Let P be a cyclic d-polytope with a vertex ordering $\boldsymbol{u}_1 < \cdots < \boldsymbol{u}_n$ that satisfies Gale's evenness condition* (2.9.4). *Then the following holds:*

(i) *For odd d, every facet of P contains $\boldsymbol{u}_1$ or $\boldsymbol{u}_n$.*
(ii) *For even d, the vertex figure of P at every vertex is a cyclic $(d-1)$-polytope.*
(iii) *For odd d, the vertex figures $P/\boldsymbol{u}_1$ and $P/\boldsymbol{u}_n$ of P at vertices $\boldsymbol{u}_1$ and $\boldsymbol{u}_n$, respectively, are cyclic $(d-1)$-polytopes.*
(iv) *If all the vertex figures of P are cyclic $(d-1)$-polytopes on $n-1$ vertices then*

$$d f_{d-1}(P) = n f_{d-2}(C(n-1, d-1)).$$

(v) *For odd $d \geqslant 5$, we have that*

$$f_{d-1}(P) = 2 f_{d-2}(C(n-1, d-1)) - f_{d-3}(C(n-2, d-2)).$$

(vi) *For $n \geqslant d+2$ and odd $d \geqslant 5$, the vertex figure $P/\boldsymbol{u}_i$ of P at some other vertex $\boldsymbol{u}_i$ is not a cyclic $(d-1)$-polytope.*
(vii) *For odd $d \geqslant 3$, P is the dual wedge of $C(n-1, d-1)$ at any vertex.*

Proof (i) If a facet of P did not contain $\boldsymbol{u}_1$ or $\boldsymbol{u}_n$, then there would be an odd number of vertices between $\boldsymbol{u}_1$ and $\boldsymbol{u}_n$, which would violate Gale's evenness condition.

(ii) This is true for $d = 2$ so assume that $d \geqslant 4$. Let $\boldsymbol{u}_i$ be a vertex of P and X a $(d-1)$-subset of $\mathcal{V}(P) \setminus \{\boldsymbol{u}_i\}$. Because P is simplicial and the $(k-1)$-faces of $P/\boldsymbol{u}_i$ are in one-to-one correspondence with the k-faces of P containing $\boldsymbol{u}_i$ (Theorem 2.7.2), we have the following.

Claim 1 The proper faces of $P/\boldsymbol{u}_i$ can be thought of as faces of P. In particular, the subset X of $\mathcal{V}(P)$ is the vertex set of a $(d-2)$-face of $P/\boldsymbol{u}_i$ if and only if the d-subset $X \cup \{\boldsymbol{u}_i\}$ of $\mathcal{V}(P)$ is the vertex set of a $(d-1)$-face F of P.

Since the graph of P is a complete graph (Proposition 2.9.3), this claim implies the next assertion.

Claim 2 The polytope $P/\boldsymbol{u}_i$ is a simplicial $(d-1)$-polytope with vertex set $\mathcal{V}(P)\backslash\{\boldsymbol{u}_i\}$.

According to Lemma 2.9.7, the ordering

$$\boldsymbol{u}_i < \boldsymbol{u}_{i+1} < \cdots < \boldsymbol{u}_n < \boldsymbol{u}_1 < \cdots < \boldsymbol{u}_{i-1} \tag{2.9.8.1}$$

is cyclic. If $X \cup \{\boldsymbol{u}_i\}$ is the vertex set of a $(d-1)$-face of P, then Gale's evenness condition on P yields an even number of elements from $X \cup \{\boldsymbol{u}_i\}$ between any two vertices $\boldsymbol{y}$ and $\boldsymbol{z}$ in $\mathcal{V}(P)\backslash(X \cup \{\boldsymbol{u}_i\})$, with respect to the ordering (2.9.8.1). As a consequence, there is an even number of vertices from X between the same two vertices $\boldsymbol{y}$ and $\boldsymbol{z}$ in $\mathcal{V}(P)\backslash(X \cup \{\boldsymbol{u}_i\})$, with respect to the ordering

$$\boldsymbol{u}_{i+1} < \cdots < \boldsymbol{u}_n < \boldsymbol{u}_1 < \cdots < \boldsymbol{u}_{i-1} \tag{2.9.8.2}$$

of the vertices of $P/\boldsymbol{u}_i$ (Claims 1 and 2). This shows that the ordering (2.9.8.2) is cyclic, implying that $P/\boldsymbol{u}_i$ is combinatorially isomorphic to a cyclic $(d-1)$-polytope on $n-1$ vertices (Proposition 2.9.6).

(iii) The reasoning is similar to that of (ii). According to (i), every facet of P contains $\boldsymbol{u}_1$ or $\boldsymbol{u}_n$. If $X \cup \{\boldsymbol{u}_1\}$ is the vertex set of a $(d-1)$-face of P, then, by Gale's evenness condition on P, there is an even number of elements from $X \cup \{\boldsymbol{u}_1\}$ between any two vertices $\boldsymbol{y}$ and $\boldsymbol{z}$ in $\mathcal{V}(P)\backslash(X \cup \{\boldsymbol{u}_1\})$, with respect to the the ordering $\boldsymbol{u}_1 < \cdots < \boldsymbol{u}_n$. As in (ii), it follows that there is an even number of vertices from X between the same two vertices $\boldsymbol{y}$ and $\boldsymbol{z}$ in $\mathcal{V}(P)\backslash(X \cup \{\boldsymbol{u}_1\})$, with respect to the ordering

$$\boldsymbol{u}_2 < \cdots < \boldsymbol{u}_n$$

of the vertices of $P/\boldsymbol{u}_1$ (see Claim 1 from the proof of (ii)). This shows that this ordering is cyclic, implying that $P/\boldsymbol{u}_1$ is combinatorially isomorphic to a cyclic $(d-1)$-polytope on $n-1$ vertices (Proposition 2.9.6). The same analysis yields that $P/\boldsymbol{u}_n$ is combinatorially isomorphic to a cyclic $(d-1)$-polytope on $n-1$ vertices.

(iv) We count the facet–vertex incidences of P in two different ways. A vertex $\boldsymbol{u}$ of P is contained in $f_{d-2}(P/\boldsymbol{u})$ facets of P, which is equal to $f_{d-2}(C(n-1,d-1))$ by assumption. Additionally, a facet contains d vertices and P has $f_{d-1}(P)$ facets. The result is now clear.

(v) Because of (iii), the vertex figures $P/\boldsymbol{u}_1$ and $P/\boldsymbol{u}_n$ are cyclic $(d-1)$-polytopes on $n-1$ vertices, and so there are $f_{d-2}(C(n-1,d-1))$ facets of P containing $\boldsymbol{u}_1$ and $f_{d-2}(C(n-1,d-1))$ facets of P containing $\boldsymbol{u}_n$. Furthermore, the number of facets of P containing both $\boldsymbol{u}_1$ and $\boldsymbol{u}_n$ coincides with the number of ways of selecting $d-2$ vertices from $\{\boldsymbol{u}_2,\dots,\boldsymbol{u}_{n-1}\}$ such that the ordering $\boldsymbol{u}_2 < \cdots < \boldsymbol{u}_{n-1}$ is cyclic; this is the same as counting the number $f_{d-3}(C(n-2,d-2))$ of $(d-3)$-faces of a cyclic $(d-2)$-polytope on $n-2$ vertices. The formula now follows.

(vi) Suppose, by way of contradiction, that the vertex figure $P/\boldsymbol{u}$ of P at every vertex $\boldsymbol{u}$ is a cyclic $(d-1)$-polytope on $n-1$ vertices. In this case, an application of (iv) to P yields that

$$d f_{d-1}(P) = n f_{d-2}(C(n-1,d-1)). \tag{2.9.8.3}$$

Moreover, as $d-1 \geqslant 4$ is even, Part (ii) gives that all vertex figures of $P/\boldsymbol{u}$ are cyclic $(d-2)$-polytopes on $n-2$ vertices. Another application of (iv) to $P/\boldsymbol{u}$ gives that

$$(d-1) f_{d-2}(C(n-1,d-1)) = (n-1) f_{d-3}(C(n-2,d-2)). \tag{2.9.8.4}$$

We solve (2.9.8.3) for $f_{d-1}(P)$ and (2.9.8.4) for $f_{d-3}(C(n-2,d-2))$, and then we put these expressions for $f_{d-1}(P)$ and $f_{d-3}(C(n-2,d-2))$ into (v) to obtain that

$$\begin{aligned}\frac{n}{d} f_{d-2}(C(n-1,d-1)) = {} & 2 f_{d-2}(C(n-1,d-1)) \\ & - \frac{d-1}{n-1} f_{d-2}(C(n-1,d-1)),\end{aligned}$$

or equivalently that

$$f_{d-2}(C(n-1,d-1)) \left(\frac{n}{d} - 2 + \frac{d-1}{n-1} \right) = 0.$$

Solving this equation amounts to solving

$$\frac{n}{d} - 2 + \frac{d-1}{n-1} = 0,$$

which reduces to $(d-n)(1+d-n) = 0$. The solutions are $n = d$ or $n = d+1$, violating our assumption of $n \geqslant d+2$.

(vii) This can be verified from Gale's evenness condition (2.9.4) on P, and so it is left to the reader. □

Cyclic Polytopes and Curves of Order d

Our initial, and the standard, presentation of cyclic polytopes uses the moment curve to realise the polytopes and unveil their properties. However, other curves could have been chosen instead, although none would beat the simplicity of the moment curve.

Denote by $\mathcal{C}[a,b]$ the space of continuous, real-valued functions defined on the interval $[a,b]$. This is a linear space over $\mathbb{R}$ with norm

$$\|\varphi\| := \max_{x\in[a,b]} |\varphi(x)|. \tag{2.9.9}$$

The moment curve is defined with the set $\{\varphi_1(x) = x, \ldots, \varphi_d(x) = x^d\}$ of functions from $\mathcal{C}[a,b]$. It turns out that polytopes combinatorially isomorphic to cyclic polytopes can be realised with sets of functions from $\mathcal{C}[a,b]$ that satisfy Haar's condition on $[a,b]$ (Timan, 1963, sec. 2.3).

Definition 2.9.10 (Haar's condition) A curve $\omega_d \colon \mathbb{R} \to \mathbb{R}^d$ defined by

$$\omega_d(x) := (\varphi_1(x), \ldots, \varphi_d(x))^t \tag{2.9.10.1}$$

satisfies *Haar's condition* if each $\varphi_i \in \mathcal{C}[a,b]$, and for every $d+1$ distinct numbers $x_1, \ldots, x_{d+1}$ in $[a,b]$ satisfying $x_1 < \cdots < x_{d+1}$, the points $\omega_d(x_1), \ldots, \omega_d(x_{d+1})$ are affinely independent in $\mathbb{R}^d$.

A curve $\omega_d \colon \mathbb{R} \to \mathbb{R}^d$ satisfying Haar's condition is said to be a *curve of order d*. Let $\omega_d \colon \mathbb{R} \to \mathbb{R}^d$ be a curve of order d defined as

$$\omega_d(x) = (\varphi_1(x), \ldots, \varphi_d(x))^t.$$

We define a d-polytope $C'(n,d)$ as the convex hull of $n \geqslant d+1$ points

$$\begin{aligned} \omega_d(x_1) &= (\varphi_1(x_1), \ldots, \varphi_d(x_1))^t, \\ &\vdots \\ \omega_d(x_n) &= (\varphi_1(x_n), \ldots, \varphi_d(x_n))^t, \end{aligned}$$

where $x_1, \ldots, x_n \in [a,b]$ and $x_1 < \cdots < x_n$.

A proof similar to that of Gale's evenness condition (Theorem 2.9.4) applies to the polytope $C'(n,d)$. As a result, the polytope $C'(n,d)$ has the same facet–vertex incidence matrix as the cyclic polytope $C(n,d)$, and so both polytopes are combinatorially isomorphic.

A result of Sturmfels (1987) states that, for every cyclic-d-polytope P of even dimension d, there exists a curve $\omega_d(x)$ of order d such that $P = \operatorname{conv}\{\omega_d(x_1), \ldots, \omega_d(x_n)\}$, for numbers $x_1 < \cdots < x_n$ in $[a,b]$. The situation is different for odd dimension $d \geqslant 3$: there are cyclic d-polytopes on $n \geqslant d+3$ vertices that do not arise from curves of order d (Cordovil and Duchet, 2000).

Neighbourly Polytopes

One of the most appealing properties of cyclic polytopes is that they are very 'neighbourly': every set of k vertices, $k \leqslant \lfloor d/2 \rfloor$, forms a $(k-1)$-face (Proposition 2.9.3(ii)). We have met this notion before. In the context of graphs, a complete graph is as 'neighbourly' as possible: every two vertices form an edge. Likewise, in the realm of polytopes, a d-simplex is as 'neighbourly' as possible: every k vertices form a proper $(k-1)$-face, for each $k \leqslant d$. In this final part, we explore the concept of 'neighbourliness'.

We say that a d-polytope P is *k-neighbourly* if every set of at most k vertices is the vertex set of a proper face of P. Proposition 2.9.3(ii) states that cyclic d-polytopes are $\lfloor d/2 \rfloor$-neighbourly. We will see that, apart from the d-simplex, no other d-polytope is k-neighbourly for $k > \lfloor d/2 \rfloor$, and so $\lfloor d/2 \rfloor$-neighbourly d-polytopes on n vertices such as $C(n,d)$ are the second best 'neighbourly' d-polytopes, and they exist for every $n \geqslant d+1$ (Proposition 2.9.3(ii)). For this reason, we call a $\lfloor d/2 \rfloor$-neighbourly d-polytope simply a *neighbourly d-polytope*; equivalently, we may say that a neighbourly d-polytope is a d-polytope with the $(\lfloor d/2 \rfloor - 1)$-skeleton of some n-simplex for $n \geqslant d+1$. Proposition 2.9.11 gathers the main properties of k-neighbourly polytopes.

Proposition 2.9.11[6] *Let P be a k-neighbourly d-polytope. Then*

(i) *every k vertices of P are affinely independent;*
(ii) $k \leqslant d$;
(iii) *P is k'-neighbourly, for each $k' \in [1..k]$;*
(iv) *if $k > \lfloor d/2 \rfloor$ then P is a d-simplex; and*
(v) *if $k = \lfloor d/2 \rfloor$ then P is $(d-2)$-simplicial, and if in addition d is even, then P is simplicial.*

2.10 Inductive Constructions of Polytopes

This section focusses on an inductive construction of the convex hull of a polytope, one in which a vertex is added at each stage. This is the so-called *beneath-beyond algorithm* of Grünbaum (1963) and Grünbaum (2003, sec. 5.2).

[6] A proof is available in Grünbaum (2003, sec. 7.1).

Let P be a d-polytope in $\mathbb{R}^d$ and let $\boldsymbol{x}$ be a point in $\mathbb{R}^d$. We say that a facet F of P is *visible* from the point $\boldsymbol{x}$ with respect to a polytope P in $\mathbb{R}^d$ if $\boldsymbol{x}$ belongs to the open halfspace determined by aff F that is disjoint from P (Fig. 2.10.1(a)). If instead $\boldsymbol{x}$ belongs to the open halfspace that contains the interior of P, we say that the facet F is *nonvisible* from $\boldsymbol{x}$ (Fig. 2.10.1(a)). Similarly, a hyperplane H, disjoint from the interior of P, is either *visible* or *nonvisible* from the point $\boldsymbol{x}$, with respect to P, depending on whether $\boldsymbol{x}$ lies in the open halfspace determined by H that is disjoint from P or in the open halfspace of H that contains the interior of P. Moreover, the point $\boldsymbol{x}$ is *beyond* a face J of P if the facets of P containing J are precisely those that are visible from $\boldsymbol{x}$ and the facets of P not containing J are all nonvisible from $\boldsymbol{x}$.

Our terminology follows that of Ziegler (1995, sec. 8.2), and it differs from that of Grünbaum (2003, sec. 5.2) in that the definitions of 'visible' and 'nonvisible' coincide with those of 'beyond' and 'beneath' in Grünbaum (2003, sec. 5.2), respectively. In addition, a facet F is visible or nonvisible from a point $\boldsymbol{x}$ in our sense if and only if $\boldsymbol{x}$ is beyond or beneath aff F, respectively, in Grünbaum's sense.

Theorem 2.10.1 (Construction of polytopes; Grünbaum [1963]) *Let P and P' be two d-polytopes in $\mathbb{R}^d$ and let $\boldsymbol{v}'$ be a vertex of P' such that $\boldsymbol{v}' \notin P$ and $P' = \operatorname{conv}(P \cup \{\boldsymbol{v}'\})$. Then the following hold:*

(i) *A face F of P is a face of P' if and only if there exists a facet of P containing F that is nonvisible from $\boldsymbol{v}'$ with respect to P.*
(ii) *If F is a face of P with $\boldsymbol{v}' \in \operatorname{aff} F$, then $F' := \operatorname{conv}(F \cup \{\boldsymbol{v}'\})$ is a face of P'.*
(iii) *If F is a face of P such that, among the facets of P containing F, there is at least one that is visible from $\boldsymbol{v}'$ (with respect to P) and at least one that is nonvisible (with respect to P), then*

$$F' := \operatorname{conv}(F \cup \{\boldsymbol{v}'\})$$

is a face of P'.
(iv) *For each face F' of P', there is a face F of P for which (i), (ii), or (iii) applies. In other words, each face of P' falls precisely in one of the above cases.*

Proof The main observation here is that every face of P' is either a face of P or the convex hull of $\boldsymbol{v}'$ and some face of P, because a hyperplane supporting P' at a face of P' other than $\boldsymbol{v}'$ also supports P.

(i) A facet J of P with supporting hyperplane H is a facet of P' with the same supporting hyperplane H if and only if J is nonvisible from $\boldsymbol{v}'$ (with

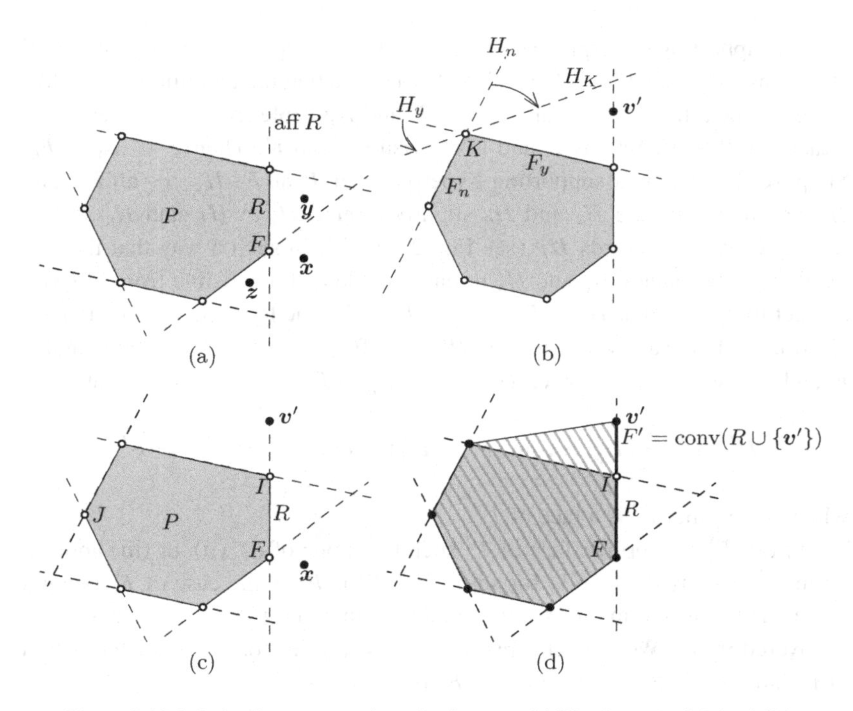

Figure 2.10.1 Inductive construction of polytopes. (a) The facet R of P is visible from the point x and nonvisible from the point z. The point x is beyond the face F but the points y and z are not. (b) Auxiliary figure for the proof of Theorem 2.10.1(iii). (c) The four cases of the proof of Theorem 2.10.1(iv). All the facets of P containing F are visible from x (Case (1)). All the facets of P containing J are nonvisible from x (Case (2)). All the facets of P containing I are visible from v' or contain v' in their affine hull and at least one such facet contains v' in their affine hull (Case (3)). All the facets of P containing F are nonvisible from v' or contain v' in their affine hull and at least one such facet contains v' in their affine hull (Case (4)). (d) $P' := \text{conv}(P \cup \{v'\})$. The polytope P is highlighted in grey and the polytope P' is highlighted in a tiling pattern.

respect to either P or P'). As a consequence, each face F of P that is in such a facet J will be a face of P'. Now consider a face F of P and P'. Then F does not contain v'. Since F is the intersection of all the facets of P' that contain it, F is in some facet J' of P' that does not contain v'. The point v' is a vertex of P', which implies that J' is nonvisible from v' with respect to P'. Hence the facet J' is facet of P that is nonvisible from v' with respect to P. This proves (i).

(ii)–(iii) Suppose that F is a face of P that satisfies (ii) or (iii). We establish that $F' := \text{conv}(F \cup \{v'\})$ is a face of P'. First, suppose that F satisfies (ii).

Then a supporting hyperplane of P at F will be a supporting hyperplane of P' at F', ensuring that F' is a face of P'. So assume that the condition (iii) holds.

Let F_y be a facet of P that contains F and is visible from $\boldsymbol{v}'$ and let F_n be a facet of P that contains F and is nonvisible from $\boldsymbol{v}'$. Then $F \subseteq F_y \cap F_n$. Suppose that H_F is a supporting hyperplane of P at F, $H_y := \operatorname{aff} F_y$, and $H_n := \operatorname{aff} F_n$. Rotate H_y and H_n slightly around $H_y \cap H_F$ and $H_n \cap H_F$, respectively, and towards H_F (see Fig. 2.10.1(b)) in such a way that the two resulting hyperplanes H'_y and H'_n remain visible and nonvisible from $\boldsymbol{v}'$ (with respect to P) and that $H'_y \cap P = H'_n \cap P = F$. The hyperplanes H'_y and H'_n allow us to define a new hyperplane $H'_F := \operatorname{aff}(\{\boldsymbol{v}'\} \cup (H'_y \cap H'_n))$ that contains $\boldsymbol{v}'$ and intersects P at F (since $H'_y \cap P = H'_n \cap P = F$). It follows that

$$H'_F \cap P' = H'_F \cap \operatorname{conv}(P \cup \{\boldsymbol{v}'\}) = \operatorname{conv}(F \cup \{\boldsymbol{v}'\}) = F',$$

which shows that F' is a face of P'.

(iv) Let F' be a proper face of P' such that none of (i), (ii), or (iii) applies. Then $F' = \operatorname{conv}(F \cup \{\boldsymbol{v}'\})$ for some face F of P, as the case of F' being a face of P is covered in (i). We can further assume that $\boldsymbol{v}' \notin \operatorname{aff} F$, as $\boldsymbol{v}' \in \operatorname{aff} F$ is covered in (ii). We (naively) list all the possibilities for the relative position of $\boldsymbol{v}'$ and the facets of P containing F (Fig. 2.10.1(c)):

(1) all the facets of P containing F are visible from $\boldsymbol{v}'$,
(2) all the facets of P containing F are nonvisible from $\boldsymbol{v}'$,
(3) all the facets of P containing F are visible from $\boldsymbol{v}'$ or contain $\boldsymbol{v}'$ in their affine hull, and at least one such facet contains $\boldsymbol{v}'$ in their affine hull (see the face I and vertex $\boldsymbol{v}'$ on Fig. 2.10.1(c)), and
(4) all the facets of P containing F are nonvisible from $\boldsymbol{v}'$ or contain $\boldsymbol{v}'$ in their affine hull, and at least one such facet contains $\boldsymbol{v}'$ in their affine hull (see the face F and vertex $\boldsymbol{v}'$ on Fig. 2.10.1(c)).

The cases (1), (2), and (3) are not real alternatives, as F' is a face of P' for which none of (i), (ii), or (iii) holds; see Fig. 2.10.1(c). Case (4) can certainly happen. In case (4), for each facet J of P that contains $\boldsymbol{v}'$ in their affine hull, we must have that $\operatorname{aff} J$ is a supporting hyperplane of P' (and of P). It must then follow that there is a proper face R of P that contains $\boldsymbol{v}'$ in the affine hull and is of the form $R = H \cap P$ for some supporting hyperplane H of P' (and of P); see Fig. 2.10.1(d). Hence (ii) applies, and $F' = \operatorname{conv}(R \cup \{\boldsymbol{v}'\})$. This completes the proof of the theorem. □

We mention that there was a mistake in the original proof of Grünbaum (1963, thm. 5.2.1); Case (4) of our proof of Theorem 2.10.1(iv) is overlooked.

However, Case (4) cannot arise if $\operatorname{conv}(F \cup \{\boldsymbol{v}'\})$ is a facet of P'; this and other consequences of Theorem 2.10.1 ensue.

Corollary 2.10.2 (Altshuler and Shemer, 1984) *Let P and P' be two d-polytopes in $\mathbb{R}^d$, and let $\boldsymbol{v}'$ be a vertex of P' such that $\boldsymbol{v}' \notin P$ and $P' = \operatorname{conv}(P \cup \{\boldsymbol{v}'\})$. Then the following hold:*

(i) *$\mathcal{V}(P') = \mathcal{V}(P) \cup \{\boldsymbol{v}'\}$ if and only if every vertex of P is in a facet of P that is nonvisible from $\boldsymbol{v}'$.*
(ii) *A facet F of P is a facet of P' if and only if it is nonvisible from $\boldsymbol{v}'$.*
(iii) *The set $\operatorname{conv}(F \cup \{\boldsymbol{v}'\})$ is a facet of P' if and only if either $\boldsymbol{v}' \in \operatorname{aff} F$ or among the facets of P containing F there is at least one that is visible from $\boldsymbol{v}'$ (with respect to P) and at least one that is nonvisible (with respect to P).*

We end this section with two applications of Theorem 2.10.1; each describes an algorithm that changes the combinatorial structure of a polytope.

Let P be a d-polytope in $\mathbb{R}^d$ and and let $\boldsymbol{v}$ be a vertex of P. Further, let $\boldsymbol{v}'$ be a point outside P such that the halfopen segment $(\boldsymbol{v}, \boldsymbol{v}']$ does not intersect any hyperplane spanned by the vertices of P. In the case that $\boldsymbol{v}$ belongs to the interior of $P' := \operatorname{conv}(P \cup \{\boldsymbol{v}'\})$, we say that P' is obtained from P by *pulling* $\boldsymbol{v}$ to $\boldsymbol{v}'$. The position of $\boldsymbol{v}'$ ensures that $\boldsymbol{v}'$ is beyond $\boldsymbol{v}$. The next result follows at once from Theorem 2.10.1.

Theorem 2.10.3 (Pulling vertices; Eggleston et al., 1964) *Let P' be a d-polytope in $\mathbb{R}^d$ obtained from a d-polytope P by pulling a vertex $\boldsymbol{v}$ of P to a vertex $\boldsymbol{v}'$ of P'. Then, for each $k \in [1 \ldots d-1]$, the k-faces of P' are as follows:*

(i) *The k-faces of P that do not contain $\boldsymbol{v}$.*
(ii) *The pyramid $\operatorname{conv}(F \cup \{\boldsymbol{v}'\})$ for each $(k-1)$-face F of P that does not contain $\boldsymbol{v}$ but belongs to a facet of P that contains $\boldsymbol{v}$.*

Moreover, $f_0(P') = f_0(P)$ and $f_k(P') \geqslant f_k(P)$ for each $k \in [1 \ldots d-1]$.

Repeated applications of Theorem 2.10.3 transform any d-polytope P into a simplicial polytope with the same number of vertices as P and at least as many faces of higher dimension. We state the result.

Theorem 2.10.4 *Let Q be a d-polytope obtained from a d-polytope P by successively pulling each of the vertices of P. Then the following hold:*

(i) *The polytope Q is a simplicial d-polytope satisfying $f_0(Q) = f_0(P)$, and $f_k(Q) \geqslant f_k(P)$ for each $k \in [1 \ldots d-1]$.*

(ii) *If some i-face of P is not a simplex, then $f_k(Q) > f_k(P)$ for each $k \in [i-1\dots d-1]$.*

'Pushing' a vertex into the interior of a polytope yields a similar outcome to that of pulling the vertex into the exterior of the polytope. Let P be a d-polytope in $\mathbb{R}^d$ and let $\boldsymbol{v}$ be a vertex of P. Further, let H_v be a hyperplane that separates $\boldsymbol{v}$ from the other vertices of P and let $\boldsymbol{v}' \in \operatorname{int} P \cap H_v$ so that $\boldsymbol{v}'$ is at Euclidean distance at most ε from $\boldsymbol{v}$ and no hyperplane spanned by vertices of P intersects the halfopen segment $(\boldsymbol{v}, \boldsymbol{v}']$. In this case, we say that the polytope $P' := \operatorname{conv}((\mathcal{V}(P)\backslash\{\boldsymbol{v}\}) \cup \{\boldsymbol{v}'\})$ is obtained from P by *pushing* $\boldsymbol{v}$ to $\boldsymbol{v}'$. The next theorem is an analogue of Theorem 2.10.3. A proof, however, does not follow from Theorem 2.10.1, and so we give one.

Theorem 2.10.5 (Pushing vertices; Klee, 1964b, sec. 2) *Let P' be a d-polytope in $\mathbb{R}^d$ obtained from a d-polytope P by pushing a vertex $\boldsymbol{v}$ of P to a point $\boldsymbol{v}'$. Then the following hold:*

(i) *The k-faces of P that do not contain $\boldsymbol{v}$ are all k-faces of P'.*
(ii) *For each pyramidal k-face F of P with apex $\boldsymbol{v}$, the pyramid*

$$\operatorname{conv}\left((F\backslash\{\boldsymbol{v}\}) \cup \{\boldsymbol{v}'\}\right)$$

is a k-face of P'.
(iii) $\mathcal{V}(P') = (\mathcal{V}(P)\backslash\{\boldsymbol{v}\}) \cup \{\boldsymbol{v}'\}$.
(iv) *For each k-face F of P that contains $\boldsymbol{v}$ but is not a pyramid with apex $\boldsymbol{v}$, the set* $\operatorname{conv}(F\backslash\{\boldsymbol{v}\})$ *is a k-face of P'.*
(v) *Each proper face of P' containing $\boldsymbol{v}'$ is a pyramid with apex $\boldsymbol{v}'$.*

Moreover, $f_0(P') = f_0(P)$ and $f_k(P') \geqslant f_k(P)$ for each $k \in [1\dots d-1]$.

Proof Let F be a k-face of P and let H_F be a hyperplane supporting P at F. As $P' \subset P$, the hyperplane H_F doesn't meet the interior of P'. It follows that $F = H_F \cap P = H_F \cap P'$ is a k-face of P' if $\boldsymbol{v} \notin F$; this proves (i). So assume that $\boldsymbol{v} \in F$.

We consider what happens when we continuously move $\boldsymbol{v}$ to $\boldsymbol{v}'$ along the segment $[\boldsymbol{v}, \boldsymbol{v}']$; at time $t = 0$ we have $\boldsymbol{v}$ and at time $t = 1$ we have $\boldsymbol{v}'$. Let $H_F(t)$ be the hyperplane obtained at time t from moving H_F together with $\boldsymbol{v}$; here, $H_F(0) = H_F$. Similarly define $P(t)$ and $F(t)$ so that $P(0) = P$, $P(1) = P'$, and $F(0) = F$. Since $F(0) = F$ is a face of $P(0) = P$, Theorem 2.3.7 ensures that

$$\operatorname{aff} \mathcal{V}(F(t)) \cap \operatorname{conv}(\mathcal{V}(P(t))\backslash\mathcal{V}(F(t))) = \emptyset \tag{2.10.5.1}$$

at time $t = 0$. If we move $\boldsymbol{v}(t)$ continuously from $t = 0$ to $t = 1$ by a sufficiently small amount, then aff $\mathcal{V}(F(t))$ and $H_F(t)$ also move continuously.

(ii) First, suppose that F is a pyramid with apex $\boldsymbol{v}$ and base R. Then $R = \operatorname{conv}(F \backslash \{\boldsymbol{v}\})$ is a $(k-1)$-face of both P and P' by (i). Additionally, $\boldsymbol{v}' \notin \operatorname{aff} \mathcal{V}(R)$ by the assumption that no hyperplane spanned by vertices of P intersects the segment $(\boldsymbol{v}, \boldsymbol{v}']$. Therefore, (2.10.5.1) becomes

$$\operatorname{aff}(\mathcal{V}(R) \cup \{\boldsymbol{v}(t)\}) \cap \operatorname{conv}(\mathcal{V}(P(t)) \backslash (\mathcal{V}(R) \cup \{\boldsymbol{v}(t)\})) = \varnothing$$

for each $t \in [0,1]$. Hence $\operatorname{conv}(\mathcal{V}(R) \cup \{\boldsymbol{v}(t)\})$ is a k-face of $P(t)$ for each $t \in [0,1]$. In the particular case $t = 1$, we have (ii).

(iii) This follows at once from (i) and (ii): every vertex of P other than $\boldsymbol{v}$ is a vertex of P' by (i), while $\boldsymbol{v}'$ is a vertex of P' by (ii).

(iv) Now suppose that F is not a pyramid with apex $\boldsymbol{v}$. Since F is a face of P, at time $t = 0$ (2.10.5.1) becomes

$$\operatorname{aff} \mathcal{V}(F) \cap \operatorname{conv}\left(\mathcal{V}(P(t)) \backslash \mathcal{V}(F)\right) = \varnothing.$$

We also have that $\operatorname{aff}(\mathcal{V}(F) \backslash \{\boldsymbol{v}\}) = \operatorname{aff} \mathcal{V}(F)$, and so $\boldsymbol{v} \in \operatorname{aff}(\mathcal{V}(F) \backslash \{\boldsymbol{v}\})$. Therefore, using the assumption that no hyperplane spanned by vertices of P intersects the segment $(\boldsymbol{v}, \boldsymbol{v}']$ and the fact that $\boldsymbol{v} \notin P(t)$ for each $t \in (0,1]$, we obtain that

$$\operatorname{aff}(\mathcal{V}(F) \backslash \{\boldsymbol{v}\}) \cap \operatorname{conv}\left(\mathcal{V}(P(t)) \backslash \left(\mathcal{V}(F) \backslash \{\boldsymbol{v}\}\right)\right) = \varnothing, \text{ for each } t \in (0,1].$$

Hence $\operatorname{conv}(F \backslash \{\boldsymbol{v}\})$ is a k-face of $P(t)$ for each $t \in (0,1]$. In the particular case $t = 1$ we have (iv).

(v) Suppose that F' is a proper face of P' that contains $\boldsymbol{v}'$ and yet is not a pyramid with apex $\boldsymbol{v}'$. If J' is a facet of P' containing F', then $\boldsymbol{v}'$ belongs to aff J', which is spanned by $\mathcal{V}(J') \cap \mathcal{V}(P)$. This contradicts the definition of pushing. □

A corollary follows at once.

Corollary 2.10.6 *Let Q be a d-polytope obtained from a d-polytope P by successively pushing each of the vertices of P. Then the following hold.*

(i) *The polytope Q is a simplicial d-polytope satisfying $f_0(Q) = f_0(P)$ and $f_k(Q) \geqslant f_k(P)$ for each $k \in [1 \dots d-1]$.*
(ii) *If some k-face of P is not a pyramid, then $f_k(Q) > f_k(P)$.*

The same idea behind the proof of Theorem 2.10.5 proves the following variation by Santos (2012).

Theorem 2.10.7 (Santos, 2012, lem. 2.2) *Let P' be a d-polytope in $\mathbb{R}^d$ obtained from a d-polytope P by pushing a vertex $\boldsymbol{v}$ of P to a point $\boldsymbol{v}'$. Then there exists a map φ from the facets of P' to the facets of P that satisfies the following:*

(i) *If F' is a facet of P' such that $\boldsymbol{v}' \in F'$, then there is a unique facet $\varphi(F')$ of P such that $(\mathcal{V}(F') \backslash \{\boldsymbol{v}'\}) \cup \{\boldsymbol{v}\} \subseteq \mathcal{V}(\varphi(F'))$.*
(ii) *If F' is a facet of P' such that $\boldsymbol{v}' \notin F'$, then there is a unique facet $\varphi(F')$ of P such that $\mathcal{V}(F') \subseteq \mathcal{V}(\varphi(F'))$.*
(iii) *The map φ sends two facets F_1' and F_2' of P' that share a ridge either to the same facet of P or to two facets $\varphi(F_1')$ and $\varphi(F_2')$ of P that share a ridge.*

Repeatedly pulling the vertices of a polytope transforms it into a simplicial polytope (Theorem 2.10.4), and so does repeatedly pushing its vertices (Corollary 2.10.6). Dually, every polytope can be transformed into a simple polytope by truncating the vertices, then the original edges, and so on up to ridges (Problem 2.15.12).

2.11 Complexes, Subdivisions, and Schlegel Diagrams

Polytopal complexes, a concept borrowed from algebraic topology, will prove useful in our study of polytopes. Among other purposes, we will use them to visualise polytopes via Schlegel diagrams, establish the existence of shellings of polytopes (Section 2.12), and prove identities such as Euler–Poincaré–Schläfli's equation for polytopes (Theorem 2.12.17). We proceed with the basic definitions related to polytopal complexes.

Definition 2.11.1 (Polytopal complex) A *polytopal complex* $\mathcal{C}$ is a finite, nonempty collection of polytopes in $\mathbb{R}^d$ that satisfies the following three conditions:

(i) the empty polytope is always in $\mathcal{C}$,
(ii) the faces of each polytope in $\mathcal{C}$ all belong to $\mathcal{C}$, and
(iii) polytopes intersect only at faces: if $P_1 \in \mathcal{C}$ and $P_2 \in \mathcal{C}$ then $P_1 \cap P_2$ is a face of both P_1 and P_2.

A complex $\mathcal{C}$ with members $\{P_1, \dots, P_n\}$ is said to be a complex *on* $\{P_1, \dots, P_n\}$. The *underlying set* of $\mathcal{C}$, denoted $\operatorname{set} \mathcal{C}$, is the set of points in $\mathbb{R}^d$ that belong to at least one polytope in $\mathcal{C}$.

If a polytope is a member of a complex $\mathcal{C}$, we say that the polytope is a *face* of the complex. The *dimension* of a complex $\mathcal{C}$ is the largest dimension of a face in $\mathcal{C}$; if $\mathcal{C}$ has dimension d we say that $\mathcal{C}$ is a *d-complex*. The set of k-faces of a complex $\mathcal{C}$ is denoted by $\mathcal{F}_k(\mathcal{C})$ and the set of all faces of $\mathcal{C}$ is denoted by $\mathcal{F}(\mathcal{C})$. Additionally, the number of k-faces is denoted by $f_k(\mathcal{C})$: $f_k(\mathcal{C}) = \#\mathcal{F}_k(\mathcal{C})$. As with the case of polytopes, we denote by $\mathcal{V}(\mathcal{C})$ the set of vertices of $\mathcal{C}$ and by $\mathcal{E}(\mathcal{C})$ the set of edges of $\mathcal{C}$. Faces of a complex of largest and second largest dimension are called *facets* and *ridges*, respectively. If each of the faces of a complex is contained in some facet, we say that the complex is *pure*.

We mention two important families of polytopal complexes. A *simplicial complex* $\mathcal{C}$ is a polytopal complex in $\mathbb{R}^d$ where all its polytopes are simplices. A *cubical complex* $\mathcal{C}$ is a polytopal complex in $\mathbb{R}^d$ where all its polytopes are cubes.

A *subcomplex* of a polytopal complex $\mathcal{C}$ is a subset of $\mathcal{C}$ that is itself a polytopal complex. A subcomplex of dimension k is a *k-subcomplex*. This book is concerned only with polytopal complexes, and so we often drop the adjective 'polytopal'. The undirected graph formed by the vertices and edges of $\mathcal{C}$, denoted by $G(\mathcal{C})$, is the *graph* of the complex $\mathcal{C}$.

For two polytonal complexes $\mathcal{C}$ and $\mathcal{C}'$, define their *intersection* $\mathcal{C} \cap \mathcal{C}'$ as the collection of polytopes in $\mathcal{C}$ and $\mathcal{C}'$, and their *union* $\mathcal{C} \cup \mathcal{C}'$ as the collection of polytopes in $\mathcal{C}$ or $\mathcal{C}'$. The intersection $\mathcal{C} \cap \mathcal{C}'$ is always a complex, and the union $\mathcal{C} \cup \mathcal{C}'$ is a complex if the intersections $P \cap P'$ with $P \in \mathcal{C}$ and $P' \in \mathcal{C}'$ are all polytopes in $\mathcal{C} \cap \mathcal{C}'$.

Complexes can be defined from a polytope P. Two basic examples are given by the complex of all faces of P, called the *complex* of P and denoted by $\mathcal{C}(P)$, and the complex of all proper faces of P, called the *boundary complex* of P and denoted by $\mathcal{B}(P)$. The *k-skeleton* $\mathcal{B}_k$ of a d-polytope P is the subcomplex formed by the faces of dimension at most k; the $(d-1)$-skeleton of P coincides with the boundary complex of P, while the 1-skeleton of P coincides with the graph $G(P)$ of P.

Similarly, a complex can be defined from a set $\{P_1, \ldots, P_n\}$ of polytopes, where each pair intersects at a common face, by forming the complex the complex $\mathcal{C}(P_1) \cup \cdots \cup \mathcal{C}(P_n)$. In this case, we say that the complex is *induced* by $\{P_1, \ldots, P_n\}$, and denote it as $\mathcal{C}(P_1 \cup \cdots \cup P_n)$.

The *face poset* $\mathcal{L}(\mathcal{C})$ of a complex $\mathcal{C}$ is the poset formed by the set of faces of $\mathcal{C}$ partially ordered by inclusion. Two complexes $\mathcal{C}$ and $\mathcal{C}'$ are *combinatorially isomorphic*, or simply *isomorphic*, if their face posets are isomorphic. For isomorphic complexes, we write $\mathcal{C} = \mathcal{C}'$.

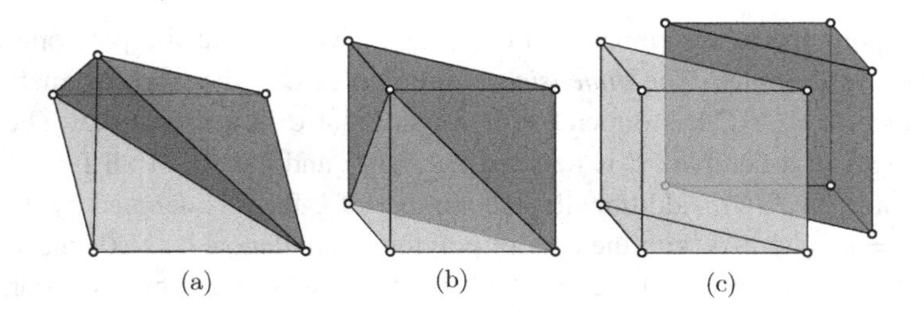

Figure 2.11.1 Triangulations of 3-polytopes. (a) A triangulation of a pyramid over a quadrangle with two 3-simplices. (b) A triangulation of a simplicial 3-prism P with three 3-simplices: first divide P into a 3-simplex and a pyramid R over a quadrangle, then triangulate R as in (a). (c) An initial step in a triangulation of a 3-cube Q with six 3-simplices: first divide Q into two simplicial 3-prims, then triangulate each prism with three 3-simplices as in (b).

Subdivisions

Polytopal subdivisions are an important kind of polytopal complexes. In general, they come into play when we want to decompose a geometric object into simpler pieces, reducing the geometry to a simpler piecewise geometry. This is the case of Delaunay subdivisions (Goodman et al., 2017, sec. 16.3, ch. 27) in computational geometry. Triangulations are the most studied subdivisions. They have found applications in computational geometry via mesh generations (Goodman et al., 2017, sec. 29.4, 29.5), in commutative algebra in connection with Gröbner bases (Sturmfels, 1996), in tropical geometry via tropical hyperplane arrangements (Ardila and Develin, 2009), and in optimisation in connection with transportation polytopes (De Loera et al., 2009).

A *polytopal subdivision* $\mathcal{S}$ of a d-polytope P in $\mathbb{R}^d$ is a pure polytopal d-complex with the same underlying set as P. A polytopal subdivision is a *triangulation* if all the polytopes in $\mathcal{S}$ are simplices; see Fig. 2.11.1. Two polytopal subdivisions are *combinatorially isomorphic* if they are combinatorially isomorphic as polytopal complexes.

Some subdivisions can be obtained from projecting a polytope in $\mathbb{R}^{d+1}$ to $\mathbb{R}^d$; these are the regular subdivisions. Suppose that a polytope Q in $\mathbb{R}^d$ is the image of a polytope P in $\mathbb{R}^{d+1}$ under the projection π that deletes the last coordinate (namely $\pi(\boldsymbol{x}, x_{d+1}) = \boldsymbol{x}$). A face F of the polytope P is a *lower face* of P if $\boldsymbol{x} - \lambda \boldsymbol{e}_{d+1} \notin P$ for each point $\boldsymbol{x} \in F$ and each $\lambda > 0$. Equivalently, F is a lower face of P if, for a vector $\boldsymbol{r} := (r_1, \ldots, r_{d+1})^t \in \mathbb{R}^{d+1}$, we have that

$$F = \{\boldsymbol{x} \in P \mid \boldsymbol{r} \cdot \boldsymbol{x} = \gamma\}, \boldsymbol{r} \cdot \boldsymbol{x} \leqslant \gamma \text{ is valid for } P, \text{ and } r_{d+1} < 0.$$

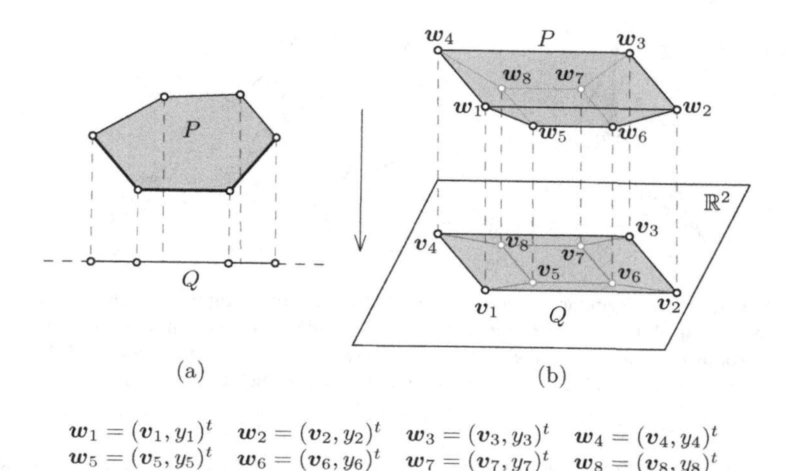

$$w_1 = (v_1, y_1)^t \quad w_2 = (v_2, y_2)^t \quad w_3 = (v_3, y_3)^t \quad w_4 = (v_4, y_4)^t$$
$$w_5 = (v_5, y_5)^t \quad w_6 = (v_6, y_6)^t \quad w_7 = (v_7, y_7)^t \quad w_8 = (v_8, y_8)^t$$

Figure 2.11.2 Regular subdivisions of polytopes. (a) A regular subdivision of a segment Q that arises from a 2-polytope P. (b) A regular subdivision of a 2-polytope $Q := \text{conv}\{v_1, \ldots, v_8\}$ that arises from a 3-cube $P := \text{conv}\{w_1, \ldots, w_8\}$; the lift vector $y := (y_1, \ldots, y_8)^t$.

This amounts to saying that F can be seen from a point far below, namely from a point $-\alpha e_{d+1}$ with sufficiently large α. The set of lower faces of P is the *lower envelope* of P. A polytopal subdivision $\mathcal{S}$ of the polytope Q in $\mathbb{R}^d$ is *regular* if it is the set of projections of all the lower faces of P:

$$\mathcal{S}(Q) = \{\pi(F) |\ F \text{ is a lower face of } P\}.$$

See Figure 2.11.2. An equivalent definition of a regular subdivision of Q in $\mathbb{R}^d$ uses a *lift vector* $y = (y_1, \ldots, y_n)^t \in \mathbb{R}^n$ to define the vertex set of P. Suppose that $Q = \text{conv}\{v_1, \ldots, v_n\}$, and for each $i \in [i \ldots n]$ let $w_i := (v_i, y_i)^t \in \mathbb{R}^{d+1}$. We say that a subdivision of Q is regular if it is combinatorially isomorphic to the lower envelope of the polytope $P := \text{conv}\{w_1, \ldots, w_n\}$; see Fig. 2.11.2(b). Being regular is not a combinatorial property; there are pairs of subdivisions that are combinatorially isomorphic and yet one is nonregular and the other is regular, as Fig. 2.11.3 and Example 2.11.2 show.

Every subdivision of a 1-polytope is regular and so are the subdivisions of a 2-polygon without interior points. It is also the case that every subdivision of a d-polytope that is the convex hull of at most $d+3$ points, vertices or otherwise, is regular (Lee, 1991). But there are subdivisions of 2-polygons with six points, vertices or otherwise, that are not regular (Fig. 2.11.3). Example 2.11.2 explores this situation.

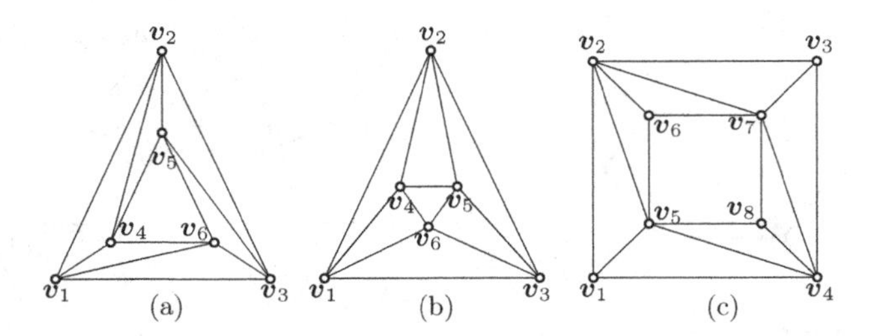

Figure 2.11.3 Regular and nonregular subdivisions of 2-polytopes with interior points. (a) A nonregular triangulation. (b) A regular triangulation that is combinatorially isomorphic to the triangulation in (a); it is the Schlegel diagram of the 3-crosspolytope. (c) A regular subdivision that is not a Schlegel diagram.

Example 2.11.2 Consider the three subdivisions of Fig. 2.11.3 and suppose that they lie in the hyperplane $x_3 = 0$ of $\mathbb{R}^3$.

(a) We show that the subdivision $\mathcal{S}_a$ of Fig. 2.11.3(a) is nonregular. Suppose that $\mathcal{S}_a$ is regular with vertex set $\{\boldsymbol{v}_1, \ldots, \boldsymbol{v}_6\}$ and lift vector $\boldsymbol{y} = (y_1, \ldots, y_6)^t$ such that $\mathcal{S}_a$ is the set of projections of the lower faces of the 3-polytope $P_a :=$ conv$\{(\boldsymbol{v}_1, y_1)^t, \ldots, (\boldsymbol{v}_6, y_6)^t\}$. By assumption, the interior triangle $\boldsymbol{v}_4\boldsymbol{v}_5\boldsymbol{v}_6$ of $\mathcal{S}_a$ is the projection of a triangular face F of P_a that lies in a plane in $\mathbb{R}^3$, and so we may assume that the three components y_4, y_5, y_6 have the same value, say zero; this may require applying an affine transformation to P, for instance a rotation, so that aff F becomes parallel to the plane $x_3 = 0$. From Fig. 2.11.3(a) we see that the segments $[\boldsymbol{v}_1, \boldsymbol{v}_2]$ and $[\boldsymbol{v}_4, \boldsymbol{v}_5]$ are parallel. This implies that if $y_1 = y_2$ then the trapezoid $\boldsymbol{v}_1\boldsymbol{v}_2\boldsymbol{v}_4\boldsymbol{v}_5$ of $\mathcal{S}_a$ would come from the 2-face $(\boldsymbol{v}_1, y_1)^t, (\boldsymbol{v}_2, y_2)^t, (\boldsymbol{v}_4, y_4)^t, (\boldsymbol{v}_5, y_5)^t$ of P_a, which disregards the existence of the edge $[\boldsymbol{v}_2, \boldsymbol{v}_4]$. Because the edge $[\boldsymbol{v}_2, \boldsymbol{v}_4]$ is present, we must have that $y_1 > y_2$. Similar reasoning on the trapezoids $\boldsymbol{v}_2\boldsymbol{v}_3\boldsymbol{v}_5\boldsymbol{v}_6$ and $\boldsymbol{v}_1\boldsymbol{v}_3\boldsymbol{v}_4\boldsymbol{v}_6$ yields that $y_2 > y_3$ and $y_3 > y_1$, respectively. Thus, we have that $y_1 > y_2 > y_3 > y_1$, a contradiction. Hence $\mathcal{S}_a$ is a nonregular subdivision.

(b) The subdivision $\mathcal{S}_b \in \mathbb{R}^2$ of Fig. 2.11.3(a) is combinatorially isomorphic to the subdivision $\mathcal{S}_a$. That $\mathcal{S}_b \in \mathbb{R}^2$ is regular follows from being a Schlegel diagram (Proposition 2.11.10).

(c) We show that the subdivision $\mathcal{S}_c \in \mathbb{R}^2$ of Fig. 2.11.3(c) is regular. Suppose that $\mathcal{S}_c$ has vertex set $\{\boldsymbol{v}_1, \ldots, \boldsymbol{v}_8\}$ and lift vector $\boldsymbol{y} = (y_1, \ldots, y_8)^t$ such that $\mathcal{S}_c$ is the set of projections of the lower faces of the 3-polytope $P_c :=$ conv$\{(\boldsymbol{v}_1, y_1)^t, \ldots, (\boldsymbol{v}_8, y_8)^t\}$. The interior quadrangle $\boldsymbol{v}_5\boldsymbol{v}_6\boldsymbol{v}_7\boldsymbol{v}_8$ of $\mathcal{S}_c$ comes from a quadrangular face of P_c and so we may assume that the four components y_5, y_6, y_7, y_8 have the same value, say zero. If we consider the trapezoid $\boldsymbol{v}_1\boldsymbol{v}_2\boldsymbol{v}_5\boldsymbol{v}_6$ and reason as in the case (a), we have that $y_1 > y_2$.

Similarly, the trapezoids $\boldsymbol{v}_2\boldsymbol{v}_3\boldsymbol{v}_6\boldsymbol{v}_7$, $\boldsymbol{v}_3\boldsymbol{v}_4\boldsymbol{v}_7\boldsymbol{v}_8$, $\boldsymbol{v}_1\boldsymbol{v}_4\boldsymbol{v}_5\boldsymbol{v}_8$ yield that $y_3 > y_2$, $y_3 > y_4$, and $y_1 > y_4$, respectively. Meeting these constraints gives a suitable lift vector, say $y_1 = 2, y_2 = 1, y_3 = 2, y_4 = 1$. The existence of the lift vector $\boldsymbol{y}$ shows that $\mathcal{S}_c$ is regular.

Schlegel Diagrams

Schlegel diagrams are regular subdivisions of a facet of a polytope that capture the combinatorics of the polytope and, as such, they will help visualise 3-polytopes and 4-polytopes.

We first define the basic elements of a Schlegel diagram. Let P be a d-polytope in $\mathbb{R}^d$ and let F be a facet of P whose affine hull is defined by the equation $\boldsymbol{r} \cdot \boldsymbol{x} = \gamma$. Choose a point $\boldsymbol{y}_F \in \mathbb{R}^d$ beyond F and consider the segment $\ell(\boldsymbol{x}) := [\boldsymbol{y}_F, \boldsymbol{x}]$ from $\boldsymbol{y}_F$ to every point $\boldsymbol{x} \in P$. We let $\varphi(\boldsymbol{x})$ be the intersection of $\ell(\boldsymbol{x})$ and F. Then

$$\varphi(\boldsymbol{x}) = \boldsymbol{y}_F + \frac{\gamma - \boldsymbol{r} \cdot \boldsymbol{y}_F}{\boldsymbol{r} \cdot \boldsymbol{x} - \boldsymbol{r} \cdot \boldsymbol{y}_F}(\boldsymbol{x} - \boldsymbol{y}_F). \tag{2.11.3}$$

We extend the function $\varphi(\boldsymbol{x})$ to proper faces of P. To do so, for each proper face J we define an affine cone C_J with apex $\boldsymbol{y}_F$ as

$$C_J := \{\boldsymbol{y}_F + \alpha(\boldsymbol{x} - \boldsymbol{y}_F) | \text{ for each } \boldsymbol{x} \in J \text{ and each } \alpha \geqslant 0\}. \tag{2.11.4}$$

With the cone C_J in place, we have that

$$\varphi(J) := C_J \cap \operatorname{aff} F. \tag{2.11.5}$$

See Fig. 2.11.4 for a depiction of the function φ and the cone C_J related to a face J of a polytope P.

A *Schlegel diagram of a polytope P based at the facet F of P*, denoted by $\mathcal{D}(P, F)$, is the image under the aforementioned function φ of all the proper faces of P other than F (see (2.11.5)):

$$\mathcal{D}(P, F) = \{\varphi(J) | \; J \in \mathcal{L}(P) \setminus \{P, F\}\}. \tag{2.11.6}$$

For simplicity, we will also refer to $\mathcal{D}(P, F)$ as the Schlegel diagram of P at F. The Schlegel diagram of P at F is a polytopal subdivision of F that captures the combinatorics of P, regardless of the point $\boldsymbol{y}_F$ that we choose beyond the facet F, as the following theorem shows.

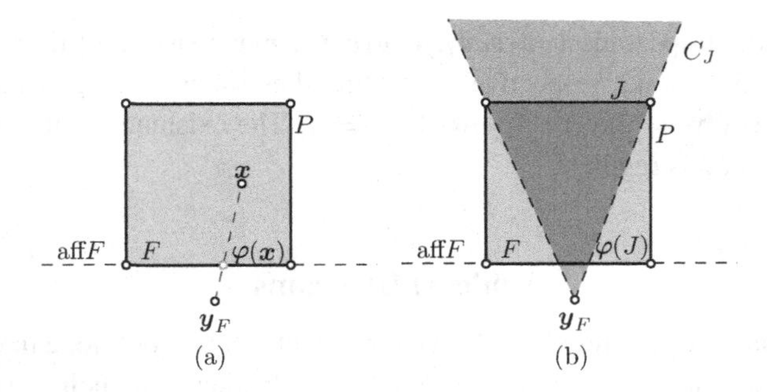

Figure 2.11.4 Definitions of the basic elements in the construction of Schlegel diagrams. (a) Definition of the function $\varphi(\boldsymbol{x})$ for each vertex $\boldsymbol{x}$ of a polytope P. (b) Definition of the cone C_J and image $\varphi(J)$ for a proper face J of P; the image $\varphi(J)$ is highlighted in bold.

Theorem 2.11.7[7] *Let P be a d-polytope in $\mathbb{R}^d$ and F a facet of it. A Schlegel diagram of P at F is a polytopal subdivision of F that is combinatorially isomorphic to the complex $\mathcal{B}(P)\backslash\{F\}$ of the proper faces of P other than F.*

Theorem 2.11.7 is illustrated in Fig. 2.11.5.

The face lattice $\mathcal{L}(P)$ of a polytope P can be readily reconstructed from the Schlegel diagram $\mathcal{D}(P,F)$ of P at a facet F of P. From $F = \operatorname{set}\mathcal{D}(P,F)$ and Theorem 2.11.7, we have that $\mathcal{D}(P,F)$ is combinatorially isomorphic to $\mathcal{B}(P)\backslash\{F\}$. We construct the facet poset $\mathcal{L}(\mathcal{D}(P,F) \cup \{F,P\}, \leqslant)$ whose faces are partially ordered by inclusion. In the poset $\mathcal{L}(\mathcal{D}(P,F) \cup \{F,P\}, \leqslant)$, a face R of $\mathcal{D}(P,F)$ satisfies $R \leqslant F$ if and only if R is a face of the facet F. We list three simple consequences of this discussion and Theorem 2.11.7.

Corollary 2.11.8 *Let P be a d-polytope in $\mathbb{R}^d$ and F a facet of it. Then the following hold.*

(i) *The underlying set of $\mathcal{D}(P,F)$ is F:* $\operatorname{set}\mathcal{D}(P,F) = F$.
(ii) *For every face R of $\mathcal{D}(P,F)$, $R \cap \operatorname{rbd} F$ is a face of F.*
(iii) *The face lattice $\mathcal{L}(P)$ of P is combinatorially isomorphic to the face poset* $\mathcal{L}(\mathcal{D}(P,F) \cup \{F,P\}, \leqslant)$.

Similar to projective transformations, duality can be applied to produce combinatorially isomorphic polytopes with prescribed properties. The application of duality relies on two observations. First, polytopes P and $\boldsymbol{x} + P$ are combinatorially isomorphic, and so if they both have zero in their interior then their duals exist and are combinatorially isomorphic. Second, if the translation

[7] A proof is available in Ziegler (1995, prop. 5.6).

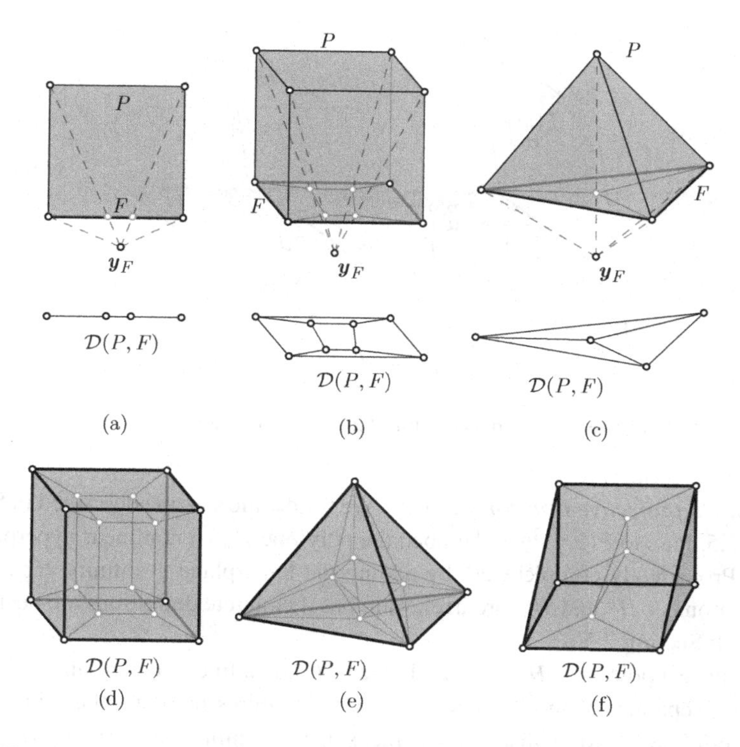

Figure 2.11.5 Schlegel diagrams of polytopes; the projection facet F has been highlighted in bold. (a) A quadrangle and one of its Schlegel diagrams. (b) A 3-cube and one of its Schlegel diagrams. (c) A 3-simplex and one of its Schlegel diagrams. (d) A Schlegel diagram of a 4-cube; the projection facet is a 3-cube. (e) A Schlegel diagram of a 4-crosspolytope; the projection facet is a 3-simplex. (f) A Schlegel diagram of the Cartesian product of a 2-simplex with another 2-simplex; the projection facet is a simplicial 3-prism.

vector $\boldsymbol{x}$ is chosen so that a vertex in $\boldsymbol{x} + P$ moves closer to $\mathbf{0}$, then the corresponding facet F in $(\boldsymbol{x}+P)^*$ moves farther from the vertices in $(\boldsymbol{x}+P)^*$ not in F (Fig. 1.11.1).

Theorem 2.11.9 *Let P be a d-polytope in $\mathbb{R}^d$ and let F be a facet of P. Then there is a d-polytope P_1 combinatorially isomorphic to P such that, for the facet F_1 of P_1 corresponding to F under this isomorphism and for every vertex $\boldsymbol{v}$ of P_1 not on F_1, the orthogonal projection of $\boldsymbol{v}$ onto F_1 lies in the relative interior of F_1.*

We offer two proofs of this result, one via projective transformations and another via duality.

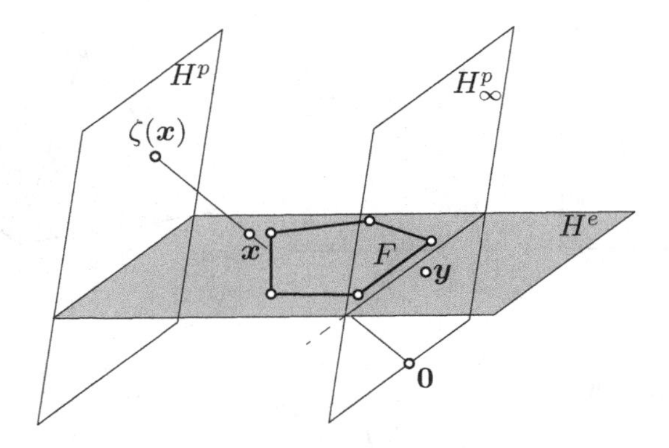

Figure 2.11.6 Projective transformation for Theorem 2.11.9.

Proof via projective transformations We follow the scheme described in Section 2.5. Pass to $\mathbb{P}(\mathbb{R}^{d+1})$ and embed the polytope P in a nonlinear hyperplane H^e. Projectively complete H^e by adding the hyperplane at infinity H^e_∞. The definitions of H^e and H^e_∞ are irrelevant, but if concrete definitions are desired, consult Section 2.5.

Choose a point $\boldsymbol{y} \in H^e$ beyond the facet F and a linear hyperplane H^p_∞ that strictly separates $\boldsymbol{y}$ and P; this latter choice requires invoking the separation theorem (1.8.5). To finalise our scheme, select a nonlinear hyperplane H^p that is parallel to H^p_∞ and admissible for P, and let ζ be a projective transformation admissible for P that fixes the points in $H^e \cap H^p$ and sends each point in $H^e \backslash (H^p \cup H^p_\infty)$ to the point in H^p lying on the same line through the origin (Fig. 2.5.1(a)); the transformation ζ maps P onto a polytope P_1 on H^p. See Fig. 2.11.6.

This transformation makes the facet $F_1 := \zeta(F)$ of P_1 very large and moves it far away from the vertices of P_1 not in F_1. Under this transformation, the polytope P_1 is contained in the pyramid $\operatorname{conv}(\{\zeta(\boldsymbol{y})\} \cup F_1)$; see Fig. 2.5.2(d). It follows that the orthogonal projection of P_1 onto $\operatorname{aff} F_1$ maps $P_1 \backslash F_1$ onto $\operatorname{rint} F_1$, as desired. □

Proof via duality Assume that $\mathbf{0} \in \operatorname{int} P$. Let $\boldsymbol{v}_1, \ldots, \boldsymbol{v}_n$ be the vertices of P^* and assume that $\boldsymbol{v}_1$ is the vertex of P^* conjugate to the facet F of P.

We translate the polytope P^* so that the vertex $\boldsymbol{v}_1$ gets closer to $\mathbf{0}$. Pick a point $\boldsymbol{y} \in \operatorname{int} P^*$ so that the hyperplane through $\boldsymbol{y}$ with normal $\boldsymbol{v}_1 - \boldsymbol{y}$ strictly separates $\boldsymbol{v}_1$ from the other vertices of P^*. Then, for every $i \in [2 \ldots n]$, we have that

$$(\boldsymbol{v}_1 - \boldsymbol{y}) \cdot \boldsymbol{v}_i < (\boldsymbol{v}_1 - \boldsymbol{y}) \cdot \boldsymbol{y} < (\boldsymbol{v}_1 - \boldsymbol{y}) \cdot \boldsymbol{v}_1. \tag{2.11.9.1}$$

Define the polytope P_1^* with vertices $\boldsymbol{v}_1 - \boldsymbol{y}, \dots, \boldsymbol{v}_n - \boldsymbol{y}$; the polytope P_1^* is equal to $P^* - \boldsymbol{y}$. We show that like P^*, P_1^* contains zero in its interior. Since $\boldsymbol{y} \in \operatorname{int} P^*$, by Theorem 1.7.6 we can write it as

$$\boldsymbol{y} = \alpha_1 \boldsymbol{v}_1 + \cdots + \alpha_n \boldsymbol{v}_n$$

for positive scalars $\alpha_1, \dots, \alpha_n$ such that $\sum_{i=1}^n \alpha_i = 1$, and so

$$\boldsymbol{0} = \alpha_1(\boldsymbol{v}_1 - \boldsymbol{y}) + \cdots + \alpha_n(\boldsymbol{v}_n - \boldsymbol{y}),$$

yielding that $\boldsymbol{0} \in \operatorname{int} P_1^*$ by Theorem 1.7.6.

Let P_1 be the dual of P_1^*. Then P_1 and P are combinatorially isomorphic; let σ be an isomorphism from $\mathcal{L}(P)$ to $\mathcal{L}(P_1)$. For each $i \in [1 \dots n]$ and each facet F_i of P_1 conjugate to the vertex $\boldsymbol{v}_i - \boldsymbol{y}$ of P_1^*, it follows that

$$\operatorname{aff} F_i = \left\{ \boldsymbol{x} \in \mathbb{R}^d \,\middle|\, (\boldsymbol{v}_i - \boldsymbol{y}) \cdot \boldsymbol{x} = 1 \right\}. \tag{2.11.9.2}$$

We show that the facet $F_1 = \sigma(F)$ of P_1 has the desired properties. Because of (2.11.9.2), for every $\boldsymbol{x} \in P_1$ and every $i \in [1 \dots n]$, we have that $(\boldsymbol{v}_i - \boldsymbol{y}) \cdot \boldsymbol{x} \leqslant 1$, and for every point $\boldsymbol{x} \in P_1 \backslash F_1$ we have that $(\boldsymbol{v}_1 - \boldsymbol{y}) \cdot \boldsymbol{x} < 1$. Consider a point $\boldsymbol{x} \in P_1 \backslash F_1$. Its projection $\boldsymbol{x}' := \pi_{\operatorname{aff} F_1}(\boldsymbol{x})$ onto aff F_1 is given by (1.4.8):

$$\boldsymbol{x}' = \boldsymbol{x} - \frac{(\boldsymbol{v}_1 - \boldsymbol{y}) \cdot \boldsymbol{x} - 1}{\|\boldsymbol{v}_1 - \boldsymbol{y}\|^2} (\boldsymbol{v}_1 - \boldsymbol{y}).$$

The point $\boldsymbol{x}'$ is in the interior of $\bigcap_{i=2}^n H_d^-(\boldsymbol{v}_i - \boldsymbol{y}, 1)$:

$$\begin{aligned}
(\boldsymbol{v}_i - \boldsymbol{y}) \cdot \boldsymbol{x}' &= (\boldsymbol{v}_i - \boldsymbol{y}) \cdot \left[\boldsymbol{x} - \frac{(\boldsymbol{v}_1 - \boldsymbol{y}) \cdot \boldsymbol{x} - 1}{\|\boldsymbol{v}_1 - \boldsymbol{y}\|^2} (\boldsymbol{v}_1 - \boldsymbol{y}) \right] \\
&= (\boldsymbol{v}_i - \boldsymbol{y}) \cdot \boldsymbol{x} - \frac{(\boldsymbol{v}_1 - \boldsymbol{y}) \cdot \boldsymbol{x} - 1}{\|\boldsymbol{v}_1 - \boldsymbol{y}\|^2} (\boldsymbol{v}_i - \boldsymbol{y}) \cdot (\boldsymbol{v}_1 - \boldsymbol{y}) \\
&\leqslant 1 - \frac{(\boldsymbol{v}_1 - \boldsymbol{y}) \cdot \boldsymbol{x} - 1}{\|\boldsymbol{v}_1 - \boldsymbol{y}\|^2} (\boldsymbol{v}_i - \boldsymbol{y}) \cdot (\boldsymbol{v}_1 - \boldsymbol{y}) \\
&< 1.
\end{aligned}$$

To see the last step, observe that

$$\frac{(\boldsymbol{v}_1 - \boldsymbol{y}) \cdot \boldsymbol{x} - 1}{\|\boldsymbol{v}_1 - \boldsymbol{y}\|^2} < 0 \text{ and that } (\boldsymbol{v}_i - \boldsymbol{y}) \cdot (\boldsymbol{v}_1 - \boldsymbol{y}) < 0 \text{ by (2.11.9.1).}$$

As $\boldsymbol{x}' \in \operatorname{aff} F_1 \cap \operatorname{int} \left(\bigcap_{i=2}^n H_d^-(\boldsymbol{v}_i - \boldsymbol{y}, 1) \right)$, we conclude that $\boldsymbol{x}' \in \operatorname{rint} F_1$, as desired. □

Thanks to Theorem 2.11.9, we can transform a polytope P with a given facet F into a combinatorially isomorphic polytope P' so that, for the facet F' of P' corresponding to F, the orthogonal projection π of each point $\boldsymbol{x} \in P'$ onto F'

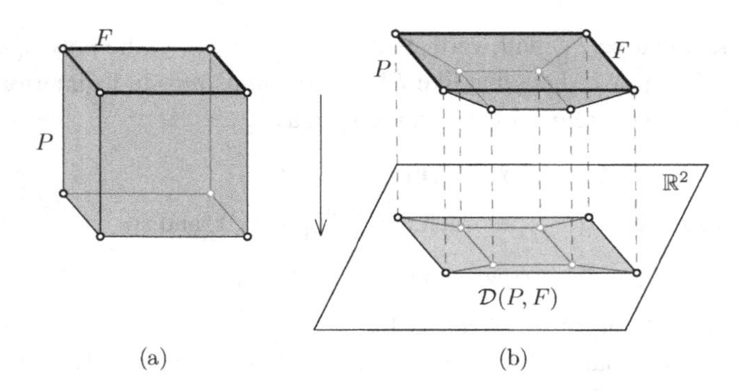

Figure 2.11.7 A Schlegel diagram of the 3-cube represented as a regular subdivision. (a) A 3-cube P with a facet F highlighted. (b) A Schlegel diagram $\mathcal{D}(P, F)$ of 3-cube P at a facet F where $\mathcal{D}(P, F)$ is combinatorially isomorphic to a regular subdivision of F.

lies in the relative interior of F'. As a result, the Schlegel diagram $\mathcal{D}(P, F)$ of P at F is combinatorially isomorphic to the Schlegel diagram $\mathcal{D}(P', F')$ of P' at F'; see Fig. 2.11.7. This construction has an important consequence.

Proposition 2.11.10 *Let P be a d-polytope, F a facet of it, and $\mathcal{D}(P, F)$ a Schlegel diagram of P at F. Then $\mathcal{D}(P, F)$ is combinatorially isomorphic to a regular subdivision of F.*

Proof We first preprocess P so that F is the orthogonal projection of P, as in Theorem 2.11.9; see also Proposition 2.1.3. It follows that $\mathcal{D}(P, F)$ is the set of projections of the lower faces of P, giving that $\mathcal{D}(P, F)$ is a regular subdivision of F. □

A Schlegel diagram of a polytope P is clearly not uniquely determined; it depends on the projection facet F and the point $\boldsymbol{y}_F$ beyond F. In view of Proposition 2.11.10 and Theorem 2.11.9, we can always assume that F is the orthogonal projection of P (see also Proposition 2.1.3).

While every Schlegel diagram can be realised as a regular subdivision (Proposition 2.11.10), not every regular subdivision is a Schlegel diagram. We use Proposition 2.11.10 to show that the regular subdivision of Fig. 2.11.3(c) is not a Schlegel diagram.

Example 2.11.11 Suppose that the regular subdivision $\mathcal{S}_c$ of Fig. 2.11.3(c) lies in the hyperplane $x_3 = 0$ of $\mathbb{R}^3$. Further suppose that $\mathcal{S}_c$ has vertex set $\{\boldsymbol{v}_1, \ldots, \boldsymbol{v}_8\}$ and lift vector $\boldsymbol{y} = (y_1, \ldots, y_8)^t$ such that $\mathcal{S}_c$ is the set of projections of the lower faces of the 3-polytope

$$P_c := \operatorname{conv}\{(\boldsymbol{v}_1, y_1)^t, \ldots, (\boldsymbol{v}_8, y_8)^t\}.$$

Suppose that $\mathcal{S}_c$ is a Schlegel diagram of P_c at a facet F with vertex set $(\boldsymbol{v}_1, y_1)^t, (\boldsymbol{v}_2, y_2)^t, (\boldsymbol{v}_3, y_3)^t, (\boldsymbol{v}_4, y_4)^t$. Then F lies in a plane of $\mathbb{R}^3$. The analysis in Example 2.11.2(c) showed that $y_1 > y_2$, $y_3 > y_2$, $y_3 > y_4$, $y_1 > y_4$; this contradicts the planarity of aff F. Hence, $\mathcal{S}_c$ is not a Schlegel diagram.

Stars, Antistars, and Links in complexes

For a polytopal complex $\mathcal{C}$, the *star* of a face F in $\mathcal{C}$, denoted $\operatorname{st}(F,\mathcal{C})$, is the subcomplex of $\mathcal{C}$ formed by all the faces containing F and their faces; the *antistar* of a face F of $\mathcal{C}$, denoted $\operatorname{ast}(F,\mathcal{C})$, is the subcomplex of $\mathcal{C}$ formed by all the faces disjoint from F; and the *link* of a face F, denoted $\operatorname{lk}(F,\mathcal{C})$, is the subcomplex of $\mathcal{C}$ formed by all the faces of $\operatorname{st}(F,\mathcal{C})$ that are disjoint from F. For a subset X of $\mathcal{V}(\mathcal{C})$, we denote by $\mathcal{C}-X$ the subcomplex of $\mathcal{C}$ formed by the faces of $\mathcal{C}$ that contain no vertex from X. It follows that $\operatorname{ast}(F,\mathcal{C}) = \mathcal{C} - \mathcal{V}(F)$ and $\operatorname{lk}(F,\mathcal{C}) = \operatorname{st}(F,\mathcal{C}) - \mathcal{V}(F)$. We define stars, antistars, and links in a polytope always with respect to $\mathcal{B}(P)$; that is why we often write $\operatorname{st}(F,P)$, $\operatorname{ast}(F,P)$, or $\operatorname{lk}(F,P)$ without explicitly stating $\mathcal{B}(P)$. Furthermore, we may also write $\operatorname{st}(F)$, $\operatorname{ast}(F)$, or $\operatorname{lk}(F)$ if P is clear from the context.

Figure 2.11.8 depicts the star and link of a vertex in the 4-cube. Let $\boldsymbol{x}$ be a vertex in a d-cube $Q(d)$ and let $\boldsymbol{x}^o$ denote the unique vertex not contained in the star of $\boldsymbol{x}$. Then the antistar of $\boldsymbol{x}$ coincides with the star of $\boldsymbol{x}^o$ and the link of $\boldsymbol{x}$ is the subcomplex $\mathcal{C}(Q(d)) - \{\boldsymbol{x}, \boldsymbol{x}^o\}$.

The star, antistar, and link of a vertex are all pure complexes. But more is true for the link.

Proposition 2.11.12 *Let P be a d-polytope. Then the link of a vertex in $\mathcal{B}(P)$ is combinatorially isomorphic to the boundary complex of a $(d-1)$-polytope.*

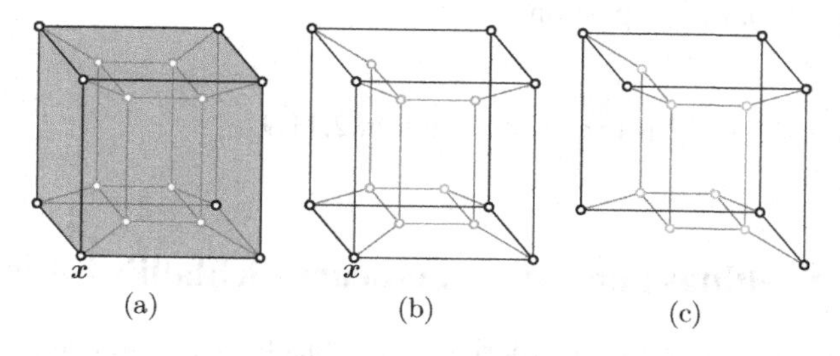

Figure 2.11.8 Complexes in the 4-cube. (a) The 4-cube with a vertex $\boldsymbol{x}$ highlighted. (b) The star of the vertex $\boldsymbol{x}$. (c) The link of the vertex $\boldsymbol{x}$.

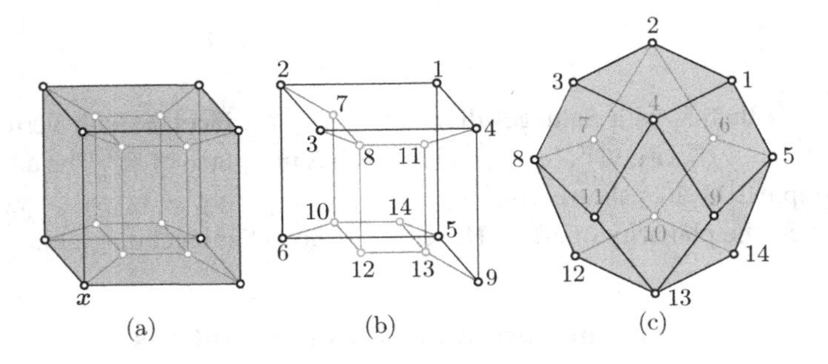

Figure 2.11.9 The link of a vertex in the 4-cube. (a) The 4-cube with a vertex $\boldsymbol{x}$ highlighted. (b) The link of the vertex $\boldsymbol{x}$ in the 4-cube. (c) The link of the vertex $\boldsymbol{x}$ as the boundary complex of the rhombic dodecahedron (Proposition 2.11.12).

In particular, for each $d \geqslant 3$, the graph of the link of a vertex is isomorphic to the graph of a $(d-1)$-polytope.

Proof Let $\boldsymbol{x}$ be a vertex of P and let $\boldsymbol{x}'$ be a point in $\mathbb{R}^d \backslash P$ beyond $\boldsymbol{x}$. Suppose $P' := \operatorname{conv}(P \cup \{\boldsymbol{x}'\})$. We could think of P' as being obtained from P by pulling the vertex $\boldsymbol{x}$ to $\boldsymbol{x}'$ (Theorem 2.10.3).

The facets in the star of $\boldsymbol{x}$ in $\mathcal{B}(P)$ are precisely those that are visible from $\boldsymbol{x}'$, and every other facet of P, which is in the antistar of $\boldsymbol{x}$, is nonvisible from $\boldsymbol{x}'$. The link of $\boldsymbol{x}$ is, by definition, the subcomplex of $\mathcal{B}(P)$ induced by the ridges of P that are contained in a facet of the star of $\boldsymbol{x}$, a facet visible from $\boldsymbol{x}'$, and a facet of the antistar of $\boldsymbol{x}$, a facet nonvisible from $\boldsymbol{x}'$. Consequently, according to Theorem 2.10.1(i) the ridges in $\operatorname{lk}(\boldsymbol{x}, \mathcal{B}(P))$ are all faces of P'. Furthermore, for every ridge $R \in \operatorname{lk}(\boldsymbol{x}, \mathcal{B}(P))$, $R' := \operatorname{conv}(R \cup \{\boldsymbol{x}'\})$ is a facet of P' (Theorem 2.10.1(iii)), a pyramid over R with apex $\boldsymbol{x}'$; and every facet in the star of $\boldsymbol{x}'$ in $\mathcal{B}(P')$ is one of these pyramids. Hence, the boundary complex of the vertex figure of P' at $\boldsymbol{x}'$ is combinatorially isomorphic to the link of $\boldsymbol{x}$ in P. Since the vertex figure is combinatorially isomorphic to a $(d-1)$-polytope (Theorem 2.7.3), the propostion follows.

□

Proposition 2.11.12 is exemplified in Fig. 2.11.9.

2.12 Shellings and Euler–Poincaré–Schläfli's Equation

A shelling is a certain linear ordering of the facets of a pure complex. While not every pure complex has a shelling, the boundary complex of every polytope has one; this is central to a number of inductive arguments on polytopes.

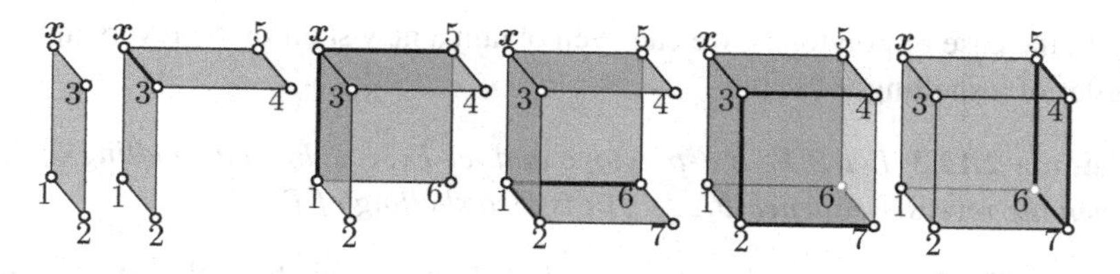

Figure 2.12.1 A shelling $F_1 = \boldsymbol{x}123, F_2 = \boldsymbol{x}345, F_3 = \boldsymbol{x}156, F_4 = 1267, F_5 = 2347, F_6 = 4567$ of the boundary complex of the 3-cube, where the facets of the star of a vertex $\boldsymbol{x}$ come first. The intersection of the current facet with the union of the previous ones is highlighted in bold.

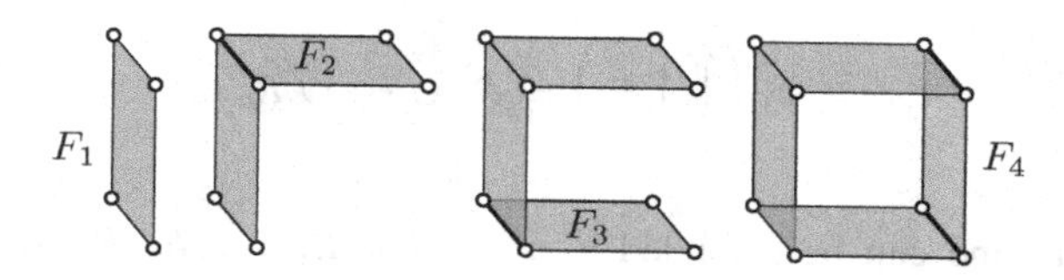

Figure 2.12.2 A sequence F_1, F_2, F_3, F_4 of facets of the boundary complex of the 3-cube that is not the beginning of any shelling of the 3-cube. The intersection of the current facet with the union of the previous ones is highlighted in bold. The intersection $F_4 \cap (F_1 \cup F_2 \cup F_3)$ is not the beginning of a shelling of F_4.

Definition 2.12.1 (Shelling) Let $\mathcal{C}$ be a pure complex. A *shelling* of $\mathcal{C}$ is a linear ordering $F_1, \ldots, F_s$ of the facets of the complex such that either $\dim \mathcal{C} = 0$, in which case the facets are vertices, or it satisfies the following:

(i) The boundary complex of F_1 has a shelling.
(ii) For $j \in [2 \ldots s]$, the intersection

$$F_j \cap \left(\bigcup_{i=1}^{j-1} F_i \right) = R_1 \cup \cdots \cup R_r$$

is nonempty and the beginning $R_1, \ldots, R_r$ of a shelling $R_1, \ldots, R_r, R_{r+1}, \ldots, R_t$ of the boundary complex of F_j.

A complex is *shellable* if it is pure and admits a shelling. Figure 2.12.1 depicts an example of a shelling, while Fig. 2.12.2 depicts a sequence of facets that is not the beginning of any shelling. Abusing terminology, we will refer to a shelling of the boundary complex of a polytope as a shelling of the polytope.

Shellings of a complex can be rearranged in number of ways.

Lemma 2.12.2 *Let $\mathcal{C}$ be a shellable complex and let $F_1, \ldots, F_r, J_{r+1}, \ldots, J_s$ be a shelling of $\mathcal{C}$. If $F_1', \ldots, F_r'$ is a different shelling of the subcomplex $\mathcal{C}(F_1 \cup \cdots \cup F_r)$ of $\mathcal{C}$ then $F_1', \ldots, F_r', J_{r+1}, \ldots, J_s$ is a different shelling of $\mathcal{C}$.*

In the case of polytopes, we can even obtain a new shelling by reversing the order of a shelling.

Lemma 2.12.3 *Let P be a d-polytope and let $F_1, \ldots, F_s$ be a shelling of P. Then the reverse sequence $F_s, \ldots, F_1$ is also shelling of P.*

Proof The lemma is plainly true for $d = 1$, and so we have the basis of an induction on d. Assume that $d \geqslant 2$. Since $F_1, \ldots, F_s$ is a shelling of P each facet F_i for $i \in [1 \ldots s]$ is shellable (Definition 2.12.1(i)–(ii)); in particular, for $j \in [2 \ldots s]$, the intersection

$$F_j \cap \left(\bigcup_{i=1}^{j-1} F_i \right) = R_1 \cup \cdots \cup R_r$$

is nonempty and can be extended to a shelling $R_1, \ldots, R_r, R_{r+1}, \ldots, R_t$ of F_j. By the induction hypothesis, the sequence $R_t, \ldots, R_{r+1}, R_r, \ldots, R_1$ is a shelling of F_j. For every $(d-2)$-face R of F_j, there is a unique facet F_ℓ such that $R = F_\ell \cap F_j$, and so the intersection

$$F_j \cap \left(\bigcup_{i=j+1}^{s} F_i \right) = R_t \cup \cdots \cup R_{r+1}$$

is nonempty and the beginning of a shelling of F_j. Since F_s is shellable, we obtain the shelling $F_s, \ldots, F_1$ of P, and so the proof of the lemma is complete. □

We extend the notion of 'general position' given in Section 1.1 from sets in $\mathbb{R}^d$ to points, lines, and line functionals in $\mathbb{R}^d$, with respect to a d-polytope.

Definition 2.12.4 (Point in general position) Let P be a d-polytope in $\mathbb{R}^d$ and let $F_1, \ldots, F_n$ be the facets of P. A point $\boldsymbol{x}$ is in *general position* with respect to P if it does not lie in any of the hyperplanes $\operatorname{aff} F_1, \ldots, \operatorname{aff} F_n$.

Definition 2.12.5 (Line in general position) Let P be a d-polytope in $\mathbb{R}^d$ and let $F_1, \ldots, F_n$ be the facets of P. A line ℓ is in *general position* with respect to P if it is not parallel to any of the hyperplanes $\operatorname{aff} F_i$ and intersects the hyperplanes $\operatorname{aff} F_i$ at pairwise distinct points.

Definition 2.12.6 (Linear functional in general position) Let P be a d-polytope in $\mathbb{R}^d$. A linear functional in $\mathbb{R}^d$ is in *general position* with respect to P if its values on the vertices of P are pairwise distinct.

We establish the existence of lines and linear functionals in general position with respect to a polytope. The proofs rely on the fact that finitely many polynomials in one variable have together only finitely many zeros.

Proposition 2.12.7[8] *Let P be a d-polytope in $\mathbb{R}^d$. Then, every nonzero vector $\boldsymbol{a}$ in $\mathbb{R}^d$ can be perturbed slightly so that the resulting vector $\boldsymbol{a}(\varepsilon)$ for some $\varepsilon > 0$ defines a line $\ell(\varepsilon) := \alpha\boldsymbol{a}(\varepsilon)$ or a linear functional $\varphi_\varepsilon(\boldsymbol{x}) := \boldsymbol{a}(\varepsilon) \cdot \boldsymbol{x}$ that is in general position with respect to P.*

Shellings in polytopes have many theoretical and computational applications. We show their existence next.

Theorem 2.12.8 (Existence of line shellings; Bruggesser and Mani, 1971) *Every polytope is shellable.*

Proof Let P be a d-polytope in $\mathbb{R}^d$ with $\boldsymbol{0}$ in its interior. We prove the existence of a particular kind of shelling of P.

> *Let $\boldsymbol{x} \in \mathbb{R}^d \setminus P$ be a point in general position with respect to P. Then there is a shelling of P in which the facets of P that are visible from $\boldsymbol{x}$ come first and the facets that are nonvisible from $\boldsymbol{x}$ come last.* (*)

The statement (*) is true for $d = 1$, and so assume that $d \geqslant 2$. As part of an induction argument on d, suppose that (*) holds for $d - 1$.

Take a line ℓ that hits the interior of P, passes through $\boldsymbol{0}$ and $\boldsymbol{x}$ (that is, $\ell = \alpha\boldsymbol{x}$ for every $\alpha \in \mathbb{R}$), and is in general position with respect to P. Orient ℓ from the interior of P to $\boldsymbol{x}$; see Fig. 2.12.3. While traversing ℓ, if you reach infinity then return to the polytope from the opposite side. Label the facets $F_1, \ldots, F_n$ of P as they become visible when traversing ℓ, and let $\boldsymbol{x}_i$ be the intersection of ℓ with aff F_i for each $i \in [1 \ldots n]$. By our choice of ℓ, the points $\boldsymbol{x}_1, \ldots, \boldsymbol{x}_n$ are all distinct. Further assume that when we traverse the line $\ell = \alpha\boldsymbol{x}$ from $\boldsymbol{0}$ to $\boldsymbol{x}$, we encounter the points $\boldsymbol{x}_1, \ldots, \boldsymbol{x}_k$, in that order before reaching infinity, and then encounter the points $\boldsymbol{x}_{k+1}, \ldots, \boldsymbol{x}_n$, again in that order, when we return to the polytope from the opposite side.

For each $i \in [1 \ldots n]$, suppose that

$$\operatorname{aff} F_i = \left\{ \boldsymbol{y} \in \mathbb{R}^d \,\middle|\, \boldsymbol{a}_i \cdot \boldsymbol{y} = 1 \right\},$$

in which case each inequality $\boldsymbol{a}_i \cdot \boldsymbol{y} \leqslant 1$ is valid for P. This implies that

$$\boldsymbol{x}_i = \frac{\boldsymbol{x}}{\boldsymbol{x} \cdot \boldsymbol{a}_i} \tag{2.12.8.1}$$

[8] A proof is available in Ziegler (1995, lem. 3.2).

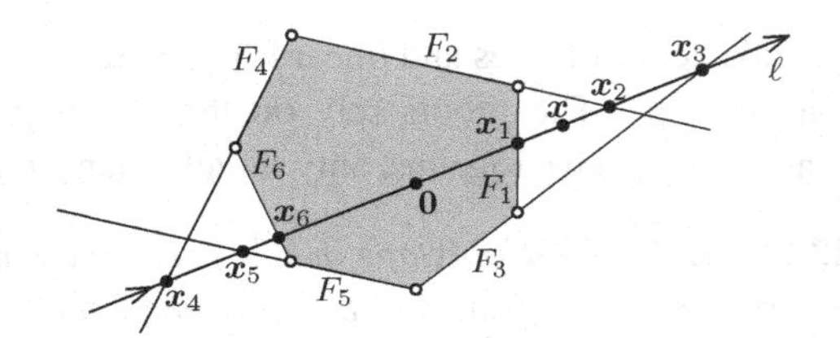

Figure 2.12.3 A line shelling $F_1, \ldots, F_6$ of the boundary complex of a polytope. The line ℓ is oriented from the interior of the polytope to a point $\boldsymbol{x}$ and, for each $i \in [1 \ldots 6]$, the point $\boldsymbol{x}_i$ is the intersection of ℓ with aff F_i.

and that

$$\boldsymbol{x} \cdot \boldsymbol{a}_1 > \boldsymbol{x} \cdot \boldsymbol{a}_2 > \cdots > \boldsymbol{x} \cdot \boldsymbol{a}_k > 0 > \boldsymbol{x} \cdot \boldsymbol{a}_{k+1} > \cdots > \boldsymbol{x} \cdot \boldsymbol{a}_n, \qquad (2.12.8.2)$$

which in turn yields that

$$\frac{1}{\boldsymbol{x} \cdot \boldsymbol{a}_{k+1}} < \cdots < \frac{1}{\boldsymbol{x} \cdot \boldsymbol{a}_n} < 0 < \frac{1}{\boldsymbol{x} \cdot \boldsymbol{a}_1} < \frac{1}{\boldsymbol{x} \cdot \boldsymbol{a}_2} < \cdots < \frac{1}{\boldsymbol{x} \cdot \boldsymbol{a}_k}. \qquad (2.12.8.3)$$

By construction, the sequence $F_1, \ldots, F_n$ is ordered so that the facets visible from $\boldsymbol{x}$ come before the facets that are nonvisible from $\boldsymbol{x}$. See Fig. 2.12.3.

It remaing to show that the sequence $F_1, \ldots, F_n$ is a shelling of P. We verify Definition 2.12.1. The facet F_1 is shellable by induction. Each point $\boldsymbol{x}_j$ with $j \in [2 \ldots n]$ is outside F_j and is in general position with respect to F_j. Suppose the facet F_j appears before reaching infinity. The intersection

$$F_j \cap \left(\bigcup_{i=1}^{j-1} F_i \right)$$

is precisely the union of the facets $R_1, \ldots, R_r$ of F_j that are visible from $\boldsymbol{x}_j$: $\boldsymbol{a}_j \cdot \boldsymbol{x}_j = 1$ and $\boldsymbol{a}_i \cdot \boldsymbol{x}_j > 1$ for each $i \in [1 \ldots j-1]$, by (2.12.8.1) and (2.12.8.2). Therefore, the induction hypothesis ensures that $R_1, \ldots, R_r$ is the beginning of a shelling of F_j, which also implies that the other facets of F_j, those nonvisible from $\boldsymbol{x}_j$, are the end of the shelling. Thus, Definition 2.12.1(ii) holds for this facet F_j. Further, suppose that the facet F_j appears after we passed infinity. The intersection

$$F_j \cap \left(\bigcup_{i=1}^{j-1} F_i \right)$$

consists precisely of the facets $R_1, \ldots, R_r$ of F_j that are nonvisible from $\boldsymbol{x}_j$: $\boldsymbol{a}_j \cdot \boldsymbol{x}_j = 1$ and $\boldsymbol{a}_i \cdot \boldsymbol{x}_j < 1$ for each $i \in [1 \ldots .j-1]$, by (2.12.8.1) and (2.12.8.2). Therefore the induction hypothesis ensures that $R_1, \ldots, R_r$ are the end of a shelling S of F_j. According to Lemma 2.12.3, the faces $R_1, \dots, R_r$ is the beginning of another shelling of F_j, the shelling obtained by reversing S. Thus, Definition 2.12.1(ii) holds for this facet F_j as well. Consequently, the two conditions of Definition 2.12.1 are met and so the order $F_1, \dots, F_n$ is a shelling of P. This completes the proof of (*), and with it, the proof of the theorem. □

Remark 2.12.9 In the proof of Theorem 2.12.8, the point $\boldsymbol{x}$ in (*) need not be in general position with respect to P. For the proof of Theorem 2.12.8 to work, what matters is that the line $\ell = \alpha\boldsymbol{x}$ intersects the interior of P and is in general position with respect to P (Definition 2.12.5), namely the points $\boldsymbol{x}_1, \dots, \boldsymbol{x}_n$ are all distinct. If $\boldsymbol{x}$ is in general condition with respect to P, these two conditions on ℓ are satisfied.

Line shellings provide a mechanism to shell a polytope in a number of ways.

Proposition 2.12.10 *A polytope P admits a shelling with any of the following prescribed conditions:*

(i) *any two facets F and F' of P can be chosen so that F is the first facet and F' is the last facet of the shelling;*
(ii) *for every vertex z of P, the star of z is the beginning of the shelling.*

Proof The two proofs are very similar; they both rely on using two points to define the line that induces the shelling.

(i) Choose a point $\boldsymbol{y}_F$ beyond the facet F and a point $\boldsymbol{y}_{F'}$ beyond the facet F', and let ℓ be the line determined by $\boldsymbol{y}_F$ and $\boldsymbol{y}_{F'}$ so that ℓ hits the interior of P and is in general position with respect to P. It follows that the points $\boldsymbol{y}_F$ and $\boldsymbol{y}_{F'}$ are in general position with respect to P. The selection of $\boldsymbol{y}_F$, $\boldsymbol{y}_{F'}$, and ℓ can be done in two steps. First, place $\boldsymbol{y}_F$ on the relative interior of F and $\boldsymbol{y}_{F'}$ on the relative interior of F' so that if they move along ℓ then they land outside P with $\boldsymbol{y}_F$ beyond the facet F and $\boldsymbol{y}_{F'}$ beyond the facet F'; this guarantees that the line ℓ hits the interior of P. Second, slightly perturb ℓ so that it becomes in general position with respect to P (Proposition 2.12.7).

Traversing the line ℓ from $\boldsymbol{y}_{F'}$ to $\boldsymbol{y}_F$ gives the desired line shelling (Theorem 2.12.8): the facet F will be the first facet visible from y_F, while if we traverse the line 'backwards', in the reverse direction, the facet F' will be the first to be visible from $\boldsymbol{y}_{F'}$.

(ii) Choose a point $\boldsymbol{x}$ beyond the vertex z and a line ℓ that passes through $\boldsymbol{x}$ and the interior of P. The point $\boldsymbol{x}$ is in general position with respect to P. If necessary, perturb ℓ so that it is in general position with respect to P (Proposition 2.12.7). Traversing the line ℓ from the interior of P to $\boldsymbol{x}$ gives a line shelling where the facets in the star of z, all visible from x, are the first facets encountered by the line ℓ (Theorem 2.12.8). □

The link of a vertex in a polytope is combinatorially isomorphic to the boundary complex of a $(d-1)$-polytope (Proposition 2.11.12), and so it is shellable by Theorem 2.12.8.

Proposition 2.12.11 *Let P be a polytope and let x be a vertex of P. Then the link of x in P is shellable.*

Shellings of a polytopal complex induce shellings of the star of a vertex.

Proposition 2.12.12 *Let $\mathcal{C}$ be a shellable $(d-1)$-complex and let x be a vertex of $\mathcal{C}$. Then the star of x in $\mathcal{C}$ is shellable. Moreover, the restriction of every shelling of $\mathcal{C}$ to the facets in the star of x is a shelling of the star of x.*

Proof The proposition is true for $d=1$, since the star of x is 0-dimensional and thus shellable by Definition 2.12.1. Assume $d \geqslant 2$. The proof proceeds by induction on d. Let $F_1, \ldots, F_n$ be a shelling S of $\mathcal{C}$, and let $F_{i_1}, \ldots, F_{i_t}$ be the restriction of S to $\operatorname{st}(x, \mathcal{C})$. Each facet in S is a $(d-1)$-polytope and, thus, it is shellable by Theorem 2.12.8; in particular, F_{i_1} is shellable. In view of Definition 2.12.1, it remains to show that, for each $p \in [2 \ldots t]$,

$$F_{i_p} \cap \left(F_{i_1} \cup \cdots \cup F_{i_{p-1}}\right) \tag{2.12.12.1}$$

is the beginning of a shelling S_{i_p} of F_{i_p}. Observe that (2.12.12.1) is a subset of $\operatorname{st}(x, F_{i_p})$.

Because $F_1, \ldots, F_n$ is a shelling of P, there is a sequence $R_1, \ldots, R_r$ of $(d-2)$-faces of F_{i_p} such that

$$F_{i_p} \cap \left(F_1 \cup F_2 \cup \cdots \cup F_{i_p - 1}\right) = R_1 \cup \cdots \cup R_r \tag{2.12.12.2}$$

is nonempty and can be extended to a shelling $S'_{i_p} := R_1, \ldots, R_r, R_{r+1}, \ldots, R_s$ of F_{i_p}. It follows that

$$\left\{F_{i_1}, \ldots, F_{i_{p-1}}\right\} \subseteq \left\{F_1, F_2, \ldots, F_{i_{p-1}}\right\}.$$

The shelling S'_{i_p} is that of a shellable $(d-2)$-complex. Hence, the induction hypothesis on $\mathcal{B}(F_{i_p})$ and S'_{i_p} implies that the restriction $R_{j_1}, \ldots, R_{j_q}$ of S'_{i_p} to the $(d-2)$-faces in the star of x in F_{i_p} is a shelling of $\operatorname{st}(x, F_{i_p})$. Let ℓ be the largest integer in $[1 \ldots q]$ such that $j_\ell \leqslant r$; see (2.12.12.2). Then $R_{j_1}, \ldots, R_{j_q}$ is a shelling of $\operatorname{st}(x, F_{i_p})$ that begins with the sequence $R_{j_1}, \ldots, R_{j_\ell}$:

$$\operatorname{st}(x, F_{i_p}) = R_{j_1} \cup \cdots \cup R_{j_\ell} \cup R_{j_{\ell+1}} \cup \cdots \cup R_{j_q}. \tag{2.12.12.3}$$

Furthermore, the $(d-2)$-faces $R_{j_1}, \ldots, R_{j_\ell}$ are precisely the $(d-2)$-faces of F_{i_p} in (2.12.12.1):

$$F_{i_p} \cap \left(F_{i_1} \cup \cdots \cup F_{i_{p-1}}\right) = R_{j_1} \cup \cdots \cup R_{j_\ell}. \tag{2.12.12.4}$$

Finally, there is a shelling S''_{i_p} of F_{i_p} that begins with the $(d-2)$-faces of $\text{st}(\boldsymbol{x}, F_{i_p})$ by Proposition 2.12.10(ii); that is, S''_{i_p} can be written as

$$S''_{i_p} = \underbrace{R_{k_1}, \ldots, R_{k_q}}_{\text{st}(\boldsymbol{x}, F_{i_p})}, R_{k_{q+1}}, \ldots, R_{k_s},$$

where $R_{k_1}, \ldots, R_{k_q}$ are the $(d-2)$-faces of $\text{st}(\boldsymbol{x}, F_{i_p})$. As a consequence of Lemma 2.12.2 and of $R_{j_1}, \ldots, R_{j_q}$ being a shelling of $\text{st}(\boldsymbol{x}, F_{i_p})$, the shelling S''_{i_p} can be rearranged to obtain the desired shelling S_{i_p} of F_{i_p}, which begins with the shelling of $\text{st}(\boldsymbol{x}, F_{i_p})$ in (2.12.12.3). That is,

$$S_{i_p} = \underbrace{R_{j_1}, \ldots, R_{j_\ell}, R_{j_{\ell+1}}, \ldots, R_{j_q}}_{\text{st}(\boldsymbol{x}, F_{i_p})}, R_{k_{q+1}}, \ldots, R_{k_s}.$$

This establishes that (2.12.12.1) is the beginning of the shelling S_{i_p} of F_{i_p}; see also (2.12.12.4). This proves the statement of the proposition. □

With the help of Lemma 2.12.2 and Proposition 2.12.12, more involved shellings of a polytope can be produced.

Proposition 2.12.13 *Let P be a polytope, $\boldsymbol{x}$ a vertex of P, and F a facet that is not in the star of $\boldsymbol{x}$ in P. Then there is a shelling of P where the facets in the star of $\boldsymbol{x}$ come first and the facet F comes last. Furthermore, any two facets F' and F'' in the star of $\boldsymbol{x}$ can be taken as the first and last facets of the shelling of the star of $\boldsymbol{x}$.*

Proof The proof is similar to that of Proposition 2.12.10. Choose a point $\boldsymbol{y}_x$ beyond the vertex $\boldsymbol{x}$ and a point $\boldsymbol{y}_F$ beyond the facet F, and let ℓ be the line determined by $\boldsymbol{y}_x$ and $\boldsymbol{y}_F$. Again, if necessary, perturb ℓ so that it becomes in general position with respect to P. Traversing the line ℓ from $\boldsymbol{y}_F$ to $\boldsymbol{y}_x$ gives a line shelling S where the facets in the star of $\boldsymbol{x}$ come first, inducing a line shelling S_i of $\text{st}(\boldsymbol{x}, P)$, and the facet F come last. Let S_f be the sequence of facets in S that follows S_i; that is, $S = S_i S_f$.

We now use Lemma 2.12.2 to modify the shelling S; we provide a different shelling S'_i of the star of $\boldsymbol{x}$ that begins with the facet F' and ends with the facet F''. This new shelling S'_i exists for the following reasons: (1) by Proposition 2.12.10(i) there exists a shelling of P that begins with the facet F' and ends with the facet F'', and (2) the restriction of this shelling of P to the star of $\boldsymbol{x}$ is a shelling of $\mathrm{st}(\boldsymbol{x}, P)$ that begins with the facet F' and ends with the facet F'' (Proposition 2.12.12); we let S_i' be this shelling of $\mathrm{st}(\boldsymbol{x}, P)$. The shelling $S_i' S_f$ is the desired shelling of P. □

The same approach in the proof of Proposition 2.12.13 gives the following.

Proposition 2.12.14 *Let P be a polytope, $\boldsymbol{x}$ a vertex of P, and F a facet in the star of $\boldsymbol{x}$ in P. Then there is a shelling of P where the facets in the star of $\boldsymbol{x}$ come last and the facet F is the last facet of the shelling.*

Euler–Poincaré–Schläfli's Equation

The polyhedral equation of Euler (1758b) is one of the earliest contributions to polytope theory. The equation relates the number of vertices, edges, and faces of a 3-polytope. According to Francese and Richeson (2007), Euler's original proof had mistakes. Schläfli (1850–52) generalised the equation to all dimensions, but his proof relied on the existence of shellings in polytopes, which was assumed but not proved at the time. As we know, the existence of shellings of polytopes was established much later by Bruggesser and Mani (1971). Poincaré (1893) attempted another proof of Schläfli's generalisation, but his proof was also erroneous, as claimed by Gruber (2007, sec. 15.2). Hence, it is fitting to call the generalisation of Euler's equation to all dimensions 'Euler–Poincaré–Schläfli's equation'.

For a complex $\mathcal{C}$ of dimension $d-1$, let $f_i := f_i(\mathcal{C})$ and define the *Euler characteristic* χ as

$$\chi(\mathcal{C}) := f_0 - f_1 + \cdots + (-1)^{d-1} f_{d-1}. \tag{2.12.15}$$

The Euler characteristic of the union $\mathcal{C} \cup \mathcal{C}'$ of complexes $\mathcal{C}$ and $\mathcal{C}'$, in case it is a complex, satisfies an attractive additivity property (can you prove it?).

Lemma 2.12.16 *Let $\mathcal{C}$ and $\mathcal{C}'$ be polytonal complexes such that $\mathcal{C} \cup \mathcal{C}'$ is a complex too. Then*

$$\chi(\mathcal{C} \cup \mathcal{C}') = \chi(\mathcal{C}) + \chi(\mathcal{C}') - \chi(\mathcal{C} \cap \mathcal{C}').$$

Theorem 2.12.17 (Euler–Poincaré–Schläfli's equation) *Let P be a d-polytope. Then*

$$\chi(\mathcal{B}(P)) = f_0 - f_1 + \cdots + (-1)^{d-1} f_{d-1} = 1 - (-1)^d,$$
$$\chi(\mathcal{C}(P)) = \chi(\mathcal{B}(P)) + (-1)^d = 1.$$

Proof Let $S := F_1, \dots, F_s$ be a shelling of P and let

$$\mathcal{C}_j := \mathcal{C}(F_1 \cup \cdots \cup F_j), \text{ for each } j \in [1 \dots s].$$

We show that

$$\chi(\mathcal{C}_j) = \begin{cases} 1, & \text{for } j \in [1 \dots s-1]; \\ 1 - (-1)^d, & \text{for } j = s. \end{cases} \tag{2.12.17.1}$$

The veracity of (2.12.17.1) implies the theorem, as $\chi(\mathcal{C}(P)) = \chi(\mathcal{B}(P)) + (-1)^d = 1$ by (2.12.15), and $\mathcal{C}_s = \mathcal{B}(P)$. We prove (2.12.17.1) by induction on d for every j.

It is clear that (2.12.17.1) is true for $d = 1$ and every $j \in [1 \dots s]$. And so we assume that $d \geqslant 2$ and that (2.12.17.1) is true for every shelling of a polytope of dimension less than d and at every step of the shelling.

For each $j \in [1 \dots s]$, (2.12.15) gives that

$$\chi(\mathcal{C}(F_j)) = \chi(\mathcal{B}(F_j)) + (-1)^{d-1},$$

and the induction hypothesis on $d - 1$ gives that

$$\chi(\mathcal{B}(F_j)) = 1 - (-1)^{d-1} \text{ and } \chi(\mathcal{C}(F_j)) = 1. \tag{2.12.17.2}$$

In particular, (2.12.17.1) holds for d and $j = 1$, as

$$\chi(\mathcal{C}_1) = \chi(\mathcal{C}(F_1)) = 1.$$

Consider $j \in [1 \dots s]$. Since $\mathcal{C}_j = \mathcal{C}_{j-1} \cup \mathcal{C}(F_j)$ and $\mathcal{C}_{j-1} \cap \mathcal{C}(F_j) = \mathcal{C}(F_j \cap (F_1 \cup \cdots \cup F_{j-1}))$, Lemma 2.12.16 together with (2.12.17.2) yields the equality

$$\begin{aligned} \chi(\mathcal{C}_j) &= \chi(\mathcal{C}_{j-1} \cup \mathcal{C}(F_j)) = \chi(\mathcal{C}_{j-1}) + \chi(\mathcal{C}(F_j)) - \chi(\mathcal{C}_{j-1} \cap \mathcal{C}(F_j)) \\ &= \chi(\mathcal{C}_{j-1}) + 1 - \chi(\mathcal{C}(F_j \cap (F_1 \cup \cdots F_{j-1}))). \end{aligned} \tag{2.12.17.3}$$

Because S is a shelling of P, the intersection

$$F_j \cap (F_1 \cup \cdots \cup F_{j-1}) = R_1 \cup \cdots \cup R_r$$

is the beginning of a shelling $R_1, \cdots, R_r, R_{r+1}, \dots, R_t$ of F_j. As a result, we rewrite (2.12.17.3) as

$$\chi(\mathcal{C}_j) = \chi(\mathcal{C}_{j-1}) + 1 - \chi(\mathcal{C}(R_1 \cup \cdots \cup R_r)). \tag{2.12.17.4}$$

From the induction hypothesis on $d - 1$, for every $r \in [1 \dots t]$ it follows that

$$\chi(\mathcal{C}(R_1 \cup \cdots \cup R_r)) = \begin{cases} 1, & \text{for } r \in [1 \dots t-1]; \\ 1 - (-1)^{d-1}, & \text{for } r = t. \end{cases} \tag{2.12.17.5}$$

If $j < s$, then $r < t$ in (2.12.17.5). And as a consequence of (2.12.17.5), for each $j \in [1 \ldots s-1]$ (2.12.17.4) becomes

$$\chi(\mathcal{C}_j) = \chi(\mathcal{C}_{j-1}) + 1 - 1 = \chi(\mathcal{C}_{j-1});$$

that is, for every $j \in [1 \ldots s-1]$, because $\chi(\mathcal{C}_1) = 1$ we get that

$$\chi(\mathcal{C}_{s-1}) = \cdots = \chi(\mathcal{C}_1) = 1. \tag{2.12.17.6}$$

If instead $j = s$, then $r = t$ in (2.12.17.5), in which case, $\mathcal{C}(R_1 \cup \cdots \cup R_r) = \mathcal{B}(F_s)$. As a result of (2.12.17.2) and (2.12.17.5), Equation (2.12.17.4) becomes

$$\chi(\mathcal{C}_s) = \chi(\mathcal{C}_{s-1}) + 1 - \left(1 - (-1)^{d-1}\right) = 1 + 1 - \left(1 - (-1)^{d-1}\right) = 1 - (-1)^d. \tag{2.12.17.7}$$

The induction is now complete, as (2.12.17.1) now follows from (2.12.17.6) and (2.12.17.7). Accordingly, the proof of the theorem is also complete. □

The case $d = 3$ of Euler–Poincaré–Schläfli's equation (2.12.17) is the famous relation of Euler (1758b,a).

Euler–Poincaré–Schläfli's equation (Theorem 2.12.17) is the unique linear equation satisfied by the f-vector of all d-polytopes. The precise meaning of this assertion ensues.

Theorem 2.12.18[9] *Let d be a positive integer, and let $\alpha_0 \in \mathbb{R}, \ldots, \alpha_d \in \mathbb{R}$. If every d-polytope P satisfies the equation*

$$\alpha_0 f_0(P) + \cdots + \alpha_{d-1} f_{d-1}(P) = \alpha_d,$$

then $\alpha_1 = \alpha_0(-1)$, $\alpha_2 = \alpha_0(-1)^2, \ldots, \alpha_{d-1} = \alpha_0(-1)^{d-1}$, and $\alpha_d = (1 - (-1)^d)\alpha_0$.

If we consider the f-vector $(f_0, \ldots, f_{d-1})$ of a d-polytope P as a point in $\mathbb{R}^d$, Theorem 2.12.18 implies that the f-vectors of all d-polytopes lie in the hyperplane

$$\left\{ \begin{pmatrix} x_1 \\ \vdots \\ x_d \end{pmatrix} \in \mathbb{R}^d \;\middle|\; \alpha_0 \begin{pmatrix} 1 \\ -1 \\ \vdots \\ (-1)^{d-1} \end{pmatrix} \cdot \begin{pmatrix} x_1 \\ \vdots \\ x_d \end{pmatrix} = \alpha_0 \left(1 - (-1)^d\right) \right\}$$

of $\mathbb{R}^d$, the so-called *Euler hyperplane*, and on no affine subspace of smaller dimension. The Euler hyperplane is affinely spanned by the f-vectors of the

[9] A proof is available in Webster (1994, thm. 3.5.3).

d-simplex and $d-1$ d-polytopes with $d+2$ vertices. For each $r \in [0 \dots d-2]$, consider the simplicial $(d-r)$-polytope $F_r := T(1) \oplus T(d-r-1)$, the direct sum of the simplices $T(1)$ and $T(d-r-1)$ (see (2.6.17)). Then consider the r-fold pyramids $\operatorname{pyr}_r(F_r)$, for $r \in [0 \dots d-2]$. It follows that the f-vectors of the d-polytopes $T(d)$ and $\operatorname{pyr}_r(F_r)$ (for $r \in [0 \dots d-2]$) form an affinely independent set in $\mathbb{R}^d$ (Problem 2.15.13).

2.13 Dehn–Sommerville's Equations

Euler–Poincaré–Schläfli's equation (Theorem 2.12.17) is the unique linear equation satisfied by the f-vectors of all d-polytopes (Theorem 2.12.18), but there are other linear equations satisfied by the f-vector of all polytopes from a certain class. This is the case of Dehn–Sommerville's (classical) equations (2.13.3), which are satisfied by the f-vector of all simplicial d-polytopes. Dehn–Sommerville's equations (Theorem 2.13.3) were first established for dimension five by Dehn (1905), and later extended to all dimensions by Sommerville (1927). As is often the case, the equations were independently rediscovered by Klee (1964). For more information on the history of the equations, consult Grünbaum (2003, sec. 9.8).

Our proof of Dehn–Sommerville's equations relies on a generalisation of Euler–Poincaré–Schläfli's equation. We denote by $\mathcal{F}_i(F,P)$ the set of i-faces in a d-polytope P containing a k-face F, for $k \in [1 \dots d]$ and $i \in [k \dots d]$. And we let $f_i(F,P) := \#\mathcal{F}_i(F,P)$.

Theorem 2.13.1 *Let P be a d-polytope in $\mathbb{R}^d$ and let F be a proper k-face of it. Then*

$$f_k(F,P) - f_{k+1}(F,P) + \cdots + (-1)^{d-k} f_d(F,P) = 0.$$

Remark 2.13.2 Euler–Poincaré–Schläfli's equation (Theorem 2.13.3) corresponds to the case $F = \varnothing$ of Theorem 2.13.1.

Proof Without loss of generality, suppose that P is a d-polytope in $\mathbb{R}^d$ that contains the origin in its interior, and let ψ be an antiisomorphism from $\mathcal{L}(P)$ to $\mathcal{L}(P^*)$. If we apply Euler–Poincaré–Schläfli's equation to the $(d-1-k)$-polytope $\psi(F)$ (Theorem 2.4.12), we get that

$$\begin{aligned} f_{-1}(\psi(F)) - f_0(\psi(F)) + \cdots + (-1)^{d-1-k} f_{d-2-k}(\psi(F)) \\ + (-1)^{d-k} f_{d-1-k}(\psi(F)) = 0. \quad (2.13.2.1) \end{aligned}$$

The number $f_{d-1-i}(\psi(F))$ of $(d-1-i)$-faces of $\psi(F)$ coincides with the number $f_i(F,P)$ of i-faces in P containing F, which is a consequence of Theorem 2.4.10 and Theorem 2.4.12. Thus (2.13.2.1) is equivalent to

$$f_d(F,P) - f_{d-1}(F,P) + \cdots + (-1)^{d-1-k} f_{k+1}(F,P) + (-1)^{d-k} f_k(F,P) = 0.$$

This proves the the theorem. □

Theorem 2.13.3 (Dehn–Sommerville's equations for simplicial polytopes) *The f-vector of a simplicial d-polytope satisfies the expression*

$$\sum_{i=k}^{d-1}(-1)^i\binom{i+1}{k+1} f_i = (-1)^{d-1} f_k,$$

for $k = -1, \ldots, d-2$.

Remark 2.13.4 Euler–Poincaré–Schläfli's equation (Theorem 2.13.3) corresponds to the case $k=-1$ of the Dehn–Sommerville equations.

Proof For a k-face F of a simplicial d-polytope P, with $k \in [-1\ldots d-2]$, Theorem 2.13.1 yields that

$$f_k(F,P) - f_{k+1}(F,P) + \cdots + (-1)^{d-k} f_d(F,P) = 0, \tag{2.13.4.1}$$

which results in

$$(-1)^k f_k(F,P) + (-1)^{k+1} f_{k+1}(F,P) + \cdots + (-1)^d f_d(F,P) = 0 \tag{2.13.4.2}$$

if we multiply (2.13.4.1) by $(-1)^k$. We run (2.13.4.2) over the set $\mathcal{F}_k(P)$ of k-faces in the polytope P, obtaining

$$\sum_{F\in\mathcal{F}_k(P)} (-1)^k f_k(F,P) + \cdots + (-1)^d f_d(F,P) = 0,$$

which is equivalent to

$$\sum_{i=k}^{d}(-1)^i \sum_{F\in\mathcal{F}_k(P)} f_i(F,P) = 0. \tag{2.13.4.3}$$

For each $k \in [-1\ldots d-2]$ and each $i \in [k\ldots d]$, the sum $\sum_{F\in\mathcal{F}_k(P)} f_i(F,P)$ counts the total number of inclusions between the k-faces and i-faces in P: a k-face F of P is contained in $f_i(F,P)$ i-faces of P, and we do this count for each k-face of P. Another way of counting these inclusions is to consider the i-faces. There are $f_i(P)$ i-faces in P, and an i-face J of P contains $\binom{i+1}{k+1}$ k-faces if $i \in [k\ldots d-1]$ (as J is a simplex) and

$f_k(P)$ k-faces if $i = d$. For each $k \in [-1 \ldots d-2]$ and each $i \in [k \ldots d]$, this analysis gives that

$$0 = \sum_{i=k}^{d}(-1)^i \sum_{F \in \mathcal{F}_k(P)} f_i(F, P) = \sum_{i=k}^{d-1}(-1)^i \binom{i+1}{k+1} f_i(P) + (-1)^d f_k(P).$$

The last equality is Dehn–Sommerville's desired equation for $k \in [-1 \ldots d - 2]$.
□

Some of Dehn–Sommerville's d equations of a simplicial d-polytope P are redundant. Consider the case $d = 3$:

$$\begin{aligned} k = -1 &: -f_{-1}(P) + f_0(P) - f_1(P) + f_2(P) = f_{-1}(P), \\ k = 0 &: f_0(P) - 2f_1(P) + 3f_2(P) = f_0(P), \\ k = 1 &: -f_1(P) + 3f_2(P) = f_1(P). \end{aligned} \tag{2.13.5}$$

The equations $k = 0$ and $k = 1$ are identical, and so the equation $k = 1$ is redundant. Furthermore, the equations $k = -1$ and $k = 0$ are irredundant. In other words, the three equations are equivalent to the the first two. Now consider the case $d = 4$:

$$\begin{aligned} k = -1 &: -f_{-1}(P) + f_0(P) - f_1(P) + f_2(P) - f_3(P) = -f_{-1}(P), \\ k = 0 &: f_0(P) - 2f_1(P) + 3f_2(P) - 4f_3 = -f_0(P), \\ k = 1 &: -f_1(P) + 3f_2(P) - 6f_3(P) = -f_1(P), \\ k = 2 &: f_2(P) - 4f_3(P) = -f_2(P). \end{aligned} \tag{2.13.6}$$

The equations $k = 1$ and $k = 2$ are identical, and the equations $k = -1$ and $k = 1$ imply the equation $k = 0$. In other words, the four equations are equivalent to the equations $k = -1$ and $k = 1$. This redundancy manifests in all dimensions.

Theorem 2.13.7 [10] *For every $d \geqslant 1$, precisely $\lfloor (d + 1)/2 \rfloor$ of Dehn–Sommerville's d equations are irredundant.*

A consequence of Theorem 2.13.7 is the existence of $\lfloor d/2 \rfloor + 1$ simplicial d-polytopes whose f-vectors form an affinely independent set in $\mathbb{R}^d$ (Problem 2.15.14). Another consequence ensues.

Corollary 2.13.8 *If we consider the f-vectors of simplicial d-polytopes as points in $\mathbb{R}^d$, then they lie in an affine subspace in $\mathbb{R}^d$ of dimension $\lfloor d/2 \rfloor$, and on no affine subspace of smaller dimension.*

[10] A proof is available in Grünbaum (2003, p. 147).

h-vectors of Simplicial Polytopes

Consider a simplicial d-polytope P and identify every face of P with its vertex set. Following Ziegler (1995, sec. 8.3), for a shelling $S := F_1, \ldots, F_s$ of the boundary complex $\mathcal{B}(P)$ of P, we define the *restriction set* R_j of the facet F_j as

$$R_j := \{\boldsymbol{v} \in \mathcal{V}(F_j) | \, (\mathcal{V}(F_j) \backslash \{\boldsymbol{v}\}) \subseteq F_i, \text{for some } i \in [1 \ldots j]\}.$$

We use the shelling to build $\mathcal{B}(P)$. For $j \in [2 \ldots s]$, suppose that we have built the subcomplex $\mathcal{C}_{j-1} := \mathcal{C}(F_1 \cup \ldots \cup F_{j-1})$ of $\mathcal{B}(P)$. When the facet F_j is added to $\mathcal{C}_{j-1}$, the faces introduced are precisely the faces X satisfying $R_j \subseteq X \subseteq F_j$. Clearly, $X \subseteq F_j$. For the other direction, if X is not new then the definition of a shelling (Definition 2.12.1) gives that $X \subseteq F_j \cap F_i = F_j \backslash \{\boldsymbol{v}\}$ for some F_i with $i < j$ and some $\boldsymbol{v} \in \mathcal{V}(F_j)$. But the definition of a restriction set yields that $\boldsymbol{v} \in R_j$, ensuring that $X \nsubseteq R_j$. We let I_j be the set of such new faces:

$$I_j := \{X \in \mathcal{B}(P) | \, R_j \subseteq X \subseteq F_j\}. \tag{2.13.9}$$

It follows that $I_1, \ldots, I_s$ is a partition of the faces of $\mathcal{B}(P)$. For $k \in [0 \ldots d]$, let $h_k(S)$ count the number of restriction sets with cardinality k:

$$h_k(S) := \#\{j | \, \#R_j = k, \text{ for } j \in [1 \ldots s]\}. \tag{2.13.10}$$

From the numbers $h_0(S), \ldots, h_d(S)$, we can recover the f-vector of P. Since P is simplicial, in the case that $\#R_j = i$, by (2.13.9) there are exactly $\binom{d-i}{k-i}$ $(k-1)$-faces in I_j, and thus

$$\begin{aligned} f_{k-1}(P) &:= \sum_{i=0}^{k} h_i(S) \binom{d-i}{k-i}, \text{ for } k = 0, \ldots, d, \\ &= \sum_{i=0}^{d} h_i(S) \binom{d-i}{k-i}. \end{aligned} \tag{2.13.11}$$

We can also obtain the number $h_k(S)$ in terms of the f-vector of P. For this, we define the two polynomials

$$\varphi_f(x) := \sum_{k=0}^{d} f_{k-1}(P) x^{d-k} \text{ and } \varphi_h(x) := \sum_{k=0}^{d} h_k(S) x^{d-k},$$

and using (2.13.11) and $(x+1)^{d-i} = \sum_{j=0}^{d-i} \binom{d-i}{j} x^{d-i-j}$ we relate them as

$$\begin{aligned}\varphi_f(x) &= \sum_{k=0}^{d} f_{k-1}(P)x^{d-k} = \sum_{k=0}^{d}\left(\sum_{i=0}^{d} h_i(S)\binom{d-i}{k-i}\right)x^{d-k}\\ &= \sum_{i=0}^{d} h_i(S)\left(\sum_{j=0}^{d-i}\binom{d-i}{j}x^{d-i-j}\right) = \sum_{i=0}^{d} h_i(S)(x+1)^{d-i} = \varphi_h(x+1).\end{aligned}$$

Equivalently, we have that $\varphi_f(x-1) = \varphi_h(x)$. This, together with the identity $(x-1)^{d-i} = \sum_{j=0}^{d-i}(-1)^j\binom{d-i}{j}x^{d-i-j}$, yields that

$$h_k(S) = \sum_{i=0}^{k}(-1)^{k-i}\binom{d-i}{k-i} f_{i-1}(P), \text{ for } k = 0,\dots,d. \tag{2.13.12}$$

From (2.13.12) it follows that the h_k numbers are independent of the shelling S, and so we will call the sequence $(h_0,\dots,h_d)$ the *h-vector* of P. The reverse shelling S' of S is also a shelling of $\mathcal{B}(P)$, where the cardinality of the restriction set R'_j of the facet F_j is $d-k$ if the cardinality of R_j in S is k, and therefore $h_{d-k}(S') = h_k(S)$, for $k \in [0\dots d]$. Since the numbers h_k are independent of any shelling of $\mathcal{B}(P)$, it follows that

$$h_{d-k} = h_k, \text{ for } k = 0,\dots,d. \tag{2.13.13}$$

With some effort, and as described in the proof of Grünbaum (2003, thm. 9.2.2), we get that (2.13.13) is another way of writing Dehn–Sommerville's equations for simplicial polytopes (Theorem 2.13.3).

Example 2.13.14 (Computation of h-vectors of simplicial polytopes) Consider the octahedron P in Fig. 3.5.2(b) and the line shelling $F_1 := 123, F_2 := 135, F_3 := 234, F_4 := 345, F_5 := 456, F_6 := 156, F_7 := 126, F_8 := 246$ of $\mathcal{B}(P)$. For this shelling we have that

$$\begin{aligned}R_1 := \emptyset, R_2 := \{5\}, R_3 := \{4\}, R_4 := \{4,5\}, R_5 := \{6\},\\ R_6 := \{1,6\}, R_7 := \{2,6\}, R_8 := \{2,4,6\}.\end{aligned}$$

Once we have the restriction sets, using (2.13.9) we get a partition of $\mathcal{B}(P)$:

$$\begin{aligned}I_1 := \{1,2,3,12,13,23,123\}, I_2 := \{5,15,35,135\}, I_3 := \{4,24,34,234\},\\ I_4 := \{45,345\}, I_5 := \{6,46,56,456\}, I_6 := \{16,156\}, I_7 := \{26,126\},\\ I_8 := \{246\}.\end{aligned}$$

The h-vector $(1,3,3,1)$ of P is computed using (2.13.10):

$$h_0 := 1, h_1 := \#\{2,3,5\}, h_2 := \#\{4,6,7\}, h_3 := \#\{8\}.$$

Ziegler (1995, ex. 8.22) also illustrates the computation of the h-vector of the octahedron by considering a shelling and its reverse shelling, both different from the shelling in Example 2.13.14.

Dehn–Sommerville's Equations for Cubical Polytopes

The f-vectors of cubical polytopes satisfy equations that are similar to Dehn–Sommerville's equations for simplicial polytopes.

Theorem 2.13.15 (Dehn–Sommerville's equations for cubical polytopes)[11] *Let P be a cubical d-polytope in $\mathbb{R}^d$. Then, apart from Euler–Poincaré–Schläfli's equation:*

$$\sum_{i=-1}^{d-1} (-1)^i f_i(P) = (-1)^{d-1} f_{-1}(P),$$

the f-vector of P satisfies

$$\sum_{i=k}^{d-1} (-1)^i 2^{i-k} \binom{i}{k} f_i(P) = (-1)^{d-1} f_k(P),$$

for $k = 0, \dots, d-2$.

The next corollary of Theorem 2.13.15 is the analogue of Corollary 2.13.8.

Corollary 2.13.16 *If we consider the f-vectors of cubical d-polytopes as points in $\mathbb{R}^d$, then they lie in an affine subspace in $\mathbb{R}^d$ of dimension $\lfloor d/2 \rfloor$, and on no affine subspace of smaller dimension.*

Problem 2.15.16 presents a family of $\lfloor d/2 \rfloor + 1$ cubical d-polytopes whose f-vectors are affinely independent in $\mathbb{R}^d$.

2.14 Gale Transforms

The Gale transform of the vertices of a polytope captures the combinatorial structure of the polytope in a very compact manner. While ideas related to Gale transforms had been exploited by a number of authors, M. A. Perles seemed to have been the first that formalised the method (Grünbaum, 2003, sec. 5.4).

An *(affine) point configuration* is a collection $X := \{\boldsymbol{x}_1, \dots, \boldsymbol{x}_n\}$ of n, not necessarily distinct, points that lies in an affine space $\mathbb{R}^d$ and that affinely

[11] A proof is available in Grünbaum (2003, sec. 9.4).

spans $\mathbb{R}^d$. Similarly, a *vector configuration* is a collection $X := \{\boldsymbol{x}_1, \ldots, \boldsymbol{x}_n\}$ of n, not necessarily distinct, vectors that lies in a linear space $\mathbb{R}^d$ and that linearly spans $\mathbb{R}^d$. In either case, often will we consider the set X as a sequence $(\boldsymbol{x}_1, \ldots, \boldsymbol{x}_n)$ or as a $d \times n$ matrix with the element $\boldsymbol{x}_i$ as the ith column. A Gale transform of a point configuration X of n points in the affine space $\mathbb{R}^d$ is a vector configuration of n vectors in the linear space $\mathbb{R}^{n-d-1}$. We next develop the machinery to compute Gale transforms.

Given a point configuration $X := \{\boldsymbol{x}_1, \ldots, \boldsymbol{x}_n\}$ in $\mathbb{R}^d$, the set $\operatorname{dep} X$ of its *affine dependences* contains all the vectors $\boldsymbol{a} \in \mathbb{R}^n$ satisfying $\boldsymbol{1}_n \cdot \boldsymbol{a} = 0$ and $X\boldsymbol{a} = \boldsymbol{0}$, where X is viewed as a $d \times n$ matrix. This is equivalent to stating that

$$\operatorname{dep} X := \left\{ \boldsymbol{a} = (a_1, \ldots, a_n)^t \,\middle|\, \sum_{i=1}^{n} a_i = 0 \text{ and } \sum_{i=1}^{n} a_i \boldsymbol{x}_i = \boldsymbol{0}_d \right\}. \qquad (2.14.1)$$

We homogenise the configuration X by associating the vector $\widehat{\boldsymbol{x}}_i := \begin{pmatrix} \boldsymbol{x}_i \\ 1 \end{pmatrix} \in \mathbb{R}^{d+1}$ with the point $\boldsymbol{x}_i \in \mathbb{R}^d$. This gives a vector configuration $\widehat{X} := \{\widehat{\boldsymbol{x}}_1, \ldots, \widehat{\boldsymbol{x}}_n\}$ in $\mathbb{R}^{d+1}$ whose set $\operatorname{ldep} \widehat{X}$ of linear dependences coincides with the set $\operatorname{dep} X$:

$$\operatorname{ldep} \widehat{X} := \left\{ \boldsymbol{a} = \begin{pmatrix} a_1 \\ \vdots \\ a_n \end{pmatrix} \,\middle|\, \widehat{X}\boldsymbol{a} = \boldsymbol{0}_{d+1}, \text{or, equivalently, } \sum_{i=1}^{n} a_i \widehat{\boldsymbol{x}}_i = \boldsymbol{0}_{d+1} \right\}. \qquad (2.14.2)$$

Remark 2.14.3 The sets $\operatorname{dep} X$ and $\operatorname{ldep} \widehat{X}$ are *symmetric about the origin* or *centrally symmetric*; that is, $\boldsymbol{a} \in \operatorname{dep} X$ if and only if $-\boldsymbol{a} \in \operatorname{dep} X$. So it often suffices to record only one element of the pair $(\boldsymbol{a}, -\boldsymbol{a})$.

For every point configuration X with $n \geqslant d + 1$ points in $\mathbb{R}^d$, the set $\operatorname{dep} X$ is a linear subspace of $\mathbb{R}^n$ of dimension $n - d - 1$. To see this, note that $\operatorname{dep} X$ coincides with the nullspace of $\widehat{X}$, where $\widehat{X}$ is viewed as a $(d + 1) \times n$ matrix. Since X affinely spans $\mathbb{R}^d$, the row vector space of $\widehat{X}$ is $(d + 1)$-dimensional. This in turn implies that $(\operatorname{row} \widehat{X})^{\perp}$, which coincides with $\operatorname{null} \widehat{X}$, is $(n-d-1)$-dimensional (Problem 1.12.5).

We now define a Gale transform of a point configuration in $\mathbb{R}^d$.

Definition 2.14.4 (Gale transform) Let $X := (\boldsymbol{x}_1, \ldots, \boldsymbol{x}_n)$ be a point configuration in $\mathbb{R}^d$, and let Y be an $n \times (n - d - 1)$ matrix whose columns $\boldsymbol{y}_1, \ldots, \boldsymbol{y}_{n-d-1}$ form a basis in $\mathbb{R}^n$ of $\operatorname{dep} X$. The *Gale transform* $\mathcal{G}$ of X is the vector configuration $\mathcal{G}(X) = (\mathcal{G}(\boldsymbol{x}_1), \ldots, \mathcal{G}(\boldsymbol{x}_n))$ in $\mathbb{R}^n$ where $\mathcal{G}(\boldsymbol{x}_i)$ is the ith row of Y, for each $i \in [1 \ldots n]$.

Regarding a Gale transform as a matrix gives rise to further observations.

Remark 2.14.5 Let $X := (\boldsymbol{x}_1, \ldots, \boldsymbol{x}_n)$ be a point configuration in $\mathbb{R}^d$, and consider the vector configuration $\widehat{X}$ as a $(d+1) \times n$ matrix and the matrix Y of Definition 2.14.4. If we regard the Gale transform $\mathcal{G}(X)$ of X as a $d \times n$ matrix with columns $\mathcal{G}(\boldsymbol{x}_1), \ldots, \mathcal{G}(\boldsymbol{x}_n)$, then Problems 1.12.4 and 1.12.5 yield that

$$Y^t = \mathcal{G}(X), \ \operatorname{null} \widehat{X} = \operatorname{col} Y = \operatorname{row} \mathcal{G}(X), \ \text{and} \ \operatorname{row} \widehat{X} = \operatorname{null} \mathcal{G}(X).$$

When computing a Gale transform of the vertex set V of a polytope, we view V as a sequence; we just fix an ordering of the elements of it. Example 2.14.6 exemplifies Definition 2.14.4.

Example 2.14.6 Consider the 3-polytope of Fig. 2.14.1, whose vertices are

$$\boldsymbol{v}_1 := (0,0,-1)^t, \ \boldsymbol{v}_2 := (1,0,-1)^t, \ \boldsymbol{v}_3 := (0,1,-1)^t, \ \boldsymbol{v}_4 := (0,0,1)^t,$$
$$\boldsymbol{v}_5 := (1,2,1)^t, \ \boldsymbol{v}_6 := (0,1,1)^t,$$

and compute a Gale transform of the sequence $V = (\boldsymbol{v}_1, \ldots, \boldsymbol{v}_6)$ according to Definition 2.14.4.

We compute a basis of dep V, or equivalently, a basis of null $\widehat{V}$. Homogenise the matrix V formed by the vertices of the polytope, obtaining $\widehat{V}$. Choose a basis of null $\widehat{V}$, for instance $(1,0,-1,-1,0,1)^t$ and $(0,1,-1,-2,-1,3)^t$, and form the 2×6 matrix Y with the vectors in the basis as columns. Then, the transpose of Y produces the Gale transform $\mathcal{G}(V)$ of V associated with the aforementioned basis. See the matrices $\widehat{V}$, Y, and $\mathcal{G}(V)$ below:

$$\widehat{V} = \begin{pmatrix} 0 & 1 & 0 & 0 & 1 & 0 \\ 0 & 0 & 1 & 0 & 2 & 1 \\ -1 & -1 & -1 & 1 & 1 & 1 \\ 1 & 1 & 1 & 1 & 1 & 1 \end{pmatrix}, \ Y = \begin{pmatrix} 1 & 0 \\ 0 & 1 \\ -1 & -1 \\ -1 & -2 \\ 0 & -1 \\ 1 & 3 \end{pmatrix}$$

and

$$\mathcal{G}(V) = \begin{pmatrix} \mathcal{G}(\boldsymbol{v}_1) & \mathcal{G}(\boldsymbol{v}_2) & \mathcal{G}(\boldsymbol{v}_3) & \mathcal{G}(\boldsymbol{v}_4) & \mathcal{G}(\boldsymbol{v}_5) & \mathcal{G}(\boldsymbol{v}_6) \\ 1 & 0 & -1 & -1 & 0 & 1 \\ 0 & 1 & -1 & -2 & -1 & 3 \end{pmatrix}$$

As expected, a Gale transform of V consists of six vectors lying in $\mathbb{R}^2$; Fig. 2.14.1(b) depicts a realisation of $\mathcal{G}(V)$.

We emphasise that a Gale transform is defined for a point configuration and not for individual points. The next proposition provides some basic properties of Gale transforms.

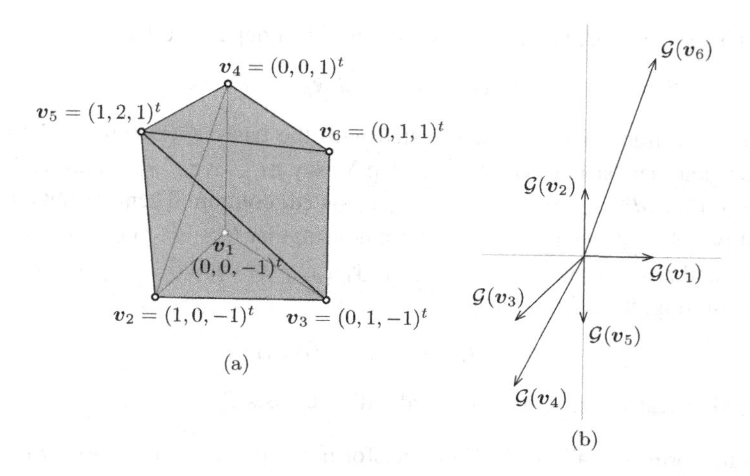

Figure 2.14.1 Computation of a Gale transform of a 3-polytope with six vertices. (a) A realisation of the polytope. (b) A realisation of a Gale transform of the polytope.

Proposition 2.14.7 *Let $X = (\boldsymbol{x}_1, \dots, \boldsymbol{x}_n)$ be a point configuration in $\mathbb{R}^d$ and let $\mathcal{G}(X) = (\mathcal{G}(\boldsymbol{x}_1), \dots, \mathcal{G}(\boldsymbol{x}_n))$ be a Gale transform of X. Then the following hold.*

(i) *The vectors of $\mathcal{G}(X)$ need not be all distinct.*
(ii) *The vectors of $\mathcal{G}(X)$ satisfy $\sum_{i=1}^n \mathcal{G}(\boldsymbol{x}_i) = \boldsymbol{0}_{n-d-1}$, and $\mathcal{G}(X)$ linearly spans $\mathbb{R}^{n-d-1}$.*
(iii) *The vectors of $\mathcal{G}(X)$ positively span $\mathbb{R}^{n-d-1}$.*
(iv) *Every open halfspace bounded by a linear hyperplane in $\mathbb{R}^{n-d-1}$ contains at least one vector from $\mathcal{G}(X)$.*
(v) *The points of X are in (affine) general position in $\mathbb{R}^d$ – that is, no $d+1$ of them lie in a hyperplane–if and only if the vectors of $\mathcal{G}(X)$ are in linear general position in $\mathbb{R}^{n-d-1}$; that is, no $n-d-1$ of them lie in a linear hyperplane.*
(vi) *A Gale transform is determined up to linear isomorphism. In other words, suppose that $Y = (\boldsymbol{y}_1, \dots, \boldsymbol{y}_{n-d-1})$ and $Z = (z_1, \dots, z_{n-d-1})$ are two bases of* dep *X so that $YA = Z$ for a nonsingular matrix A. And suppose that $\mathcal{G}_Y$ and $\mathcal{G}_Z$ are the Gale transforms of X associated with Y and Z (Definition 2.14.4). Then, for each $i \in [1 \dots n]$, we have that*

$$\mathcal{G}_Z(\boldsymbol{x}_i) = A^t\, \mathcal{G}_Y(\boldsymbol{x}_i).$$

Proof We provide a proof of (vi), and leave the rest to the reader.

Let $Y := (y_1, \ldots, y_{n-d-1})$ be a basis in $\mathbb{R}^n$ of dep X and let

$$(\mathcal{G}(x_1), \ldots, \mathcal{G}(x_n))$$

be the Gale transform of X associated with the basis Y (Definition 2.14.4). Choose another basis in $\mathbb{R}^{n-d-1}$ of dep X, say $z_1, \ldots, z_{n-d-1}$, and let Z be the $n \times (n-d-1)$ matrix having z_i as its ith column. Then we obtain the equality $YA = Z$ where $A = (a_{i,j})$ is a nonsingular $(n-d-1) \times (n-d-1)$ matrix and $z_i = a_{1,i} y_1 + \cdots + a_{n-d-1,i} y_{n-d-1}$ for each $i \in [1 \ldots n-d-1]$. In this setting, the sequence

$$(A^t \mathcal{G}(x_1), \ldots, A^t \mathcal{G}(x_n))$$

is the Gale transform of X associated with the basis Z. □

Henceforth, we talk of *the* Gale transform of a point configuration, with the understanding stated in Proposition 2.14.7(vi).

While we have stressed that the elements of a Gale transform should be understood as vectors in a linear space $\mathbb{R}^m$, sometimes we need to interpret them as points in the affine space $\mathbb{R}^m$ so that we can carry out operations on them such as convex hulls. Theorem 2.14.8 exemplifies this situation and shows how to query a Gale transform.

Theorem 2.14.8[12] *Let* $V := \{v_1, \ldots, v_n\}$ *be a point configuration in* $\mathbb{R}^d$ *and let* $\mathcal{G}(V) := (\mathcal{G}(v_1), \ldots, \mathcal{G}(v_n))$ *be the Gale transform of* V. *Moreover, let* $W \subset V$ *and let* $\mathcal{G}(W)$ *be the restriction of* $\mathcal{G}(V)$ *to the points in* W. *The set* conv W *is a proper face of the* d*-polytope* conv V *if and only if* $\mathbf{0}_{n-d-1} \in \operatorname{rint}(\operatorname{conv}(\mathcal{G}(V) \setminus \mathcal{G}(W)))$.

We test Theorem 2.14.8 on the 3-polytope of Fig. 2.14.1.

Example 2.14.9 Let $V := \{v_1, \ldots, v_6\}$ be the vertex set of the 3-polytope P of Fig. 2.14.1, and let $\mathcal{G}(V) := (\mathcal{G}(v_1), \ldots, \mathcal{G}(v_6))$ be the Gale transform of V. Applying Theorem 2.14.8 to $\mathcal{G}(V)$ yields the following.

(i) The vertex sets of the edges of P incident with v_1 are $W_1 := \{v_1, v_2\}$, $W_2 := \{v_1, v_4\}$, and $W_3 := \{v_1, v_3\}$, as $\mathbf{0}_2 \in \operatorname{rint}(\operatorname{conv}(\mathcal{G}(V) \setminus \mathcal{G}(W_i)))$ for $i = 1, 2, 3$. It also follows that the sets $\{v_1, v_5\}$ and $\{v_1, v_6\}$ are not edges of P.

(ii) The vertex sets of the 2-faces of P incident with v_1 are $W_4 := \{v_1, v_2, v_3\}$, $W_5 := \{v_1, v_2, v_4\}$, and $W_6 := \{v_1, v_3, v_4, v_6\}$, as $\mathbf{0}_2 \in \operatorname{rint}(\operatorname{conv}(\mathcal{G}(V) \setminus \mathcal{G}(W_i)))$ for $i = 4, 5, 6$.

[12] A proof is available in Webster (1994, thm. 3.6.6).

The power of the Gale transform (and Theorem 2.14.8) comes to light when dealing with d-polytopes that have $d+2$, $d+3$, or $d+4$ vertices, since in these cases the transforms lie in $\mathbb{R}$, $\mathbb{R}^2$, and $\mathbb{R}^3$, respectively. Its power is also visible when studying pyramids and simplicial polytopes.

Theorem 2.14.10 [13] *Let P be a polytope with vertex set V and let $\mathcal{G}(V)$ be the Gale transform of V. Then P is a pyramid with apex $\boldsymbol{v}$ if and only if $\mathcal{G}(\boldsymbol{v}) = \mathbf{0}$. Furthermore, the Gale transform of the base* $\operatorname{conv}(V \setminus \{\boldsymbol{v}\})$ *of P is $\mathcal{G}(V) \setminus \{\mathcal{G}(\boldsymbol{v})\}$.*

A corollary immediately ensues.

Corollary 2.14.11 *Let P be a polytope with vertex set V and let $\mathcal{G}(V)$ be the Gale transform of V. Then P is an r-fold pyramid with apices $\boldsymbol{v}_1, \ldots, \boldsymbol{v}_r$ if and only if $\mathcal{G}(\boldsymbol{v}_1) = \cdots = \mathcal{G}(\boldsymbol{v}_r) = \mathbf{0}$.*

A d-simplex is a d-fold pyramid over a 0-simplex, which combined with Proposition 2.14.7 yields the following.

Corollary 2.14.12 *Let $\{\boldsymbol{v}_1, \ldots, \boldsymbol{v}_{d+1}\}$ be the vertex set of a d-simplex. Then $\mathcal{G}(\boldsymbol{v}_1) = \cdots = \mathcal{G}(\boldsymbol{v}_{d+1}) = \mathbf{0}$.*

As in the case of pyramids, Gale transforms of simplicial polytopes can be easily characterised.

Theorem 2.14.13 [14] *Let $\mathcal{G}$ be the Gale transform of a d-polytope on n vertices. The polytope is simplicial if and only if $\mathbf{0} \notin$* $\operatorname{rint}(\operatorname{conv}(\mathcal{G} \cap H))$ *for every linear hyperplane H in $\mathbb{R}^{n-d-1}$.*

The computation described in Definition 2.14.4 can certainly be applied to a point configuration X in $\mathbb{R}^d$ that does not represent the vertices of a polytope. So how can we tell if the sequence X is in convex position? Theorem 2.14.14 provides an answer.

Theorem 2.14.14 [15] *A sequence $\mathcal{G}(V) := (\mathcal{G}(\boldsymbol{v}_1), \ldots, \mathcal{G}(\boldsymbol{v}_n))$ of vectors in $\mathbb{R}^{n-d-1}$ is the Gale transform of a d-polytope (other than the simplex) with vertex set V if and only if*

(i) $\sum_{i=1}^{n} \mathcal{G}(\boldsymbol{v}_i) = \mathbf{0}_{n-d-1}$ *and,*
(ii) *every open halfspace bounded by a linear hyperplane in $\mathbb{R}^{n-d-1}$ contains at least two vectors of $\mathcal{G}(V)$.*

[13] A proof is available in McMullen and Shephard (1971, thm. 3).
[14] A proof is available in Webster (1994, thm. 3.6.9).
[15] A proof is available in Webster (1994, thm. 3.6.8).

Projectively isomorphic polytopes can be told from their corresponding Gale transforms, and so can affinely isomorphic polytopes. Two polytopes P and P' are *affinely isomorphic* if there is an affine isomorphism ϱ such that $\varrho(P) = P'$.

Theorem 2.14.15 (Grünbaum 2003, thms. 5.4.6–4.7) *Let P and P' be two polytopes, let $V = \{\boldsymbol{v}_1, \ldots, \boldsymbol{v}_n\}$ and $V' = \{\boldsymbol{v}'_1, \ldots, \boldsymbol{v}'_n\}$ be their respective vertex sets, and let $(\mathcal{G}(\boldsymbol{v}_1), \ldots, \mathcal{G}(\boldsymbol{v}_n))$ and $(\mathcal{G}(\boldsymbol{v}'_1), \ldots, \mathcal{G}(\boldsymbol{v}'_n))$ be their respective Gale transforms. Then the following hold.*

(i) *There is an affine isomorphism ϱ between P and P' with $\varrho(\boldsymbol{v}_i) = \boldsymbol{v}'_i$ if and only if there is a nonsingular matrix A such that $\mathcal{G}(\boldsymbol{v}'_i) = A\,\mathcal{G}(\boldsymbol{v}_i)$ for each $i \in [1 \ldots n]$.*
(ii) *There is a projective isomorphism ζ between P and P' that is admissible for P and satisfies $\zeta(\boldsymbol{v}_i) = \boldsymbol{v}'_i$ if and only if there is a nonsingular matrix A and positive scalars $\alpha_1, \ldots, \alpha_n$ such that $\mathcal{G}(\boldsymbol{v}'_i) = \alpha_i A\,\mathcal{G}(\boldsymbol{v}_i)$ for each $i \in [1 \ldots n]$.*

If the vectors of the Gale transform $\mathcal{G}$ of a polytope P are multiplied by positive scalars, then we obtain a vector configuration $\mathcal{S}$ that will produce the same face lattice as $\mathcal{G}$ when we apply Theorem 2.14.8 to it. But $\mathcal{S}$ may not satisfy all the properties of a Gale transform (Proposition 2.14.7); in particular, the sum of vectors of $\mathcal{S}$ may not result in the zero vector. We say that a *Gale diagram* of a d-polytope on n vertices is a vector configuration $\mathcal{S}$ with n vectors in $\mathbb{R}^{n-d-1}$ that produces the face lattice of the polytope when we apply Theorem 2.14.8 to it. It follows that Gale diagrams generalise Gale transforms. In this context, we say that two Gale diagrams of polytopes are *isomorphic* if they produce isomorphic face lattices.

It is customary to *normalise* the vectors of a Gale diagram and call the resulting configuration a *standard Gale diagram*, as done in McMullen and Shephard (1971, p. 138). For Grünbaum (2003, sec. 5.4), however, every Gale diagram is a standard Gale diagram. From a Gale diagram

$$\mathcal{G}(X) := (\mathcal{G}(\boldsymbol{x}_1), \ldots, \mathcal{G}(\boldsymbol{x}_n)),$$

a standard Gale diagram $\mathcal{G}'(X)$ can be obtained as follows:

$$\mathcal{G}'(X) := \begin{cases} \mathcal{G}(\boldsymbol{x}_i), & \text{if } \mathcal{G}(\boldsymbol{x}_i) = \mathbf{0}; \\ \frac{\mathcal{G}(\boldsymbol{x}_i)}{\|\boldsymbol{x}_i\|}, & \text{otherwise.} \end{cases} \tag{2.14.16}$$

The new configuration $\mathcal{G}'$ of vectors is a subset of $\{\mathbf{0}\} \cup \mathbb{S}^{n-d-2}$. And by virtue of Theorem 2.14.8, the sequence $\mathcal{G}'(X)$ is endowed with the same combinatorial properties of $\mathcal{G}(X)$. In drawing the diagram $\mathcal{G}'$ on $\{\mathbf{0}\} \cup \mathbb{S}^{n-d-2}$,

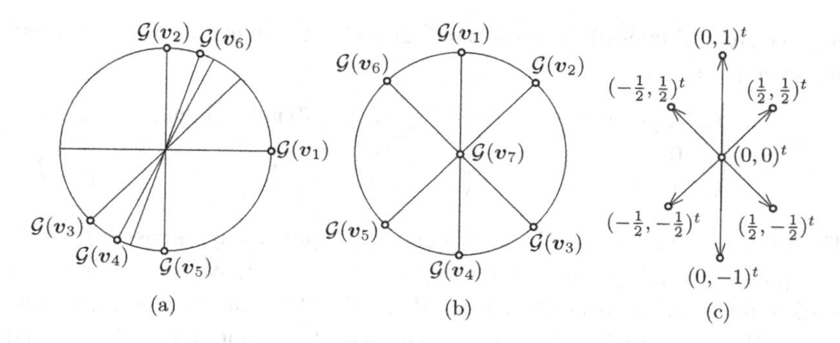

Figure 2.14.2 Gale diagrams of polytopes. (a) The standard Gale diagram corresponding to the Gale transform of Fig. 2.14.1(b). (b) The standard Gale diagram of a pyramidal 4-polytope. (c) A Gale transform isomorphic to the Gale diagram in (b).

we draw the sphere and extend each vector so that it becomes a diameter of the sphere, as in Fig. 2.14.2(a)–(b). In what follows, we normalise our Gale diagrams.

When we are interested only in the face lattice that results from applying Theorem 2.14.8 to a Gale transform, we will resort to Gale diagrams without specifying concrete vectors. The next result is the combinatorial equivalent of Theorem 2.14.15.

Theorem 2.14.17 *Let P and P' be two polytopes with vertex sets V and V', respectively, and let $\mathcal{G}(V)$ and $\mathcal{G}(V')$ be their respective Gale diagrams. The polytopes are combinatorially isomorphic if and only if their Gale diagrams are isomorphic.*

We have learnt how to construct the Gale transform of a polytope given its vertices (Definition 2.14.4), and how to produce the face lattice of a polytope P from its Gale diagram (Theorem 2.14.8). But what about if we want a realisation of a polytope that is combinatorially isomorphic to P; Example 2.14.18 demonstrates how to do so.

Example 2.14.18 Produce a realisation of a polytope P that is combinatorially isomorphic to a polytope whose Gale diagram $\mathcal{G}$ appears in Fig. 2.14.2(b). Let $V := \mathcal{V}(P)$.

Figure 2.14.2(b) first reveals that P has $n = 7$ vertices and dimension $d = 4$, as the Gale diagram lies in $\{\mathbf{0}\} \cup \mathbb{S}$. Moreover, P is a pyramid because $\mathcal{G}(v_7) = \mathbf{0}$ (Theorem 2.14.10). To produce a realisation of P, we produce a Gale transform $\mathcal{G}'$ isomorphic to $\mathcal{G}$ by associating concrete vectors in $\mathbb{R}^2$ that are nonzero multiples of the vectors in $\mathcal{G}$ and whose sum is $\mathbf{0}_2$. Possible vectors are given below.

$$\mathcal{G}'(V) = \begin{pmatrix} \mathcal{G}'(v_1) & \mathcal{G}'(v_2) & \mathcal{G}'(v_3) & \mathcal{G}'(v_4) & \mathcal{G}'(v_5) & \mathcal{G}'(v_6) & \mathcal{G}'(v_7) \\ 0 & \frac{1}{2} & \frac{1}{2} & 0 & -\frac{1}{2} & -\frac{1}{2} & 0 \\ 1 & \frac{1}{2} & -\frac{1}{2} & -1 & -\frac{1}{2} & \frac{1}{2} & 0 \end{pmatrix}.$$

The rows of $\mathcal{G}'(V)$ form a basis of dep V or, equivalently, a basis of null $\widehat{V}$ or, equivalently, a basis of $(\text{row } \widehat{V})^{\perp}$ by Problem 1.12.5. We are looking for a basis of row $\widehat{V}$ or, equivalently, a basis B of null $\mathcal{G}'(V)$ that includes the all-one vector (Remark 2.14.5). After finding the basis B, the matrix $\widehat{V}$ is constructed by placing the vectors of B as rows, with the all-one vector as the last row. The matrices V and $\widehat{V}$ are given below, from left to right.

$$\begin{pmatrix} v_1 & v_2 & v_3 & v_4 & v_5 & v_6 & v_7 \\ -1 & 1 & 0 & 0 & 0 & 1 & 0 \\ 0 & 1 & 0 & 0 & 1 & 0 & 0 \\ 1 & 0 & 0 & 1 & 0 & 0 & 0 \\ 1 & -1 & 1 & 0 & 0 & 0 & 0 \end{pmatrix}, \begin{pmatrix} -1 & 1 & 0 & 0 & 0 & 1 & 0 \\ 0 & 1 & 0 & 0 & 1 & 0 & 0 \\ 1 & 0 & 0 & 1 & 0 & 0 & 0 \\ 1 & -1 & 1 & 0 & 0 & 0 & 0 \\ 1 & 1 & 1 & 1 & 1 & 1 & 1 \end{pmatrix}.$$

Applications of Gale Diagrams

Gale diagrams facilitate the description and enumeration of d-polytopes with $d + 2$ and $d + 3$ vertices. The case of $d + 2$ vertices is not as complicated as the case of $d + 3$ vertices (Fusy, 2006), which makes it appropriate for a neat application of Gale diagrams. See Fig. 2.14.3.

Theorem 2.14.19 [16] *There are precisely $\lfloor d^2/4 \rfloor$ nonisomorphic d-polytopes with $d + 2$ vertices. Among these, the $\lfloor d/2 \rfloor$ simplicial d-polytopes are direct sums of simplices, namely $T(r) \oplus T(s)$ with $r, s \geqslant 1$ and $r + s = d$. The remaining nonsimplicial d-polytopes are t-fold pyramids over $T(r) \oplus T(s)$ with $r, s, t \geqslant 1$ and $r + s + t = d$.*

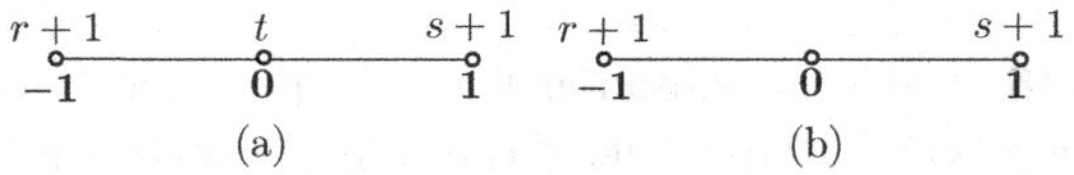

Figure 2.14.3 Gale diagrams of d-polytopes with $d + 2$ vertices. (a) The Gale diagrams corresponding to nonsimplicial polytopes. (b) The Gale diagrams corresponding to simplicial polytopes.

[16] A proof is available in Webster (1994, thm. 3.6.10).

Affine Gale Diagrams

As Theorem 2.14.8 demonstrates, the combinatorial structure of a d-polytope on n vertices can be read off from its Gale transform on $\mathbb{R}^{n-d-1}$. It turns out that this can also be done from a point configuration on $\mathbb{R}^{n-d-2}$, from affine Gale diagrams. This reduction in dimension means that if $n-d-1=3$, then the Gale transform would lie in $\mathbb{R}^3$, but an affine Gale diagram would lie in $\mathbb{R}^2$. Sturmfels (1988) seems to have been the first to utilise affine Gale diagrams.

Affine Gale diagrams are central projections of Gale transforms onto nonlinear hyperplanes. Let $V := \{\boldsymbol{v}_1, \ldots, \boldsymbol{v}_n\}$ be the vertex set of a d-polytope conv V and let $\mathcal{G}$ be the Gale transform of V. We choose a nonlinear hyperplane $H := \{\boldsymbol{x} \in \mathbb{R}^{n-d-1} \mid \boldsymbol{a} \cdot \boldsymbol{x} = 1\}$ not parallel to any vector in $\mathcal{G}$; that is, $\boldsymbol{a} \cdot \mathcal{G}(\boldsymbol{v}_i) \neq 0$ for each $i \in [1 \ldots n]$. Centrally project each nonzero vector $\mathcal{G}(\boldsymbol{v}_i)$ of $\mathcal{G}$ onto a point $\mathcal{G}_a(\boldsymbol{v}_i)$ in H: the vector $\mathcal{G}(\boldsymbol{v}_i)$ is mapped to the intersection $\mathcal{G}_a(\boldsymbol{v}_i)$ of H and a line through $\boldsymbol{0}$ and $\mathcal{G}(\boldsymbol{v}_i)$. Notationally, we have that

$$\mathcal{G}_a(\boldsymbol{v}_i) := \frac{\mathcal{G}(\boldsymbol{v}_i)}{\boldsymbol{a} \cdot \mathcal{G}(\boldsymbol{v}_i)}, \text{ for each } i \in [1 \ldots n]. \tag{2.14.20}$$

The *affine Gale diagram* $\mathcal{G}_a$ of V is the point configuration

$$\mathcal{G}_a(V) = (\mathcal{G}_a(\boldsymbol{v}_1), \ldots, \mathcal{G}_a(\boldsymbol{v}_n)). \tag{2.14.21}$$

We need to distinguish between a point $\mathcal{G}_a(\boldsymbol{v}_i)$ obtained from a vector $\mathcal{G}(\boldsymbol{v}_i)$ directed towards H, one satisfying $\boldsymbol{a} \cdot \mathcal{G}(\boldsymbol{v}_i) > 0$, and a point obtained from vectors $\mathcal{G}(\boldsymbol{v}_i)$ directed away from H, one satisfying $\boldsymbol{a} \cdot \mathcal{G}(\boldsymbol{v}_i) < 0$. Call the former point *positive*, and if a differentiation is necessary, denote it $\mathcal{G}_a^+(\boldsymbol{v}_i)$; and call the latter point *negative* and denote it $\mathcal{G}_a^-(\boldsymbol{v}_i)$, if necessary. The set of positive points of $\mathcal{G}_a(V)$ will be denoted by $\mathcal{G}_a^+(V)$, while the set of negative points will be denoted by $\mathcal{G}_a^-(V)$. A zero vector $\mathcal{G}(\boldsymbol{v}_i)$ has no central projection onto H and so is a *special point*. To realise the reduction in dimension, we need an isomorphism σ between H and $\mathbb{R}^{n-d-2}$; since $\boldsymbol{a} = (a_1, \ldots, a_{n-d-1})^t$ is nonzero, there exists $a_i \neq 0$, which implies that the projection 'deleting' the i coordinate of each $\mathcal{G}_a(\boldsymbol{v}_i)$ does the trick. For the sake of simplicity, we also denote by $\mathcal{G}_a(\boldsymbol{v}_i)$ the point $\sigma(\mathcal{G}_a(\boldsymbol{v}_i))$ and denote by $\mathcal{G}_a(V)$ the set of these points in $\mathbb{R}^{n-d-2}$. Finally, to depict $\mathcal{G}_a(V)$ in $\mathbb{R}^{n-d-2}$, we draw the positive points with black dots, the negative points with white dots, and specify the number of special points by drawing a grey point and a number. We exemplify the construction of an affine Gale diagram for the Gale transform in Fig. 2.14.1(b).

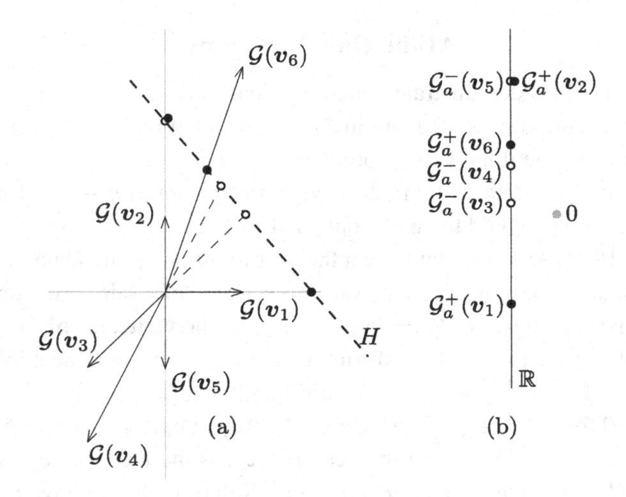

Figure 2.14.4 An affine Gale diagram from a Gale transform. (a) A realisation of the Gale transform $\mathcal{G}$ in Fig. 2.14.1(b), a nonlinear hyperplane H, and the central projection of the vectors in $\mathcal{G}$ onto H. (b) A realisation of an affine Gale diagram $\mathcal{G}_a$ for $\mathcal{G}$.

Example 2.14.22 Consider the Gale transform in Fig. 2.14.1(b):

$$\mathcal{G}(V) = \begin{pmatrix} \mathcal{G}(\boldsymbol{v}_1) & \mathcal{G}(\boldsymbol{v}_2) & \mathcal{G}(\boldsymbol{v}_3) & \mathcal{G}(\boldsymbol{v}_4) & \mathcal{G}(\boldsymbol{v}_5) & \mathcal{G}(\boldsymbol{v}_6) \\ 1 & 0 & -1 & -1 & 0 & 1 \\ 0 & 1 & -1 & -2 & -1 & 3 \end{pmatrix}.$$

We choose $\boldsymbol{a} := (1/2, 1/2)^t$, compute the affine Gale diagram

$$\mathcal{G}_a(\boldsymbol{v}_1) := \frac{\mathcal{G}(\boldsymbol{v}_1)}{\boldsymbol{a}\cdot\mathcal{G}(\boldsymbol{v}_1)} = \begin{pmatrix}2\\0\end{pmatrix},\ \mathcal{G}_a(\boldsymbol{v}_2) := \frac{\mathcal{G}(\boldsymbol{v}_2)}{\boldsymbol{a}\cdot\mathcal{G}(\boldsymbol{v}_2)} = \begin{pmatrix}0\\2\end{pmatrix},$$

$$\mathcal{G}_a(\boldsymbol{v}_3) := \frac{\mathcal{G}(\boldsymbol{v}_3)}{\boldsymbol{a}\cdot\mathcal{G}(\boldsymbol{v}_3)} = \begin{pmatrix}1\\1\end{pmatrix},\ \mathcal{G}_a(\boldsymbol{v}_4) := \frac{\mathcal{G}(\boldsymbol{v}_4)}{\boldsymbol{a}\cdot\mathcal{G}(\boldsymbol{v}_4)} = \begin{pmatrix}2/3\\4/3\end{pmatrix},$$

$$\mathcal{G}_a(\boldsymbol{v}_5) := \frac{\mathcal{G}(\boldsymbol{v}_5)}{\boldsymbol{a}\cdot\mathcal{G}(\boldsymbol{v}_5)} = \begin{pmatrix}0\\2\end{pmatrix},\ \mathcal{G}_a(\boldsymbol{v}_6) := \frac{\mathcal{G}(\boldsymbol{v}_6)}{\boldsymbol{a}\cdot\mathcal{G}(\boldsymbol{v}_6)} = \begin{pmatrix}1/2\\3/2\end{pmatrix},$$

and project the points onto the $x_2 = 0$ hyperplane to produce

$$\mathcal{G}_a(\boldsymbol{v}_1) := 2,\ \mathcal{G}_a(\boldsymbol{v}_2) := 0,\ \mathcal{G}_a(\boldsymbol{v}_3) := 1,\ \mathcal{G}_a(\boldsymbol{v}_4) := 2/3,\ \mathcal{G}_a(\boldsymbol{v}_5) := 0,$$
$$\mathcal{G}_a(\boldsymbol{v}_6) := 1/2.$$

There is no special point in $\mathcal{G}_a$. As expected, an affine Gale diagram of V consists of six points lying in $\mathbb{R}^1$; Figure 2.14.4(b) depicts $\mathcal{G}_a(V)$.

Affine Gale diagrams convey the same combinatorial information as Gale transforms and Gale diagrams, as we now demonstrate.

Theorem 2.14.23 *Let $V := \{\boldsymbol{v}_1, \ldots, \boldsymbol{v}_n\}$ be a point configuration in $\mathbb{R}^d$ and $\mathcal{G}_a(V) := (\mathcal{G}_a(\boldsymbol{v}_1), \ldots, \mathcal{G}_a(\boldsymbol{v}_n))$ an affine Gale diagram of V with no special points. Moreover, let $W \subset V$ and let $\mathcal{G}_a(W)$ be the restriction of $\mathcal{G}_a(V)$ to the points in W. The set* conv W *is a proper face of the d-polytope* conv V *if and only if*

$$\operatorname{rint}(\operatorname{conv}(\mathcal{G}_a^+(V)\setminus\mathcal{G}_a^+(W))) \cap \operatorname{rint}(\operatorname{conv}(\mathcal{G}_a^-(V)\setminus\mathcal{G}_a^-(W))) \neq \emptyset.$$

Proof It suffices that the condition of this theorem is equivalent to the condition of Theorem 2.14.8. Let $\mathcal{G}(V) := \{\mathcal{G}(\boldsymbol{v}_1), \ldots, \mathcal{G}(\boldsymbol{v}_n)\}$ be the Gale transform and $H := \{\boldsymbol{x} \in \mathbb{R}^{n-d-1} \mid \boldsymbol{a}\cdot\boldsymbol{x} = 1\}$ the hyperplane in $\mathbb{R}^{n-d-1}$ used in the computation of $\mathcal{G}_a$. Assume that $\mathcal{G}_a(V)$ lies in $H \cap \mathbb{R}^{n-d-1}$. Further, without loss of generality assume that $\mathcal{G}(V)\setminus\mathcal{G}(W) = \{\boldsymbol{v}_1, \ldots, \boldsymbol{v}_k\}$. We partition the set $I = \{1, \ldots, k\}$ into two subsets I^+ and I^- according to the sign of $\boldsymbol{a}\cdot\mathcal{G}(\boldsymbol{v}_i)$. That is,

$$I^+ := \{i \in I \mid \boldsymbol{a}\cdot\mathcal{G}(\boldsymbol{v}_i) > 0\} \text{ and } I^- := \{j \in I \mid \boldsymbol{a}\cdot\mathcal{G}(\boldsymbol{v}_j) < 0\}.$$

Suppose that $\boldsymbol{0}_{n-d-1} \in \operatorname{rint}(\operatorname{conv}(\mathcal{G}(V)\setminus\mathcal{G}(W)))$. By Theorem 1.7.6, this is equivalent to the existence of positive scalars $\alpha_1, \ldots, \alpha_k$ satisfying $\sum_{i=1}^k \alpha_i = 1$ and $\boldsymbol{0}_{n-d-1} = \sum_{i=1}^k \alpha_i\,\mathcal{G}(\boldsymbol{v}_i)$, which in turn implies that

$$\sum_{i\in I^+} \alpha_i\,\mathcal{G}(\boldsymbol{v}_i) = -\sum_{j\in I^-} \alpha_j\,\mathcal{G}(\boldsymbol{v}_j) \tag{2.14.23.1}$$

and that

$$\sum_{i\in I^+} \alpha_i \boldsymbol{a}\cdot\mathcal{G}(\boldsymbol{v}_i) = -\sum_{j\in I^-} \alpha_j \boldsymbol{a}\cdot\mathcal{G}(\boldsymbol{v}_j). \tag{2.14.23.2}$$

From (2.14.23.2), it is clear that both I^+ and I^- are nonempty. We show the existence of a point

$$\boldsymbol{x} \in \operatorname{rint}(\operatorname{conv}(\mathcal{G}_a^+(V)\setminus\mathcal{G}_a^+(W))) \cap \operatorname{rint}(\operatorname{conv}(\mathcal{G}_a^-(V)\setminus\mathcal{G}_a^-(W))),$$

which would settle the first direction.

Let $\alpha := \sum_{i\in I^+} \alpha_i \boldsymbol{a}\cdot\mathcal{G}(\boldsymbol{v}_i)$ and let

$$\boldsymbol{x} := \sum_{i\in I^+} \frac{\alpha_i \boldsymbol{a}\cdot\mathcal{G}(\boldsymbol{v}_i)}{\alpha}\mathcal{G}_a(\boldsymbol{v}_i) = \sum_{i\in I^+} \frac{\alpha_i \boldsymbol{a}\cdot\mathcal{G}(\boldsymbol{v}_i)}{\alpha}\,\frac{\mathcal{G}(\boldsymbol{v}_i)}{\boldsymbol{a}\cdot\mathcal{G}(\boldsymbol{v}_i)}. \tag{2.14.23.3}$$

Since $\boldsymbol{a}\cdot\mathcal{G}(\boldsymbol{v}_i) > 0$ for $i \in I^+$ and $\alpha_i > 0$, we find that $\alpha_i\boldsymbol{a}\cdot\mathcal{G}(\boldsymbol{v}_i)/\alpha > 0$ and $\alpha > 0$, which combined with $\sum_{i\in I^+}(\alpha_i\boldsymbol{a}\cdot\mathcal{G}(\boldsymbol{v}_i))/\alpha = 1$ give that

$$\boldsymbol{x} \in \operatorname{rint}(\operatorname{conv}(\mathcal{G}_a^+(V)\setminus\mathcal{G}_a^+(W))).$$

On the other hand, $\alpha = -\sum_{j\in I^-}\alpha_j \boldsymbol{a}\cdot\mathcal{G}(\boldsymbol{v}_j)$ by (2.14.23.2). So from (2.14.23.1) and (2.14.23.3), it follows that

$$\boldsymbol{x} = \sum_{j\in I^-}\frac{(-\alpha_j)\boldsymbol{a}\cdot\mathcal{G}(\boldsymbol{v}_j)}{\alpha}\mathcal{G}_a(\boldsymbol{v}_j) = \sum_{j\in I^-}\frac{(-\alpha_j)\boldsymbol{a}\cdot\mathcal{G}(\boldsymbol{v}_j)}{\alpha}\frac{\mathcal{G}(\boldsymbol{v}_j)}{\boldsymbol{a}\cdot\mathcal{G}(\boldsymbol{v}_j)},$$

which implies $\boldsymbol{x} \in \operatorname{rint}(\operatorname{conv}(\mathcal{G}_a^-(V)\setminus\mathcal{G}_a^-(W)))$, as desired.

For the other direction, suppose that there exists a point $\boldsymbol{x} \in H$ satisfying

$$\boldsymbol{x} \in \operatorname{rint}(\operatorname{conv}(\mathcal{G}_a^+(V)\setminus\mathcal{G}_a^+(W))) \cap \operatorname{rint}(\operatorname{conv}(\mathcal{G}_a^-(V)\setminus\mathcal{G}_a^-(W))).$$

Then

$$\boldsymbol{x} = \sum_{i\in I^+}\beta_i\,\mathcal{G}_a(\boldsymbol{v}_i),\ \text{with } \beta_i > 0 \text{ for } i \in I^+ \text{ and } \sum_{i\in I^+}\beta_i = 1,$$
$$\boldsymbol{x} = \sum_{j\in I^-}\beta_j\,\mathcal{G}_a(\boldsymbol{v}_j),\ \text{with } \beta_j > 0 \text{ for } j \in I^- \text{ and } \sum_{j\in I^-}\beta_j = 1,$$

which yields that

$$\begin{aligned}\boldsymbol{0}_{n-d-1} &= \sum_{i\in I^+}\beta_i\,\mathcal{G}_a(\boldsymbol{v}_i) + \sum_{j\in I^-}(-\beta_j)\,\mathcal{G}_a(\boldsymbol{v}_j)\\ &= \sum_{i\in I^+}\beta_i\frac{\mathcal{G}(\boldsymbol{v}_i)}{\boldsymbol{a}\cdot\mathcal{G}(\boldsymbol{v}_i)} + \sum_{j\in I^-}(-\beta_j)\frac{\mathcal{G}(\boldsymbol{v}_j)}{\boldsymbol{a}\cdot\mathcal{G}(\boldsymbol{v}_j)}.\end{aligned}$$

Since $\beta_i/(\boldsymbol{a}\cdot\mathcal{G}(\boldsymbol{v}_i)) > 0$ for $i \in I^+$ and $(-\beta_j)/(\boldsymbol{a}\cdot\mathcal{G}(\boldsymbol{v}_j)) > 0$ for $j \in I^-$, it follows that

$$\boldsymbol{0}_{n-d-1} \in \operatorname{rint}(\operatorname{conv}\{\mathcal{G}(\boldsymbol{v}_1), \dots, \mathcal{G}(\boldsymbol{v}_k)\}) = \operatorname{rint}(\operatorname{conv}(\mathcal{G}(V)\setminus\mathcal{G}(W))).$$

If σ is an isomorphism from H to $\mathbb{R}^{n-d-2}$, then it is clear that any positive combination involving $\boldsymbol{x}$ and points in $\mathcal{G}_a$ will remain valid under σ. This completes the proof of the theorem. □

We test Theorem 2.14.23 with the polytope in Fig. 2.14.1.

Example 2.14.24 Let $V := \{\boldsymbol{v}_1, \dots, \boldsymbol{v}_6\}$ be the vertex set of the 3-polytope P of Fig. 2.14.1 and let $\mathcal{G}_a(V) := (\mathcal{G}_a(\boldsymbol{v}_1), \dots, \mathcal{G}(\boldsymbol{v}_6))$ be an affine Gale diagram of V. We have that $\mathcal{G}_a^+(V) := \{\mathcal{G}_a^+(\boldsymbol{v}_1), \mathcal{G}_a^+(\boldsymbol{v}_2), \mathcal{G}_a^+(\boldsymbol{v}_6)\}$ and $\mathcal{G}_a^-(V) := \{\mathcal{G}_a^-(\boldsymbol{v}_3), \mathcal{G}_a^-(\boldsymbol{v}_4), \mathcal{G}_a^-(\boldsymbol{v}_5)\}$. Applying Theorem 2.14.23 to $\mathcal{G}_a(V)$ yields the following.

(i) The sets $W_1 := \{\boldsymbol{v}_1, \boldsymbol{v}_2\}$, $W_2 := \{\boldsymbol{v}_1, \boldsymbol{v}_4\}$, and $W_3 := \{\boldsymbol{v}_1, \boldsymbol{v}_3\}$ are edges of P, as

$$\operatorname{rint}(\operatorname{conv}(\mathcal{G}_a^+(V)\setminus\mathcal{G}_a^+(W_i))) \cap \operatorname{rint}(\operatorname{conv}(\mathcal{G}_a^-(V)\setminus\mathcal{G}_a^-(W_i))) \neq \emptyset$$

for $i = 1, 2, 3$. It also follows that the set $W_1' := \{v_1, v_6\}$ is not an edge of P since

$$\mathrm{rint}(\mathrm{conv}(\mathcal{G}_a^+(V)\backslash \mathcal{G}_a^+(W_1'))) \cap \mathrm{rint}(\mathrm{conv}(\mathcal{G}_a^-(V)\backslash \mathcal{G}_a^-(W_1'))) = \emptyset.$$

However, observe that

$$\mathrm{conv}(\mathcal{G}_a^+(V)\backslash \mathcal{G}_a^+(W_1')) \cap \mathrm{conv}(\mathcal{G}_a^-(V)\backslash \mathcal{G}_a^-(W_1')) \neq \emptyset.$$

(ii) The set $W_6 := \{v_1, v_3, v_4, v_6\}$ is a 2-face of P, as

$$\mathrm{rint}(\mathrm{conv}(\mathcal{G}_a^+(V)\backslash \mathcal{G}_a^+(W_6))) \cap \mathrm{rint}(\mathrm{conv}(\mathcal{G}_a^-(V)\backslash \mathcal{G}_a^-(W_6))) \neq \emptyset.$$

We now give the translation of Theorem 2.14.14 to affine Gale diagrams

Theorem 2.14.25[17] *A sequence $\mathcal{G}_a(V) := (\mathcal{G}_a(v_1), \dots, \mathcal{G}_a(v_n))$ of points in $\mathbb{R}^{n-d-2}$ is an affine Gale diagram of a d-polytope (other than a pyramid) with vertex set V if and only if, for any hyperplane H spanned by some of the points in $\mathcal{G}_a(V)$ and for each open halfspace determined by H, the number of positive points on this halfspace plus the number of negative points on the other open halfspace is at least two.*

Verify this theorem with the affine Gale diagram in Fig. 2.14.4.

Theorems 2.14.23 and 2.14.25 are stated for a polytope P with no special points. This is not a limitation, since adding k special points amounts to taking a k-fold pyramid Q over P, and the combinatorics of Q is determined by that of P and the number k.

2.15 Problems

2.15.1 Let P and Q be polytopes in $\mathbb{R}^d$, let $\alpha \in \mathbb{R}$, and let K be an affine subspace of $\mathbb{R}^d$. Prove that $P + Q$, $P \cap Q$, $P \cap K$, and αP are all polytopes in $\mathbb{R}^d$.

2.15.2 Let P be a polytope and let F be a facet of it. Suppose there are exactly two vertices outside F. Prove that these vertices must be adjacent.

2.15.3 Let P be a d-polytope, F an h-face of P, and F_0 a proper k-face of F with $-1 \leqslant k < h \leqslant d$. Prove that there exists a $(d - h + k)$-face F_1 of P such that $F_0 = F_1 \cap F$ and $P = \mathrm{conv}(F \cup F_1)$.

[17] A proof is available in Ziegler (1995, thm. 6.1.9).

2.15.4 Let P be a d-polytope, F_j a j-face of P, and F_i an i-face of P such that $-1 \leqslant i < j \leqslant d$. Prove that the lattice $\mathcal{L}(F_j/F_i)$ is a sublattice of the lattice $\mathcal{L}(P)$.

2.15.5 Prove that a d-polytope is a d-simplex if and only if it has $d+1$ facets.

2.15.6 (Pyramids) Let P be a d-polytope. Prove the following.

(i) If $P = F * \boldsymbol{v}$ is a pyramid, then F is a pyramid with apex $\boldsymbol{w} \neq \boldsymbol{v}$ if and only if P is a pyramid with apex $\boldsymbol{w}$.
(ii) A d-polytope is a pyramid over r distinct facets if and only if it is an r-fold pyramid.

2.15.7 Let P and P' be two d-polytopes with a facet F of P projectively isomorphic to a facet F' of P'. Prove that there exists a projective transformation ζ such that $\operatorname{conv}(P \cup \zeta(P'))$ is a realisation of $P\#_F P'$.

2.15.8 Prove that a polytope P is k-simplicial if and only if P^* is k-simple.

2.15.9 (McMullen, 1976) Prove that the connected sum of two polytopes along simplex facets is always possible. This amounts to proving that a simplex is projectively isomorphic to any other realisation of a simplex.

2.15.10 Prove that if we stack over a facet F of a polytope P, then the conjugate vertex of F in the dual polytope P^* of P gets truncated, and vice versa.

2.15.11 Prove that if we perform the wedge of a polytope P at a facet F of it, then we are performing the dual wedge of the dual polytope P^* at the conjugate vertex of F in P^*, and vice versa.

2.15.12 (Ewald and Shephard, 1974) A polytope can be made simple by truncating the vertices, then the original edges, and so on up to the ridges of the polytope.

2.15.13 Consider the simplicial $(d-r)$-polytope $F_r := T(1) \oplus T(d-r-1)$, and the r-fold pyramids $\operatorname{pyr}_r(F_r)$, for $r \in [0 \ldots d-2]$. Prove that the f-vectors of the d-polytopes $T(d)$ and $\operatorname{pyr}_r(F_r)$ (for $r \in [0 \ldots d-2]$) form an affinely independent set in $\mathbb{R}^d$.

2.15.14 Prove that the f-vectors of the following sets of simplicial d-polytopes form a set of affinely independent vectors in $\mathbb{R}^d$.

(i) The d-simplex and the $\lfloor d/2 \rfloor$ simplicial d-polytopes $T(r) \oplus T(s)$ with $r, s \geqslant 1$ and $r + s = d$.
(ii) The cyclic d-polytopes $C(n,d), C(n+1,d), \ldots, C(n + \lfloor d/2 \rfloor, d)$ with $n \geqslant d+1$.

2.15.15 (Cuboids; Grünbaum, 2003, sec. 4.6) The *cuboid* $Q(d,0)$ is combinatorially isomorphic to the d-cube $Q(d)$. For $r \in [0 \ldots d]$, the *cuboid* $Q(d,r)$ is obtained by 'pasting together' two cuboids $Q(d,k-1)$ along a common cuboid $Q(d-1,k-1)$. This operation requires that the two cuboids $Q(d,k-1)$ are deformed beforehand. Prove the following.

(i) $f_k(Q(d,0)) = 2^{d-1}\binom{d}{k}$.
(ii) $f_k(Q(d,r)) = 2f_k(Q(d,r-1)) - f_k(Q(d-,r-1))$, for $k \in [0 \ldots d - 1 - r]$.
(iii) For $k, r \geqslant 0$ and $k \in [0 \ldots d-1-r]$, it holds that

$$f_k(Q(d,r)) = \sum_{i=0}^{r} \binom{r}{i}\binom{d-i}{k} 2^{d+r-k-2i}.$$

2.15.16 Prove that the cubical d-polytopes $Q(d,0), \ldots, Q(d, \lfloor d/2 \rfloor)$ defined in Problem 2.15.15 form a set of affinely independent vectors in $\mathbb{R}^d$.

2.15.17 Let P be a d-polytope in $\mathbb{R}^d$ and let $\mathcal{G}$ be a Gale diagram of P. Prove that if $\boldsymbol{v}$ is a vertex of P, then $\mathcal{G} \setminus \{\mathcal{G}(\boldsymbol{v})\}$ is a Gale diagram, but not necessarily the Gale transform, of the vertex figure $P/\boldsymbol{v}$ of P at $\boldsymbol{v}$.

2.15.18 Produce realisations of polytopes combinatorially isomorphic to the polytopes whose Gale diagrams appear in Fig. 2.15.1.

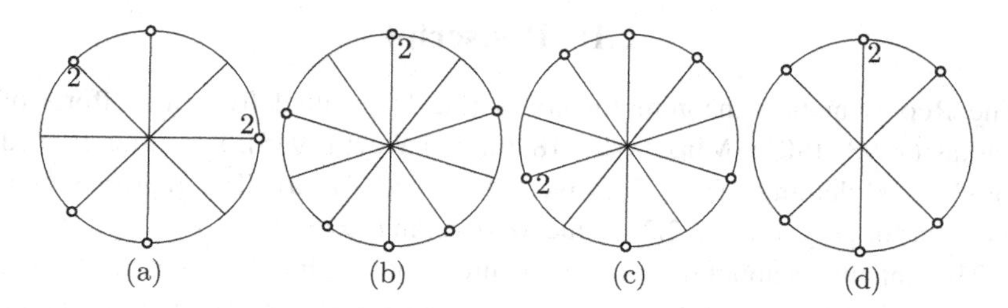

Figure 2.15.1 Gale diagrams of four 4-polytopes with seven vertices; the labels state the number of vectors in the corresponding position.

2.15.19 (Projectively unique polytopes; McMullen, 1976) A polytope P is *projectively unique* if every polytope combinatorially isomorphic to P is projectively isomorphic to P. Prove the following.

(i) The d-simplex is projectively unique.
(ii) Every d-polytope with $d + 2$ vertices is projectively unique.
(iii) The dual polytope of a projectively unique polytope is also projectively unique.
(iv) The join of two polytopes P_1 and P_2 is projectively unique if and only if both P_1 and P_2 are projectively unique.

2.15.20* (Perfect shellings; Ziegler, 1995, ex. 8.9) Let $F_1, \ldots, F_s$ be a shelling of a polytope. For each $i \in [1 \ldots s]$, order the $(d-2)$-faces of F_i as they appear in this list: $F_1 \cap F_i, F_2 \cap F_i, F_{i-1} \cap F_i, F_{i+1} \cap F_i, \ldots, F_s \cap F_i$. If, for each $i \in [1 \ldots s]$, this ordering of the $(d-2)$-faces of F_i is a shelling of F_i, then we say that the shelling is *perfect*.

(i) Does every polytope have a perfect shelling?
(ii) Does every simple polytope have a perfect shelling?
(iii) Does every cubical polytope have a perfect shelling?

Ziegler attributes the first question to Gil Kalai. It is known that simplicial polytopes, duals of cyclic polytopes, d-cubes, and 3-polytopes all have perfect shellings.

2.16 Postscript

The Representation theorem for cones (2.2.1) resulted from the efforts of Farkas (1898, 1901), Minkowski (1896), and Weyl (1935). It is often proved via Farkas' lemma, as in Schrijver (1986, cor. 7.1a). The representation theorem for polyhedra (2.2.2) is due to Motzkin (1936).

The statements about the facial structure of polyhedra (Theorem 2.3.1) and, more generally, those in Section 2.3, are standard and scattered across many texts on convexity; for instance see Webster (1994, sec. 3.2), Brøndsted (1983, sec. 2.8), and Lauritzen (2013, ch. 4). Our proofs, while standard, were inspired by the presentation in Webster (1994, sec. 3.2) and Brøndsted (1983, sec. 2.8). The proof of Proposition 2.3.2 follows the same ideas as that of the second part of the proof of Lauritzen (2013, prop. 4.3). The proof for the sufficiency part of the characterisation of faces of polytopes in Theorem 2.3.7 is inspired by ideas from the proof of Webster (1994, thm. 3.1.4).

The section on preprocessing has the spirit of Ziegler (1995, sec. 2.6), while trying to have the appeal and concreteness of Yaglom (1973); consult Ziegler (1995, sec. 2.6) for the formulas that we did not provide. The presentation of the embedding of affine spaces into projective spaces follow that in Berger (2009, ch. 5) and Gallier (2011, sec. 5.6).

Sections 2.7 on face figures and 2.8 on simple and simplicial polytopes are based on the excellent accounts of Brøndsted (1983, sec. 2.11–2.12). The proof by duality of Theorem 2.11.9 is based on the proof by Brøndsted (1983, thm. 2.11.10), while the proof that, for $d \geqslant 3$, a d-simplex is the only simple and simplicial polytope (Theorem 2.8.8) is inspired by that of Brøndsted (1983, thm. 2.11.19).

The material of Section 2.9 on cyclic and neighbourly polytopes is fairly standard. Our presentation is similar to those in Grünbaum (2003, sec. 4.7,7.1), Webster (1994, sec. 3.4), and Brøndsted (1983, sec. 2.13).

The proof of the inductive construction of polytopes offered in Theorem 2.10.1 is based on the original proof of Grünbaum (1963, thm. 5.2.1). As we stated after the proof of Theorem 2.10.1, the original proof of Grünbaum (1963, thm. 5.2.1) is slightly incorrect. The same mistake is carried over in the proof of McMullen and Shephard (1971, thm. 2.22). This mistake was first noted by M. A. Perles, as acknowledged by Altshuler and Shemer (1984). This inductive construction is often described as the *beneath-beyond* algorithm and plays an important role in the computation of convex hulls in computational geometry; see, for instance, Edelsbrunner (2012, sec. 8.4) and Preparata and Shamos (1985, sec. 3.4.2).

In many settings, the operations of pulling and pushing vertices allow us to focus on simplicial polytopes. This is the case with the upper bound theorem of McMullen (1970), which states that the cyclic d-polytope on n vertices has the largest number of faces among the d-polytopes with that number of vertices. The process of pulling vertices first appeared in Eggleston et al. (1964, sec. 2), while the process of pushing vertices was first announced in Klee (1964b, sec. 2). Our proof of Theorem 2.10.5 is based on that of Matoušek (2002, lem. 5.5.4) and Santos (2012, lem. 2.2).

The presentation of polytopal complexes, subdivisions, and Schlegel diagrams is similar to that of Ziegler (1995, sec. 5.1,5.2). Schlegel diagrams first appeared in Schlegel (1883), but Sommerville (1958) seems to be the first to exploit their use on polytopes. Lee (1991) showed that the regularity of a subdivision of a polytope conv V can be tested via the Gale transform of the set V. We did not go into algorithmic aspects of subdivisions; they are well covered in De Loera et al. (2010, sec. 8.2). We just remark that checking whether a subdivision is regular is equivalent to the feasibility of a linear program (De Loera et al., 2010, sec. 8.2).

Proposition 2.12.12 states that stars of vertices of shellable complexes are shellable. This is due to Courdurier (2006), and our proof follows his. It is also the case that links of vertices in shellable polytopal complexes are shellable (Courdurier, 2006); this generalises Proposition 2.12.11, which gives that links of vertices in boundary complexes of polytopes are shellable.

Our proof of Euler–Poincaré–Schläfli's equation (Theorem 2.12.17) is inspired by that of Gruber (2007, thm. 15.5). The proof of Dehn–Sommerville's equations is somehow standard: it relies on the generalisation of Euler–Poincaré–Schläfli's equation stated in Theorem 2.13.1; see, for instance, Webster (1994, thm. 3.5.4) or McMullen and Shephard (1971, thm. 2.4.19). The derivation of the h-vector of simplicial polytopes from shellings is explained in more detail in Ziegler (1995, sec. 8.3); our description aims to summarise his. Dehn–Sommerville's equations for cubical polytopes seem to have first appeared in Grünbaum (2003, thm. 9.4.1).

The section on Gale transforms (Section 2.14) is based on the presentations in McMullen and Shephard (1971, ch. 3) and Webster (1994, sec. 3.6).

3

Polytopal Graphs

The graph of a polytope is formed by its vertices and edges. This graph as an abstract is often viewed graph and analysed using methods from graph theory. A *polytopal graph* is simply a graph of a polytope. Appendix C reviews the relevant graph-theoretical prerequisites.

Graphs of three-dimensional polytopes are planar; therefore, we review the basic topological background necessary to study graphs embedded in a topological space, with an emphasis on plane-embedded graphs. We then explore properties of polytopal graphs.

We analyse acyclic orientations of graphs of polytopes in Section 3.5; these orientations are closely related to the shelling orders of the corresponding dual polytopes. This relationship has important applications in the reconstruction of polytopes from their graphs (Chapter 5). We also examine convex realisations of 3-connected planar graphs, including a proof of a celebrated theorem of Tutte (1963). A nontrivial graph G is defined as r*-connected*, for $r \geqslant 0$, if it contains more than r vertices and no two vertices of G can be separated by fewer than r other vertices. Section 3.8 covers Steinitz's characterisation of graphs of 3-polytopes: they are precisely the 3-connected planar graphs.

The graph of a 3-polytope contains a subdivision of the graph of the 3-simplex, namely K^4. Section 3.9 extends this result to every dimension, showing that the graph of a d-polytope contains a subdivision of K^{d+1}. Since K^5 is the 1-skeleton of a 4-simplex, the nonplanarity of K^5 is a special case of a theorem of Flores (1934) and Van Kampen (1932), which states that the d-skeleton of the $(2d + 2)$-simplex cannot be embedded in $\mathbb{R}^{2d}$; these topics are discussed in Section 3.10.

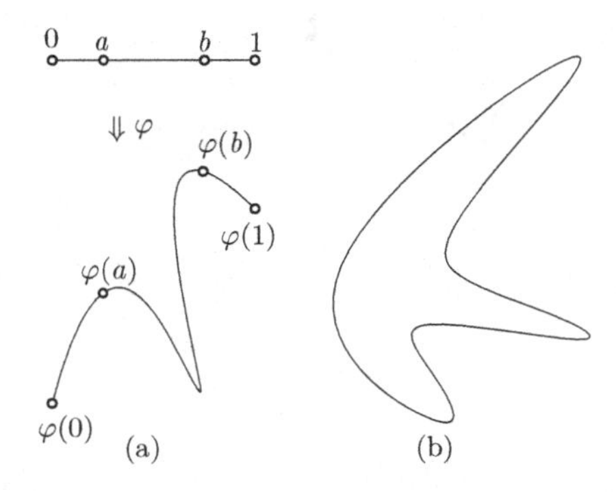

Figure 3.1.1 Curves in $\mathbb{R}^2$. (a) A simple curve in $\mathbb{R}^2$. (b) A simple closed curve in $\mathbb{R}^2$.

3.1 Realisations of Graphs

Graphs can be embedded, or realised, in a topological space. Let X be a topological space. A *curve* in X is the image of a continuous function φ that maps the interval $[0,1]$ into X. A curve is *simple* if it does not intersect itself; that is, if φ is injective. It is *closed* if $\varphi(0) = \varphi(1)$. And a *simple closed curve* is a closed curve that does not intersect itself; that is, it is a space homeomorphic to the one-dimensional unit sphere. See Fig. 3.1.1. The image $\varphi([a,b])$ for $0 \leqslant a < b \leqslant 1$ is a *segment* of the curve from $\varphi(a)$ to $\varphi(b)$.

A graph G is *embedded* in a topological space X if every vertex of G is represented by a distinct point of X and every edge is represented by a distinct simple curve in X connecting the endvertices of the edge. In this representation, the endpoints of a curve represent the endvertices of the corresponding edge, and the interiors of the curves are mutually disjoint and do not meet the points representing vertices.

An *embedding* of an abstract graph G in a topological space X is an isomorphism from G to a graph G' embedded in X. The graph G' is a *realisation* of G in X. If there is an embedding of G into X, then we say that G *can be embedded* or *realised* in X. It turns out that every realisation of a graph is a subset of $\mathbb{R}^2$ or $\mathbb{R}^3$. In fact, every graph can be realised in $\mathbb{R}^3$ in such a way that every edge is represented by a line segment (Corollary 3.10.3).

A *geometric graph* is a realisation in $\mathbb{R}^d$ of a graph G where the edges in G correspond to line segments, and no three vertices are collinear. Accordingly, for an edge $e := \boldsymbol{x}\boldsymbol{y}$ of a geometric graph, we often write $e = [\boldsymbol{x}, \boldsymbol{y}]$ to highlight its geometric nature. The graph of a polytope in $\mathbb{R}^d$ is a geometric graph in $\mathbb{R}^d$.

A topological space X is *arcwise connected* if every two distinct points in X are joined by a simple curve in X. As an example of arcwise connected spaces,

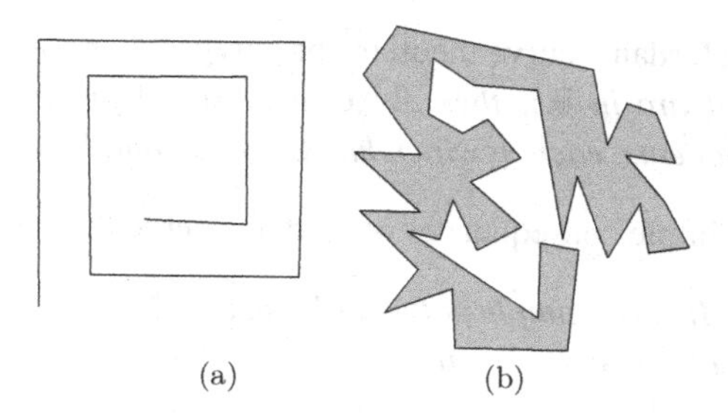

Figure 3.1.2 Polygonal arcs in $\mathbb{R}^2$. (a) A simple polygonal arc in $\mathbb{R}^2$. (b) A simple, closed polygonal arc in $\mathbb{R}^2$.

we have $\mathbb{R}^d$ for each $d \geqslant 1$. Arcwise connectivity can be naturally extended to subsets of X. A maximal arcwise-connected subset of the topological space X is a *component* of X, and a *face* of a subset C of X is an arcwise-connected component of $X \backslash C$.

In the following we focus on graphs embedded in the plane, namely *plane graphs*. A graph is *planar* if it can be realised in $\mathbb{R}^2$ as a plane graph; this plane graph is a *planar realisation* of the planar graph.

A graph G embedded in $\mathbb{R}^2$ partitions $\mathbb{R}^2 \backslash G$ into a number of faces of G. The *boundary* of a face F is the boundary of the arcwise-connected component F in the usual topological sense. We may regard the boundary of a face of a graph as a subgraph of the graph. In this setting, a face is said to be *incident* with the vertices and edges of its boundary. The plane graph G is bounded in the sense that it lies in a sufficiently large ball B of $\mathbb{R}^2$, and so it has exactly one unbounded face, namely the face that contains $\mathbb{R}^2 \backslash B$. We call the unbounded face of G the *outer face* of G, and call the other faces the *inner faces* of G.

The set $\mathbb{R}^2$ is arcwise connected, and so every two points of $\mathbb{R}^2$ can be connected by a simple curve. It is also true that that this connection can be made by a polygonal arc; this is what is needed to show that planar graphs can be realised in $\mathbb{R}^2$ with edges represented by simple polygonal arcs rather than simple curves. A *polygonal arc* is a curve in $\mathbb{R}^2$ that is the union of finitely many line segments in $\mathbb{R}^2$ (Fig. 3.1.2(a)). Similar to general curves, we have simple polygonal arcs and closed polygonal arcs (Fig. 3.1.2(b)).

Proposition 3.1.1 [1] *Every planar graph admits a planar realisation such that all edges are simple polygonal arcs.*

To work with plane graphs with edges as simple polygonal arcs (Proposition 3.1.1), the main topological result is a special case of the curve theorem of Jordan (1882).

[1] A proof is available in Mohar and Thomassen (2001, lem. 2.1.1).

Theorem 3.1.2 (Jordan's curve theorem for polygonal arcs)[2] *If C is a simple, closed polygonal arc in $\mathbb{R}^2$, then $\mathbb{R}^2 \backslash C$ consists of precisely two arcwise-connected components, each of which has C as its boundary.*

We give two simple consequences of Jordan's curve theorem.

Corollary 3.1.3 *If L is a simple polygonal arc, then $\mathbb{R}^2 \backslash L$ consists of precisely one arcwise-connected component.*

Theorem 3.1.4[3] *The graphs K^5 and $K^{3,3}$ are both nonplanar.*

While planar graphs can be represented in $\mathbb{R}^2$ with edges as simple polygonal arcs (Proposition 3.1.1), henceforth we will realise them with edges represented by line segments, which is always possible by virtue of a theorem of Fáry (1948).

Theorem 3.1.5 (Fáry, 1948) *Every planar graph admits a planar realisation such that all edges are line segments.*

Planar Graphs in the Sphere

Here, we show that a graph can be embedded in the plane if and only if it can be embedded in the two-dimensional sphere. For this, we use the stereographic projection.

Consider the d-dimensional unit sphere $\mathbb{S}^d$ in $\mathbb{R}^{d+1}$, the point $\boldsymbol{e}_{d+1} = (0, \ldots, 1)^t$ of $\mathbb{S}^d$, and a hyperplane

$$H := \left\{ \boldsymbol{x} = (x_1, \ldots, x_{d+1})^t \in \mathbb{R}^{d+1} \,\middle|\, x_{d+1} = 0 \right\}$$

in $\mathbb{R}^{d+1}$. The *stereographic projection* $\pi_e \colon \mathbb{S}^d \backslash \{\boldsymbol{e}_{d+1}\} \to H$ maps a point $\boldsymbol{x} = (x_1, \ldots, x_{d+1})^t$ of $\mathbb{S}^d \backslash \{\boldsymbol{e}_{d+1}\}$ to the point $\boldsymbol{y}$ of H where the line between $\boldsymbol{x}$ and $\boldsymbol{e}_{d+1}$ meets H. See Fig. 3.1.3. We can readily produce expressions for $\boldsymbol{x}$ and $\boldsymbol{y}$ in terms of each other:

$$\boldsymbol{y} = \frac{1}{1 - x_{d+1}} \boldsymbol{x} - \frac{x_{d+1}}{1 - x_{d+1}} \boldsymbol{e}_{d+1}, \; \boldsymbol{x} = \frac{\|\boldsymbol{y}\|^2 - 1}{\|\boldsymbol{y}\|^2 + 1} \boldsymbol{e}_{d+1} + \frac{2}{\|\boldsymbol{y}\|^2 + 1} \boldsymbol{y}. \quad (3.1.6)$$

The stereographic projection π_e is a homeomorphism from $\mathbb{S}^d \backslash \{\boldsymbol{e}_{d+1}\}$ onto H. And since H is isomorphic to $\mathbb{R}^d$, we have the following.

[2] A proof is available in Mohar and Thomassen (2001, lem. 2.1.3).
[3] A proof is available in Mohar and Thomassen (2001, lem. 2.1.7).

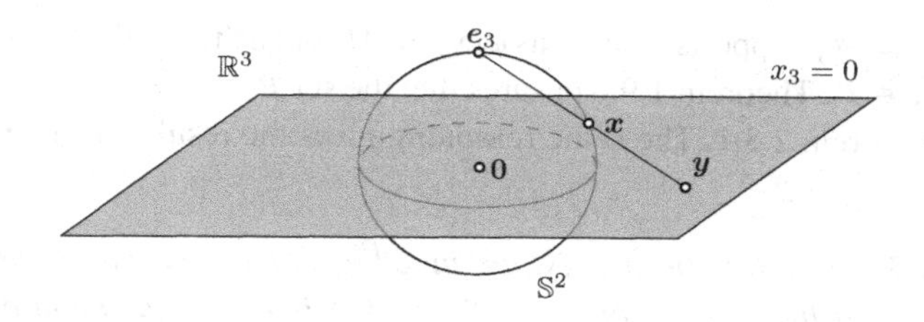

Figure 3.1.3 Stereographic projections in $\mathbb{R}^3$.

Proposition 3.1.7 *The unit sphere $\mathbb{S}^d$ with a point removed is homeomorphic to $\mathbb{R}^d$.*

We can now prove the promised result.

Theorem 3.1.8 *A graph can be embedded in the plane if and only if it can be embedded in the two-dimensional sphere.*

Proof Suppose that a graph G is embedded in the two-dimensional unit sphere. Without loss of generality, further suppose that the point $\boldsymbol{e}_3$ is not in G. Then $\mathbb{S}^2 \backslash \{\boldsymbol{e}_3\}$ is homeomorphic to the plane (Proposition 3.1.7). It also follows that the image of G under the stereographic projection is a planar realisation of G.

Suppose that G is a plane graph. Then the inverse of the stereographic projection will give a realisation of G in the sphere. □

The stereographic projection and Proposition 3.1.7 can be used to obtain planar realisations of planar graphs with some desired properties (Problem 3.11.1).

3.2 Basic Properties of Polytopal Graphs

The first two results are about the extreme values of linear functionals over a polytope; these are often seen in the context of linear programming.

Theorem 3.2.1 *Let P be a nonempty polyhedron in $\mathbb{R}^d$, and let φ be a linear functional in $\mathbb{R}^d$. Then the maximum of φ over P is attained at a face of P. Similarly, the minimum of φ over P is attained at a face of P.*

Proof Let F be the set of points of P that maximises φ over P, and say that $\varphi(z) = M$ for every $z \in F$; see Proposition B.3.1. The hyperplane $H := \{\boldsymbol{x} \in$

$\mathbb{R}^d \mid \varphi(x) = M\}$ supports P at F, as $\varphi(x) \leqslant M$ for each $x \in P$ and $\varphi(x) = M$ for some $x \in P$. Theorem 1.9.6 ensures that the set $P \cap H = F$ is a face of P; see also Theorem 2.3.1. The same reasoning gives the result for the minimum of φ. □

Theorem 3.2.2 *Let P be a polytope in $\mathbb{R}^d$ and let v be any vertex in P. Suppose φ is a linear functional in $\mathbb{R}^d$. If $\varphi(v)$ is not a maximum of φ over P then there is a neighbour u of v with $\varphi(u) > \varphi(v)$. Similarly, if $\varphi(v)$ is not a minimum of φ over P then there is a neighbour u of v with $\varphi(u) < \varphi(v)$.*

Proof We prove the contrapositive of the assertion related to the maximum of φ. Let $u_1, \ldots, u_n$ be the neighbours of v in P. Suppose that, for every neighbour u_i of v ($i \in [1 \ldots n]$), we have that $\varphi(u_i) \leqslant \varphi(v)$.

Let w be any point in P. The polytope P is contained in the affine cone based at v and spanned by the neighbours of v (Theorem 2.7.6), and so there are nonnegative scalars $\alpha_1, \ldots, \alpha_n$ such that

$$w = v + \sum_{i=1}^{n} \alpha_i (u_i - v).$$

The linearity of φ now yields that

$$\varphi(w) = \varphi(v) + \sum_{i=1}^{n} \alpha_i (\varphi(u_i) - \varphi(v)).$$

The assumption $\varphi(u_i) \leqslant \varphi(v)$ for each $i \in [1 \ldots n]$ then gives that

$$\alpha_i (\varphi(u_i) - \varphi(v)) \leqslant 0 \text{ for each } i \in [1 \ldots n],$$

which produces the desired conclusion $\varphi(w) \leqslant \varphi(v)$. □

For a linear functional φ in $\mathbb{R}^d$ and a graph Z whose vertices are points in $\mathbb{R}^d$, we say that a path $x_1 \ldots x_n$ in Z is *φ-increasing* if, for each $i \in [1 \ldots n-1]$, we have that $\varphi(x_i) < \varphi(x_{i+1})$, and it is *$\varphi$-decreasing* if, for each $i \in [1 \ldots n-1]$, we have that $\varphi(x_i) > \varphi(x_{i+1})$. Similarly, we define *$\varphi$-nonincreasing paths* and *φ-nondecreasing paths*. Repeated applications of Theorem 3.2.2 yield the following.

Corollary 3.2.3 *Let P be a polytope in $\mathbb{R}^d$ and let v be any vertex in P. Suppose φ is any linear functional in $\mathbb{R}^d$. Then, there is an φ-increasing path in $G(P)$ from v to some vertex of P where φ achieves its maximum over P and there is an φ-decreasing path in $G(P)$ from v to some vertex of P where φ achieves its minimum over P.*

We proceed with a corollary of Theorem 3.2.2 that follows from the continuity of a linear functional on $\mathbb{R}^d$ (Problem 1.12.8).

Corollary 3.2.4 *Let $\alpha \in \mathbb{R}$ and let P be a polytope in $\mathbb{R}^d$. Suppose that φ is a linear functional in $\mathbb{R}^d$. Then there is a point $\boldsymbol{x}$ in P satisfying $\varphi(\boldsymbol{x}) = \alpha$ if and only if*

$$\min\{\varphi(\boldsymbol{v}) |\ \boldsymbol{v} \in \mathcal{V}(P)\} \leqslant \alpha \leqslant \max\{\varphi(\boldsymbol{v}) |\ \boldsymbol{v} \in \mathcal{V}(P)\}.$$

The next result can be found in Sallee (1967, sec. 3) and Balinski (1961). For a graph G and subsets X, Y of the vertex set $\mathcal{V}(G)$ of G, we call a path $L := x_1 \ldots x_n$ in G an $X - Y$ *path* if $\mathcal{V}(L) \cap X = \{x_1\}$ and $\mathcal{V}(L) \cap Y = \{x_n\}$. We write $x - Y$ path instead of $\{x\} - Y$ path and, likewise, write $X - y$ path instead of $X - \{y\}$.

Lemma 3.2.5 *Let P be a d-polytope in $\mathbb{R}^d$, and let $\varphi\colon \mathbb{R}^d \to \mathbb{R}$ be a linear functional in general position with respect to P. Suppose that $\varphi(\boldsymbol{x}) > 0$ for some $\boldsymbol{x} \in P$. If $\boldsymbol{u}$ and $\boldsymbol{v}$ are vertices of P with $\varphi(\boldsymbol{u}) \geqslant 0$ and $\varphi(\boldsymbol{v}) \geqslant 0$, then there exists a $\boldsymbol{u} - \boldsymbol{v}$ path $\boldsymbol{x}_0\boldsymbol{x}_1 \ldots \boldsymbol{x}_n$ with $\boldsymbol{x}_0 = \boldsymbol{u}$ and $\boldsymbol{x}_n = \boldsymbol{v}$ such that $\varphi(\boldsymbol{x}_i) > 0$, for each $i \in [1 \ldots n-1]$.*

Proof Starting at the vertex $\boldsymbol{u}$, we use Corollary 3.2.3 to produce a φ-increasing directed path L_u from $\boldsymbol{u}$ to some vertex $\boldsymbol{u}_m$ of P for which $\varphi(\boldsymbol{u}_m)$ is the maximum of φ on P. Similarly, there is a φ-increasing directed path L_v from the vertex $\boldsymbol{v}$ to some vertex $\boldsymbol{v}_m$ of P for which $\varphi(\boldsymbol{v}_m)$ is the maximum of φ on P. Since φ is in general position with respect to P, we must have that $\boldsymbol{u}_m = \boldsymbol{v}_m$. The desired $\boldsymbol{u} - \boldsymbol{v}$ path is obtained by concatenating the paths L_u and L_v. □

The first consequence of Lemma 3.2.5 is that graphs of polytopes are connected. A stronger result is even possible. If G is a graph and $X \subseteq \mathcal{V}(G)$, then $G[X]$ denotes the subgraph of G induced by the set X and $G - X$ denotes the subgraph $G[\mathcal{V}(G) \setminus X]$ of G; we prefer $G - G'$ to $G - \mathcal{V}(G')$, in case that G' is a subgraph of G. In case $X \subseteq V$ and $X = \{x\}$, we prefer $G - x$ to $G - \{x\}$.

Theorem 3.2.6 (Connectivity of polytopes) *Let P be a polytope, F a proper face of P, and $X \subseteq \mathcal{V}(F)$. Then the subgraph $G(P) - X$ is connected. In particular, $G(P)$ is connected.*

Proof Let φ be a linear functional that vanishes on the face F. Further suppose that $\varphi(\boldsymbol{x}) > 0$ for some $\boldsymbol{x} \in P$. The function φ certainly exists as it can be taken to define a hyperplane supporting P at F. Perturb φ slightly so that it becomes in general position with respect to P.

Let $\boldsymbol{u}$ and $\boldsymbol{v}$ be vertices in $\mathcal{V}(P)\backslash X$. Then $\varphi(\boldsymbol{u}) \geqslant 0$ and $\varphi(\boldsymbol{v}) \geqslant 0$, and Lemma 3.2.5 ensures the existence of a $\boldsymbol{u}-\boldsymbol{v}$ path in P whose interior vertices are not in F. This establishes that $G(P) - X$ is connected. Taking $X = \emptyset$ gives that $G(P)$ is connected. □

Lemma 3.2.5 can accomodate linear functionals that are not in general position with respect to polytopes once we have the connectivity of the graph of a polytope presented in Theorem 3.2.6.

Lemma 3.2.7 *Let P be a d-polytope in $\mathbb{R}^d$, and let $\varphi\colon \mathbb{R}^d \to \mathbb{R}$ be a linear functional. Suppose that $\varphi(\boldsymbol{x}) > 0$ for some $\boldsymbol{x} \in P$. If $\boldsymbol{u}$ and $\boldsymbol{v}$ are vertices of P with $\varphi(\boldsymbol{u}) \geqslant 0$ and $\varphi(\boldsymbol{v}) \geqslant 0$, then there exists a $\boldsymbol{u}-\boldsymbol{v}$ path $\boldsymbol{x}_0\boldsymbol{x}_1 \ldots \boldsymbol{x}_n$ with $\boldsymbol{x}_0 = \boldsymbol{u}$ and $\boldsymbol{x}_n = \boldsymbol{v}$ such that $\varphi(\boldsymbol{x}_i) > 0$, for each $i \in [1 \ldots n-1]$.*

Graphs of 3-Polytopes

We now turn our attention to graphs of 3-polytopes.

The *radial projection* π_r with centre at the origin maps a point $\boldsymbol{x}$ in $\mathbb{R}^d\backslash\{\boldsymbol{0}\}$ onto the point $\boldsymbol{x}/\|\boldsymbol{x}\|$ of the $(d-1)$-dimensional unit sphere $\mathbb{S}^{d-1}$; that is

$$\pi_r(\boldsymbol{x}) = \frac{\boldsymbol{x}}{\|\boldsymbol{x}\|}. \tag{3.2.8}$$

Let P be a d-polytope in $\mathbb{R}^d$ with $\boldsymbol{0}$ in its interior. Think of the polytope as being inscribed in the sphere (Fig. 3.2.1(a)). Then the radial projection from set $\mathcal{B}(P)$ onto $\mathbb{S}^{d-1}$ is a homeomorphism (Problem 3.11.2). Every interior point in the closed unit d-ball $B_d[\boldsymbol{0}, 1]$ lies in a segment $[\boldsymbol{0}, \boldsymbol{y}]$ for some point $\boldsymbol{y}$ in the boundary of $B_d[\boldsymbol{0}, 1]$, namely in $\mathbb{S}^{d-1}$. As a result, the radial projection from the boundary of P to the boundary of $B_d[\boldsymbol{0}, 1]$ can be extended radially to a homeomorphism of their interiors:

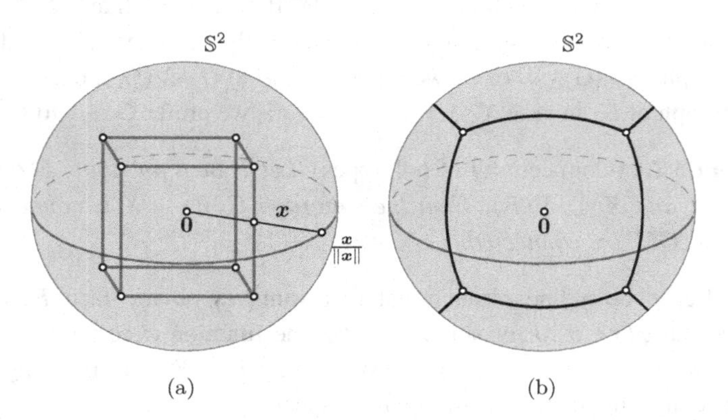

Figure 3.2.1 Radial projection in $\mathbb{R}^3$. (a) The radial projection from the boundary of a 3-cube to the boundary of $\mathbb{S}^2$. (b) A spherical cube.

$$\pi_r^*(\boldsymbol{x}) := \begin{cases} \|\boldsymbol{x}\|\pi_r^{-1}\left(\frac{\boldsymbol{x}}{\|\boldsymbol{x}\|}\right), & \text{if } \boldsymbol{x} \neq \mathbf{0}; \\ \mathbf{0}, & \text{otherwise.} \end{cases} \tag{3.2.9}$$

It follows that the function $\pi_r^*(\boldsymbol{x})$ is a homeomorphism from the closed unit d-ball in $\mathbb{R}^d$ onto the underlying set of the complex of P. We have just established the following.

Proposition 3.2.10 *The underlying set of the boundary complex of a d-polytope is homeomorphic to the unit $(d-1)$-sphere, and that of the complex of the d-polytope is homeomorphic to the closed unit d-ball.*

In view of Proposition 3.2.10, we say that a *polytopal $(d-1)$-sphere* is a polytopal $(d-1)$-complex $\mathcal{C}$ where set $\mathcal{C}$ is homeomorphic to the unit $(d-1)$-sphere and $\mathcal{C}$ is isomorphic to the boundary complex of a d-polytope.

The radial projection also yields a simple proof that graphs of 3-polytopes can be embedded in the two-dimensional sphere $\mathbb{S}^2$, which means that these graphs are planar (Theorem 3.1.8). In fact, very attractive planar realisations of such graphs are available.

A *convex realisation* of a graph is a planar realisation of the graph in which each edge is represented by a line segment and each inner face is a convex polygon. However, convex realisations are not available for all 2-connected planar graphs as exemplified in Fig. 3.2.2(c). But they are, however, available for graphs of 3-polytopes.

Proposition 3.2.11 *The graph of a 3-polytope is planar and 3-connected. Furthermore, the graph admits a convex realisation.*

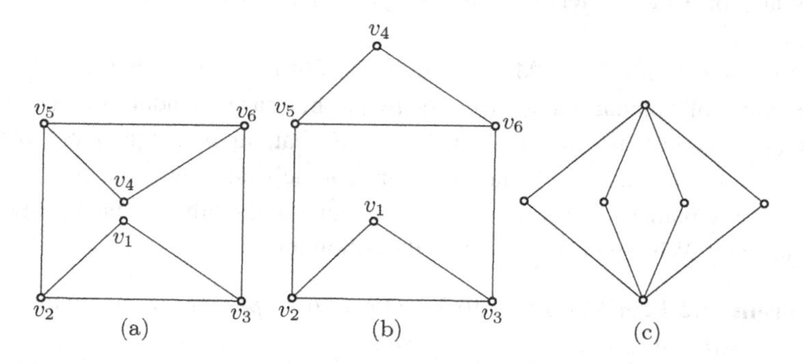

Figure 3.2.2 Planar realisations of 2-connected planar graphs. (a)–(b) Two inequivalent planar realisations of a 2-connected planar graph G. (a) A planar realisation of G with no facial cycle of length five. (b) A planar realisation of G with a facial cycle of length five. (c) A graph not admitting a convex realisation.

Proof Let P be a 3-polytope in $\mathbb{R}^3$. We give two proofs of the planarity of $G(P)$, with the latter giving the convex realisation.

First proof via radial projections Assume that P contains the origin in its interior and think of P as inscribed in the two-dimensional sphere $\mathbb{S}^2$ (Fig. 3.2.1(a)). Consider a radial projection from set $\mathcal{B}(P)$ to $\mathbb{S}^2$. Then the vertices, edges, and faces of the polytope are projected onto points, 'spherical' arcs, and 'spherical' polygons in the sphere, respectively; see Fig. 3.2.1(b). This gives a realisation of $G(P)$ in the sphere. □

Second proof via Schlegel diagrams Let F be a facet of P. A Schlegel diagram of P at F is a polytopal subdivision that is combinatorially isomorphic to the complex $\mathcal{B}(P)\backslash\{F\}$ (Theorem 2.11.7). This gives a convex realisation of $G(P)$ in which every vertex of P not on F is in the relative interior of F; see Fig. 2.11.5. □

The fact that $G(P)$ is 3-connected is a special case of Balinski's theorem (4.1.1) on the d-connectivity of graphs of d-polytopes. However, once we have the planarity of $G(P)$, a proof along the lines of that of Theorem C.7.6 is readily available. This completes the proof of the lemma. □

We end this part with two theorems, one from Whitney (1933) and one from Tutte (1963). Faces of a 2-connected plane graph G are all bounded by cycles (Proposition C.7.4). But it still may be the case that a cycle in one planar realisation of G is the boundary of a face, while in another realisation it is not; see Fig. 3.2.2(a)–(b). This observation suggests that, in some sense, two such realisations are not 'equivalent'. In plane graphs, a cycle that forms the boundary of a face is termed a *facial cycle*; if it doesn't, it is referred to as *nonfacial*.

Following Bondy and Murty (2008, p. 266), we say that two planar realisations of a planar graph are *equivalent* if their face boundaries, considered as sets of edges, are identical. This means that, in two equivalent planar realisations of a graph, the facial cycles are uniquely determined: a cycle that is facial in one planar realisation will also be facial in the other planar realisation of the graph. Whitney (1933) proved the following.

Theorem 3.2.12 (Whitney, 1933) *Every two planar realisations of a 3-connected planar graph are equivalent.*

A modern proof of this result relies on a theorem of Tutte (1963) that characterises the facial cycles in a 3-connected graph. A cycle in a graph is *nonseparating* if removing its vertices does not disconnect the graph; otherwise the cycle is described as *separating*.

Theorem 3.2.13 (Tutte, 1963)[4] *A cycle of a 3-connected plane graph is facial if and only if it is nonseparating.*

Whitney's theorem is now a corollary of Tutte's theorem.

Proof of Whitney's theorem (3.2.12) Let G be a 3-connected planar graph. According to Tutte's theorem (3.2.13), the facial cycles of a planar realisation of G are precisely those that are nonseparating. Being nonseparating is a combinatorial property that does not depend on the realisation of G, and so the facial cycles of G must be the same in every planar realisation of G. □

We remark that Whitney's theorem is also a corollary of Steinitz's theorem (3.8.1). According to Steinitz's theorem, a 3-connected planar graph G is the graph of a 3-polytope P, and the facial cycles of a planar realisation of G correspond to the boundaries of the 2-faces of the polytope. The boundary of a 2-face F of P is formed by the edges contained in F. The incidence between the edges and 2-faces of P is captured by the face lattice of the polytope, thus making it a combinatorial property of P. Hence, the facial cycles of G must be the same in every planar realisation of G.

3.3 Strongly Connected Complexes

A pure complex $\mathcal{C}$ is *strongly connected* if every pair of facets F and F' is connected by a path $F_1 \dots F_n$ of facets in $\mathcal{C}$ such that $F_i \cap F_{i+1}$ is a ridge of $\mathcal{C}$ for $i \in [1 \dots n-1]$, $F_1 = F$, and $F_n = F'$; we say that such a path is a *$(d-1, d-2)$-path*, or a *facet-ridge path* if the dimensions of the faces can be deduced from the context.

The definition of strongly connected complexes yields the connectivity of their graphs.

Proposition 3.3.1 *For $d \geqslant 1$, the graph of a strongly connected d-complex is connected.*

Proof Let $\mathcal{C}$ be a strongly connected d-complex, and let $\boldsymbol{u}$ and $\boldsymbol{v}$ be two vertices in $\mathcal{C}$. Since $\mathcal{C}$ is pure, there are facets $F_{\boldsymbol{u}}$ and $F_{\boldsymbol{v}}$ of $\mathcal{C}$ such that $\boldsymbol{u} \in F_{\boldsymbol{u}}$ and $\boldsymbol{v} \in F_{\boldsymbol{v}}$. And because $\mathcal{C}$ is strongly connected, these facets $F_{\boldsymbol{u}}$ and $F_{\boldsymbol{v}}$ are connected by a $(d-1, d-2)$-path in $\mathcal{C}$. It is now clear that this facet-ridge path gives a $\boldsymbol{u} - \boldsymbol{v}$ path in $G(\mathcal{C})$. □

[4] A proof is available in Diestel (2017, prop. 4.2.7).

Sallee (1967, cor. 2.11, thm. 3.5) showed that the star and antistar of a vertex in a d-polytope are strongly connected $(d-1)$-complexes.

Proposition 3.3.2 *Let P be a d-polytope. Then the boundary complex of P as well as the star and antistar of a vertex in $\mathcal{B}(P)$ are all strongly connected $(d-1)$-complexes.*

Proof Let ψ define the natural antiisomorphism from the face lattice of P to the face lattice of its dual P^*. Since the graph of P^* is connected (Theorem 3.2.6), we have that $\mathcal{B}(P)$ is a strongly connected $(d-1)$-complex. The facets in the star $\mathcal{S}_x$ of a vertex $\boldsymbol{x}$ in $\mathcal{B}(P)$ correspond to the vertices in the facet $\psi(\boldsymbol{x})$ in P^*. The existence a facet-ridge path in $\mathcal{S}_x$ between any two facets F_1 and F_2 of $\mathcal{S}_x$ amounts to the existence of a vertex-edge path in $\psi(\boldsymbol{x})$ between the vertices $\psi(F_1)$ and $\psi(F_2)$ of $\psi(\boldsymbol{x})$. That $\mathcal{S}_x$ is a strongly connected $(d-1)$-complex now follows from the connectivity of the graph of $\psi(\boldsymbol{x})$. The facets in the antistar $\mathcal{A}_x$ of $\boldsymbol{x}$ correspond to the vertices of P^* that are not in $\psi(\boldsymbol{x})$. That is, if F_1 and F_2 are any two facets of $\mathcal{A}_x$, then $\psi(F_1), \psi(F_2) \in \mathcal{V}(P^*)\backslash\mathcal{V}(\psi(\boldsymbol{x}))$. The existence of a facet-ridge path between F_1 and F_2 in $\mathcal{A}_x$ amounts to the existence of an vertex-edge path between $\psi(F_1)$ and $\psi(F_2)$ in the subgraph $G(P^*) - \mathcal{V}(\psi(\boldsymbol{x}))$ of $G(P^*)$. The removal of the vertices of a facet does not disconnect the graph of a polytope (Theorem 3.2.6), wherefrom it follows that $G(P^*) - \mathcal{V}(\psi(\boldsymbol{x}))$ is connected, as desired. □

We now study the connectivity of polytopal complexes generated by a shelling of a polytope.

Proposition 3.3.3 *Let P be a d-polytope in $\mathbb{R}^d$ with the origin in its interior, P^* the dual polytope of P, and ψ an antiisomorphism from $\mathcal{L}(P)$ to $\mathcal{L}(P^*)$. Suppose that $F_1, \ldots, F_s$ is a shelling of P. Then the subgraph of $G(P^*)$ induced by the vertices $\psi(F_1), \ldots, \psi(F_i)$ is connected, for each $i \in [1 \ldots s]$.*

Proof The proof is similar to that of Proposition C.4.1. Suppose that the vertices $\psi(F_1), \ldots, \psi(F_i)$ have already been chosen. By an inductive argument on i, the subgraph $G[\psi(F_1), \ldots, \psi(F_i)]$ of $G(P^*)$ is connected. Since $F_1, \ldots, F_s$ is a shelling of P, it follows that

$$F_{i+1} \cap (F_1 \cup \cdots \cup F_i)$$

is a union of $(d-2)$-faces of F_{i+1}; in particular, $F_{i+1} \cap F_\ell$ is a ridge for some $\ell \in [1 \ldots i]$. This means that the vertex $\psi(F_{i+1})$ is adjacent to the vertex $\psi(F_\ell)$ in $G[\psi(F_1), \ldots, \psi(F_i)]$, which yields the connectivity of $G[\psi(F_1), \ldots, \psi(F_{i+1})]$. The induction is now complete. □

In Proposition 3.3.3, we remark that $G(P^*) = G[\psi(F_1), \dots, \psi(F_s)]$, implying that $G(P^*)$ is connected and giving another proof of the connectivity of polytopal graphs.

An immediate consequence of Proposition 3.3.3 is stated next.

Proposition 3.3.4 *Let P be a d-polytope, and let $F_1, \dots, F_s$ be a shelling of P. Then the complex $\mathcal{C}(F_1 \cup \dots \cup F_i)$ is a strongly connected $(d-1)$-complex, for each $i \in [1 \dots s]$.*

Proof Let P^* be the dual polytope of P and let ψ be an antiisomorphism from $\mathcal{L}(P)$ to $\mathcal{L}(P^*)$. That the complex $\mathcal{C}(F_1 \cup \dots \cup F_i)$ is strongly connected amounts to saying that the subgraph of $G(P^*)$ induced by the vertices $\psi(F_i), \dots, \psi(F_i)$ is connected, for each $i \in [1 \dots s]$; the latter follows from Proposition 3.3.3. □

3.4 Excess Degree of a Polytope

By the *degree* of a vertex $\boldsymbol{v}$ in a d-polytope P we mean the degree of the vertex in the graph $G(P)$ of P. We know that $\deg_P \boldsymbol{v} \geqslant d$ (Theorem 2.4.14) and that there are d-polytopes whose vertices all have degree d, the simple d-polytopes. It is therefore natural to define the *excess degree* ξ of the vertex $\boldsymbol{v}$ as

$$\xi_P(\boldsymbol{v}) = \deg_P \boldsymbol{v} - d, \tag{3.4.1}$$

and the *excess degree* of P as

$$\xi(P) = \sum_{\boldsymbol{u} \in \mathcal{V}(P)} \xi_P(\boldsymbol{u}) = 2f_1(P) - df_0(P). \tag{3.4.2}$$

When referring to the excess degree of a vertex, we mean the excess degree in the polytope unless we make explicit we are referring to the excess degree in a face of the polytope.

The definition of excess degree gives at once the following.

Remark 3.4.3 Let P be a d-polytope.

(i) The excess degree of a vertex $\boldsymbol{v}$ in a facet F, $\xi_F(\boldsymbol{v}) = \deg_F \boldsymbol{v} - (d-1)$, is not larger than its excess degree in P, $\xi_P(\boldsymbol{v}) = \deg_P \boldsymbol{v} - d$.
(ii) A vertex of P has null excess degree if and only if it is contained in precisely d facets if and only if it is contained in precisely d edges.
(iii) An edge in P incident with a vertex of excess degree zero is contained in precisely $d-1$ facets.
(iv) If d is even, then only even excess degrees are possible for P.
(v) P is simple if and only if its excess degree is zero.

The main result on excess degrees is due to Pineda-Villavicencio et al. (2018).

Theorem 3.4.4 (Excess degree theorem) *The smallest excess degrees of a d-polytope are 0 and $d-2$*

The excess degree theorem clearly rules out the existence of d-polytopes with excess degree in the interval $[1 \dots d-3]$. This has proven valuable in characterising d-polytopes with a minimum number of edges or faces (Pineda-Villavicencio et al., 2018, 2022; Pineda-Villavicencio and Yost, 2022) and in characterising the Minkowski decomposable d-polytopes with $2d+1$ vertices (Theorem 6.3.2). The theorem has also been instrumental in determining all pairs (f_0, f_1) for which there exists a 5-polytope with f_0 vertices and f_1 edges (Theorem 8.1.6).

We proceed with lemmas required to prove the excess degree theorem. The first lemma shows that the excess degree is increasing with respect to faces.

Lemma 3.4.5 *The excess degree of a nonsimple polytope is larger than the excess degree of any of its facets.*

Proof Let F be a facet of a d-polytope P. If $\xi(F) = 0$, then we are done. So we assume that $\xi(F) > 0$, and let $\boldsymbol{u}$ be a vertex in F incident with more than $d-1$ edges in F. If a neighbour $\boldsymbol{x} \in P \backslash F$ of $\boldsymbol{u}$ is incident with more than d edges in P, then we are also done. Thus we can assume that $\boldsymbol{x}$ is a simple vertex in P, implying that the edge $\boldsymbol{xu}$ is contained in precisely $d-1$ facets (Remark 3.4.3). Since $\boldsymbol{u}$ is contained in at least $d+1$ facets (Remark 3.4.3), it follows that $\boldsymbol{u}$ has two neighbours outside F, ensuring that $\xi(P) > \xi(F)$, as desired. □

A small excess degree in a polytope imposes strong constraints on the intersection of facets.

Lemma 3.4.6 *Let P be a d-polytope, let F and F' be two distinct nondisjoint facets in P, and let $\ell :=$ dim $F \cap F'$. Then the following hold.*

(i) *Every vertex in $F \cap F'$ has excess degree at least $d-2-\ell$ in P.*
(ii) *The excess degree of P is at least* $\max\{\xi(F), \xi(F')\} + (d-2-\ell)(\ell+1)$.

Proof (i) Set $R := F \cap F'$. For each vertex $\boldsymbol{u} \in R$, the numbers of neighbours of $\boldsymbol{u}$ in R, $F \backslash R$, and $F' \backslash R$ are at least ℓ, $d-1-\ell$, and $d-1-\ell$, respectively. Thus, the degree of $\boldsymbol{u}$ in $F \cup F'$ is at least $2d-2-\ell$, which settles (i).

(ii) There are at least $\ell + 1$ vertices in R, and so the vertices in $F \cap F'$, and therefore in $F \cup F'$, contribute to $\xi(P)$ at least $(\ell + 1)(d - 2 - \ell)$. If F and F' are simple polytopes, we get equality. Hence, in any case

$$\xi(P) \geqslant \max\{\xi(F), \xi(F')\} + (\ell + 1)(d - 2 - \ell),$$

as desired. □

We need a further lemma on polytopes whose nondisjoint facets intersect at ridges.

Lemma 3.4.7 *Let P be a polytope in which every pair of nondisjoint facets intersect at a ridge. If every facet of P is simple, then so is P.*

Proof Let $d := \dim P$, let $\boldsymbol{u}$ be a vertex of P, and let F be a facet of P containing $\boldsymbol{u}$. Then F is simple. By the lemma hypothesis, the facets in P containing $\boldsymbol{u}$ are F and the facets of P that intersect F at a ridge containing $\boldsymbol{u}$, $1 + d - 1$ in total (since $\xi_F(\boldsymbol{u}) = 0$). The vertex $\boldsymbol{u}$ was arbitrary, implying that P is a simple polytope. □

We are now ready to prove the excess degree theorem.

Proof of the excess degree theorem (3.4.4) A 3-simplex has null excess degree, and a square pyramid has excess degree one. Let P be a nonsimple d-polytope. Thus, we can start an induction argument on d, with $d = 3$ as the basis case. Assume that $d \geqslant 4$.

If a pair of nondisjoint facets of P intersect at a face of dimension $\ell \leqslant d-3$, then $\xi(P) \geqslant (d - 2 - \ell)(\ell + 1)$ by Lemma 3.4.6, which, for $d \geqslant 4$, yields that $\xi(P) \geqslant d - 2$. Thus, we can assume that every pair of nondisjoint facets of P intersect at a ridge. Since P is nonsimple, Lemma 3.4.7 ensures that at least one facet F of P is also nonsimple. And by the induction hypothesis on F, we have that $\xi(F) \geqslant d - 3$. The conclusion $\xi(P) \geqslant d - 2$ now follows from the inequality $\xi(P) > \xi(F)$ (Lemma 3.4.5). □

We know that there are polytopes with excess degree zero, the simple polytopes, and that for each $d \geqslant 3$ there is a d-polytope with excess degree $d - 2$, for instance, a $(d - 2)$-fold pyramid over a quadrangle. We also know that, for each $d \geqslant 4$, the cyclic d-polytope on $d + 2$ vertices has excess degree $d + 2$. However, d-polytopes with excess degree $d - 1$ exist only in dimensions three and five (Pineda-Villavicencio et al., 2018). Whereas in the case of even d the excess degree $d - 1$ is not possible by a parity argument (Remark 3.4.3), in the case of odd $d \geqslant 7$ the value $d - 1$ is possible but is missing. Pineda-Villavicencio et al. (2018) showed that, for each dimension, the number of possible and missing excess degrees is finite. Let $\Xi(d)$ be the set of possible excess degrees of all d-polytopes.

Theorem 3.4.8 *If d is even, then $\Xi(d)$ contains every even integer in the interval $[d\sqrt{d},\infty]$. If instead d is odd, then $\Xi(d)$ contains every integer in the interval $[d\sqrt{2d},\infty]$.*

3.5 Acyclic Orientations and Shellings

An *orientation* of a graph is a digraph that arises from the graph by directing every edge from one of its ends to the other. The *reverse orientation* of an orientation O is another orientation that is obtained by reversing each arc of O: O contains the arc (x,y) if and only if its reverse orientation contains the arc (y,x). An orientation is *acyclic* if it has no directed cycles; a *directed* path or *directed* cycle is an orientation of a path or cycle in which there is a directed edge from each vertex to its successor in the sequence. If an orientation is acyclic, so is its reverse orientation.

We relate dual graphs and dual polytopes.

Proposition 3.5.1 *Let P be a 3-polytope and let P^* be its dual polytope. Then the graph $G(P^*)$ of P^* is isomorphic to the dual graph of $G(P)$.*

Proof Since the face lattice of P^* is isomorphic to the opposite lattice of the face lattice of P (Corollary 2.4.11), the proposition follows at once from the definition of dual plane graphs (Section C.7). □

We define the *dual graph* $G^*(P)$ of a polytope P as the graph whose vertices are the facets of P and whose edges are the pairs of facets that intersect at a ridge of P; in this context, two facets are *adjacent* if they share a ridge. By virtue of Proposition 3.5.1, for a 3-polytope P, we have that

$$G^*(P) = (G(P))^* = G(P^*).$$

The same proof of Proposition 3.5.1 yields the following.

Proposition 3.5.2 *The dual graph of a polytope P is isomorphic to the graph of the dual polytope of P.*

A shelling $F_1,\dots,F_s$ of the facets of a polytope P with $\mathbf{0} \in \operatorname{int} P$ gives rise to an acyclic orientation of its dual graph, the graph of P^*. Let ψ be an antiisomorphism from $\mathcal{L}(P)$ to $\mathcal{L}(P^*)$. An edge $\psi(F_i)\psi(F_j)$ in $G(P^*)$ ensures that the facets F_i and F_j share a ridge. We obtain an orientation of

$G(P^*)$ by orienting the edge from $\psi(F_j)$ to $\psi(F_i)$ whenever F_i comes before F_j in the shelling. The shelling also causes a linear ordering $\psi(F_1), \ldots, \psi(F_s)$ of the vertices of $G(P^*)$.

Following Kalai (1988b), an acyclic orientation of a graph of a polytope P is *good* if the subgraph of every nonempty face F of P has a unique *sink* (a vertex of outdegree zero); otherwise it is *bad*. The orientation of $G(P)$ induced by a shelling of P^* is a good orientation.

Proposition 3.5.3 *Let P be a d-polytope in $\mathbb{R}^d$ with the origin in its interior, P^* the dual polytope of P, and ψ an antiisomorphism from $\mathcal{L}(P)$ to $\mathcal{L}(P^*)$. Then a shelling of P gives rise to an orientation O of the graph of P^* in which, for every nonempty face F of P^*, the restriction of O to $G(F)$ has a unique sink and a unique* source *(a vertex of indegree zero).*

Proof Consider the orientation O of $G(P^*)$ induced by a shelling $J_1, \ldots, J_n$ of P. The orientation is clearly acyclic, which implies that, for each nonempty face F of P^*, the restriction $O|_F$ of O to $G(F)$ is acyclic. We show that $O|_F$ has a unique sink. The case $\dim F \leqslant 1$ is trivial, so assume $\dim F \geqslant 2$.

Let $\boldsymbol{x}_1 := \psi(J_1), \ldots, \boldsymbol{x}_n := \psi(J_n)$, and let $\boldsymbol{x}_{i_1}, \ldots, \boldsymbol{x}_{i_m}$ be the restriction of the ordering $\boldsymbol{x}_1, \ldots, \boldsymbol{x}_n$ to the vertices of $G(F)$. Then $\boldsymbol{x}_{i_1}$ is a sink of $O|_F$. Suppose there is another sink $\boldsymbol{x}_{i_p}$, for $p > 1$. It follows that the set

$$J_{i_p} \cap \left(J_{i_1} \cup \cdots \cup J_{i_{p-1}}\right) \tag{3.5.3.1}$$

contains no $(d-2)$ faces, but contains the face $\psi^{-1}(F)$ of P. The facets J_j of P with $\boldsymbol{x}_j \notin F$ do not contain the face $\psi^{-1}(F)$. Therefore, the set

$$\mathcal{C} := J_{i_p} \cap \left(J_1 \cup \cdots \cup J_{i_1} \cup \cdots \cup J_{i_{p-1}} \cup \cdots \cup J_{i_p-1}\right) \tag{3.5.3.2}$$

contains the face $\psi^{-1}(F)$ but a maximal face in $\mathcal{C}$ containing $\psi^{-1}(F)$ is not a $(d-2)$-face. However, since $J_1, \ldots, J_n$ is a shelling of P, $\mathcal{C}$ is a union of $(d-2)$-faces. This contradiction concludes that $O|_F$ has a unique sink.

Additionally, the orientation $O|_F$ has a unique source. The reverse sequence $F_n, \ldots, F_1$ of the shelling $F_n, \ldots, F_1$ is also a shelling (Lemma 2.12.3), which gives rise to another orientation O' of $G(P^*)$, the reverse orientation of O. From our previous reasoning, the restriction $O'|_F$ of O' to $G(F)$ has a unique sink. A sink of $O'|_F$ is a source of $O|_F$, and so $O|_F$ has a unique source. □

Since every shelling of a polytope produces a good orientation in the graph of the dual polytope (Proposition 3.5.3), the shellings in Propositions 2.12.13 and 2.12.14 give good orientations with additional constraints. First, define an *initial set* with respect to some orientation as a set of vertices such that no arc is directed from a vertex not in the set to a vertex in the set; the vertices

$\boldsymbol{x}_4, \boldsymbol{x}_5, \boldsymbol{x}_6$ in Fig. 3.5.1(a) form an initial set. Similarly, a *final set* with respect to some orientation is a set of vertices such that no arc is directed from a vertex in the set to a vertex not in the set; the vertices $\boldsymbol{x}_1, \boldsymbol{x}_2, \boldsymbol{x}_3$ in Fig. 3.5.1(a) form a final set.

Corollary 3.5.4 *Let P be a polytope, $\boldsymbol{v}$ a vertex of P, and F a facet of P. Then $G(P)$ admits the following good orientations.*

(i) *If F does not contain $\boldsymbol{v}$ then there is a good orientation of $G(P)$ having $\mathcal{V}(F)$ as a final set and $\boldsymbol{v}$ as the unique source.*
(ii) *If F contains $\boldsymbol{v}$ then there is a good orientation of $G(P)$ having $\mathcal{V}(F)$ as an initial set and $\boldsymbol{v}$ as the unique source.*

Proof Let ψ be an antiisomorphism from $\mathcal{L}(P)$ to the face lattice $\mathcal{L}(P^*)$ of the dual polytope P^* of P. Suppose that $J := \psi(\boldsymbol{v})$ and $\boldsymbol{x} := \psi(F)$. According to Proposition 2.12.13, there is a shelling of P^* where the facets of $\operatorname{st}(\boldsymbol{x}, P^*)$ come first and the facet J comes last. In view of Proposition 3.5.3, this shelling gives rise to a good orientation of $G(P)$ that satisfies Part (i). Furthermore, from Proposition 2.12.14 follows the existence of a shelling of P^* where the facets of $\operatorname{st}(\boldsymbol{x}, P^*)$ come last and the facet J is the last facet of the shelling. This shelling produces a good orientation of $G(P)$ that matches the description in Part (ii). □

A linear functional φ in general position with respect to a polytope P in $\mathbb{R}^d$ (Definition 2.12.6) gives a linear ordering $<'$ of the vertices of $G(P)$ where $\boldsymbol{x}_i <' \boldsymbol{x}_j$ if $\varphi(\boldsymbol{x}_i) > \varphi(\boldsymbol{x}_j)$ for $\boldsymbol{x}_i, \boldsymbol{x}_j \in \mathcal{V}(P)$. The ordering $\varphi(\boldsymbol{x}_1) > \cdots > \varphi(\boldsymbol{x}_n)$ induces an acyclic orientation O_φ of $G(P)$ when we orient the edge $\boldsymbol{x}_i\boldsymbol{x}_j$ of $G(P)$ from $\boldsymbol{x}_j$ to $\boldsymbol{x}_i$ if $\boldsymbol{x}_i <' \boldsymbol{x}_j$ in the ordering. More generally, we have the following.

Remark 3.5.5 Any linear ordering $x_1, \ldots, x_n$ of the vertices of a graph G induces an acyclic orientation of G when we orient the edge $x_i x_j$ of G from x_j to x_i if x_i precedes x_j in the ordering.

The next result is also straightforward.

Lemma 3.5.6 *Let P be a d-polytope in $\mathbb{R}^d$, let φ be a linear functional in general position with respect to P, and let O_φ be the orientation of $G(P)$ induced by φ. Then, for every nonempty face F of P, the restriction of O to $G(F)$ has a unique sink and a unique source.*

Lemma 3.5.6 motivates the characterisation of orientations that arise from linear functionals. An orientation O of the graph of a polytope P in $\mathbb{R}^d$ is an

LP orientation if there exists a realisation of P and a linear functional φ in general position with respect to p such that the orientation induced by φ on $G(P)$ is precisely O.

Bruggesser–Mani's shelling theorem (2.12.8) can be dualised with the use of LP orientations.

Theorem 3.5.7 (Danaraj and Klee, 1974, lem. 2.1) *Let P be a d-polytope in $\mathbb{R}^d$ with the origin in its interior, P^* the dual polytope of P, and ψ an antiisomorphism from $\mathcal{L}(P)$ to $\mathcal{L}(P^*)$. If $\boldsymbol{x}_1, \ldots, \boldsymbol{x}_n$ is an ordering of $\mathcal{V}(P^*)$ induced by an LP orientation, then $\psi(\boldsymbol{x}_1), \ldots, \psi(\boldsymbol{x}_n)$ is a line shelling of P.*

Proof We reduce the proof of this theorem to the proof of Bruggesser–Mani's shelling theorem (2.12.8). Let φ be the linear functional in general position with respect to P^* that induced the ordering $\boldsymbol{x}_1, \ldots, \boldsymbol{x}_n$. We need concrete expressions for φ and each facet $\psi(\boldsymbol{x}_i)$ of P. By Theorems 2.3.1 and 2.4.3, for each $i \in [1 \ldots n]$ we have that

$$\psi(\boldsymbol{x}_i) = \{\boldsymbol{z} \in P \mid \boldsymbol{x}_i \cdot \boldsymbol{z} = 1\}.$$

Suppose $\varphi(\boldsymbol{z}) := \boldsymbol{y} \cdot \boldsymbol{z}$ for some $\boldsymbol{y} \in \mathbb{R}^d$ and every $\boldsymbol{z} \in \mathbb{R}^d$. Without loss of generality, assume that φ is nonzero on $\{\boldsymbol{x}_1, \ldots, \boldsymbol{x}_n\}$ and that

$$\boldsymbol{y} \cdot \boldsymbol{x}_1 > \boldsymbol{y} \cdot \boldsymbol{x}_2 > \cdots > \boldsymbol{y} \cdot \boldsymbol{x}_k > 0 > \boldsymbol{y} \cdot \boldsymbol{x}_{k+1} > \cdots > \boldsymbol{y} \cdot \boldsymbol{x}_n. \quad (3.5.7.1)$$

For each $i \in [1 \ldots n]$, define

$$\boldsymbol{y}_i := \frac{\boldsymbol{y}}{\boldsymbol{y} \cdot \boldsymbol{x}_i}.$$

It follows that the point $\boldsymbol{y}_i$ lies in aff $\psi(\boldsymbol{x}_i)$. Furthermore, because of (3.5.7.1), we have that

$$\frac{1}{\boldsymbol{y} \cdot \boldsymbol{x}_{k+1}} < \cdots < \frac{1}{\boldsymbol{y} \cdot \boldsymbol{x}_n} < 0 < \frac{1}{\boldsymbol{y} \cdot \boldsymbol{x}_1} < \frac{1}{\boldsymbol{y} \cdot \boldsymbol{x}_2} < \cdots < \frac{1}{\boldsymbol{y} \cdot \boldsymbol{x}_k}, \quad (3.5.7.2)$$

whence we get that the points $\boldsymbol{y}_i$ are all pairwise distinct.

Consider the line $\ell := \alpha \boldsymbol{y}$ for every $\alpha \in \mathbb{R}$. Then, for each $i \in [1 \ldots n]$, the point $\boldsymbol{y}_i$ lies in ℓ, and, thus, $\boldsymbol{y}_i$ is in the intersection of ℓ and aff $\psi(\boldsymbol{x}_i)$. Because of (3.5.7.2), when we traverse the line ℓ from the origin, which is in the interior of P, to $\boldsymbol{y}$, we encounter the points $\boldsymbol{y}_1, \ldots, \boldsymbol{y}_k$, in that order, before reaching infinity, and then encounter the points $\boldsymbol{y}_{k+1}, \ldots, \boldsymbol{y}_n$, again in that order, when we return to the polytope from the opposite side. We now have the same setting as in the proof of Theorem 2.12.8; see also Remark 2.12.9. As a consequence, we find that $\psi(\boldsymbol{x}_1), \ldots, \psi(\boldsymbol{x}_n)$ is a line shelling of P. □

The converse of Theorem 3.5.7 is also true.

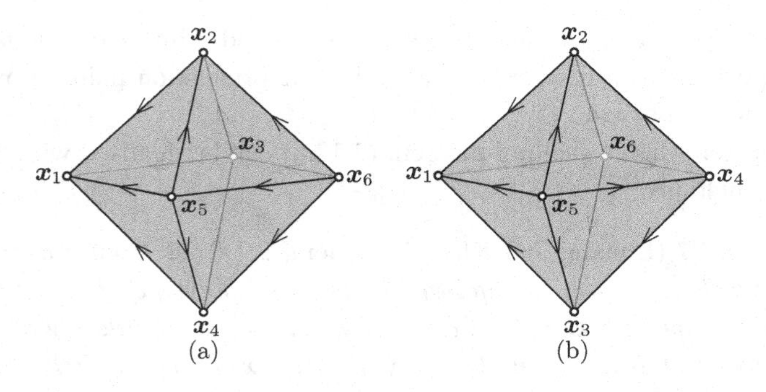

Figure 3.5.1 Acyclic orientations of the graph of the octahedron that lead to sequences $F_1 := \psi(\boldsymbol{x}_1), \ldots, F_6 := \psi(\boldsymbol{x}_6)$ of facets of the 3-cube, where ψ is the relevant antiisomorphism. (a) An orientation that leads to the line shelling of the 3-cube depicted in Fig. 2.12.1: the vertex $\boldsymbol{x}_1$ is the unique sink, the vertex $\boldsymbol{x}_6$ is the unique source, and there are three independent paths from $\boldsymbol{x}_1$ to $\boldsymbol{x}_6$ (Theorem 3.5.9). (b) An orientation that leads to the sequence of facets in the 3-cube shown in Fig. 2.12.2, which is the beginning of no shelling of the 3-cube: the vertex $\boldsymbol{x}_1$ is the unique sink, and the vertices $\boldsymbol{x}_5$ and $\boldsymbol{x}_6$ are two sources (Theorem 3.5.10).

Theorem 3.5.8 (Ziegler, 1995, ex. 8.10) *Let P be a d-polytope in $\mathbb{R}^d$ with the origin in its interior, and P^* the dual polytope of P. An orientation O on the graph of P is an LP orientation if and only if there exists a line shelling of P^* that induces the orientation O.*

According to Theorem 3.5.7, for a polytope P, some linear orderings of the vertices of $G(P^*)$ translate into line shellings of P. In the case of 3-polytopes, there is a neat characterisation of this phenomenon.

Theorem 3.5.9 (Mihalisin and Klee, 2000, thm. 3.6, cor. 3.7) *Let P be a 3-polytope in $\mathbb{R}^3$ with the origin in its interior, let P^* be the dual polytope of P, and let ψ be an antiisomorphism from $\mathcal{L}(P)$ to $\mathcal{L}(P^*)$. Suppose that $\boldsymbol{x}_1, \ldots, \boldsymbol{x}_n$ is a linear ordering of $\mathcal{V}(P^*)$. Then $\psi(\boldsymbol{x}_1), \ldots, \psi(\boldsymbol{x}_n)$ is a line shelling of P if and only if the acyclic orientation of $G(P^*)$ induced by the ordering has a unique sink, a unique source, and three independent paths from the source to the sink.*

The acyclic orientation of Fig. 3.5.1(a) yields the line shelling of the 3-cube depicted in Fig. 2.12.1, while the acyclic orientation of Fig. 3.5.2(c) yields a line shelling of the octahedron. Also, in case of a 3-polytopes, there is also a neat characterisation of shellings in general.

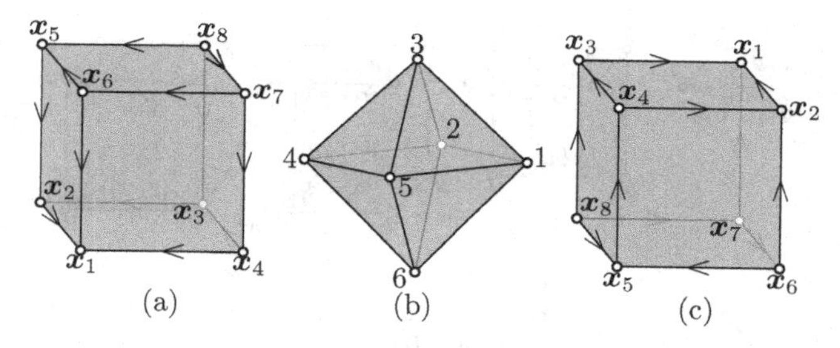

Figure 3.5.2 Acyclic orientations of the graph of the 3-cube that leads to shellings of the octahedron. (a) An orientation of the 3-cube that leads to a nonline shelling: the vertex $\boldsymbol{x}_1$ is the unique sink, the vertex $\boldsymbol{x}_8$ is the unique source, but there are no three independent paths from $\boldsymbol{x}_1$ to $\boldsymbol{x}_8$ (Theorem 3.5.9). The corresponding nonline shelling is $F_1 := \psi(\boldsymbol{x}_1) = 246, F_2 := \psi(\boldsymbol{x}_2) = 126, F_3 := \psi(\boldsymbol{x}_3) = 156, F_4 := \psi(\boldsymbol{x}_4) = 456, F_5 := \psi(\boldsymbol{x}_5) = 123, F_6 := \psi(\boldsymbol{x}_6) = 234, F_7 := \psi(\boldsymbol{x}_7) = 345, F_8 := \psi(\boldsymbol{x}_8) = 135$; the vertex labelling is depicted in (b). (b) The octahedron with vertex labels. (c) An orientation of the 3-cube that leads to a line shelling: the vertex $\boldsymbol{x}_1$ is the unique sink, the vertex $\boldsymbol{x}_8$ is the unique source, and there are three independent paths from $\boldsymbol{x}_1$ to $\boldsymbol{x}_8$ (Theorem 3.5.9). The corresponding line shelling is $F_1 := \psi(\boldsymbol{x}_1) = 123, F_2 := \psi(\boldsymbol{x}_2) = 135, F_3 := \psi(\boldsymbol{x}_3) = 234, F_4 := \psi(\boldsymbol{x}_4) = 345, F_5 := \psi(\boldsymbol{x}_5) = 456, F_6 := \psi(\boldsymbol{x}_6) = 156, F_7 := \psi(\boldsymbol{x}_7) = 126, F_8 := \psi(\boldsymbol{x}_8) = 246$; the vertex labelling is depicted in (b). Here, ψ is the relevant antiisomorphism.

Theorem 3.5.10 (Ishizeki and Takeuchi, 1999, prop. 1.2) *Let P be a 3-polytope in $\mathbb{R}^3$ with the origin in its interior, let P^* be the dual polytope of P, and let ψ be an antiisomorphism from $\mathcal{L}(P)$ to $\mathcal{L}(P^*)$. Suppose that $\boldsymbol{x}_1, \dots, \boldsymbol{x}_n$ is an ordering of $\mathcal{V}(P^*)$. Then $\psi(\boldsymbol{x}_1), \dots, \psi(\boldsymbol{x}_n)$ is a shelling of P if and only if the acyclic orientation of $G(P^*)$ induced by the ordering has a unique sink and a unique source.*

The acyclic orientation of Fig. 3.5.1(b) yields the sequence of facets in the 3-cube shown in Fig. 2.12.2, which leads to no shelling of the 3-cube.

One consequence of Theorem 3.5.9 and Theorem 3.5.10 is the existence of nonline shellings of 3-polytopes. We depict a nonline shelling of the octahedron in Fig. 3.5.2, which was discovered by Smilansky (1990). Smilansky (1990) noted that nonline shellings of d-polytopes with $d \geqslant 4$ were known to exist; see Develin (2004) for examples of nonline shellings of crosspolytopes.

Another consequence of Theorem 3.5.9 and Theorem 3.5.10 is that every good orientation of a graph of a 3-polytope leads to a shelling of the dual polytope. This does not extend to higher dimensions. Develin (2004) showed that there exist good orientations of graphs of d-crosspolytopes, with $d \geqslant 4$, that have d independent paths from the source to the sink, but lead to no

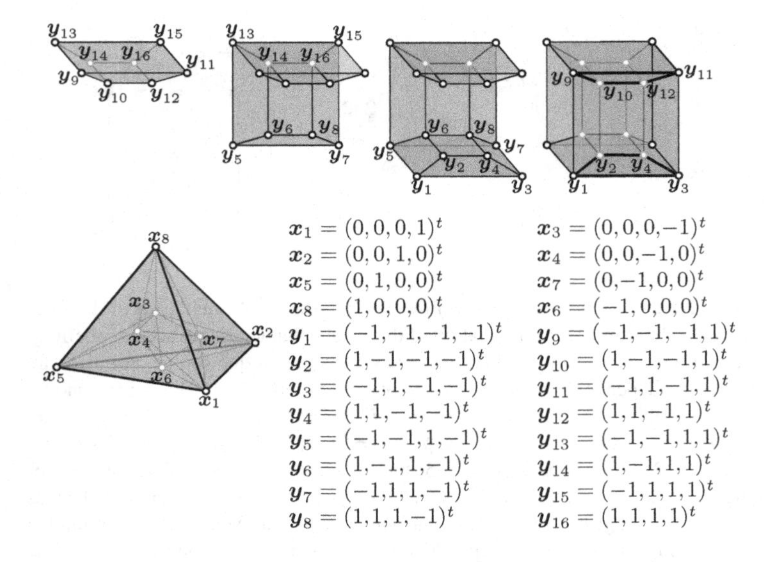

Figure 3.5.3 A linear ordering of vertices of the 4-crosspolytope that lead to a nonshelling sequence $F_1 := \psi(\boldsymbol{x}_1), \ldots, F_8 := \psi(\boldsymbol{x}_8)$ of facets of the 4-cube; here, ψ is the relevant antiisomorphism. The current facet in the sequence has its vertices labelled. The intersection $F_4 \cap (F_1 \cup F_2 \cup F_3)$, highlighted in bold, is not the beginning of a shelling of F_4.

shelling of the d-cube; we present one such orientation of the 4-crosspolytope in Fig. 3.5.3. Avis and Moriyama (2009, prop. 7) gave an example with fewer vertices than the 4-crosspolytope.

h-Vectors of Simple Polytopes

The (good) orientations of the graph of a simple polytope P induced by linear functionals in general position give rise to line shellings of the dual polytope (Theorem 3.5.7). We will establish a relation between the k-faces of P and the vertices of indegree k in those orientations. This is the dual approach to the derivation of the h-vectors of simplicial polytopes in Section 2.13.

Consider the graph of a simple d-polytope P and a good orientation O of $G(P)$ induced by a linear functional φ in general position. Let $h_i(O)$ denote the number of vertices with indegree i in O. By virtue of Theorem 2.8.7, a simple vertex $\boldsymbol{w}$ and every k of its neighbours define a unique k-face of P. Thus, if $\boldsymbol{w}$ has indegree i in O, then it is a sink in $\binom{i}{k}$ k-faces of P. Because O is a good orientation, each nonempty k-face of P has a unique sink in O. These observations yield an expression for the number f_k of k-faces of P:

$$
\begin{aligned}
f_k(P) &:= \sum_{i=k}^{d} h_i(O)\binom{i}{k}, \quad \text{for } k = 0, \dots, d, \\
&= \sum_{i=0}^{d} h_i(O)\binom{i}{k}.
\end{aligned} \tag{3.5.11}
$$

As in the case of simplicial polytopes (Section 2.13), we can also obtain the number $h_k(O)$ in terms of the f-vector of P. For this, we define the two polynomials

$$\zeta_f(x) := \sum_{k=0}^{d} f_k(P)x^k \text{ and } \zeta_h(x) := \sum_{k=0}^{d} h_k(O)x^k,$$

and we relate them using (3.5.11),

$$
\begin{aligned}
\zeta_f(x) &= \sum_{k=0}^{d} f_k(P)x^k = \sum_{k=0}^{d}\left(\sum_{i=0}^{d} h_i(O)\binom{i}{k}\right)x^k = \sum_{i=0}^{d} h_i(O)\left(\sum_{k=0}^{d}\binom{i}{k}x^k\right) \\
&= \sum_{i=0}^{d} h_i(O)\left(\sum_{k=0}^{i}\binom{i}{k}x^k\right) = \sum_{i=0}^{d} h_i(O)(x+1)^i = \zeta_h(x+1).
\end{aligned}
$$

Equivalently, we have that $\zeta_f(x-1) = \zeta_h(x)$. It is now clear that

$$h_k(O) := \sum_{i=k}^{d}(-1)^{i-k}\binom{i}{k} f_i(P), \text{ for } k = 0, \dots, d. \tag{3.5.12}$$

From (3.5.12) it follows that the $h_k(O)$ numbers are independent of the good orientation O, and so we will call the sequence $(h_0, \dots, h_d)$ the *h-vector* of P. The reverse orientation O' of O is also a good orientation of $G(P)$ where the vertices of indegree $d-k$ correspond to the vertices of indegree k in O, and therefore, $h_{d-k}(O') = h_k(O)$ for $k \in [0 \dots d]$. Since the numbers h_k are independent of any good orientation of $G(P)$, we get the Dehn–Sommerville equations for simple polytopes.

Theorem 3.5.13 (Dehn–Sommerville's equations for simple polytopes) *The h-vector of a simple d-polytope satisfies the equality*

$$h_k = h_{d-k}, \textit{ for } k = 0, \dots, d.$$

This is the dual version of Dehn–Sommerville's equations for simplicial polytopes (Theorem 2.13.3).

Example 3.5.14 (Computation of h-vectors of simple polytopes) Consider the LP orientation O of the graph of the cube in Fig. 3.5.2(c). From this orientation we get the h-vector $(1,3,3,1)$ of the cube: $h_0 := \#\{\boldsymbol{x}_8\}, h_1 := \#\{\boldsymbol{x}_4, \boldsymbol{x}_6, \boldsymbol{x}_7\}, h_2 := \#\{\boldsymbol{x}_2, \boldsymbol{x}_3, \boldsymbol{x}_5\}$, $h_3 := \#\{\boldsymbol{x}_1\}$.

3.6 Cycles in Polytopes

We study cycles of graphs of polytopes, including their cycle space. Recall that in a plane graph, we refer to a cycle that is the boundary of a face as a facial cycle; otherwise the cycle is nonfacial. In the context of polytopes, a *facial cycle* is defined as a cycle that forms the boundary of a 2-face of a polytope; otherwise the cycle is *nonfacial.*

As the graph of a 3-polytope is planar and 3-connected (Proposition 3.2.11), the next proposition is relevant to our discussion on cycles in polytopes.

Proposition 3.6.1 *Every 3-connected plane graph has a vertex of degree three or a triangular face.*

Proof Let G be a 3-connected plane graph with no triangular face. Suppose that G has f_0 vertices, f_1 edges, and f_2 faces. Since every face has four or more edges and since every edge is contained in exactly two faces, we have that

$$4f_2 \leqslant 2f_1. \tag{3.6.1.1}$$

Euler's equation (C.7.1) (see also Euler–Poincaré–Schläfli's equation (2.12.17)) gives that

$$2f_0 - 2f_1 + 2f_2 = 4,$$

or, equivalently, that

$$2f_0 - f_1 = 4 - 2f_2 + f_1. \tag{3.6.1.2}$$

Combining (3.6.1.1) and (3.6.1.2), we get that

$$f_1 \leqslant 2f_0 - 4.$$

This now gives that G has a vertex of degree three, as if every vertex had degree at least four then $f_1 \geqslant 2f_0$. □

The next result states that the girth of a 3-polytope is at most five, a consequence of the fact that a planar graph has a vertex of degree at most five (Proposition C.7.2); the *girth* of a graph G with at least one cycle, denoted $g(G)$, is the length of a shortest cycle in G,

Proposition 3.6.2 *Every d-polytope with $d \geqslant 3$ has a facial cycle of length at most five.*

Proof Let P be a 3-polytope that contains the origin in its interior. Then, the dual polytope P^* is also a 3-polytope (Proposition 2.4.7). From Proposition 3.2.11 it now follows that $G(P^*)$ is planar, and so by Proposition C.7.2, it has a vertex of degree at most five. Thus, the 2-face of P corresponding to this vertex is bounded by a cycle of length at most five.

Every d-polytope with $d \geqslant 3$ has a 3-face, which, according to the previous discussion, has a facial cycle of length at most five. Hence the d-polytope has girth at most five. □

A corollary immediately ensues.

Corollary 3.6.3 *The girth of a d-polytope with $d \geqslant 3$ is at most five.*

There are 3-polytopes and 4-polytopes whose facial cycles have all length at least five. The dodecahedron and the 120-cell are examples of such polytopes in dimensions three and four, respectively. Kalai (1990) proved that this cannot happen in dimensions greater than four.

Theorem 3.6.4 (Kalai, 1990) *Every d-polytope with $d \geqslant 5$ has a facial cycle of length three or four.*

The proof of Kalai (1990) for Theorem 3.6.4 is computer assisted and uses inequalities for the flag numbers of 5-polytopes, a generalisation of face numbers (Chapter 8). The equivalent statement for simple polytopes is an easy application of Dehn–Sommerville's equations for simple polytopes (Theorem 3.5.13).

Theorem 3.6.5 *Every simple d-polytope with $d \geqslant 5$ has a facial cycle of length three or four.*

Proof Let P be a simple 5-polytope, and let $(f_0, f_1, f_2, f_3, f_4)$ be the f-vector of P. If we combine the equality $5f_0 = 2f_1$ with Dehn–Sommerville's equations for simple polytopes (Theorem 3.5.13) (the cases $h_2 = h_3$ and $h_1 = h_4$), then we obtain that

$$f_0 = 2f_3 - 6f_4 + 12, \ f_2 = 4f_3 - 10f_4 + 20. \tag{3.6.5.1}$$

Each vertex of P is contained in exactly $\binom{5}{2}$ 2-faces (Theorem 2.8.6), and by way of contradiction let us suppose that each 2-face of P contains at least five vertices. Then $10f_0 \geqslant 5f_2$. Combining this inequality with (3.6.5.1) we get that

$$2f_0 = 4f_3 - 12f_4 + 24 \geqslant 4f_3 - 10f_4 + 20 = f_2,$$

or, equivalently, that $2 \geqslant f_4$, contradicting the fact that $f_4 \geqslant 6$ (Theorem 2.4.13). Hence P must contain a facial cycle of length at most four.

Every simple d-polytope with $d \geqslant 5$ has a 5-face, which, according to our discussion, has a facial cycle of length at most four. The theorem is now proved. □

Facial cycles are induced cycles. It may, however, be the case that not every induced cycle in a polytope is a facial cycle; for instance, the cycle in the base of a 3-bipyramid is induced but not facial. In the case of simple polytopes, every 'small' induced cycle is facial.

Theorem 3.6.6 *Every induced cycle of length at most five in a simple polytope is facial.*

Proof Let P be a simple d-polytope. Let C be an induced cycle of P.

First, suppose that $\mathcal{V}(C) = \{\boldsymbol{x}_1, \boldsymbol{x}_2, \boldsymbol{x}_3\}$. Then $\boldsymbol{x}_1$ and the edges $\boldsymbol{x}_1\boldsymbol{x}_2$ and $\boldsymbol{x}_1\boldsymbol{x}_3$ are contained in a 2-face F of P (Theorem 2.8.6). By the convexity of F, the edge $\boldsymbol{x}_2\boldsymbol{x}_3$ must be part of F, implying that $\mathcal{V}(F) = \mathcal{V}(C)$.

Now suppose that $C = \boldsymbol{x}_1\boldsymbol{x}_2\boldsymbol{x}_3\boldsymbol{x}_4\boldsymbol{x}_1$. Then the vertex $\boldsymbol{x}_1$ and the edges $\boldsymbol{x}_1\boldsymbol{x}_2$ and $\boldsymbol{x}_1\boldsymbol{x}_4$ are contained in a 2-face F_1 (Theorem 2.8.6), and the vertex $\boldsymbol{x}_3$ and the edges $\boldsymbol{x}_3\boldsymbol{x}_2$ and $\boldsymbol{x}_3\boldsymbol{x}_4$ are contained in a 2-face F_3. It follows that $\{\boldsymbol{x}_2, \boldsymbol{x}_4\} \subseteq F_1 \cap F_3$. If $F_1 \neq F_3$ then $\boldsymbol{x}_2\boldsymbol{x}_4$ must be an edge of P, contradicting the fact that C is induced. Hence $F_1 = F_3$.

Finally, suppose that C is a 5-cycle. First, consider the case of P being a 3-polytope. By way of contradiction, suppose that C is nonfacial. The graph of a 3-polytope is 3-connected (Proposition 3.2.11). According to Tutte's theorem (3.2.13), the removal of the vertices C disconnects $G(P)$. Let G' and G'' be two components of $G - \mathcal{V}(C)$. Because G is 3-connected, and thus 3-edge-connected, there must be at least three edges from G' to C and at least three edges from G'' to C. But then, there must exist a vertex in C of degree at least four in $G(P)$, a contradiction. Hence C is facial in P.

For the case $d \geqslant 4$, it suffices to show that C must be contained in a 3-face of P, which would give that C is facial in P. Let $C = \boldsymbol{x}_1\boldsymbol{x}_2\boldsymbol{x}_3\boldsymbol{x}_4\boldsymbol{x}_5\boldsymbol{x}_1$. Then $\boldsymbol{x}_1$ and the edges $\boldsymbol{x}_1\boldsymbol{x}_5$ and $\boldsymbol{x}_1\boldsymbol{x}_2$ are contained in a 2-face F_1 of P. It may happen that $\boldsymbol{x}_3 \in F_1$, in which case the edges $\boldsymbol{x}_1\boldsymbol{x}_5$, $\boldsymbol{x}_1\boldsymbol{x}_2$, and $\boldsymbol{x}_2\boldsymbol{x}_3$ are contained in a 3-face of P; otherwise the vertex $\boldsymbol{x}_2$, the edges $\boldsymbol{x}_2\boldsymbol{x}_3$ and $\boldsymbol{x}_1\boldsymbol{x}_2$, and the other edge incident with $\boldsymbol{x}_2$ in F_1 are contained in a 3-face. Thus, in any case the edges $\boldsymbol{x}_1\boldsymbol{x}_5$, $\boldsymbol{x}_1\boldsymbol{x}_2$, and $\boldsymbol{x}_2\boldsymbol{x}_3$ are contained in a 3-face J of P. The vertex $\boldsymbol{x}_4$ and the edges $\boldsymbol{x}_4\boldsymbol{x}_3$ and $\boldsymbol{x}_4\boldsymbol{x}_5$ are contained in a 2-face F_4. If $F_4 \nsubseteq J$, then $F_4 \cap J = \{\boldsymbol{x}_3, \boldsymbol{x}_5\}$ must be the edge $\boldsymbol{x}_3\boldsymbol{x}_5$, contradicting the fact that C is induced. Hence C must be contained in J, concluding the proof of the case and the theorem. □

Theorem 3.6.6 is best possible in the sense that there are induced 6-cycles in simple polytopes that are not facial cycles; for instance, the 3-cube has one such 6-cycle.

Kalai (1990) obtained other results on the existence of certain faces in simple polytopes.

Cycle Space of Polytopal Graphs

In view of the planarity of graphs of 3-polytopes (Proposition 3.2.11), we start with results on plane graphs.

Let G be a graph and let G' and G'' be two spanning subgraphs of G. The *symmetric difference* of two spanning subgraphs G' and G'' of G is the spanning subgraph $G' \triangle G''$ of G whose edge set is the symmetric difference $\mathcal{E}(G') \triangle \mathcal{E}(G'')$ of $\mathcal{E}(G')$ and $\mathcal{E}(G'')$. The spanning subgraph G' of G is an *even subgraph* if every vertex of G' has even degree. Every cycle in G can be regarded as an even subgraph if enough isolated vertices of G are added. We collect some basic properties of even subgraphs next.

Proposition 3.6.7 (Even subgraphs)[5] *The following hold.*

(i) *The symmetric difference of two even subgraphs of a graph is an even subgraph of the graph.*
(ii) *A subgraph of a graph is an even subgraph if and only if it is an edge-disjoint union of cycles viewed as even subgraphs.*

A consequence of Proposition 3.6.7(i) is that the set of all even subgraphs of a graph G forms a linear space $\mathcal{Z}(G)$ over the field $GF(2)$, the 2-element field, with respect to the symmetric difference of spanning subgraphs. The linear space $\mathcal{Z}(G)$ is called the *cycle space* of G. And a consequence of Proposition 3.6.7(ii) is that the cycles of G, viewed as even subgraphs, span $\mathcal{Z}(G)$. A collection B of cycles in G (viewed as even subgraphs) is a *2-basis* of G if it is a basis of $\mathcal{Z}(G)$ and every edge of G is contained in at most two elements of B. MacLane (1937) characterised planar graphs in terms of 2-bases.

[5] A proof is available in Diestel (2017, prop. 1.9.1).

Theorem 3.6.8 (MacLane, 1937) *A graph is planar if and only if its cycle space has a 2-basis.*

Proofs of MacLane's theorem via Kuratowski's theorem (C.8.1) can be found in Diestel (2017, thm. 4.5.1), and Mohar and Thomassen (2001, thm. 2.4.5).

Let G be a 2-connected planar graph and let G' be a planar realisation of G. Then every face of G' is bounded by a facial cycle (Proposition C.7.4) and every edge of G' lies in at most two facial cycles. Moreover, every cycle C in $\mathcal{Z}(G')$ is the symmetric difference of facial cycles (regarded as even subgraphs). To see this, observe that all the edges in the interior of C occur in two facial cycles in the interior of C, and so cancel themselves out. This assertion will be used later in the book, and so we turn it into a proposition.

Proposition 3.6.9 *Every cycle in a 2-connected plane graph is the symmetric difference of the facial cycles contained in the interior of the cycle.*

Proposition 3.6.9 gives a characterisation of the 2-connected planar graphs that are bipartite.

Corollary 3.6.10 *A 2-connected plane graph is bipartite if and only if every facial cycle is bipartite.*

Proof If a 2-connected plane graph is bipartite, then every facial cycle must have even length and so it is bipartite (Proposition C.4.7). If a 2-connected plane graph G is nonbipartite, then G has a cycle C of odd length (Proposition C.4.7). By Proposition 3.6.9, the cycle C (as an even subgraph) is the symmetric difference of facial cycles of G. Accordingly, one of these facial cycles must have odd length. □

Also from Proposition 3.6.9, it follows that the facial cycles of a planar representation G' of a graph G form a 2-basis of G. Combining this discussion with MacLane's theorem and Tutte (1963, thm. 2.6) on the cycle space of 3-connected graphs, we obtain another planar criterion for 3-connected planar graphs.

Theorem 3.6.11 (Kelmans, 1980)[6] *A 3-connected graph is planar if and only if every edge belongs to exactly two nonseparating cycles.*

Both Proposition 3.6.9 and Corollary 3.6.10 immediately extend to graphs of 3-polytopes (Proposition 3.2.11). The next theorem shows that they extend

[6] A proof is available in Diestel (2017, thm. 4.5.2).

to graphs of polytopes of all dimensions. The proof of the theorem is a neat application of shellings.

Theorem 3.6.12 *For $d \geqslant 2$, every even subgraph in the graph of a d-polytope is the symmetric difference of facial cycles. In particular, every cycle in the graph of a d-polytope is the symmetric difference of facial cycles.*

Proof For $d = 2$, the theorem is trivially true. So assume that $d \geqslant 3$ and proceed by induction on d.

Let $S := F_1, \ldots, F_s$ be a shelling of P, let

$$\mathcal{C}_n := \mathcal{C}(F_1 \cup \cdots \cup F_n),$$

and let G_n be the graph of $\mathcal{C}_n$. For $i \in [1 \ldots s-1]$, each facet F_i is a $(d-1)$-polytope, and so the induction hypothesis ensures that the even subgraphs in each $G(F_i)$ are generated by facial cycles in such a graph.

Since all the edges of P lie in G_{s-1} – and, in particular, $G_{s-1} = G$ – it suffices to prove that every even subgraph in G_n is spanned by the facial cycles of G_n, for each $n \in [1 \ldots s-1]$. We further proceed by induction on n. The case $n = 1$ holds because $G_1 = G(F_1)$. We then assume that each even subgraph of G_{n-1}, where $1 \leqslant n-1 \leqslant s-2$, is spanned by the facial cycles in G_{n-1} and that some *problematic* even subgraph in G_n is not. Because $G_n = G_{n-1} \cup G(F_n)$, each problematic even subgraph in G_n is contained in neither G_{n-1} nor $G(F_n)$. Let C be a problematic even subgraph in G_n with a smallest number of edges in $G_{n-1} \backslash G(F_n)$ among the problematic even subgraphs in G_n. Each even subgraph is the symmetric difference of cycles (Proposition 3.6.7); and consequently, we have that C is a cycle of G_n.

Think of C as a cycle directed from $G_{n-1} \backslash G(F_n)$ to $G(F_n)$, starting at a vertex in $G_{n-1} \backslash G(F_n)$. The cycle C intersects $G_{n-1} \cap G(F_n)$ in the distinct vertices $\boldsymbol{x}_1, \ldots, \boldsymbol{x}_j$ found in this order as we traverse C. That is, $\boldsymbol{x}_1$ is the first vertex on C that touches $G_{n-1} \cap G(F_n)$, $\boldsymbol{x}_j$ is the last vertex on C that touches $G_{n-1} \cap G(F_n)$, and the directed subpath $L := \boldsymbol{x}_j C \boldsymbol{x}_1$ of C lies in $G_{n-1} \backslash G(F_n)$ except for $\boldsymbol{x}_1$ and $\boldsymbol{x}_j$. If $v(G_{n-1} \cap G(F_n) \cap C) = 1$, then C would be a union of cycles with a cycle in G_{n-1} and a cycle in $G(F_n)$, a contradiction to C being a cycle. Therefore $j \geqslant 2$.

The graph $G_{n-1} \cap G(F_n)$ is connected. Since S is a shelling, the complex $\mathcal{B}(F_n) \cap \mathcal{C}_{n-1}$ is the beginning of a shelling of F_n, which implies that $\mathcal{B}(F_n) \cap \mathcal{C}_{n-1}$ is a strongly connected $(d-2)$-complex (Proposition 3.3.4); the connectivity of $G_{n-1} \cap G(F_n)$ now follows from Proposition 3.3.1.

From the connectivity of $G_{n-1} \cap G(F_n)$ follows the existence of an $\boldsymbol{x}_1 - \boldsymbol{x}_j$ path M in $G_{n-1} \cap G(F_n)$. Concatenating the paths L and M, we form a cycle

C_1 in G_{n-1}. By the induction hypothesis on n, this cycle C_1 is the symmetric difference of facial cycles in G_{n-1}. Let L' be the directed subpath of C from $\boldsymbol{x}_1$ to $\boldsymbol{x}_j$. Then $C = L \cup L'$ and $\mathcal{V}(L) \cap \mathcal{V}(L') = \{\boldsymbol{x}_1, \boldsymbol{x}_j\}$. In addition, let $W := L' \triangle M$; here, we understand L' and M as spanning subgraphs of G_n (see Section C.1). It follows that W is an even subgraph of G, since $\boldsymbol{x}_1$ and $\boldsymbol{x}_j$ each have degree one in both L' and W, and every other vertex in W has even degree in both L' and W. It is also the case that W has fewer edges in $G_{n-1} \backslash G(F_n)$ than C, and so it is the symmetric difference of facial cycles in G_n.

The cycle C is the symmetric difference of C_1 and W and, as a consequence, it is the symmetric difference of facial cycles in G_n. This contradiction ensures that the cycle C does not exist, which amounts to saying that every even subgraph in G_n is spanned by facial cycles in G_n. Hence the induction is complete, and so is the proof of the theorem. □

As a first corollary of Theorem 3.6.12 we get the following.

Corollary 3.6.13 *For $d \geqslant 2$, the cycle space of the graph of a d-polytope is spanned by the facial cycles of the polytope.*

As a second corollary of Theorem 3.6.12, we obtain a characterisation of bipartite polytopal graphs due to Blind and Blind (1994).

Corollary 3.6.14 (Blind and Blind, 1994) *For $d \geqslant 2$, the graph of a d-polytope is bipartite if and only if every 2-face of the polytope is bipartite.*

Proof With the use of Theorem 3.6.12, proceed, mutatis mutandis, as in the proof of Corollary 3.6.10. □

3.7 Tutte's Realisation Theorem

We present a theorem of Tutte (1963) that produces a concrete convex realisation in $\mathbb{R}^2$ of a 3-connected planar graph G. Under some circumstances, this convex realisation yields a Schlegel diagram of a 3-polytope whose graph is G; we will use these realisations to prove Steinitz's theorem (3.8.1).

A convex polygon P is *strictly convex* if, for every edge $\boldsymbol{uv}$ of P, all the vertices of $\mathcal{V}(P) \backslash \{\boldsymbol{u}, \boldsymbol{v}\}$ are on one open halfspace of the line aff$\{\boldsymbol{u}, \boldsymbol{v}\}$ in aff P.

Let X be a graph with a vertex set $\mathcal{V}(X) = \{1, \ldots, n\}$. We assign points or positions $\boldsymbol{p}_1, \ldots, \boldsymbol{p}_n$ in $\mathbb{R}^2$ to the vertices of X to form a graph G in $\mathbb{R}^2$ whose vertices are $\boldsymbol{p}_1, \ldots, \boldsymbol{p}_n$ and whose edges are precisely the pairs $\boldsymbol{p}_i\boldsymbol{p}_j$ such that

ij is an edge of X. That is, there is an isomorphism σ between the graphs X and G. We join adjacent vertices of G by line segments. The graph G, or the pair $(X, \boldsymbol{p})$ where $\boldsymbol{p} := (\boldsymbol{p}_1, \ldots, \boldsymbol{p}_n)$, is said to be a *framework* of X in $\mathbb{R}^2$; here, we use the notion of 'framework' and not that of 'realisation' (Section 3.1) because two distinct vertices of G may be represented by the same point in $\mathbb{R}^2$, which is not allowed for realisations, and edges may intersect at points other than their endpoints, which is not allowed for realisations either. We will return to the notion of framework in connection to the rigidity of graphs in Section 8.3.

Let G be a framework in $\mathbb{R}^2$ of a graph X and let $\mathcal{V}(G) = \{\boldsymbol{p}_1, \ldots, \boldsymbol{p}_n\}$. Suppose that positive weights have been placed on the edges of X. Thanks to the bijection σ, we often think that the weight c_{ij} of the edge ij of X has also been placed on the edge $\boldsymbol{p}_i\boldsymbol{p}_j$ of G; here, we assume that $c_{ij} = c_{ji}$, since we deal only with (undirected) graphs. Suppose that $\boldsymbol{p}_i = (x_i, y_i)^t$ (for $i \in [1 \ldots n]$). We say that a vertex $\boldsymbol{p}_\ell$ of G is in *equilibrium* provided that it is in the convex hull of its neighbours in G, namely provided that

$$\begin{aligned} &\sum_{\ell j \in \mathcal{E}(X)} c_{\ell j}(\boldsymbol{p}_\ell - \boldsymbol{p}_j) = \boldsymbol{0}_2; \text{ or equivalently that} \\ &\sum_{\ell j \in \mathcal{E}(X)} c_{\ell j}(x_\ell - x_j) = 0, \quad \sum_{\ell j \in \mathcal{E}(X)} c_{\ell j}(y_\ell - y_j) = 0. \end{aligned} \tag{3.7.1}$$

The faces of a 3-connected planar graph X are uniquely determined, by Whitney's theorem (3.2.12). So we can talk of faces and facial cycles of X, independent of a planar realisation of X. Similarly, we can talk of the corresponding faces and facial cycles in a framework in $\mathbb{R}^2$ of X.

Definition 3.7.2 (Tutte realisation) Let X be a 3-connected planar graph, let F be a face of X, and let B be the facial cycle of F. A framework G of X in $\mathbb{R}^2$ is a *Tutte realisation* (with respect to F) if

(i) the facial cycle C in G corresponding to B is the boundary of a strictly convex polygon in $\mathbb{R}^2$, and
(ii) every vertex in $\mathcal{V}(G)\backslash\mathcal{V}(C)$ is in equilibrium.

Figure 3.7.1 illustrates the difference between a framework that is a Tutte realisation and a framework that is not. In a Tutte realisation G, we think of the vertices of the aforementioned facial cycle C as being nailed to the plane and the edges with one endvertex in $\mathcal{V}(G)\backslash\mathcal{V}(C)$ as being rubber bands. Then we let the configuration find its equilibrium. Tutte's realisation theorem reads as follows.

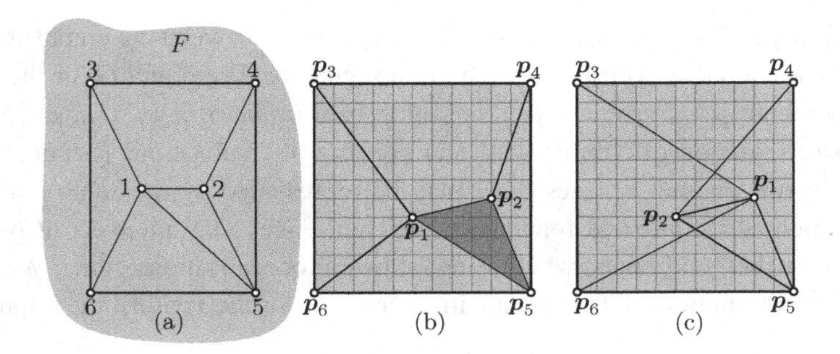

Figure 3.7.1 Two frameworks of a 3-connected plane graph X_1. (a) A planar realisation of X_1 with outer face F. (b) A framework G_1 of X_1 with respect to F that is a Tutte realisation. The boundary polygon of G_1 is depicted in grey, while its boundary cycle $\boldsymbol{p}_3\boldsymbol{p}_4\boldsymbol{p}_5\boldsymbol{p}_6\boldsymbol{p}_3$ is depicted in heavier lines. The Tutte realisation G_1 has five interior faces, including a face highlighted in dark grey, and one exterior face. (c) A framework G_1' of X_1 with respect to F that is not a Tutte realisation; for instance, the vertex $\boldsymbol{p}_2$ is not in equilibrium. The boundary polygon of G_1' is depicted in grey, while its boundary cycle $\boldsymbol{p}_3\boldsymbol{p}_4\boldsymbol{p}_5\boldsymbol{p}_6\boldsymbol{p}_3$ is depicted in heavier lines.

Theorem 3.7.3 (Tutte, 1963) *Let X be a 3-connected planar graph. Every Tutte realisation of X with respect to any face of X is a convex realisation of X in which the edges are represented by line segments and the faces are strictly convex polygons.*

From now to the end of this section, we let X be a 3-connected planar graph with positive weights c_{ij} on its edges, let $\mathcal{V}(X) = \{1, \ldots, n\}$, let G be a framework in $\mathbb{R}^2$ of X, and let $\mathcal{V}(G) = \{\boldsymbol{p}_1, \ldots, \boldsymbol{p}_n\}$. Additionally, we let F represent a fixed face of X. The facial cycle C in G corresponding to the facial cycle B of F is the *boundary cycle* of G (Fig. 3.7.1(b)–(c)). The vertices in C are the *boundary vertices* of G, while the vertices in $\mathcal{V}(G)\backslash\mathcal{V}(C)$ are the *interior vertices* of G. Likewise, the *boundary edges* of G are those of C, and the *interior edges* of G are those containing at least one interior vertex. We also find it convenient to define the *interior edges* of X as those not in B. An *interior face* of G is a face of G other than the one bounded by C; the face of G bounded by C is its *exterior face*; see Fig. 3.7.1(b).

Assume that the boundary vertices $\boldsymbol{p}_{r+1}, \ldots, \boldsymbol{p}_n$ of the framework G are predetermined and form the boundary of a strictly convex polygon K in $\mathbb{R}^2$; this is clearly always possible. The polygon K is the *boundary polygon* of G (Fig. 3.7.1(b)–(c)). We show that the points of the interior vertices $\boldsymbol{p}_1, \ldots, \boldsymbol{p}_r$ of G can be chosen so that G is a Tutte realisation of X.

Lemma 3.7.4 *There exist unique interior vertices $p_1, \dots, p_r$ in $\mathbb{R}^2$ for the framework G that make it a Tutte realisation of X.*

Proof For $i \in [1 \dots n]$, suppose that $p_i = (x_i, y_i)^t$, and for $j \in [r+1 \dots n]$ suppose that the points p_j are known. Further, let $\mathcal{E}'(X) := \mathcal{E}(X) \backslash \mathcal{E}(F)$ and let $\varphi \colon \mathbb{R}^{2r} \to \mathbb{R}$ be defined as

$$\varphi(x_1, \dots, x_r, y_1, \dots, y_r) := \frac{1}{2} \sum_{ij \in \mathcal{E}'(X)} c_{ij} \left((x_i - x_j)^2 + (y_i - y_j)^2 \right), \tag{3.7.4.1}$$

$$= \frac{1}{2} \sum_{ij \in \mathcal{E}'(X)} c_{ij} \| p_i - p_j \|^2.$$

We let the values of the interior vertices of G vary until they reach a configuration that minimises φ.

A quadratic function is convex, and the function φ is a sum of quadratic functions. Thus φ is convex. Being *convex* in $\mathbb{R}^d$ means that $\varphi(\alpha u + (1 - \alpha) w) \leqslant \alpha \varphi(u) + (1 - \alpha) \varphi(w)$ for every $\alpha \in (0, 1)$ and every two distinct points u and w in $\mathbb{R}^d$. Because X is connected, it is even true that φ is *strictly convex* in $\mathbb{R}^d$, namely that $\varphi(\alpha u + (1 - \alpha) w) < \alpha \varphi(u) + (1 - \alpha) \varphi(w)$ for every $\alpha \in (0, 1)$ and every two distinct points u and w in $\mathbb{R}^d$.

Claim 1 The function φ is strictly convex in $\mathbb{R}^{2r}$.

Proof of claim Consider two different points $x' := (x'_1, \dots, x'_r, y'_1, \dots, y'_r)^t$ and $x'' := (x''_1, \dots, x''_r, y''_1, \dots, y''_r)^t$ in $\mathbb{R}^{2r}$. For an edge $p_k p_\ell$ of G, define the function:

$$\zeta(x_1, \dots, x_r, y_1, \dots, y_r) := \frac{1}{2} \sum_{\substack{ij \in \mathcal{E}'(X) \\ ij \neq k\ell}} c_{ij} \left((x_i - x_j)^2 + (y_i - y_j)^2 \right)$$

$$+ \frac{1}{2} c_{k\ell} (y_k - y_\ell)^2.$$

The function ζ is a sum of quadratic functions, and so ζ is convex (as $c_{ij} > 0$ for each edge ij of X).

Further, let $I := \{ i \in [1 \dots r] \mid x'_i \neq x''_i \}$. Since $x' \neq x''$, we have that $I \neq \varnothing$. Suppose that L is a shortest path in G from the set $\{ p_i \mid i \in I \}$ to the boundary vertices of G, which exists because G is connected. We consider two cases according to the length of L.

First, suppose that $L = p_k p_\ell$ with $k \in I$ and $p_\ell \in C$. Let

$$\varrho(x) := \frac{1}{2} c_{k\ell} (x - x_\ell)^2.$$

Then

$$\zeta(x_1, \ldots, x_r, y_1, \ldots, y_r) = \varphi(x_1, \ldots, x_r, y_1, \ldots, y_r) - \varrho(x_k).$$

The function ϱ is strictly convex in $\mathbb{R}$ because its second derivative is positive, for every $x \in \mathbb{R}$ (Problem C.10.21); here, x_ℓ is a constant. Moreover, since ζ is convex, for every $\alpha \in (0, 1)$ it follows that

$$\zeta(\alpha\boldsymbol{x}' + (1-\alpha)\boldsymbol{x}'') \leqslant \alpha\zeta(\boldsymbol{x}') + (1-\alpha)\zeta(\boldsymbol{x}''),$$

and since $x'_k \neq x''_k$ (as $k \in I$) and ϱ is strictly convex it follows that

$$\varrho(\alpha x'_k + (1-\alpha)x''_k) < \alpha\varrho(x'_k) + (1-\alpha)\varrho(x''_k).$$

As a consequence of these two inequalities, we have that

$$\begin{aligned}\varphi(\alpha\boldsymbol{x}' + (1-\alpha)\boldsymbol{x}'') &= \zeta(\alpha\boldsymbol{x}' + (1-\alpha)\boldsymbol{x}'') + \varrho(\alpha x'_k + (1-\alpha)x''_k)\\ &< \alpha\zeta(\boldsymbol{x}') + (1-\alpha)\zeta(\boldsymbol{x}'') + \alpha\varrho(x'_k) + (1-\alpha)\varrho(x''_k)\\ &= \alpha\varphi(\boldsymbol{x}') + (1-\alpha)\varphi(\boldsymbol{x}'').\end{aligned}$$

This proves that φ is strictly convex in this case.

Now suppose that the path L has length greater than one. Then there is an edge $\boldsymbol{p}_k\boldsymbol{p}_\ell$ where $\boldsymbol{p}_k$ and $\boldsymbol{p}_\ell$ are interior vertices, and $k \in I$ and $\ell \notin I$. It follows that $x'_k \neq x''_k$ and $x'_\ell = x''_\ell$. Let

$$\eta(u, v) := \frac{1}{2}c_{k\ell}(u - v)^2.$$

The function η is convex but not strictly convex. However, if we fix the variable v, then it is strictly convex in u as the second derivative is positive for every u (Problem C.10.21); this kind of function is sometimes called *coordinate-wise strictly convex*. Thus, for every different $u', u'' \in \mathbb{R}$ and every $\beta \in (0, 1)$, we have that

$$\eta(\beta u' + (1-\beta)u'', v) < \beta\eta(u', v) + (1-\beta)\eta(u'', v).$$

And because $x'_k \neq x''_k$ and $\alpha x'_\ell + (1-\alpha)x''_\ell = x'_\ell = x''_\ell$, this equation implies that

$$\eta(\alpha x'_k + (1-\alpha)x''_k, \alpha x'_\ell + (1-\alpha)x''_\ell) < \alpha\eta(x'_k, x'_\ell) + (1-\alpha)\eta(x''_k, x''_\ell). \quad (3.7.4.2)$$

Furthermore, it is true that

$$\zeta(x_1, \ldots, x_r, y_1, \ldots, y_r) = \varphi(x_1, \ldots, x_r, y_1, \ldots, y_r) - \eta(x_k, x_\ell).$$

By virtue of this equality, the inequality (3.7.4.2), and the convexity of ζ, we have that, for every $\alpha \in (0,1)$,

$$\begin{aligned}\varphi(\alpha\boldsymbol{x}' + (1-\alpha)\boldsymbol{x}'') &= \zeta(\alpha\boldsymbol{x}' + (1-\alpha)\boldsymbol{x}'') \\ &\quad + \eta(\alpha x_k' + (1-\alpha)x_k'', \alpha x_\ell' + (1-\alpha)x_\ell'') \\ &= \zeta(\alpha\boldsymbol{x}' + (1-\alpha)\boldsymbol{x}'') + \eta(\alpha x_k' + (1-\alpha)x_k'', x_\ell'') \\ &< \alpha\zeta(\boldsymbol{x}') + (1-\alpha)\zeta(\boldsymbol{x}'') + \alpha\eta(x_k', x_\ell') \\ &\quad + (1-\alpha)\eta(x_k'', x_\ell'') \\ &= \alpha\varphi(\boldsymbol{x}') + (1-\alpha)\varphi(\boldsymbol{x}'').\end{aligned}$$

This inequality implies that φ is strictly convex in this case as well. □

Because φ is nonnegative and strictly convex in $\mathbb{R}^{2r}$ (by Claim 1), it attains a unique minimum on $\mathbb{R}^{2r}$ (Problem C.10.22). At this minimum point, for every $i \in [1 \dots r]$ the partial derivative of φ with respect to x_i is zero:

$$\frac{\partial\varphi}{\partial x_i} = \sum_{ij \in \mathcal{E}'(X)} c_{ij}(x_i - x_j) = 0, \quad \frac{\partial\varphi}{\partial y_i} = \sum_{ij \in \mathcal{E}'(X)} c_{ij}(y_i - y_j) = 0. \quad (3.7.4.3)$$

This is precisely the equilibrium condition for the interior vertices; see (3.7.1). Hence, if the values of the interior vertices of G are computed according to (3.7.4.3), then the framework G becomes a Tutte realisation of X. □

We give two examples of the computation of the equilibrium positions of interior vertices using (3.7.4.3).

Example 3.7.5 Let X_1 be the 3-connected plane graph of Fig. 3.7.1(a) and let X_2 be the 3-connected plane graph of Fig. 3.7.2(a); both graphs have weights

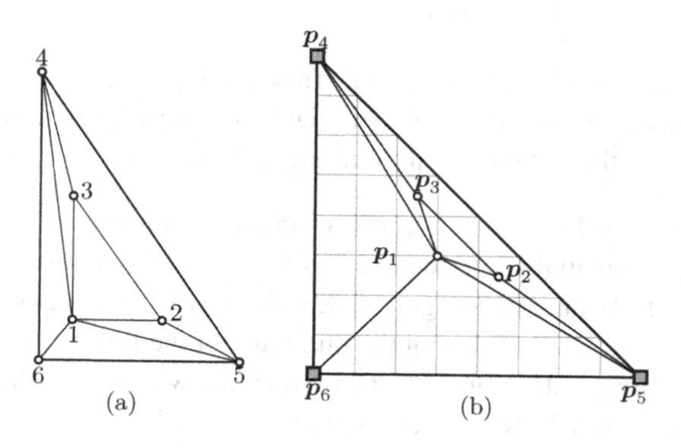

Figure 3.7.2 A Tutte realisation of a 3-connected plane graph X_2. (a) A planar realisation of X_2. (b) A Tutte realisation of the planar realisation in (a).

equal to one on all the interior edges. We compute Tutte realisations of X_1 and X_2, respectively.

(i) Let G_1 represent a framework of X_1 with boundary vertices $\boldsymbol{p}_3 = (x_3, y_3)^t = (0,1)^t$, $\boldsymbol{p}_4 = (x_4, y_4)^t = (1,1)^t$, $\boldsymbol{p}_5 = (x_5, y_5)^t = (1,0)^t$, and $\boldsymbol{p}_6 = (x_6, y_6)^t = (0,0)^t$. We compute the values $\boldsymbol{p}_1 = (x_1, y_1)^t$ and $\boldsymbol{p}_2 = (x_2, y_2)^t$ of the interior vertices of G_1 so that G_1 becomes a Tutte realisation of X_1. Consider the equilibrium of each interior vertex:

$$\boldsymbol{p}_1\colon x_1 - x_2 + x_1 - x_3 + x_1 - x_5 + x_1 - x_6 = 0,$$
$$\boldsymbol{p}_1\colon y_1 - y_2 + y_1 - y_3 + y_1 - y_5 + y_1 - y_6 = 0,$$
$$\boldsymbol{p}_2\colon x_2 - x_1 + x_2 - x_4 + x_2 - x_5 = 0,$$
$$\boldsymbol{p}_2\colon y_2 - y_1 + y_2 - y_4 + y_2 - y_5 = 0.$$

And then solve for x_1, y_1, x_2, y_2. This gives $\boldsymbol{p}_1 = (5/11, 4/11)^t$, $\boldsymbol{p}_2 = (9/11, 5/11)^t$. With coordinates for all the vertices of G_1, we draw the edges of G_1 with line segments, obtaining the realisation in Fig. 3.7.1(b).

(ii) Let G_2 be a framework of X_2 with boundary vertices $\boldsymbol{p}_4 = (x_4, y_4)^t = (0,1)^t$, $\boldsymbol{p}_5 = (x_5, y_5)^t = (1,0)^t$, and $\boldsymbol{p}_6 = (x_6, y_6)^t = (0,0)^t$. We compute the values $\boldsymbol{p}_1 = (x_1, y_1)^t$, $\boldsymbol{p}_2 = (x_2, y_2)^t$, and $\boldsymbol{p}_3 = (x_3, y_3)^t$ of the interior vertices of G_2. Consider the equilibrium of each interior vertex:

$$\boldsymbol{p}_1\colon x_1 - x_2 + x_1 - x_3 + x_1 - x_4 + x_1 - x_5 + x_1 - x_6 = 0,$$
$$\boldsymbol{p}_1\colon y_1 - y_2 + y_1 - y_3 + y_1 - y_4 + y_1 - y_5 + y_1 - y_6 = 0,$$
$$\boldsymbol{p}_2\colon x_2 - x_1 + x_2 - x_3 + x_2 - x_5 = 0,$$
$$\boldsymbol{p}_2\colon y_2 - y_1 + y_2 - y_3 + y_2 - y_5 = 0,$$
$$\boldsymbol{p}_3\colon x_3 - x_1 + x_3 - x_2 + x_3 - x_4 = 0,$$
$$\boldsymbol{p}_3\colon y_3 - y_1 + y_3 - y_2 + y_3 - y_4 = 0.$$

And then solve for $x_1, y_1, x_2, y_2, x_3, y_3$. This gives $\boldsymbol{p}_1 = (3/8, 3/8)^t$, $\boldsymbol{p}_2 = (9/16, 5/16)^t$, and $\boldsymbol{p}_3 = (5/16, 9/16)^t$. Drawing the edges of G_2 with line segments, we get the Tutte realisation in Fig. 3.7.2(b).

With the help of Lemma 3.7.4, and specifically equations (3.7.4.3), given positive weights on all the interior edges of the graph X (and of the framework G), and given the boundary polygon of G, we can vary the values of the interior vertices $\boldsymbol{p}_1, \ldots, \boldsymbol{p}_r$ of G so that, at the minimum of the function φ in (3.7.4.1), G becomes a Tutte realisation of X. Consequently, we henceforth assume that the framework G is a Tutte realisation of the graph X.

In a framework of a 3-connected planar graph, we may have a facial cycle that is not a simple closed curve in $\mathbb{R}^2$. For instance, the facial cycle

$\boldsymbol{p}_1\boldsymbol{p}_2\boldsymbol{p}_4\boldsymbol{p}_3\boldsymbol{p}_1$ in Fig. 3.7.1(c) is not a simple closed curve in $\mathbb{R}^2$, while the facial cycle $\boldsymbol{p}_1\boldsymbol{p}_5\boldsymbol{p}_6\boldsymbol{p}_1$ is. The definition of a Tutte realisation does not require the facial cycles of the framework G to all be simple closed curves in $\mathbb{R}^2$. However, Tutte's realisation theorem (3.7.3) ensures this and more. To prove the theorem, we need to exclude six possible degeneracies:

Degeneracy 1 An interior vertex of G does not lie in the interior of the boundary polygon of G.
Degeneracy 2 An interior vertex of G is collinear with all its neighbours.
Degeneracy 3 An edge ij of X is mapped to a single point $\boldsymbol{p}_i = \boldsymbol{p}_j$ in G.
Degeneracy 4 A facial cycle of G is not a simple closed curve in $\mathbb{R}^2$ or does not bound a face that is a strictly convex polygon.
Degeneracy 5 Two faces in G overlap.
Degeneracy 6 Two edges cross each other.

We rule out these degeneracies in order.

Proposition 3.7.6 (Excluding Degeneracy 1) *All the interior vertices of G lie in the interior of the boundary polygon of G.*

Proof Take an interior vertex $\boldsymbol{p}_\ell$ of G. The vertex $\boldsymbol{p}_\ell$ is in equilibrium (see (3.7.1)), and so $\boldsymbol{p}_\ell$ can be expressed as the convex hull of its neighbours in G:

$$\boldsymbol{p}_\ell = \frac{1}{\sum_{\boldsymbol{p}_i \in \mathcal{N}(\boldsymbol{p}_\ell)} c_{\ell i}} \sum_{\boldsymbol{p}_i \in \mathcal{N}(\boldsymbol{p}_\ell)} c_{\ell i} \boldsymbol{p}_i. \tag{3.7.6.1}$$

Use this expression to remove $\boldsymbol{p}_\ell$ from the system (3.7.4.3). Removing $\boldsymbol{p}_\ell$ from this system adds edges between the neighbours of $\boldsymbol{p}_\ell$. For instance, suppose that $\boldsymbol{p}_r$ is a neighbour of $\boldsymbol{p}_\ell$. Then the equilibrium equation for $\boldsymbol{p}_r$ reads as:

$$\sum_{\substack{\boldsymbol{p}_j \in \mathcal{N}(\boldsymbol{p}_r) \\ j \neq \ell}} c_{rj}(\boldsymbol{p}_r - \boldsymbol{p}_j) + c_{r\ell}(\boldsymbol{p}_r - \boldsymbol{p}_\ell) = \boldsymbol{0}_2. \tag{3.7.6.2}$$

We then let $\beta := \sum_{\boldsymbol{p}_i \in \mathcal{N}(\boldsymbol{p}_\ell)} c_{\ell i}$ and replace $\boldsymbol{p}_\ell$ in (3.7.6.2) with (3.7.6.1), which yields the following:

$$\sum_{\substack{\boldsymbol{p}_j \in \mathcal{N}(\boldsymbol{p}_r) \\ j \neq \ell}} c_{rj}(\boldsymbol{p}_r - \boldsymbol{p}_j) + c_{r\ell}\boldsymbol{p}_r - c_{r\ell}\left(\sum_{\boldsymbol{p}_i \in \mathcal{N}(\boldsymbol{p}_\ell)} \frac{c_{\ell i}}{\beta}\boldsymbol{p}_i\right) = \boldsymbol{0}_2,$$

$$\sum_{\substack{\boldsymbol{p}_j \in \mathcal{N}(\boldsymbol{p}_r) \\ j \neq \ell}} c_{rj}(\boldsymbol{p}_r - \boldsymbol{p}_j) + \sum_{\boldsymbol{p}_i \in \mathcal{N}(\boldsymbol{p}_\ell)} \frac{c_{r\ell}c_{\ell i}}{\beta}(\boldsymbol{p}_r - \boldsymbol{p}_i) = \boldsymbol{0}_2. \tag{3.7.6.3}$$

To get the last equality we used that $\sum_{p_i \in \mathcal{N}(p_\ell)} c_{\ell i}/\beta = 1$. The equality (3.7.6.3) means that p_r has become a neighbour of all the neighbours of p_ℓ that were not already neighbours of p_r, and so p_r is now a convex combination of its neighbours in $G - p_\ell$. Thus, if, for each neighbour p_r of p_ℓ, we replace (3.7.6.2) with (3.7.6.3) in the system (3.7.4.3), we obtain a new system that expresses the equilibrium condition for the interior vertices of G other than p_ℓ.

It is now clear that eliminating all but one interior vertex p_s leaves a constraint that expresses this vertex as a convex combination of the boundary vertices of G. As a consequence, p_s lies in the interior of K. There was nothing special about p_s, and so the lemma holds for all interior vertices. □

We now prepare the ground to deal with the second degeneracy.

Lemma 3.7.7 *Let p_ℓ be a vertex of G, and let ℓ be a line that passes through it and the interior of the boundary polygon of G. If p_ℓ is a boundary vertex, then there are neighbours of p_ℓ on both open halfspaces determined by ℓ. If instead p_ℓ is an interior vertex of G, then either all neighbours of p_ℓ are on ℓ or there are neighbours of p_ℓ on both open halfspaces determined by ℓ.*

Proof For each $x \in \mathbb{R}^2$, let φ be a linear functional in $\mathbb{R}^2$ such that the line ℓ is defined as $\varphi(x) = \varphi(p_\ell)$. If p_ℓ is a boundary vertex, then p_ℓ is a vertex of a strictly convex polytope, therefrom the lemma follows. So assume that p_ℓ is an interior vertex of G.

The vertex p_ℓ is in equilibrium, and so letting $\beta := \sum_{p_i \in \mathcal{N}(p_\ell)} c_{\ell i}$ we get that

$$p_\ell = \sum_{p_i \in \mathcal{N}(p_\ell)} \frac{c_{\ell i}}{\beta} p_i.$$

This expression, the equality $\sum_{p_i \in \mathcal{N}(p_\ell)} c_{\ell i}/\beta = 1$, and the linearity of φ yield that

$$\begin{aligned} 0 = \varphi(p_\ell - p_\ell) &= \varphi\left(\sum_{p_i \in \mathcal{N}(p_\ell)} \frac{c_{\ell i}}{\beta} p_\ell - \sum_{p_i \in \mathcal{N}(p_\ell)} \frac{c_{\ell i}}{\beta} p_i\right) \\ &= \sum_{p_i \in \mathcal{N}(p_\ell)} \frac{c_{\ell j}}{\beta} \varphi(p_\ell - p_i). \end{aligned}$$

If some summand $\varphi(p_\ell - p_r)$ in this last expression is positive, then some other summand $\varphi(p_\ell - p_s)$ must be negative. This proves the lemma. □

The next lemma is the key to exclude the collinearity of an interior vertex of G and its neighbours.

Lemma 3.7.8 *Let ℓ be a line in $\mathbb{R}^2$ intersecting the interior of the boundary polygon K of G, and let G^+ and G^- be the induced subgraphs of G whose vertices lie in the two open halfspaces determined by ℓ, respectively. Then both G^+ and G^- are connected.*

Proof For each $\boldsymbol{x} \in \mathbb{R}^2$ and some $\alpha \in \mathbb{R}$, let φ be a linear functional in $\mathbb{R}^2$ such that the line ℓ is given as $\varphi(\boldsymbol{x}) = \alpha$. Because of the boundary vertices of G, we have that $\mathcal{V}(G^+)$ and $\mathcal{V}(G^-)$ are nonempty. Without loss of generality, assume that the vertices of G^+ satisfy $\varphi(\boldsymbol{x}) > \alpha$. We prove the statement for G^+.

Let $\boldsymbol{p}_m$ be the vertex of G with maximum φ-value among the vertices of G, and let $\boldsymbol{p}_u$ be a vertex of G^+. Then $\boldsymbol{p}_m \in G^+$. Furthermore, by Proposition 3.7.6, $\boldsymbol{p}_m$ must be a boundary vertex of G. We prove the existence of a φ-nondecreasing (directed) path in G^+ from $\boldsymbol{p}_u$ to $\boldsymbol{p}_m$. If $\varphi(\boldsymbol{p}_u) = \varphi(\boldsymbol{p}_m)$, then either $\boldsymbol{p}_u = \boldsymbol{p}_m$ or $\boldsymbol{p}_u\boldsymbol{p}_m$ is an edge of K (and of G^+) because K is a strictly convex polygon. Thus assume that $\varphi(\boldsymbol{p}_m) > \varphi(\boldsymbol{p}_u)$.

Suppose there is a φ-nondecreasing path from every vertex with φ-value larger than $\varphi(\boldsymbol{p}_u)$ to $\boldsymbol{p}_m$. Let U be the set of vertices of G that are connected to $\boldsymbol{p}_u$ by a path L in G such that all the vertices of L have the same φ-value as $\boldsymbol{p}_u$; trivially, $\boldsymbol{p}_u \in U$ and $U \subseteq \mathcal{V}(G^+)$. Because G is a connected, there must exist a vertex $\boldsymbol{p}_v \in U$ with a neighbour not in U; otherwise there would not be a path in G from $\boldsymbol{p}_u$ to $\boldsymbol{p}_m$. Since $\boldsymbol{p}_v \in U$, we have a φ-nondecreasing $\boldsymbol{p}_u - \boldsymbol{p}_v$ path L_{uv} in G^+.

Moreover, we consider a line ℓ_v defined as $\varphi(\boldsymbol{x}) = \varphi(\boldsymbol{p}_v) = \varphi(\boldsymbol{p}_u)$, which passes through $\boldsymbol{p}_v$ and the interior of K. Then, from $\boldsymbol{p}_v$ having a neighbour not on ℓ_v follows the existence of a neighbour $\boldsymbol{p}_w$ of $\boldsymbol{p}_v$ satisfying $\varphi(\boldsymbol{p}_w) > \varphi(\boldsymbol{p}_v) = \varphi(\boldsymbol{p}_u)$ (Lemma 3.7.7). By our assumption, there is a φ-nondecreasing path L_m from $\boldsymbol{p}_w$ to $\boldsymbol{p}_m$. Concatenating the path L_{uv}, the edge $\boldsymbol{p}_v\boldsymbol{p}_w$, and the path L_m gives a φ-nondecreasing path from $\boldsymbol{p}_u$ to $\boldsymbol{p}_m$, as desired.

Proving the statement for G^- is analogous, and so the proof of the lemma is concluded. □

Disallowing the second degeneracy is now at hand.

Proposition 3.7.9 (Excluding Degeneracy 2) *No vertex of G is collinear with all its neighbours.*

Proof The statement is true for the boundary vertices of G, as they are vertices of a strictly convex polygon. By way of contradiction, suppose that an interior vertex $\boldsymbol{p}_u$ of G is collinear with all its neighbours; let ℓ be that line. Suppose that G^+ and G^- are the induced subgraphs of G whose vertices lie in the two open halfspaces determined by ℓ, respectively. The existence

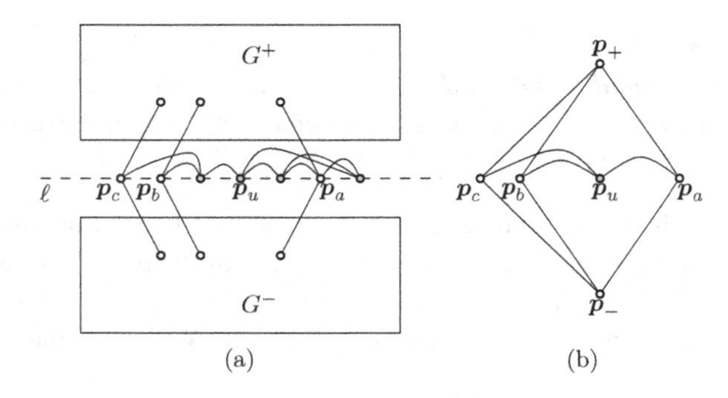

Figure 3.7.3 Auxiliary figure for the proof of Proposition 3.7.9. (a) A line ℓ and the subgraphs G^+ and G^- of G whose vertices lie in the two open halfspaces of ℓ. The vertices of U are drawn in light grey. (b) A $K^{3,3}$ minor in G.

of boundary vertices of G guarantees that G^+ and G^- are both nonempty subgraphs.

Let U represent the set of vertices of G that are connected to $\boldsymbol{p}_u$ by a path and whose neighbours are all on ℓ; trivially, $\boldsymbol{p}_u \in U$ but the boundary vertices of G are not all in U. Besides, let W be the set of vertices on ℓ that are neighbours of vertices in U, but are themselves not in U. From Lemma 3.7.7 it follows that every vertex in W has a neighbour in both G^+ and G^-. Removing the vertices of W disconnects the vertices in U from the rest of G. And since G is 3-connected, we have that $\#W \geqslant 3$.

Consider three distinct vertices $\boldsymbol{p}_a, \boldsymbol{p}_b, \boldsymbol{p}_c$ in W. See Fig. 3.7.3(a). In view that both G^+ and G^- are connected (Lemma 3.7.8), we can contract G^+ to a vertex $\boldsymbol{p}_+$ and G^- to a vertex $\boldsymbol{p}_-$. Moreover, we contract the set U to the vertex $\boldsymbol{p}_u$. In this contracted graph G', the vertices $\boldsymbol{p}_a, \boldsymbol{p}_b, \boldsymbol{p}_c$ are all adjacent to $\boldsymbol{p}_+$ and $\boldsymbol{p}_-$; see Fig. 3.7.3(b). It follows that the minor G' of G contains $K^{3,3}$ as a subgraph, where the partite sets are $\{\boldsymbol{p}_a, \boldsymbol{p}_b, \boldsymbol{p}_c\}$ and $\{\boldsymbol{p}_u, \boldsymbol{p}_+, \boldsymbol{p}_-\}$. This contradicts the planarity of G (Wagner's theorem (C.8.3)), and thus proves the proposition. □

It remains to deal with the edges, facial cycles, and faces of G. Spielman (2019, ch. 15) credits the next lemma to Jim Geelen. This lemma also appears in Lovász (2019, p. 25).

Lemma 3.7.10 (Geelen's lemma) *Let $\boldsymbol{p}_u\boldsymbol{p}_v$ be an interior edge of G, and let C_a and C_b be two facial cycles of G that share the edge $\boldsymbol{p}_u\boldsymbol{p}_v$. Further suppose that L is a path in G that is disjoint from $\{\boldsymbol{p}_u, \boldsymbol{p}_v\}$ and goes from a vertex in $\mathcal{V}(C_a)\backslash\{\boldsymbol{p}_u, \boldsymbol{p}_v\}$ to a vertex in $\mathcal{V}(C_b)\backslash\{\boldsymbol{p}_u, \boldsymbol{p}_v\}$. Then every $\boldsymbol{p}_u - \boldsymbol{p}_v$ path in G either consists of the edge $\boldsymbol{p}_u\boldsymbol{p}_v$ or contains a vertex of L.*

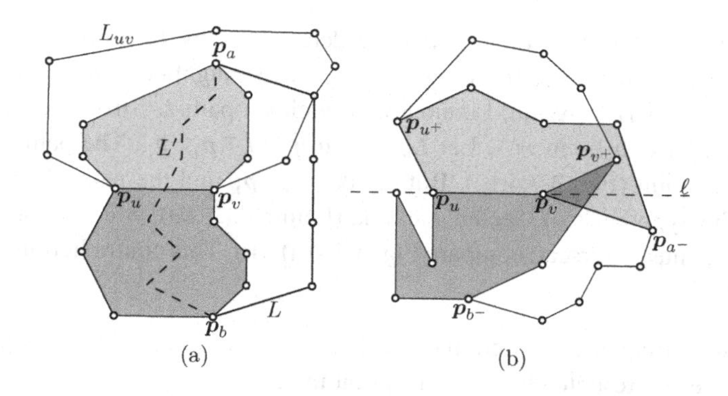

Figure 3.7.4 Auxiliary figure for the proofs of Lemmas 3.7.10 and 3.7.11.

Proof Let Y be a planar realisation of X and let J_a and J_b be the faces of Y whose facial cycles are C_a and C_b, respectively; here, we again used that the faces of X are uniquely determined by Whitney's realisation theorem (3.2.12). Suppose that L is a path in Y from a vertex $\boldsymbol{p}_a$ in $\mathcal{V}(C_a)\backslash\{\boldsymbol{p}_u, \boldsymbol{p}_v\}$ to a vertex $\boldsymbol{p}_b$ in $\mathcal{V}(C_b)\backslash\{\boldsymbol{p}_u, \boldsymbol{p}_v\}$.

Let L' be a simple polygonal arc from $\boldsymbol{p}_a$ to $\boldsymbol{p}_b$ that lies inside $J_a \cup J_b$ and passes through the edge $\boldsymbol{p}_u\boldsymbol{p}_v$ once. Then we may assume that the simple closed polygonal arc $M := \boldsymbol{p}_a L \boldsymbol{p}_b L' \boldsymbol{p}_a$ contains $\boldsymbol{p}_v$ in the interior and $\boldsymbol{p}_u$ in the exterior; see Fig. 3.7.4(a). By Jordan's curve theorem (3.1.2), every $\boldsymbol{p}_u - \boldsymbol{p}_v$ path L_{uv} in Y crosses M, which implies that L_{uv} either is the edge $\boldsymbol{p}_u\boldsymbol{p}_v$ or contains a vertex of L. That the path L_{uv} either consists of the edge $\boldsymbol{p}_u\boldsymbol{p}_v$ or contains a vertex of L is a combinatorial property of X, independent of the planar realisation Y. Hence, the lemma holds. □

Lemma 3.7.11 *Let $\boldsymbol{p}_u\boldsymbol{p}_v$ be an interior edge of G and let C_a and C_b be the facial cycles of G that share the edge $\boldsymbol{p}_u\boldsymbol{p}_v$. Then the vertices in $\mathcal{V}(C_a)\backslash\{\boldsymbol{p}_u, \boldsymbol{p}_v\}$ and $\mathcal{V}(C_b)\backslash\{\boldsymbol{p}_u, \boldsymbol{p}_v\}$ lie in opposite open halfspaces of a line ℓ through $\boldsymbol{p}_u$ and $\boldsymbol{p}_v$.*

Proof By way of contradiction, suppose that the lemma is false. Then we may assume that there are vertices $\boldsymbol{p}_a \in \mathcal{V}(C_a)\backslash\{\boldsymbol{p}_u, \boldsymbol{p}_v\}$ and $\boldsymbol{p}_b \in \mathcal{V}(C_b)\backslash\{\boldsymbol{p}_u, \boldsymbol{p}_v\}$ such that, for each $i = a, b$, $\boldsymbol{p}_i$ is on ℓ or on an open halfspace H^- determined by ℓ.

If $\boldsymbol{p}_a$ lies in ℓ, then there is a neighbour $\boldsymbol{p}_{a^-}$ of $\boldsymbol{p}_a$ in H^- (Proposition 3.7.9); otherwise, $\boldsymbol{p}_a$ lies in H^- and we let $\boldsymbol{p}_{a^-} := \boldsymbol{p}_a$. Similarly, if $\boldsymbol{p}_b$ lies in ℓ then some neighbour $\boldsymbol{p}_{b^-}$ of it lies in H^-; else, $\boldsymbol{p}_b$ lies in H^- and we let $\boldsymbol{p}_{b^-} := \boldsymbol{p}_b$. In any case, from Lemma 3.7.8 follows the existence of a path L' from $\boldsymbol{p}_{a^-}$ to $\boldsymbol{p}_{b^-}$ that has all its vertices in H^-. Let $L := \boldsymbol{p}_a\boldsymbol{p}_{a^-}L'\boldsymbol{p}_{b^-}\boldsymbol{p}_b$.

Let H^+ be the other open halfspace determined by ℓ. Proposition 3.7.9 also ensures that the vertices $\boldsymbol{p}_u$ and $\boldsymbol{p}_v$ have neighbours $\boldsymbol{p}_{u^+}$ and $\boldsymbol{p}_{v^+}$, respectively, in H^+. Again, Lemma 3.7.8 yields a path L^+ from p_{u^+} to p_{v^+} whose vertices all lie in H^+. Let $L_{uv} := \boldsymbol{p}_u \boldsymbol{p}_{u^+} L^+ \boldsymbol{p}_{v^+} \boldsymbol{p}_v$. The paths L and L_{uv} are disjoint (Fig. 3.7.4(b)). But the edge $\boldsymbol{p}_u \boldsymbol{p}_v$ and the paths L and L_{uv} satisfy the hypothesis of Geelen's lemma (Lemma 3.7.10), which implies that L and L_{uv} must intersect; compare Fig. 3.7.4(a)–(b). This contradiction proves the lemma. □

A first consequence of Lemma 3.7.11 is the nonexistence of edges joining two vertices represented by the same point in $\mathbb{R}^2$.

Proposition 3.7.12 (Excluding Degeneracy 3) *No interior edge of G is represented by a single point in* $\mathbb{R}^2$.

Proof Suppose that the edge ij of X maps to the edge $\boldsymbol{p}_i \boldsymbol{p}_j$ of G with $\boldsymbol{p}_i = \boldsymbol{p}_j$. Let C_a be a facial cycle that contains $\boldsymbol{p}_i \boldsymbol{p}_j$, and let ℓ be a line through $\boldsymbol{p}_i$ and a vertex $\boldsymbol{p}_u$ in C_a other than $\boldsymbol{p}_i$ and $\boldsymbol{p}_j$. Suppose that C_b is the other facial cycle of G containing the edge $\boldsymbol{p}_i \boldsymbol{p}_j$. When we apply Lemma 3.7.11 to the edge $\boldsymbol{p}_i \boldsymbol{p}_j$, and the facial cycles C_a and C_b, we get that the vertices in $\mathcal{V}(C_a) \backslash \{\boldsymbol{p}_i, \boldsymbol{p}_j\}$ and $\mathcal{V}(C_b) \backslash \{\boldsymbol{p}_i, \boldsymbol{p}_j\}$ lie in opposite open halfspaces determined by ℓ. But this contradicts the fact that $\boldsymbol{p}_u \in \mathcal{V}(C_a) \backslash \{\boldsymbol{p}_i, \boldsymbol{p}_j\}$ and $\boldsymbol{p}_u$ is on ℓ. □

A second consequence of Lemma 3.7.11 is that the faces of X are mapped onto strictly convex polygons of G.

Proposition 3.7.13 (Excluding Degeneracy 4) *Every facial cycle of G bounds a face that is a strictly convex polygon.*

Proof If the proposition did not hold for some facial cycle C_a, then we could find an interior edge $\boldsymbol{p}_u \boldsymbol{p}_v$ in C_a so that the edge $\boldsymbol{p}_u \boldsymbol{p}_v$, the facial cycle C_a, and the other facial cycle C_b containing the edge $\boldsymbol{p}_u \boldsymbol{p}_v$ would contradict Lemma 3.7.11. Thus the proposition follows. □

Finally, we exclude the existence of overlapping faces and crossing edges in G.

Proposition 3.7.14 (Excluding Degeneracies 5 and 6) *The interiors of any two faces in G are disjoint. Similarly, the interiors of any two edges in G are disjoint.*

Proof We first prove the statement for the faces. Let J be an interior face of G, and let $\boldsymbol{y}$ be a point in the interior of J. We show that no other face of G contains $\boldsymbol{y}$.

Let ℓ be a line through $\boldsymbol{y}$ that passes through no vertex of G or the intersection of two edges of G. For each point $\boldsymbol{x}$ of ℓ, we count the number $\tau(\boldsymbol{x})$ of faces of G that contain it in their interiors. We move along ℓ, from outside the boundary polygon K of G to $\boldsymbol{y}$. We first have that the points of ℓ outside K are contained only in the interior of the exterior face of G (the one with C as a facial cycle). Thus, $\tau(\boldsymbol{x}) = 1$ for these points $\boldsymbol{x}$ of ℓ. As we enter the interior of K, the number τ does not change unless we cross an edge of G. And when we cross an edge, we leave a face of G and enter another face of G, thanks to Lemma 3.7.11. Thus, this number $\tau(\boldsymbol{x})$ is constant for every $\boldsymbol{x} \in \ell$ that is not on an edge of G; in particular, $\tau(\boldsymbol{y}) = 1$. This proves that the statement related to the faces of G.

We now consider two edges intersecting at an interior point $\boldsymbol{z}$ of them. Then there is a point $\boldsymbol{y}$ very close to $\boldsymbol{z}$ that lies in the interiors of two faces. The existence of $\boldsymbol{y}$ contradicts the first part of the proposition. Hence we conclude the proof of the proposition. □

We are finally ready to prove Tutte's realisation theorem (3.7.3).

Proof of Tutte's realisation theorem (3.7.3) We choose a face F of X and, according to Lemma 3.7.4, produce a Tutte realisation G of X with respect to F. From Propositions 3.7.6, 3.7.9 and 3.7.12 to 3.7.14, it follows that G is a convex realisation of X in which the faces are strictly convex polygons. □

Fáry's theorem (3.1.5) follows from the existence of Tutte realisations of 3-connected planar graphs. If a plane graph G is 3-connected, then a Tutte realisation of it depicts all edges as line segments. If, instead, G is not 3-connected, then we add to G as many edges as possible to obtain a maximal plane graph G'. Maximal plane graphs are 3-connected according to Theorem C.7.6, which ensures that G' admits a Tutte realisation G''. Removing the edges $\mathcal{E}(G')\backslash\mathcal{E}(G)$ from G'' yields a planar realisation of G where all the edges are represented by line segments.

Determining a Tutte realisation of a 3-connected planar graph involves solving the system (3.7.4.3) of linear equations. Thus, if we choose appropriate integer coordinates for the boundary vertices of the Tutte realisation and weights equal to one on all the interior edges of the planar graph, then we can obtain a Tutte realisation with rational coordinates, and thus with integer coordinates (Richter-Gebert, 2006, sec. 13.2). This implies that every 3-connected planar graph can be embedded in a two-dimensional grid; see, for instance, Richter-Gebert (2006, lem. 13.2.2), for a Tutte realisation of a 3-connected planar graph with n vertices in a grid of size singly exponential in n^2.

A Tutte realisation is a *strictly convex realisation*, a planar realisation in which each face is a strictly convex polygon. For such realisations we have realisations in grids of quadratic size.

Theorem 3.7.15 (Bárány and Rote, 2006) *Every 3-connected planar graph with n vertices admits a strictly convex realisation in a grid of size $O(n^2)$.*

If the faces of a 3-connected planar graph are not required to be strictly convex polygons, then a convex realisation of the graph can be realised in a $O(n) \times O(n)$ grid (Bonichon et al., 2007).

3.8 Steinitz's Theorem

This section is devoted to a seminal result of Steinitz (1922) on graphs of 3-polytopes.

Theorem 3.8.1 (Steinitz, 1922) *A graph is the graph of a 3-polytope if and only if it is planar and 3-connected.*

One direction of Steinitz's theorem was established in Proposition 3.2.11 and so we now deal with the second, harder part. Our proof relies on Tutte's realisation theorem (3.7.3) and a construction first suggested by Maxwell (1864, 1869) and later expanded by Cremona (1890).

We let X be a 3-connected planar graph with vertex set $\mathcal{V}(X) = \{1, \dots, n\}$ and positive weights c_{ij} placed on its edges. Further, let F be a fixed face of X; we can talk about the faces of X because, by Whitney's theorem (3.2.12), in every planar realisation of X its faces are bounded by the same facial cycles. By Tutte's realisation theorem, a Tutte realisation G of X in the plane with respect to F is a convex realisation whose faces are strictly convex polygons. Let $\mathcal{V}(G) = \{\boldsymbol{v}_1, \dots, \boldsymbol{v}_n\}$ be the distinct points in the plane representing the vertices of G, let K be the boundary polygon of G with $\mathcal{V}(K) = \{\boldsymbol{v}_{r+1}, \dots, \boldsymbol{v}_n\}$, and let C be the boundary cycle of G. Then $\mathcal{V}(C) = \{\boldsymbol{v}_{r+1}, \dots, \boldsymbol{v}_n\}$. Recall that the vertices and edges of C are the boundary vertices and boundary edges of G, respectively, while the interior vertices of G are the vertices in $\mathcal{V}(G) \backslash \mathcal{V}(C)$ and the interior edges of G are those containing at least one interior vertex; these definitions were first given in Section 3.7. In addition, recall that an interior face of G is a face of G other than the one bounded by C, and that the face of G bounded by C is its exterior face.

We assume that the Tutte realisation G has been placed in the plane H_G of $\mathbb{R}^3$ with equation $z = 1$:

$$H_G := \left\{(x, y, z)^t \in \mathbb{R}^3 \,\middle|\, z = 1\right\}; \tag{3.8.2}$$

that is, each vertex $\boldsymbol{v}_i = (x_i, y_i)^t$ of G in $\mathbb{R}^2$ becomes the vertex $\boldsymbol{u}_i = (\boldsymbol{v}_i, 1)^t$ of G in $\mathbb{R}^3$.

The plan of the proof is first to associate vectors to the interior faces of G, which will then be used to define vertices and normal vectors for the faces of a 3-polytope P whose graph is isomorphic to G. These normal vectors will be defined with the cross product of vectors.

The *cross product* $\boldsymbol{a} \times \boldsymbol{b}$ of two vectors $\boldsymbol{a} = (a_1, a_2, a_3)^t$ and $\boldsymbol{b} = (b_1, b_2, b_3)^t$ in $\mathbb{R}^3$ is the vector

$$\boldsymbol{a} \times \boldsymbol{b} := (a_2b_3 - a_3b_2, a_3b_1 - a_1b_3, a_1b_2 - a_2b_1)^t. \tag{3.8.3}$$

The direction of $\boldsymbol{a} \times \boldsymbol{b}$ is given by the *right-hand rule*: if the fingers of your right hand curl in the direction of a rotation from $\boldsymbol{a}$ to $\boldsymbol{b}$, then your thumb points in the direction of $\boldsymbol{a} \times \boldsymbol{b}$. The next remark is plain.

Remark 3.8.4 Let $\boldsymbol{a}, \boldsymbol{b} \in \mathbb{R}^3$. Then

(i) the vector $\boldsymbol{a} \times \boldsymbol{b}$ is orthogonal to both $\boldsymbol{a}$ and $\boldsymbol{b}$,
(ii) $\boldsymbol{a} \times \boldsymbol{b} = -\boldsymbol{b} \times \boldsymbol{a}$, and
(iii) two nonzero vectors $\boldsymbol{a}$ and $\boldsymbol{b}$ are parallel if and only if $\boldsymbol{a} \times \boldsymbol{b} = \boldsymbol{0}$.

Let m be the number of interior faces of G. To each interior face F_k ($k \in [1 \dots m]$) of G we associate a vector $\boldsymbol{f}_k \in \mathbb{R}^3$ as follows. If we orient an edge $\boldsymbol{u}_i\boldsymbol{u}_j$ of G from $\boldsymbol{u}_i$ to $\boldsymbol{u}_j$ then we let $\boldsymbol{f}_\ell$ and $\boldsymbol{f}_r$ be the vectors associated to the faces on the left and the right of $\boldsymbol{u}_i\boldsymbol{u}_j$, respectively. We single out one interior face of G, say F_1 and assign the vector $\boldsymbol{f}_1 := \boldsymbol{0}_3$ to it. Then the formula

$$\boldsymbol{f}_r := c_{ij}(\boldsymbol{u}_i \times \boldsymbol{u}_j) + \boldsymbol{f}_\ell \tag{3.8.5}$$

allows us to compute the other vectors $\boldsymbol{f}$ by moving from interior face to interior face of G.

For faces $F_{k_1}, \dots, F_{k_r}$ of G, let us call a directed path $F_{k_1} \dots F_{k_r}$ from the face F_{k_1} to the face F_{k_r} a *face-edge path* of G if $F_{k_i} \cap F_{k_{i+1}}$ is an edge of G, for each $i \in [1 \dots r-1]$; a closed face-edge path is a *face-edge cycle* of G. Let G^* be the dual graph of G (Section C.7). It follows that all the vectors of interior faces can be computed by traversing face-edge paths in G according to a BFS-tree in G^* that starts at F_1 and disregards the exterior face of G (see Fig. 3.8.1(a)); recall that we can obtain the standard planar realisation of the dual graph G^* of G by placing each vertex f^* of G^* inside the corresponding face f of G and then drawing an edge e^* between the vertices f^* and g^* of G^* so that it crosses the edge e shared by the faces f and g of G (see Section 3.1 and Fig. C.7.2).

We say two face-edge paths are *independent* if they share no inner face.

Lemma 3.8.6 *There are three independent face-edge paths between any two distinct faces of G.*

Proof The dual graph G^* of G is 3-connected (Problem C.10.19) and so there are three independent vertex-edge paths between any two distinct vertices of G^* (Corollary C.6.6). Face-edge paths in G correspond to (vertex-edge) paths in G^*, and therefore the lemma holds. □

We show that no matter what path we take to get to an interior face F_k of G, we always get the same vector $\boldsymbol{f}_k$ for F_k.

Lemma 3.8.7 *The vectors $\boldsymbol{f}_1, \ldots, \boldsymbol{f}_m$ are well defined.*

Proof Let F be an interior face of G with vector $\boldsymbol{f}$. For an edge $\boldsymbol{u}_i\boldsymbol{u}_j$ of G directed from $\boldsymbol{u}_i$ to $\boldsymbol{u}_j$, we can consistently compute the left vector $\boldsymbol{f}_\ell$ from the right $\boldsymbol{f}_r$ or compute the right vector from the left vector; this is so because $c_{ij} = c_{ji}$ and

$$\boldsymbol{f}_r = c_{ij}(\boldsymbol{u}_i \times \boldsymbol{u}_j) + \boldsymbol{f}_\ell, \ \boldsymbol{f}_\ell = c_{ji}(\boldsymbol{u}_j \times \boldsymbol{u}_i) + \boldsymbol{f}_r. \tag{3.8.7.1}$$

Thus, if we know the vector of a face sharing an edge with F, then we can consistently compute the vector $\boldsymbol{f}$.

We now consider the star $\mathcal{C}$ of a vertex $\boldsymbol{u}_j$ in G. We prove that any two different face-edge paths in $\mathcal{C}$ from a face F' to a face F'' of G yield the same vector $\boldsymbol{f}''$ for F''. Without loss of generality, we can assume that these face-edge paths share no face except for F' and F''; see also Lemma 3.8.6. Let $F_{i_1}, \ldots, F_{i_a}, F_{i_{a+1}}, \ldots, F_{i_b}$ be the faces of $\mathcal{C}$ in clockwise order. It suffices to prove that a face-edge cycle L consisting of a face-edge path $L' := F_{i_1} \ldots F_{i_a}$ in $\mathcal{C}$ from F' to F'' followed by the face-edge path $L'' := F_{i_a}F_{i_{a+1}} \ldots F_{i_b}F_{i_1}$ in $\mathcal{C}$ from F'' to F' yields the same 'consistent' value of $\boldsymbol{f}' = \boldsymbol{f}_{i_1}$; see Fig. 3.8.1(b).

Suppose that $\boldsymbol{u}_{i_1}, \ldots, \boldsymbol{u}_{i_a}, \boldsymbol{u}_{i_{a+1}}, \ldots, \boldsymbol{u}_{i_b}$ are the neighbours of $\boldsymbol{u}_j$, also in clockwise order, so that the edge $\boldsymbol{u}_j\boldsymbol{u}_{i_k}$ is the intersection of F_{i_k} and $F_{i_{k+1}}$, for $k \in [1 \ldots b-1]$, and $\boldsymbol{u}_j\boldsymbol{u}_{i_b} = F_{i_b} \cap F_{i_1}$. For simplicity, for an edge $\boldsymbol{u}_j\boldsymbol{u}_{i_x}$ we let $c'_x := c_{ji_x}$. By traversing the cycle L and using (3.8.5), we get the following:

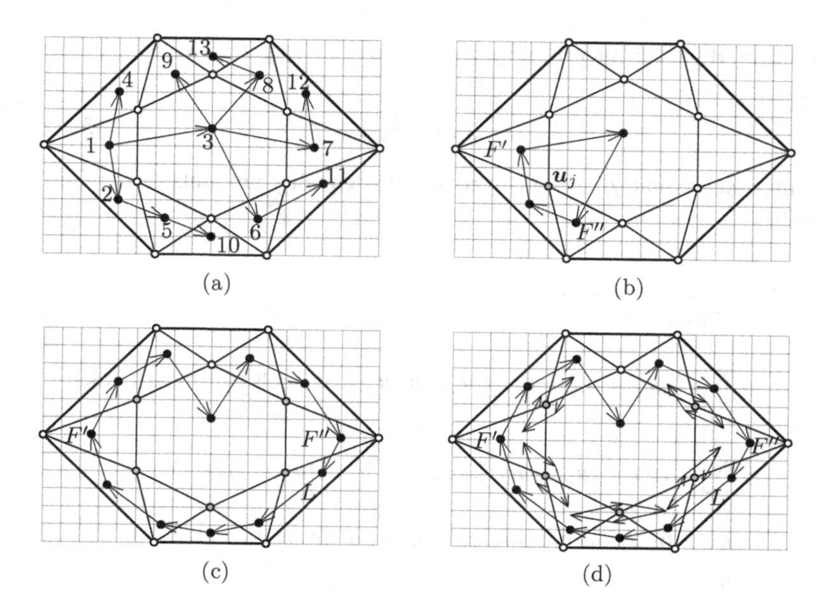

Figure 3.8.1 Face-edge paths in a Tutte realisation G of a 3-connected planar graph. (a) A BFS-tree traversing the faces of G; the labels represent the order in which the faces are added to the tree. (b) A face-edge cycle in G around the star of the vertex $\boldsymbol{u}_j$ in G. (c) A face-edge cycle L in G enclosing five vertices; the enclosed vertices are drawn in grey. (d) Computing the vector $\boldsymbol{f}'$ of a face F' by moving along a cycle L; this computation is carried out by traversing the stars of the vertices enclosed by L.

$$
\begin{aligned}
\boldsymbol{f}_{i_2} &= c'_1(\boldsymbol{u}_j \times \boldsymbol{u}_{i_1}) + \boldsymbol{f}_{i_1},\\
\boldsymbol{f}_{i_3} &= c'_2(\boldsymbol{u}_j \times \boldsymbol{u}_{i_2}) + \boldsymbol{f}_{i_2},\\
&\vdots\\
\boldsymbol{f}_{i_a} &= c'_{a-1}(\boldsymbol{u}_j \times \boldsymbol{u}_{i_{a-1}}) + \boldsymbol{f}_{i_{a-1}},\\
\boldsymbol{f}_{i_{a+1}} &= c'_a(\boldsymbol{u}_j \times \boldsymbol{u}_{i_a}) + \boldsymbol{f}_{i_a},\\
&\vdots\\
\boldsymbol{f}_{i_b} &= c'_{b-1}(\boldsymbol{u}_j \times \boldsymbol{u}_{i_{b-1}}) + \boldsymbol{f}_{i_{b-1}}.
\end{aligned}
$$

By summing both sides of these equalities, we get that

$$
\boldsymbol{f}_{i_1} = \boldsymbol{f}_{i_b} - \sum_{k=1}^{b-1} c'_k(\boldsymbol{u}_j \times \boldsymbol{u}_{i_k}). \tag{3.8.7.2}
$$

Moreover, if we consider the edge $\boldsymbol{u}_j\boldsymbol{u}_{i_b}$, then (3.8.5) gives a vector $\boldsymbol{f}'_{i_1}$ for the face F_{i_1}:

$$\boldsymbol{f}'_{i_1} = c'_b(\boldsymbol{u}_j \times \boldsymbol{u}_{i_b}) + \boldsymbol{f}_{i_b}. \tag{3.8.7.3}$$

We must show that $\boldsymbol{f}_{i_1} = \boldsymbol{f}'_{i_1}$. Proving this equality amounts to proving the equality

$$\begin{aligned}
\mathbf{0} &= \sum_{k=1}^{b} c'_k(\boldsymbol{u}_j \times \boldsymbol{u}_{i_k}) \\
&= \sum_{k=1}^{b} c'_k(\boldsymbol{u}_j \times \boldsymbol{u}_{i_k} - \boldsymbol{u}_j \times \boldsymbol{u}_j), \text{ as } \boldsymbol{u}_j \times \boldsymbol{u}_j = \mathbf{0} \text{ by Remark 3.8.4,} \\
&= \sum_{k=1}^{b} c'_k(\boldsymbol{u}_j \times (\boldsymbol{u}_{i_k} - \boldsymbol{u}_j)) = \boldsymbol{u}_j \times \sum_{k=1}^{b} c'_k(\boldsymbol{u}_{i_k} - \boldsymbol{u}_j) \\
&= \boldsymbol{u}_j \times \mathbf{0}, \text{ by the equilibrium (3.7.1) of an interior vertex of } G, \\
&= \mathbf{0}.
\end{aligned}$$

Thus we have consistency when moving along face-edge paths in a star of a vertex. As a consequence, if we know the vector of a face in a star that contains F, then we can consistently compute the vector $\boldsymbol{f}$.

Finally, for distinct interior faces F' and F'' and for each (directed) face-edge cycle L in G consisting of a face-edge path L' from F' to F'' followed by a face-edge path L'' from F'' to F', it suffices to prove that the vector $\boldsymbol{f}'$ of F' is well defined along the face-edge cycle L. A vertex of G is *enclosed* by the face-edge cycle L in G if the vertex is in the interior of the cycle when the cycle is viewed as a simple closed curve connecting points in the interior of the faces in L; see Fig. 3.8.1(c)–(d). The vectors for the faces in L can be computed from the vectors for the faces in the stars of the vertices in G enclosed by L.

If we have only one enclosed vertex, then the face-edge cycle L lies in the star of a vertex, which was done in the proof when we considered the case of the start of a vertex $\boldsymbol{u}_j$ in G. If more vertices are enclosed, then we move from the star $\mathcal{S}_1$ of an enclosed vertex to the star $\mathcal{S}_2$ of another enclosed vertex so that $\mathcal{S}_1$ and $\mathcal{S}_2$ share a face, until all the stars of enclosed vertices are covered; see Fig. 3.8.1(d) for an example of this computation. This approach guarantees the consistency of the computations and completes the proof of the lemma. □

Remark 3.8.8 In a Tutte realisation A, there is a dual interpretation for the computation of a face vector along a face-edge cycle L by traversing the stars of the vertices enclosed by L. The face-edge cycle L corresponds to a cycle

L^* in the dual graph A^* of A, and the stars of the vertices enclosed by L correspond to the facial cycles of A^* contained in the interior of L^*, whose symmetric difference is L^* (Proposition 3.6.9).

The vectors $\boldsymbol{f}_1, \ldots, \boldsymbol{f}_m$ are now used to produce heights for the vertices of G and supporting hyperplanes for the faces of a 3-polytope with graph G. For each point $\boldsymbol{x}$ in the boundary polygon K, define a piecewise linear function $\varphi\colon K \to \mathbb{R}$ as

$$\varphi(\boldsymbol{x}) := \begin{cases} \boldsymbol{f}_1 \cdot \boldsymbol{x}, & \text{if } \boldsymbol{x} \in F_1; \\ & \vdots \\ \boldsymbol{f}_m \cdot \boldsymbol{x}, & \text{if } \boldsymbol{x} \in F_m. \end{cases} \tag{3.8.9}$$

We must show that there is no local inconsistency in the definition of φ: for each point $\boldsymbol{x}$ on an edge of G, the value $\varphi(\boldsymbol{x})$ is the same whether we use the vector of the left face of the edge or the vector of the right face.

Lemma 3.8.10 *The function φ is well defined.*

Proof Let $e := \boldsymbol{u}_i\boldsymbol{u}_j$ be an edge of G, oriented from $\boldsymbol{u}_i$ to $\boldsymbol{u}_j$, and let $\boldsymbol{f}_\ell$ and $\boldsymbol{f}_r$ be the left and right faces of e, respectively. For each point $\boldsymbol{x} \in e$ we show that $\boldsymbol{f}_r \cdot \boldsymbol{x} = \boldsymbol{f}_\ell \cdot \boldsymbol{x}$, as it should. Since $\boldsymbol{x} \in e$, we can write as $\boldsymbol{x} = \alpha\boldsymbol{u}_i + (1-\alpha)\boldsymbol{u}_j$ for $\alpha \in [0, 1]$, in which case we have that

$$\begin{aligned}
\boldsymbol{f}_r \cdot \boldsymbol{x} &= \boldsymbol{f}_r \cdot (\alpha\boldsymbol{u}_i + (1-\alpha)\boldsymbol{u}_j) \\
&= \alpha\boldsymbol{f}_r \cdot \boldsymbol{u}_i + (1-\alpha)\boldsymbol{f}_r \cdot \boldsymbol{u}_j, \text{ by the linearity of the dot product,} \\
&= \alpha(c_{ij}(\boldsymbol{u}_i \times \boldsymbol{u}_j) + \boldsymbol{f}_\ell) \cdot \boldsymbol{u}_i + (1-\alpha)(c_{ij}(\boldsymbol{u}_i \times \boldsymbol{u}_j) + \boldsymbol{f}_\ell) \cdot \boldsymbol{u}_j \\
&= \alpha c_{ij}(\boldsymbol{u}_i \times \boldsymbol{u}_j) \cdot \boldsymbol{u}_i + \alpha\boldsymbol{f}_\ell \cdot \boldsymbol{u}_i + (1-\alpha)c_{ij}(\boldsymbol{u}_i \times \boldsymbol{u}_j) \cdot \boldsymbol{u}_j \\
&\quad + (1-\alpha)\boldsymbol{f}_\ell \cdot \boldsymbol{u}_j \\
&= \alpha\boldsymbol{f}_\ell \cdot \boldsymbol{u}_i + (1-\alpha)\boldsymbol{f}_\ell \cdot \boldsymbol{u}_j \\
&= \boldsymbol{f}_\ell \cdot (\alpha\boldsymbol{u}_i + (1-\alpha)\boldsymbol{u}_j) \\
&= \boldsymbol{f}_\ell \cdot \boldsymbol{x}.
\end{aligned}$$

Here, we used the fact that $(\boldsymbol{u}_i \times \boldsymbol{u}_j) \cdot \boldsymbol{u}_i = (\boldsymbol{u}_i \times \boldsymbol{u}_j) \cdot \boldsymbol{u}_j = 0$. The lemma now follows. □

By virtue of Lemma 3.8.10, we can use the function φ to produce heights for the vertices of G. For each interior face F_k of G with $k \in [1 \ldots m]$ and for each vertex $\boldsymbol{u}_i = (x_i, y_i, 1)^t$ of G contained in F_k, we define the point $\boldsymbol{w}_i := (x_i, y_i, \boldsymbol{f}_k \cdot \boldsymbol{u}_i)^t$ and the set

$$J_k := \operatorname{conv}\left\{ \boldsymbol{w}_i = \begin{pmatrix} x_i \\ y_i \\ \boldsymbol{f}_k \cdot \boldsymbol{u}_i \end{pmatrix} \middle| \text{ for each } \boldsymbol{u}_i = \begin{pmatrix} x_i \\ y_i \\ 1 \end{pmatrix} \in \mathcal{V}(F_k) \right\}. \quad (3.8.11)$$

Let

$$\mathcal{J} := \{J_1, \ldots, J_m\}.$$

In this way, we have a correspondence between the inner faces of G and the elements of $\mathcal{J}$ so that the face F_k of G is mapped to the set J_k of $\mathcal{J}$.

We show that the vertices of a set J_k lie in a common plane of $\mathbb{R}^3$, and so J_k is a convex polygon in $\mathbb{R}^3$. For each $k \in [1 \ldots m]$, from the vector $\boldsymbol{f}_k$ associated with the face F_k of G we derive another vector $\boldsymbol{g}_k$ as

$$\boldsymbol{f}_k = \begin{pmatrix} f_{k1} \\ f_{k2} \\ f_{k3} \end{pmatrix} \mapsto \begin{pmatrix} f_{k1} \\ f_{k2} \\ -1 \end{pmatrix} =: \boldsymbol{g}_k. \quad (3.8.12)$$

The next lemma explains the relevance of these new vectors $\boldsymbol{g}_k$.

Lemma 3.8.13 *Let $k \in [1 \ldots m]$ and let $\mathcal{V}(F_k) = \{\boldsymbol{u}_1, \ldots, \boldsymbol{u}_p\}$. Then* aff J_k *is a plane in $\mathbb{R}^3$ with equation $\boldsymbol{g}_k \cdot \boldsymbol{x} = \alpha_k$ for some $\alpha_k \in \mathbb{R}$ and J_k is a strictly convex polygon whose vertices are $\boldsymbol{w}_1, \ldots, \boldsymbol{w}_p$.*

Proof Let $\boldsymbol{w}_a$ and $\boldsymbol{w}_b$ represent two vertices in J_k. Suppose that $\boldsymbol{f}_k = (f_{k1}, f_{k2}, f_{k3})^t$. Then $\boldsymbol{g}_k = (f_{k1}, f_{k2}, -1)^t$ by (3.8.12). For the first part, it suffices to show that $\boldsymbol{g}_k \cdot \boldsymbol{w}_a - \boldsymbol{g}_k \cdot \boldsymbol{w}_b = 0$.

$$\begin{aligned}
\boldsymbol{g}_k \cdot \boldsymbol{w}_a - \boldsymbol{g}_k \cdot \boldsymbol{w}_b &= \boldsymbol{g}_k \cdot (\boldsymbol{w}_a - \boldsymbol{w}_b) \\
&= \boldsymbol{g}_k \cdot (x_a - x_b, y_a - y_b, \boldsymbol{f}_k \cdot (\boldsymbol{u}_a - \boldsymbol{u}_b))^t \\
&= \boldsymbol{g}_k \cdot (x_a - x_b, y_a - y_b, \boldsymbol{f}_k \cdot (x_a - x_b, y_a - y_b, 0)^t)^t \\
&= \boldsymbol{g}_k \cdot (x_a - x_b, y_a - y_b, f_{k1}(x_a - x_b) + f_{k2}(y_a - y_b))^t \\
&= f_{k1}(x_a - x_b) + f_{k2}(y_a - y_b) - (f_{k1}(x_a - x_b) \\
&\quad + f_{k2}(y_a - y_b)) \\
&= 0.
\end{aligned}$$

Since the face F_k of G is a strictly convex polygon with vertices $\boldsymbol{u}_1, \ldots, \boldsymbol{u}_p$ by Tutte's realisation theorem (3.7.3), it follows that J_k is a strictly convex polygon with vertices $\boldsymbol{w}_1, \ldots, \boldsymbol{w}_p$. This completes the proof of the lemma. □

By Lemma 3.8.13, for each $k \in [1 \ldots m]$ we can assume that

$$\operatorname{aff} J_k := \left\{ \boldsymbol{x} \in \mathbb{R}^3 \middle| \boldsymbol{g}_k \cdot \boldsymbol{x} = \alpha_k \right\}, \text{ for some } \alpha_k \in \mathbb{R}. \quad (3.8.14)$$

The next step is to prove that, for faces F_ℓ and F_r of G incident with an edge $\boldsymbol{u}_i\boldsymbol{u}_j$ of G, the corresponding polygons J_ℓ and J_r bend along the edge $J_i \cap J_j = \boldsymbol{w}_i\boldsymbol{w}_j$ in the 'right way'. We say that J_ℓ is the *left polygon* of $\boldsymbol{w}_i\boldsymbol{w}_j$, while J_r is the *right polygon* of $\boldsymbol{w}_i\boldsymbol{w}_j$.

Lemma 3.8.15 *Let J_ℓ and J_r be two polygons in $\mathcal{J}$ that intersect at an edge $\boldsymbol{w}_i\boldsymbol{w}_j := J_\ell \cap J_r$ such that J_ℓ and J_r are the left and right polygons of $\boldsymbol{w}_i\boldsymbol{w}_j$. Then for every point $z \in J_\ell \backslash J_r$, we have that $\boldsymbol{g}_r \cdot z > \alpha_r$.*

Proof The lemma will hold if, for every vertex $\boldsymbol{w}_\ell$ of J_ℓ other than $\boldsymbol{w}_i$ and $\boldsymbol{w}_j$, we have that $\boldsymbol{g}_r \cdot \boldsymbol{w}_\ell > \alpha_r$; this is what we prove next. Suppose that $\boldsymbol{f}_k = (f_{k1}, f_{k2}, f_{k3})^t$. Then $\boldsymbol{g}_k = (f_{k1}, f_{k2}, -1)^t$ by (3.8.12).

$$
\begin{aligned}
\boldsymbol{g}_r \cdot \boldsymbol{w}_\ell - \alpha_r &= \boldsymbol{g}_r \cdot \boldsymbol{w}_\ell - \boldsymbol{g}_r \cdot \boldsymbol{w}_i, \text{ as } \boldsymbol{w}_i \in J_r,\\
&= \boldsymbol{g}_r \cdot (\boldsymbol{w}_\ell - \boldsymbol{w}_i) - \boldsymbol{g}_\ell \cdot (\boldsymbol{w}_\ell - \boldsymbol{w}_i), \text{ as } \boldsymbol{w}_\ell, \boldsymbol{w}_i \in J_\ell,\\
&= (\boldsymbol{g}_r - \boldsymbol{g}_\ell) \cdot (\boldsymbol{w}_\ell - \boldsymbol{w}_i)\\
&= (f_{r1} - f_{\ell 1}, f_{r2} - f_{\ell 2}, 0)^t \cdot (x_\ell - x_i, y_\ell - y_i, \boldsymbol{f}_\ell \cdot (\boldsymbol{u}_\ell - \boldsymbol{u}_i))^t\\
&= (\boldsymbol{f}_r - \boldsymbol{f}_\ell) \cdot (\boldsymbol{u}_\ell - \boldsymbol{u}_i), \text{ by the definition of } \boldsymbol{f}_\ell, \boldsymbol{f}_i, \boldsymbol{u}_\ell, \boldsymbol{u}_i,\\
&= c_{ij}(\boldsymbol{u}_i \times \boldsymbol{u}_j) \cdot (\boldsymbol{u}_\ell - \boldsymbol{u}_i), \text{ by the definition of } \boldsymbol{f}_r, \boldsymbol{f}_\ell,\\
&= c_{ij}(\boldsymbol{u}_i \times \boldsymbol{u}_j) \cdot \boldsymbol{u}_\ell - c_{ij}(\boldsymbol{u}_i \times \boldsymbol{u}_j) \cdot \boldsymbol{u}_i\\
&= c_{ij}(\boldsymbol{u}_i \times \boldsymbol{u}_j) \cdot \boldsymbol{u}_\ell, \text{ since } (\boldsymbol{u}_i \times \boldsymbol{u}_j) \cdot \boldsymbol{u}_i = 0,\\
&> 0.
\end{aligned}
$$

The product $(\boldsymbol{u}_i \times \boldsymbol{u}_j) \cdot \boldsymbol{u}_\ell$, often called the *scalar triple product*, is positive when $\boldsymbol{u}_\ell$ and $\boldsymbol{u}_i \times \boldsymbol{u}_j$ are on the same side of the hyperplane in $\mathbb{R}^3$ defined by the vectors $\boldsymbol{u}_i$ and $\boldsymbol{u}_j$. In our case, we have that $(\boldsymbol{u}_i \times \boldsymbol{u}_j) \cdot \boldsymbol{u}_\ell > 0$ because $c_{ij} > 0$ and $\boldsymbol{u}_i, \boldsymbol{u}_j, \boldsymbol{u}_\ell$ (considered as points) lie in H_G and the point $\boldsymbol{u}_\ell$ lies to the left of the line $\operatorname{aff}\{\boldsymbol{u}_i, \boldsymbol{u}_j\}$. This proves the lemma. □

A face-edge path $L := J_{i_1} \ldots J_{i_r}$ from a polygon $J_a = J_{i_1}$ to a polygon $J_b = J_{i_r}$ gives rise to a face-edge path $L_G := F_{i_1} \ldots F_{i_r}$ in G from the face $F_a = F_{i_1}$ of G to the face $F_b = F_{i_r}$ of G such that, for each $k \in [1 \ldots r]$, the polygon J_{i_k} of L corresponds to the inner face F_{i_k} of L_G.

Lemma 3.8.16 *Let J_a and J_b be any two polygons in $\mathcal{J}$. Then, for every point $z \in J_a \backslash J_b$, we have that $\boldsymbol{g}_b \cdot z > \alpha_b$.*

Proof We consider a face-edge path $L := J_a \ldots J_{b-1} J_b$ from J_a to J_b. While traversing the face-edge path L, we go from the left polygon to the right polygon of the corresponding edge; for instance, J_{b-1} is the left polygon of $J_{b-1} \cap J_b$, while J_b is the right polygon. We prove the lemma by induction on the *length* of a face-edge path from J_a to some other polygon. If the length of the face-edge path is one, then Lemma 3.8.15 settles the argument. Suppose that the lemma is true for a face J_c connected to J_a by a face-edge path of length less than that of L.

If $z \in J_{b-1}$ then $g_{b-1} \cdot z = \alpha_{b-1}$, else $z \in J_a \backslash J_{b-1}$ and the induction hypothesis gives that $g_{b-1} \cdot z > \alpha_{b-1}$. Thus, in any case, if we let w_b be a vertex of the edge $J_{b-1} \cap J_b$, then we have that

$$g_{b-1} \cdot (z - w_b) \geqslant 0. \tag{3.8.16.1}$$

The face-edge path L gives rise to a corresponding face-edge path L_G from the face F_a of G to the face F_b of G. If $z = (z_1, z_2, z_3)^t$ and we let $z' := (z_1, z_2, 1)$, then $z' \in F_a$. It follows that z' lies in the open halfspace in H_G determined by $\operatorname{aff}(F_{b-1} \cap F_b)$ and containing the vertices of G in $F_{b-1} \backslash F_b$. We reason as in Lemma 3.8.15.

$$\begin{aligned}
g_b \cdot z - \alpha_b &= g_b \cdot z - g_b \cdot w_b \\
&\geqslant g_b \cdot (z - w_b) - g_{b-1} \cdot (z - w_b), \text{ by (3.8.16.1)}, \\
&\geqslant (g_b - g_{b-1}) \cdot (z - w_b) \\
&\geqslant (f_b - f_{b-1}) \cdot (z' - u_b), \text{ by the definition of } w_b, z', \\
&\geqslant (f_b - f_{b-1}) \cdot z' - (f_b - f_{b-1}) \cdot u_b \\
&\geqslant (f_b - f_{b-1}) \cdot z', \text{ since } (f_b - f_{b-1}) \cdot u_b = 0 \\
&= c_{b-1,b}(u_{b-1} \times u_b) \cdot z', \text{ by the definition of } f_{b-1}, f_b, \\
&> 0.
\end{aligned}$$

The product $c_{b-1,b}(u_{b-1} \times u_b) \cdot z'$ is positive because $c_{b-1,b} > 0$, and z' and $u_{b-1} \times u_b$ are on the same side of the hyperplane in $\mathbb{R}^3$ defined by the vectors u_{b-1} and u_b. □

We now let P be the convex hull of the n points $w_1, \ldots, w_n$:

$$P = \operatorname{conv}\{w_1, \ldots, w_n\}. \tag{3.8.17}$$

By virtue of Lemmas 3.8.13 and 3.8.16, P is contained in the H-polyhedron

$$Q = \left\{ x \in \mathbb{R}^3 \,\middle|\, g_1 \cdot x \geqslant \alpha_1, \ldots, g_m \cdot x \geqslant \alpha_m \right\}. \tag{3.8.18}$$

If the vertices of the polytope P arising from the boundary polygon K of G lie in a plane of $\mathbb{R}^3$, say with equation $g_0 \cdot x = \alpha_0$, then adding the inequality $g_0 \cdot x \geqslant \alpha_0$ to Q yields an H-description of P. Moreover, in this case, P has graph G. We have just established the following.

Proposition 3.8.19 *A 3-connected planar graph with a triangular face is the graph of a 3-polytope.*

Proof Let X be a 3-connected planar graph with a triangular face F and let G be a Tutte realisation of X with respect to F. Suppose that G is embedded in the plane H_G of $\mathbb{R}^3$ with equation $z = 1$ (3.8.2) and that $\mathcal{V}(G) = \{\boldsymbol{u}_1, \ldots, \boldsymbol{u}_n\}$.

We now follow the lifting process described above. Produce the vectors $\boldsymbol{f}_1, \ldots, \boldsymbol{f}_m$ associated with each interior face of G (3.8.5), and using these vectors lift the vertices of G to obtain the vertices $\boldsymbol{w}_1, \ldots, \boldsymbol{w}_n$ of a 3-polytope P; see (3.8.11) and (3.8.17). We also use the vectors $\boldsymbol{f}_1, \ldots, \boldsymbol{f}_m$ to obtain vectors $\boldsymbol{g}_1, \ldots, \boldsymbol{g}_m$ as in (3.8.12). Since F is a triangle, the vertices of P lifted from the vertices of the boundary polygon of G lie in a plane, say with equation $\boldsymbol{g}_0 \cdot \boldsymbol{x} = \alpha_0$. As a consequence, the graph G, and thus X, is the graph of the polytope P whose H-description is $\{\boldsymbol{x} \in \mathbb{R}^3 \mid \boldsymbol{g}_0 \cdot \boldsymbol{x} \geqslant \alpha_0, \boldsymbol{g}_1 \cdot \boldsymbol{x} \geqslant \alpha_1, \ldots, \boldsymbol{g}_m \cdot \boldsymbol{x} \geqslant \alpha_m\}$. □

From the lifting to three dimensions of a Tutte realisation G with respect to a triangular face F, we get that this Tutte realisation is a Schlegel diagram based at F of a 3-polytope with graph G. It is also true that every Schlegel diagram of a 3-polytope P based at a face, not necessarily triangular, can be obtained as a Tutte realisation of the graph of P with respect to this face; see Pestenjak (1999) and Pisanski and Žitnik (2009) for further details.

Proof of Steinitz's theorem (3.8.1) The theorem now follows from Propositions 3.2.11, 3.6.1 and 3.8.19. Let X be a 3-connected planar graph and let G be a Tutte realisation of X.

The graph G contains a triangular face or a vertex of degree three (Proposition 3.6.1). If G has a triangular face, then the existence of a polytope with graph G is ensured by Proposition 3.8.19. Otherwise, G has vertex of degree three. This implies that the dual graph G^* of G has a triangular face. The graph G^* is planar and it is 3-connected by Problem C.10.19. Thus, there is a 3-polytope Q with graph G^* according to Proposition 3.8.19. From Proposition 3.5.1 it now follows that the dual polytope P of Q has graph $(G^*)^*$, which is isomorphic to G. Hence, in any case, we have a polytope with graph G, which completes the proof of the theorem. □

We give an example of the construction of a 3-polytope P from a Tutte realisation of a 3-connected planar graph.

Example 3.8.20 We revisit the Tutte realisations G_1 and G_2 computed in Example 3.7.5 and depicted in Fig. 3.7.1(b) and Fig. 3.7.2(b). Recall that both

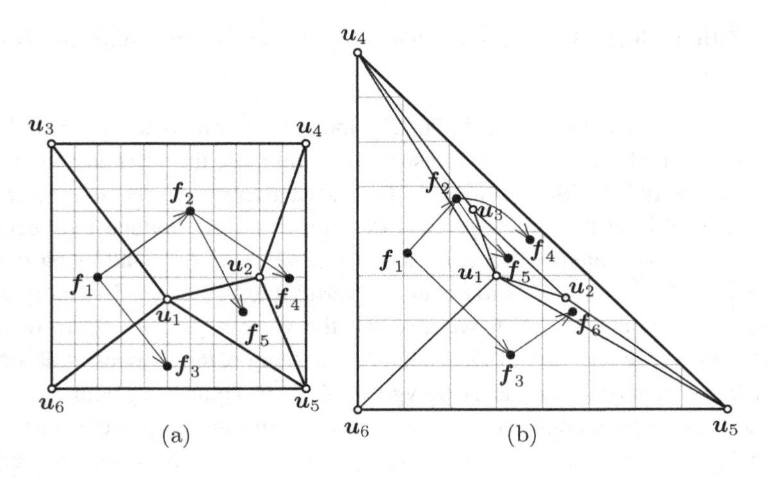

Figure 3.8.2 Tutte realisations of 3-connected plane graphs with BFS-trees traversing the faces of the graphs.

graphs have weights equal to one on all their interior edges. We assume that both Tutte realisations G_1 and G_2 have been placed in the plane H_G of $\mathbb{R}^3$ with equation $z = 1$.

(i) The vertices of G_1 are $\boldsymbol{u}_1 = (5/11, 4/11, 1)^t$, $\boldsymbol{u}_2 = (9/11, 5/11, 1)^t$, $\boldsymbol{u}_3 = (0, 1, 1)^t$, $\boldsymbol{u}_4 = (1, 1, 1)^t$, $\boldsymbol{u}_5 = (1, 0, 1)^t$, and $\boldsymbol{u}_6 = (0, 0, 1)^t$. With the BFS-tree in Fig. 3.8.2(a), we produce the vectors $\boldsymbol{f}_1, \boldsymbol{f}_2, \boldsymbol{f}_3, \boldsymbol{f}_4, \boldsymbol{f}_5$ associated with each interior face of G_1 (3.8.5), and with these vectors we compute the vectors $\boldsymbol{g}_1, \boldsymbol{g}_2, \boldsymbol{g}_3, \boldsymbol{g}_4, \boldsymbol{g}_5$ as in (3.8.12):

$$\boldsymbol{f}_1 = (0,0,0)^t\,, \qquad \boldsymbol{g}_1 = (0,0,-1)^t\,,$$

$$\boldsymbol{f}_2 = \boldsymbol{u}_1 \times \boldsymbol{u}_3 = \left(-\frac{7}{11}, -\frac{5}{11}, \frac{5}{11}\right)^t, \qquad \boldsymbol{g}_2 = \left(-\frac{7}{11}, -\frac{5}{11}, -1\right)^t,$$

$$\boldsymbol{f}_3 = \boldsymbol{u}_6 \times \boldsymbol{u}_1 = \left(-\frac{4}{11}, \frac{5}{11}, 0\right)^t, \qquad \boldsymbol{g}_3 = \left(-\frac{4}{11}, \frac{5}{11}, -1\right)^t,$$

$$\boldsymbol{f}_4 = \boldsymbol{u}_2 \times \boldsymbol{u}_4 + \boldsymbol{f}_2 = \left(-\frac{13}{11}, -\frac{3}{11}, \frac{9}{11}\right)^t, \qquad \boldsymbol{g}_4 = \left(-\frac{13}{11}, -\frac{3}{11}, -1\right)^t,$$

$$\boldsymbol{f}_5 = \boldsymbol{u}_1 \times \boldsymbol{u}_2 + \boldsymbol{f}_2 = \left(-\frac{8}{11}, -\frac{1}{11}, \frac{4}{11}\right)^t, \qquad \boldsymbol{g}_5 = \left(-\frac{8}{11}, -\frac{1}{11}, -1\right)^t.$$

With the vectors $\boldsymbol{f}_1, \boldsymbol{f}_2, \boldsymbol{f}_3, \boldsymbol{f}_4, \boldsymbol{f}_5$ at hand, we proceed to compute the vertices $\boldsymbol{w}_1, \boldsymbol{w}_2, \boldsymbol{w}_3, \boldsymbol{w}_4, \boldsymbol{w}_5$ as in (3.8.11):

$$
\begin{aligned}
\boldsymbol{w}_1 &= \left(\frac{5}{11}, \frac{4}{11}, \boldsymbol{f}_1 \cdot \boldsymbol{u}_1\right)^t = \left(\frac{5}{11}, \frac{4}{11}, 0\right)^t, \\
\boldsymbol{w}_2 &= \left(\frac{9}{11}, \frac{5}{11}, \boldsymbol{f}_2 \cdot \boldsymbol{u}_2\right)^t = \left(\frac{9}{11}, \frac{5}{11}, -\frac{3}{11}\right)^t, \\
\boldsymbol{w}_3 &= (0, 1, \boldsymbol{f}_1 \cdot \boldsymbol{u}_3)^t = (0, 1, 0)^t, \\
\boldsymbol{w}_4 &= (1, 1, \boldsymbol{f}_2 \cdot \boldsymbol{u}_4)^t = \left(1, 1, -\frac{7}{11}\right)^t, \\
\boldsymbol{w}_5 &= (1, 0, \boldsymbol{f}_3 \cdot \boldsymbol{u}_5)^t = \left(1, 0, -\frac{4}{11}\right)^t, \\
\boldsymbol{w}_6 &= (0, 0, \boldsymbol{f}_1 \cdot \boldsymbol{u}_6)^t = (0, 0, 0)^t.
\end{aligned}
$$

Moreover, we produce scalars $\alpha_1, \alpha_2, \alpha_3, \alpha_4, \alpha_5$ so that, for each $i \in [1 \ldots 5]$, $\boldsymbol{g}_i \cdot \boldsymbol{x} = \alpha_i$ is the equation of the plane aff J_i; see (3.8.11). To do so, for each $i \in [1 \ldots 5]$ we pick a vertex $\boldsymbol{w}_j$ in J_i. Refer to Fig. 3.8.2(a).

$$
\begin{aligned}
\alpha_1 &= \boldsymbol{g}_1 \cdot \boldsymbol{w}_1 = 0, \\
\alpha_2 &= \boldsymbol{g}_2 \cdot \boldsymbol{w}_1 = -5/11, \qquad \alpha_4 = \boldsymbol{g}_4 \cdot \boldsymbol{w}_2 = -9/11, \\
\alpha_3 &= \boldsymbol{g}_3 \cdot \boldsymbol{w}_1 = 0, \qquad \alpha_5 = \boldsymbol{g}_5 \cdot \boldsymbol{w}_2 = -4/11.
\end{aligned}
$$

Finally, we let $Q_1 := \{\boldsymbol{x} \in \mathbb{R}^3 \mid \boldsymbol{g}_1 \cdot \boldsymbol{x} \geqslant \alpha_1, \ldots, \boldsymbol{g}_5 \cdot \boldsymbol{x} \geqslant \alpha_5\}$ and let $P_1 := \operatorname{conv}\{\boldsymbol{w}_1, \ldots, \boldsymbol{w}_6\}$. An H-description of P_1 is given as

$$
P_1 = \left\{\boldsymbol{x} \in \mathbb{R}^3 \,\middle|\, \boldsymbol{g}_1 \cdot \boldsymbol{x} \geqslant \alpha_1, \ldots, \boldsymbol{g}_5 \cdot \boldsymbol{x} \geqslant \alpha_5, \boldsymbol{h}_1 \cdot \boldsymbol{x} \geqslant \beta_1, \boldsymbol{h}_2 \cdot \boldsymbol{x} \geqslant \beta_2\right\},
$$

where $\boldsymbol{h}_1 = (4, 3, 11)^t$, $\boldsymbol{h}_2 = (7, 0, 11)^t$, and $\beta_1 = \beta_2 = 0$. It follows that the polytope P_1 is contained in the H-polyhedron Q_1 and that the vertices in the boundary polygon of G_1 does not lie in a plane. Thus, the polytope P_1 does not have G_1 as a graph. In this case, if we follow the same procedure with a triangular boundary polygon of G_1, then the resulting polytope would have G_1 as its graph.

(ii) The vertices of G_2 are $\boldsymbol{u}_1 = (3/8, 3/8, 1)^t$, $\boldsymbol{u}_2 = (9/16, 5/16, 1)^t$, $\boldsymbol{u}_3 = (5/16, 9/16, 1)^t$, $\boldsymbol{u}_4 = (0, 1, 1)^t$, $\boldsymbol{u}_5 = (1, 0, 1)^t$, and $\boldsymbol{u}_6 = (0, 0, 1)^t$. Using the BFS-tree in Fig. 3.8.2(b), we produce the vectors $\boldsymbol{f}_1, \boldsymbol{f}_2, \boldsymbol{f}_3, \boldsymbol{f}_4, \boldsymbol{f}_5, \boldsymbol{f}_6$ associated with each interior face of G_2 (3.8.5), and with these vectors we compute the vectors $\boldsymbol{g}_1, \boldsymbol{g}_2, \boldsymbol{g}_3, \boldsymbol{g}_4, \boldsymbol{g}_5, \boldsymbol{g}_6$ as in (3.8.12):

$$\boldsymbol{f}_1 = (0,0,0)^t, \qquad \boldsymbol{g}_1 = (0,0,-1)^t,$$

$$\boldsymbol{f}_2 = \boldsymbol{u}_1 \times \boldsymbol{u}_4 = \left(-\frac{5}{8}, -\frac{3}{8}, \frac{3}{8}\right)^t, \qquad \boldsymbol{g}_2 = \left(-\frac{5}{8}, -\frac{3}{8}, -1\right)^t,$$

$$\boldsymbol{f}_3 = \boldsymbol{u}_6 \times \boldsymbol{u}_1 = \left(-\frac{3}{8}, \frac{3}{8}, 0\right)^t, \qquad \boldsymbol{g}_3 = \left(-\frac{3}{8}, \frac{3}{8}, -1\right)^t,$$

$$\boldsymbol{f}_4 = \boldsymbol{u}_3 \times \boldsymbol{u}_4 + \boldsymbol{f}_2 = \left(-\frac{17}{16}, -\frac{11}{16}, \frac{11}{16}\right)^t, \qquad \boldsymbol{g}_4 = \left(-\frac{17}{16}, -\frac{11}{16}, -1\right)^t,$$

$$\boldsymbol{f}_5 = \boldsymbol{u}_1 \times \boldsymbol{u}_3 + \boldsymbol{f}_2 = \left(-\frac{13}{16}, -\frac{7}{16}, \frac{15}{32}\right)^t, \qquad \boldsymbol{g}_5 = \left(-\frac{13}{16}, -\frac{7}{16}, -1\right)^t,$$

$$\boldsymbol{f}_6 = \boldsymbol{u}_5 \times \boldsymbol{u}_1 + \boldsymbol{f}_3 = \left(-\frac{3}{4}, -\frac{1}{4}, \frac{3}{8}\right)^t, \qquad \boldsymbol{g}_6 = \left(-\frac{3}{4}, -\frac{1}{4}, -1\right)^t.$$

With the vectors $\boldsymbol{f}_1, \boldsymbol{f}_2, \boldsymbol{f}_3, \boldsymbol{f}_4, \boldsymbol{f}_5, \boldsymbol{f}_6$ at hand, we proceed to compute the vertices $\boldsymbol{w}_1, \boldsymbol{w}_2, \boldsymbol{w}_3, \boldsymbol{w}_4, \boldsymbol{w}_5, \boldsymbol{w}_6$ as in (3.8.11):

$$\begin{aligned}
\boldsymbol{w}_1 &= \left(\frac{3}{8}, \frac{3}{8}, \boldsymbol{f}_1 \cdot \boldsymbol{u}_1\right)^t = \left(\frac{3}{8}, \frac{3}{8}, 0\right)^t,\\
\boldsymbol{w}_2 &= \left(\frac{9}{16}, \frac{5}{16}, \boldsymbol{f}_4 \cdot \boldsymbol{u}_2\right)^t = \left(\frac{9}{16}, \frac{5}{16}, -\frac{1}{8}\right)^t,\\
\boldsymbol{w}_3 &= (\frac{5}{16}, \frac{9}{16}, \boldsymbol{f}_4 \cdot \boldsymbol{u}_3)^t = \left(\frac{5}{16}, \frac{9}{16}, -\frac{1}{32}\right)^t,\\
\boldsymbol{w}_4 &= (0,1,\boldsymbol{f}_1 \cdot \boldsymbol{u}_4)^t = (0,1,0)^t,\\
\boldsymbol{w}_5 &= (1,0,\boldsymbol{f}_3 \cdot \boldsymbol{u}_5)^t = \left(1,0,-\frac{3}{8}\right)^t,\\
\boldsymbol{w}_6 &= (0,0,\boldsymbol{f}_1 \cdot \boldsymbol{u}_6)^t = (0,0,0)^t.
\end{aligned}$$

In addition, we produce scalars $\alpha_1,\alpha_2,\alpha_3,\alpha_4,\alpha_5,\alpha_6$ so that, for each $i \in [1 \ldots 6]$, $\boldsymbol{g}_i \cdot \boldsymbol{x} = \alpha_i$ is the equation of the plane aff J_i; see (3.8.11). To do so, for each $i \in [1 \ldots 6]$ we pick a vertex $\boldsymbol{w}_j$ in J_i. Refer to Fig. 3.8.2(b).

$$\begin{aligned}
\alpha_1 &= \boldsymbol{g}_1 \cdot \boldsymbol{w}_1 = 0, & \alpha_4 &= \boldsymbol{g}_4 \cdot \boldsymbol{w}_2 = -11/16,\\
\alpha_2 &= \boldsymbol{g}_2 \cdot \boldsymbol{w}_1 = -3/8, & \alpha_5 &= \boldsymbol{g}_5 \cdot \boldsymbol{w}_1 = -15/32,\\
\alpha_3 &= \boldsymbol{g}_3 \cdot \boldsymbol{w}_1 = 0, & \alpha_6 &= \boldsymbol{g}_6 \cdot \boldsymbol{w}_1 = -3/8.
\end{aligned}$$

Finally, we let $Q_2 := \{\boldsymbol{x} \in \mathbb{R}^3 \mid \boldsymbol{g}_1 \cdot \boldsymbol{x} \geqslant \alpha_1, \ldots, \boldsymbol{g}_6 \cdot \boldsymbol{x} \geqslant \alpha_6\}$ and let $P_2 := \operatorname{conv}\{\boldsymbol{w}_1, \ldots, \boldsymbol{w}_6\}$. An H-description of P_2 is given as

$$P_2 = \left\{\boldsymbol{x} \in \mathbb{R}^3 \,\middle|\, \boldsymbol{g}_0 \cdot \boldsymbol{x} \geqslant \alpha_0, \boldsymbol{g}_1 \cdot \boldsymbol{x} \geqslant \alpha_1, \ldots, \boldsymbol{g}_6 \cdot \boldsymbol{x} \geqslant \alpha_6\right\},$$

where $g_0 = (3,0,8)^t$ and $\alpha_0 = 0$. If we add the inequality $g_0 \cdot x \geqslant 0$ to Q_2, we get an H-description of P_2. It follows that the graph G_2 is isomorphic to the graph of P_2.

Proofs of Steinitz's Theorem and their Consequences

According to Ziegler (2007), the proofs of the harder part of Steinitz's theorem (3.8.1) can be roughly categorised into three types: the *Cremona–Maxwell–Tutte proofs* (proofs similar to the one we provided), the *Steinitz proofs* (proofs similar to the original proof of Steinitz [1922]), and the *Andreev–Koebe–Thurston proofs*, which are based on circle packings of planar graphs.

Cremona–Maxwell–Tutte proofs rely on lifting to $\mathbb{R}^3$ a strictly convex realisation of a 3-connected planar graph. This approach was presented in Hopcroft and Kahn (1992), Richter-Gebert (2006, sec. 3.1), and Lovász (2019, ch. 3). From Theorem 3.7.15 we know that strictly convex realisations of 3-connected planar graphs with n vertices can be embedded in two-dimensional grids of size $O\left(n^2\right)$. And it turns out that the lifting operation can be also achieved with integer coordinates for the z coordinate (Richter-Gebert, 2006, sec. 13.2). Thus, all combinatorial types of 3-polytopes can be realised in a three-dimensional grid. Richter-Gebert (2006, sec. 13.2) gives a very accessible proof for a realisation in a $O\left(2^{18n^2}\right) \times O\left(2^{18n^2}\right) \times O(2^{18n^2})$ grid. The current best algorithms give grids of size $O(2^{cn})$, for some constant c.

Theorem 3.8.21 (Ribó Mor et al., 2011) *Every combinatorial type of a 3-polytope with n vertices can be realised in a three-dimensional grid of size* $O\left(2^{7.55n}\right)$.

For special classes of polytopes such as simplicial polytopes and polytopes with a triangular or a quadrilateral face, better bounds for the grid size are possible (Ribó Mor et al., 2011).

A polytope is *rational* if it is combinatorially isomorphic to a polytope whose vertices have rational coordinates; otherwise it is *nonrational*. Under suitable coordinate scaling, every polytope whose vertices have rational coordinates can be converted into a *lattice polytope*, a polytope whose vertices have integral coordinates. Thus, a nonquantitative version of Theorem 3.8.21 could read as that every combinatorial type of a 3-polytope can be realised as a lattice polytope. But this doesn't extend to higher dimensions. We already know that in dimension four we have nonrational polytopes (Richter-Gebert, 2006, thm. 9.2.1). For a simpler construction of a nonrational polytope, albeit of dimension at least eight, we refer the reader to Grünbaum (2003, thm. 5.5.4) and Ziegler (1995, sec. 6.5).

Steinitz proofs follow induction arguments on the number of edges of a 3-connected planar graph; examples of these proofs are given in Grünbaum (2003, sec. 13.1), Barnette and Grünbaum (1969), and Ziegler (1995, ch. 4). The proofs in Grünbaum (2003, sec. 13.1) and Ziegler (1995, ch. 4) can be carried out in rational coordinates, and so they also give that the combinatorial types of 3-polytopes can be realised as lattice polytopes. As we have seen, when realising a 3-connected planar graph as a 3-polytope in $\mathbb{R}^3$, we have a choice for the realisation: for instance, we could ask for rational vertices (Theorem 3.8.21). It is also possible to ask for the shape of a facet to be a preassigned convex polygon in $\mathbb{R}^3$; this is a consequence of the proof in Grünbaum (2003, sec. 13.1), according to Barnette and Grünbaum (1970). Whereas preassigning the shape of facet is possible for all the d-polytopes with at most $d+3$ vertices (Grünbaum, 2003, ex. 6.5.3), this does not extend to 4-polytopes, as demonstrated in Barnette (1987b), Sturmfels (1988, sec. 5), and Ziegler (1995, ex. 6.22). Other corollaries of Steinitz proofs can be found in Ziegler (1995, sec. 4.4) and Grünbaum (2003, sec. 13.2).

A *circle packing* of a planar graph with n vertices is a family of n nonoverlapping circles in $\mathbb{R}^2$ where the centre of the circle C_i represents the vertex v_i of the graph and two circles C_i and C_j are tangent to each other if and only if $v_i v_j$ is an edge of the graph. The edge $v_i v_j$ is represented by a line segment between the centres of C_i and C_j, through $C_i \cap C_j$. See Fig. 3.8.3(a). The existence of a circle packing of a plane triangulation was first established by Koebe (1936) and was later rediscovered by Andreev (1970a,b), Thurston (1980, sec. 13.6), Marden and Rodin (1990), and de Verdière (1991). It was later observed that some of these proofs are applicable to the general planar case (Sachs, 1994).

Theorem 3.8.22 (Andreev–Koebe–Thurston's theorem) *Every planar graph admits a circle packing.*

Modern proofs of this theorem can be found in Lovász (2019, ch. 5), Mohar and Thomassen (2001, sec. 2.8), and Pach and Agarwal (1995, thm. 8.1).

Section 1.3 presented a model of a real projective space where the space decomposes into an affine part, a nonlinear hyperplane, and a projective part given by the hyperplane at infinity (see (2.5.1)). This model can be readily extended to the complex numbers. In the one-dimensional case $\mathbb{P}(\mathbb{C}^1)$, also called the *extended complex plane*, we consider $\mathbb{R}^2$ as the complex plane $\mathbb{C}$ and add one additional point, the *point p_∞ at infinity*; that is, $\mathbb{P}\left(\mathbb{C}^1\right) = \mathbb{C} \cup \{p_\infty\}$. We use $\mathbb{P}\left(\mathbb{C}^1\right)$ to obtain additional circle packings of a plane graph.

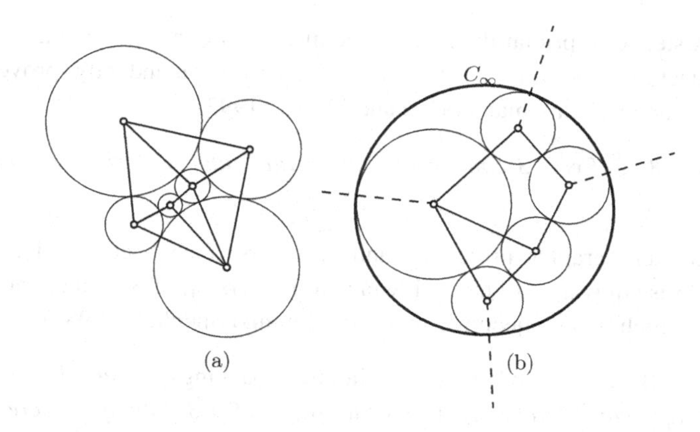

Figure 3.8.3 Circle packings of planar graphs. (a) A circle packing in $\mathbb{R}^2$ of the plane graph G in Fig. 3.7.1(a). (b) A circle packing in $\mathbb{P}(\mathbb{C}^1)$ of the dual graph of G; the cycle C_∞ is highlighted in bold.

In a circle packing of a plane graph G in $\mathbb{P}(\mathbb{C}^1)$, we have the choice of singling out a vertex u of G and representing it by the point p_∞ at infinity. The point p_∞ is the 'centre' of a new cycle C_∞ that behaves differently from the other cycles: the other cycles are contained in its interior. We then construct a circle packing in $\mathbb{R}^2$ of the graph $G - u$. Finally, the edge $v_i u$ is in G if and only if the cycles C_i and C_∞ are tangent to each other; we represent this edge by a ray from the centre of C_i to the centre of C_∞. If we want to be specific, we say that this is a *circle packing in* $\mathbb{P}(\mathbb{C}^1)$ *of* G *with respect to the vertex* u. See Fig. 3.8.3(b).

Simultaneous circle packings in $\mathbb{P}(\mathbb{C}^1)$ of a 3-connected plane graph G and its dual graph G^* can be used to prove Steinitz's theorem. We denote by f_∞ the unbounded face of G, which is a vertex of G^*. A *primal-dual circle packing* of a 3-connected plane graph G consists of a circle packing $\{C_i \mid v_i \in \mathcal{V}(G)\}$ of G and a circle packing $\{D_i \mid f_i \in \mathcal{V}(G^*)\}$ of G^* with respect to f_∞ such that

(i) for every pair of dual edges $e = v_i v_j \in \mathcal{E}(G)$ and $e^* = f_i f_j \in \mathcal{E}(G^*)$ (the edge e is shared by the faces f_i and f_j of G), the cycles C_i and C_j corresponding to v_i and v_j are tangent at the same point as the cycles D_i and D_j corresponding to f_i and f_j; and
(ii) the line through the centres of C_i and C_j is perpendicular to the line through the centres of D_i and D_j.

In other words, we have overlaid the two circle packings in $\mathbb{P}(\mathbb{C}^1)$ such that dual edges cross at a right angle, and no other edges cross.

The existence of primal-dual circle packings of 3-connected planar graphs was conjectured by Tutte (1963, sec. 13), and independently proved by Brightwell and Scheinerman (1993) and Mohar (1993).

Theorem 3.8.23 *Every 3-connected planar graph admits a primal-dual circle packing.*

Via the stereographic projection and its inverse (see Section 3.1), Theorem 3.8.23 is equivalent to the following strong version of Steinitz's theorem, a version which was conjectured by Grünbaum and Shephard (1987).

Theorem 3.8.24 (Steinitz's theorem via circle packings) *Every 3-connected planar graph can be represented as the graph of a 3-polytope where every edge of the polytope is tangent to the two-dimensional unit sphere.*

Textbook proofs of these last two theorems can be found in Lovász (2019, ch. 5) and Mohar and Thomassen (2001, sec. 2.8).

3.9 Refinements

Refinements extend notions of minors from graphs to polytopal complexes and, like minors in graph theory, they have a number of applications in polytope theory. The main result of this section is that polytopes are refinements of simplices. We first revisit minors of graphs.

Minors of Graphs

Let e be an edge of a graph G, and let G/e denote the graph obtained from G by removing the edge e, identifying its ends with a new vertex v_e, and replacing all the multiple edges with single edges that have the same ends; we say that G/e is obtained by the *contraction of the edge e* (Fig. 3.9.1). If X is a subgraph or simply a set of edges of a graph G, we denote by G/X the graph obtained from G by successively contracting the edges of X; it is clear that the order in which we contract the edges of X does not affect the resulting graph G/X.

A *minor* of a graph G is any graph obtained from G by contracting edges and deleting vertices and edges. Alternatively, a graph X is a minor of G if G has a subgraph Y whose vertex set can be partitioned into connected subgraphs G_u such that there is an edge between u and v in X if and only if there is an edge between G_u and G_v in Y (Fig. 3.9.2); the vertex sets of the subgraphs G_u are called *branch sets*. A graph Y is a *subdivision* of a graph X if it can be obtained from X by *subdividing* edges of X, inserting vertices of degree

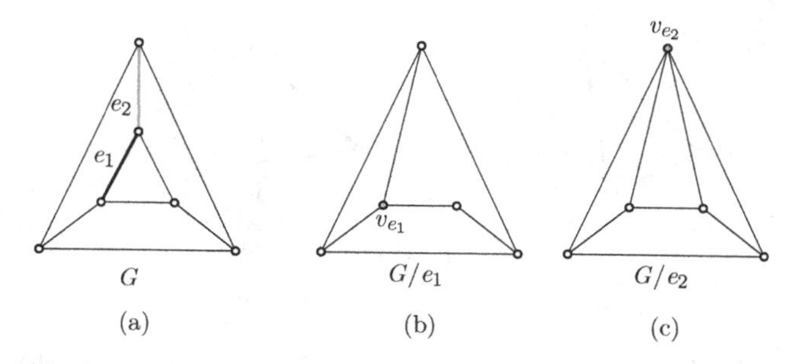

Figure 3.9.1 Edge contractions in a graph G. (a) A 3-connected graph G with a noncontractible edge e_1 and a contractible edge e_2 for definitions of contractible and noncontractible edges in graphs, please refer to Section C.5. (b) A 2-connected graph G/e_1 obtained by the contraction of the edge e_1. (c) A 3-connected graph G/e_2 obtained by the contraction of the edge e_2.

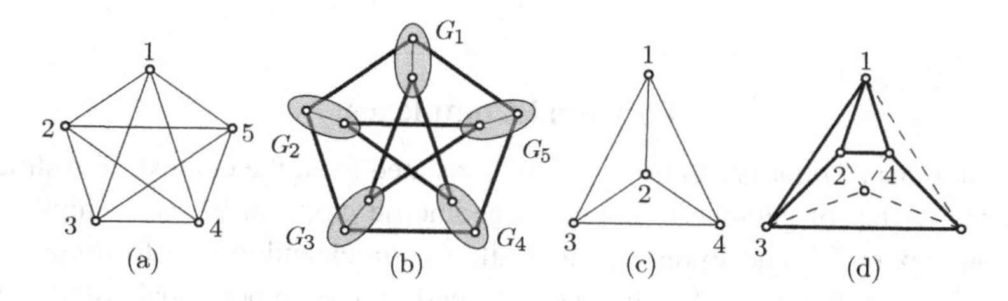

Figure 3.9.2 Minors. (a) K_5. (b) The Petersen graph has K_5 as a minor. (c) K_4. (d) The octahedron has K_4 as a topological minor.

two on the edges. The original vertices of X are the *branch vertices* of the subdivision and the new vertices are its *subdividing vertices*. A graph X is a *topological minor* of G if G contains a subdivision of X (Fig. 3.9.2). Minors and topological minors are closely related.

Proposition 3.9.1[7] *In a graph, the following hold.*

(i) *Every topological minor is also a minor of the graph.*
(ii) *Every minor of maximum degree at most three is also a topological minor of the graph.*

[7] A proof is available in Diestel (2017, prop. 1.7.3).

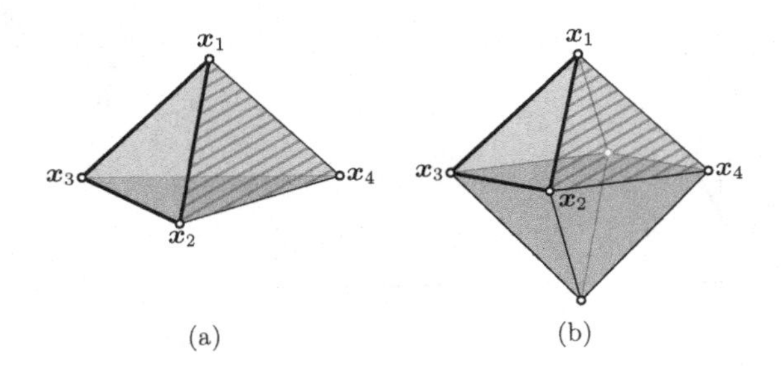

Figure 3.9.3 Refinement of a simplex. (a) The 3-simplex. (b) The octahedron as a refinement of the 3-simplex. The inverse image of the 2-face $\boldsymbol{x}_2\boldsymbol{x}_3\boldsymbol{x}_4$ of the simplex is the strongly connected 2-subcomplex of the octahedron that is highlighted in dark grey, and the 2-faces $\boldsymbol{x}_1\boldsymbol{x}_2\boldsymbol{x}_3$ and $\boldsymbol{x}_1\boldsymbol{x}_2\boldsymbol{x}_4$ of the octahedron are principal faces of the refinement homeomorphism: $\sigma^{-1}(\boldsymbol{x}_1\boldsymbol{x}_2\boldsymbol{x}_3) = \boldsymbol{x}_1\boldsymbol{x}_2\boldsymbol{x}_3$ and $\sigma^{-1}(\boldsymbol{x}_1\boldsymbol{x}_2\boldsymbol{x}_4) = \boldsymbol{x}_1\boldsymbol{x}_2\boldsymbol{x}_4$.

Polytopal Complexes

The notions of minors can be naturally translated from the context of abstract graphs to that of geometric graphs. A geometric graph in $\mathbb{R}^d$ is a simplicial 1-complex in $\mathbb{R}^d$, and so these graphs allow us to extend minors to simplicial 1-complexes. Refinements allow us to extend minors to polytopal complexes of higher dimension.

A $(d-1)$-complex $\mathcal{C}$ is a *refinement* of a $(d-1)$-complex $\mathcal{D}$ if there is a homeomorphism σ from $\operatorname{set}\mathcal{C}$ onto $\operatorname{set}\mathcal{D}$ such that the inverse image σ^{-1} of every k-face F of $\mathcal{D}$, with $k \in [0 \dots d-1]$, is the underlying set of a subcomplex $\mathcal{C}_F$ of $\mathcal{C}$: $\sigma^{-1}(F) = \operatorname{set}\mathcal{C}_F$. If you recall, the underlying set of a complex $\mathcal{C}$, denoted $\operatorname{set}\mathcal{C}$, is the set of points in $\mathbb{R}^d$ that belong to at least one polytope in the complex (Section 2.11). The bijection σ is called a *refinement homeomorphism*. A homeomorphism preserves the dimensions of the spaces, whence it follows that the complex $\mathcal{C}_F$ is a strongly connected k-complex. If $\sigma^{-1}(F)$ is a face of $\mathcal{C}$, then we say that $\sigma^{-1}(F)$ is a *principal face* of the refinement. A polytope P is a refinement of a polytope Q if there is a refinement homeomorphism from $\operatorname{set}\mathcal{B}(P)$ onto $\operatorname{set}\mathcal{B}(Q)$. Naturally, every refinement homeomorphism from $\operatorname{set}\mathcal{B}(P)$ onto $\operatorname{set}\mathcal{B}(Q)$ extends to one from $\operatorname{set}\mathcal{C}(P)$ onto $\operatorname{set}\mathcal{C}(Q)$; Figure 3.9.3 shows the octahedron as a refinement of the 3-simplex. A geometric graph G, viewed as a polytopal 1-complex, is a refinement of a geometric graph G', viewed as a polytopal 1-complex, if G is a subdivision of G'.

We state a simple consequence of the definition of a refinement.

Proposition 3.9.2 *If a $(d-1)$-complex $\mathcal{C}$ is a refinement of a $(d-1)$-complex $\mathcal{D}$, then the f-vector of $\mathcal{C}$ is componentwise not less than that of $\mathcal{D}$. Moreover, $f_i(\mathcal{C}) = f_i(\mathcal{D})$ for each $i \in [0 \ldots d-1]$ if and only if $\mathcal{C}$ and $\mathcal{D}$ are combinatorially isomorphic.*

The composition of refinement homeomorphisms is another refinement homeomorphism.

Proposition 3.9.3 *If a complex $\mathcal{C}$ is a refinement of a complex $\mathcal{D}$ and the complex $\mathcal{D}$ is a refinement of a complex $\mathcal{E}$, then $\mathcal{C}$ is a refinement of $\mathcal{E}$.*

A consequence of Steinitz's theorem (3.8.1) is that a 3-polytope P is a refinement of a 3-polytope P' if and only if $G(P)$ contains a subdivision of $G(P')$. While the octahedron is a refinement of the 3-simplex, the graph of the octahedron is not a refinement of the graph of the 3-simplex (Fig. 3.9.3). It is, however, true that the graph of the octahedron contains a subdivision of the graph of the 3-simplex; in other words, the graph of the 3-simplex is a topological minor of the graph of the octahedron. These observations extend to higher dimensions via refinements of polytopes.

We proceed with a common way to construct refinements.

Lemma 3.9.4 *Let P be a d-pyramid with apex $\boldsymbol{x}_0$ and base Q and let P' be another d-pyramid with apex $\boldsymbol{x}'_0$ and base Q'. Then every refinement homeomorphism $\sigma_Q\colon \operatorname{set}\mathcal{C}(Q) \to \operatorname{set}\mathcal{C}(Q')$ can be extended to a refinement homeomorphism $\sigma_P\colon \operatorname{set}\mathcal{C}(P) \to \operatorname{set}\mathcal{C}(P')$ where the principal faces of σ_P are the principal faces of σ_Q plus the pyramids with apex $\boldsymbol{x}_0$ over the principal faces of σ_Q.*

Proof For each point $\boldsymbol{x} \in P$ other than $\boldsymbol{x}_0$, let $\varphi(\boldsymbol{x})$ be the point at the intersection of Q and the line $\operatorname{aff}\{\boldsymbol{x}_0, \boldsymbol{x}\}$. It is plain that, for each point $\boldsymbol{x} \in P$ other than $\boldsymbol{x}_0$, there is a number $n(\boldsymbol{x}) \in (0,1]$ such that

$$\boldsymbol{x} = n(\boldsymbol{x})\varphi(\boldsymbol{x}) + (1 - n(\boldsymbol{x}))\boldsymbol{x}_0.$$

Construct σ_P from σ_Q as follows:

$$\sigma_P(\boldsymbol{x}) := \begin{cases} \boldsymbol{x}'_0, & \text{if } \boldsymbol{x} = \boldsymbol{x}_0; \\ n(\boldsymbol{x})\sigma_Q(\varphi(\boldsymbol{x})) + (1 - n(\boldsymbol{x}))\boldsymbol{x}'_0, & \text{otherwise.} \end{cases}$$

That σ_P is a refinement homeomorphism from $\operatorname{set}\mathcal{C}(P)$ to $\operatorname{set}\mathcal{C}(P')$ now follows from σ_Q being a refinement homeomorphism from $\operatorname{set}\mathcal{C}(Q)$ to $\operatorname{set}\mathcal{C}(Q')$. We have that σ_P coincides with σ_Q on $\operatorname{set}\mathcal{C}(Q)$, implying that the principal faces of σ_Q remain principal faces of σ_P. The other principal faces of σ_P must be pyramids over principal faces of σ_Q. The proof is complete. □

We next show how to use Lemma 3.9.4 and radial projections to produce the most important result on refinements (Corollary 3.9.6).

Theorem 3.9.5 (Grünbaum, 2003, ex. 11.1.3) *Suppose that P is a d-polytope in $\mathbb{R}^d$ and*

$$\emptyset = F_{-1} \subset F_0 \subset \cdots \subset F_d = P$$

is a complete flag of faces of P. Then there is a refinement homeomorphism σ from set $\mathcal{C}(P)$ *to* set $\mathcal{C}(T)$ *for a simplex T in $\mathbb{R}^d$ such that the following are satisfied.*

(i) *The complete flag of faces of P is mapped facewise onto the complete flag of faces of T; in particular, for every $i \in [-1 \dots d]$, F_i is a principal face of the refinement; and*
(ii) *for every $i \in [-1 \dots d]$ and every $k \in [i \dots d]$, if J is a k-face of T containing $\sigma(F_i)$, then $\sigma^{-1}(J)$ is the underlying set of a k-subcomplex of P whose k-faces all contain F_i.*
(iii) *The principal vertices of σ are affinely independent in $\mathbb{R}^d$.*

Proof We prove all the parts concurrently. We induct on the dimension d. The case $d = 1$ is trivial, so assume that $d \geqslant 2$.

(i) Let $\boldsymbol{y} := F_0$, let H be a hyperplane in $\mathbb{R}^d$ that strictly separates $\boldsymbol{y}$ from the other vertices of P, and let H^- be the closed halfspace in $\mathbb{R}^d$ that is determined by H and contains $\boldsymbol{y}$. The intersection $P \cap H$ yields the vertex figure $P/\boldsymbol{y}$ of P at $\boldsymbol{y}$ and the intersection $P \cap H^-$ yields a pyramid over $P/\boldsymbol{y}$.

Let T^0 be a $(d-1)$-simplex. The induction hypothesis on $d-1$ gives a refinement homeomorphism

$$\sigma_y\colon\ \operatorname{set}\mathcal{C}(P \cap H) \to \operatorname{set}\mathcal{C}(T^0)$$

that satisfies (i), (ii), and (iii) for the complete flag of faces of $P \cap H$

$$\emptyset = F_0 \cap H \subset \cdots \subset F_d \cap H = P \cap H. \tag{3.9.5.1}$$

From Part (i) we get that this flag is mapped facewise onto a complete flag of faces of T^0

$$\emptyset := T_0^0 \subset \cdots \subset T_d^0 = T^0, \tag{3.9.5.2}$$

say $T_i^0 := \operatorname{conv}\{\boldsymbol{x}_1, \dots, \boldsymbol{x}_i\}$, for $i \in [1 \dots d]$.

Let $T_i := \operatorname{conv}(T_i^0 \cup \{\boldsymbol{x}_0\})$ be an i-simplex in $\mathbb{R}^d$ for some point $\boldsymbol{x}_0 \in \mathbb{R}^d$ and each $i \in [0 \dots d]$, and let $T := T_d$. By Lemma 3.9.4, since $F_i \cap H^- =$

$\mathrm{conv}((F_i \cap H) \cup \{y\})$ (for $i \in [0 \ldots d]$), the refinement homeomorphism σ_y can be extended to a new refinement homeomorphism

$$\sigma_a\colon \operatorname{set}\mathcal{C}(P \cap H^-) \to \operatorname{set}\mathcal{C}(T)$$

in such a way that the complete flag of faces of $P \cap H^-$

$$\emptyset = F_{-1} \cap H^- \subset F_0 \cap H^- \subset \cdots \subset F_{d-1} \cap H^- \subset F_d \cap H^- = P \cap H^-$$

is mapped facewise onto the complete flag of faces of T

$$\emptyset := T_{-1} \subset T_0 \subset \cdots \subset T_{d-1} \subset T_d = T.$$

Finally, via the radial projection centred at y we produce a refinement homeomorphism

$$\sigma_b\colon \operatorname{set}\mathcal{C}(P) \to \operatorname{set}\mathcal{C}(P \cap H^-)$$

under which $\sigma_b(F_i) = F_i \cap H^-$, for $i \in [0 \ldots d]$. For every point $x \in P \setminus \{y\}$, the line $\ell_x := \mathrm{aff}\{y, x\}$ intersects the underlying set of $\mathrm{ast}(y, \mathcal{B}(P))$ at a unique point $\alpha(x)$. Hence, x lies in the segment $[y, \alpha(x)]$ for a unique number $n(x) \in (0, 1]$:

$$x = n(x)\alpha(x) + (1 - n(x))y.$$

The same line ℓ_x intersects the hyperplane H at a unique point $\beta(x)$. These lines define a radial projection with centre y. Define σ_b as

$$\sigma_b(x) := \begin{cases} y, & \text{if } x = y; \\ n(x)\beta(x) + (1 - n(x))y, & \text{otherwise.} \end{cases}$$

Under the refinement homeomorphism σ_b, each face in the complete flag of faces

$$\emptyset = F_{-1} \subset F_0 \subset \cdots \subset F_d = P$$

is principal. The desired refinement homeomorphism $\sigma\colon \operatorname{set}\mathcal{C}(P) \to \operatorname{set}\mathcal{C}(T)$ is given by the composition $\sigma_a \circ \sigma_b$.

(ii) We continue with the same notation and refinement homeomorphisms as in (i). By the proof of Part (i), the statement (ii) holds for the refinement homeomorphism $\sigma_y\colon \operatorname{set}\mathcal{C}(P \cap H) \to \operatorname{set}\mathcal{C}(T^0)$ and the complete flags of faces in (3.9.5.1) and (3.9.5.2).

Let J be a k-face of T that contains the face $\sigma(F_i) = T_i$. Then $x_0 \in J$. First, we let $J^0 := \mathrm{conv}(\mathcal{V}(J) \setminus \{x_0\})$. Then J^0 is in T^0 and contains the face $T_i^0 = \sigma_y(F_i \cap H)$. Thus, by the induction hypothesis, the inverse image $\sigma_y^{-1}(J^0)$ is the underlying set of a $(k-1)$-subcomplex of $P \cap H$ whose $(k-1)$-faces all contain $F_i \cap H$. By construction, the inverse image σ_a^{-1} of J is a strongly connected k-subcomplex of $P \cap H^-$ whose k-faces all contain $F_i \cap H^-$;

see also Lemma 3.9.4. Finally, under the refinement homeomorphism σ_b the inverse image of a k-face of $P \cap H^-$ that contains $F_i \cap H^-$ is a k-face of P that contains F_i. As a consequence, the inverse image of J under $\sigma = \sigma_a \circ \sigma_b$ is a strongly connected k-subcomplex in P that satisfies (ii).

(iii) By the proof of Part (i), we find that the principal vertices of σ_y are affinely independent in $\mathbb{R}^d$. This implies the same for the principal vertices of σ. □

An immediate consequence of Theorem 3.9.5 is the following.

Corollary 3.9.6 (Grünbaum, 1965) *Every d-polytope is a refinement of the d-simplex.*

Corollary 3.9.6 trivially extends to skeletons.

Corollary 3.9.7 *For every d and every $k \in [1 \ldots d-1]$, the k-skeleton of each d-polytope contains a refinement of the k-skeleton of the d-simplex.*

We zoom in on the case $k = 1$ of Corollary 3.9.7.

Corollary 3.9.8 *Let P be a d-polytope and let $\boldsymbol{y}$ be a vertex of P. Then $G(P)$ contains a subdivision of K^{d+1} where the branch vertices are the vertex $\boldsymbol{y}$ and a subset of its neighbours.*

We say that a subdivision as in Corollary 3.9.8 is *rooted* at the vertex y of a graph G; that is, a subdivision in G whose branch vertices are in $\{y\} \cup \mathcal{N}_G(y)$. Another corollary of Theorem 3.9.5 is the following.

Corollary 3.9.9 *For $d \geqslant 4$, the graph of a d-polytope is nonplanar.*

Proof For a d-polytope P, Corollary 3.9.8 ensures that $G(P)$ contains a subdivision of K^{d+1}. This means that the graph of a 4-polytope contains a subdivision of K^5, which establishes that such a graph is nonplanar according to Kuratowski's theorem (C.8.1). For $d \geqslant 4$, every d-polytope contains a 4-face, and thus its graph contains a subdivision of K^5. It is now clear that $G(P)$ is nonplanar by Kuratowski's theorem. □

We now mention some additional results on refinements and present others in the next subsection and the problems.

A polytope P in $\mathbb{R}^d$ is *centrally symmetric* if, for each point $\boldsymbol{x} \in P$, we have that $-\boldsymbol{x} \in P$. The d-crosspolytope is an example, since it can be realised as $\operatorname{conv}\{\boldsymbol{e}_1, -\boldsymbol{e}_1, \ldots, \boldsymbol{e}_d, -\boldsymbol{e}_d\}$. Grünbaum (2003, sec. 11.3) conjectured that the graph of a centrally symmetric simplicial d-polytope contains a refinement of the complete graph on $\lfloor 3d/2 \rfloor$ vertices. Halin (1966) showed that this conjecture is true for the crosspolytope, while Larman and Mani (1970) showed that it is true for $d = 4$.

The boundary complex of the d-crosspolytope is a *flag complex*, a simplicial complex where every set of pairwise adjacent vertices is a face of the complex. Athanasiadis (2011) proved that the graph of a flag $(d - 1)$-complex contains a refinement of the graph of the d-crosspolytope.

Refinements of 3-Polytopes

Existing research on minors in 3-polytopes focuses primarily on the octahedron and the cube as representative minors. Hereafter, if the graph of a polytope is r-connected, we also refer to the polytope itself as *r-connected*.

Let us start with the octahedron, the smallest 4-connected 3-polytope. Barnette (1987a) established that 5-connected 3-polytopes are refinements of the octahedron. In terms of connectivity, this is best possible because there are 4-connected 3-polytopes, such as antiprisms other than the octahedron, that are not refinements of the octahedron; an *antiprism* in $\mathbb{R}^3$ is the convex hull of two disjoint copies F_0 and F_1 of a polygon positioned in parallel hyperplanes so that the 2-faces of P other than F_0 and F_1 are triangles formed by the convex hull of a vertex of F_i and an edge of F_{1-i}, for $i = 0, 1$. The octahedron, as a simplicial 3-polytope with minimum degree at least four, naturally prompts the question of whether larger members of the class of simplicial 3-polytopes with a minimum degree at least four are refinements of it. A positive answer was given by Batagelj (1989, thm. 2).

Graphs of antiprisms do, however, contain the graph of the octahedron as a minor (Maharry, 1999). In fact, the graph of every 4-connected 3-polytope contains the graph of the octahedron as a minor (Maharry, 1999, thm. 5.1).

We continue with results about the cube. Because the cube has maximum degree three, it suffices to look for polytopes whose graph contains the cube as a topological minor, since containing the cube as a minor implies containing it as a topological minor (Proposition 3.9.1). There are 3-polytopes with at least eight vertices that are not a refinement of the cube; the connected sum of two octahedra along a triangular face is one such polytope (see Problem 3.11.5). However, every 4-connected 3-polytope with at least eight vertices is a refinement of the cube (Maharry, 2000, cor. 4.5, thm. 8.2), and so is the dual of a 4-connected 3-polytope (Barnette, 1973a, lem. 8). The cube is the smallest cubical 3-polytope, and the other cubical 3-polytopes are refinements of it (Brinkmann et al., 2005).

3.10 Embeddability of Simplicial Complexes

This section extends both the nonplanarity of K^5 and Fáry's theorem (3.1.5) to simplicial complexes of all dimensions.

We define abstract simplicial complexes, which feature heavily in combinatorics, topology, and computer science. Let X be a finite set and let $\mathcal{K}$ be a family of subsets of X. The family $\mathcal{K}$ is an *abstract simplicial complex* if it is closed under taking sets: if $S_1 \in \mathcal{K}$ and $S_2 \subseteq S_1$ then $S_2 \in \mathcal{K}$. The set X is the *vertex set* of $\mathcal{K}$ and is denoted by $\mathcal{V}(\mathcal{K})$, and its elements are the *vertices* of $\mathcal{K}$. The sets in $\mathcal{K}$ are called *abstract simplices* or *faces*. The *dimension* of an abstract simplex S is $\#S - 1$, and the *dimension* of an abstract simplicial complex $\mathcal{K}$ is the largest dimension of an abstract simplex in $\mathcal{K}$; faces of the largest and second largest dimension are called *facets* and *ridges*, respectively. If $\mathcal{K}$ has dimension d, then we say that $\mathcal{K}$ is an *abstract simplicial d-complex*. An abstract graph is an abstract simplicial 1-complex.

A *subcomplex* of an abstract simplicial d-complex $\mathcal{K}$ is a subset of $\mathcal{K}$ that is itself an abstract simplicial complex. The *k-skeleton* of $\mathcal{K}$ is the subcomplex formed by the abstract simplices of dimension at most $k \leqslant d$. The 1-skeleton of $\mathcal{K}$ is the *graph* $G(\mathcal{K})$ of $\mathcal{K}$.

Isomorphisms of graphs (see Section C.1) naturally generalise to isomorphisms of abstract simplicial complexes. A *simplicial map* from an abstract simplicial complex $\mathcal{K}$ to an abstract simplicial complex $\mathcal{K}'$ is a map $\varphi\colon \mathcal{V}(\mathcal{K}) \to \mathcal{V}(\mathcal{K}')$ that maps abstract simplices to abstract simplices: $\varphi(S) \in \mathcal{K}'$ for every $S \in \mathcal{K}$. An *isomorphism* from $\mathcal{K}$ to $\mathcal{K}'$ is a bijective simplicial map; in this case, we say that $\mathcal{K}$ and $\mathcal{K}'$ are *isomorphic*.

A simplicial complex $\mathcal{C}$ in some $\mathbb{R}^d$ is a *realisation* in $\mathbb{R}^d$ of an abstract simplicial complex $\mathcal{K}$ if there is a bijection $\sigma : \mathcal{V}(\mathcal{K}) \to \mathcal{V}(\mathcal{C})$ such that, for any subset $\{x_1, \dots, x_n\}$ of $\mathcal{V}(\mathcal{K})$, we have that

$$\{x_1, \dots, x_n\} \in \mathcal{K} \text{ if and only if } \operatorname{conv}\{\sigma(x_1), \dots, \sigma(x_n)\} \in \mathcal{C}.$$

Moreover, with each simplicial complex $\mathcal{C}'$ we can associate an abstract simplicial complex $\mathcal{K}'$. The elements of $\mathcal{V}(\mathcal{K}')$ are the vertices of $\mathcal{V}(\mathcal{C}')$, and the abstract simplices in $\mathcal{K}'$ are the vertex sets of the simplices of $\mathcal{C}'$.

It is well known that that every graph has a realisation in $\mathbb{R}^3$ in which the edges are represented by line segments (Corollary 3.10.3). Alternatively, this can be rephrased as 'every abstract simplicial 1-complex can be realised as a simplicial 1-complex in $\mathbb{R}^3$'. This latter interpretation can be extended to abstract simplicial d-complexes. First, we need a result on affine independent sets.

Lemma 3.10.1 *Let $\{\boldsymbol{x}_1, \dots, \boldsymbol{x}_n\}$ be an affinely independent set in $\mathbb{R}^d$ and let $I := \{i_1, \dots, i_r\}$, $J := \{j_1, \dots, j_s\}$, and $K := \{k_1, \dots, k_t\}$ be subsets of $\{1, \dots, n\}$ such that $K := I \cap J$. Then*

$$\operatorname{conv} I \cap \operatorname{conv} J = \operatorname{conv} K.$$

Proof The set $\operatorname{conv}\{\boldsymbol{x}_{k_1}, \ldots, \boldsymbol{x}_{k_t}\}$ is a subset of $\operatorname{conv}\{\boldsymbol{x}_{i_1}, \ldots, \boldsymbol{x}_{i_r}\} \cap \operatorname{conv}\{\boldsymbol{x}_{j_1}, \ldots, \boldsymbol{x}_{j_s}\}$. We prove the other direction. Let

$$\boldsymbol{y} \in \operatorname{conv}\{\boldsymbol{x}_{i_1}, \ldots, \boldsymbol{x}_{i_r}\} \cap \operatorname{conv}\{\boldsymbol{x}_{j_1}, \ldots, \boldsymbol{x}_{j_s}\}.$$

Because $\boldsymbol{y} \in \operatorname{conv}\{\boldsymbol{x}_{i_1}, \ldots, \boldsymbol{x}_{i_r}\}$, there are nonnegative scalars $\alpha_{i_1}, \ldots, \alpha_{i_r}$ such that $\sum_{\ell \in I} \alpha_\ell = 1$ and

$$\boldsymbol{y} = \sum_{\ell \in I} \alpha_\ell \boldsymbol{x}_\ell, \tag{3.10.1.1}$$

and because $\boldsymbol{y} \in \operatorname{conv}\{\boldsymbol{x}_{j_1}, \ldots, \boldsymbol{x}_{j_s}\}$, there are nonnegative scalars $\beta_{j_1}, \ldots, \beta_{j_s}$ such that $\sum_{\ell \in J} \beta_\ell = 1$ and

$$\boldsymbol{y} = \sum_{\ell \in J} \beta_\ell \boldsymbol{x}_\ell. \tag{3.10.1.2}$$

Combining (3.10.1.1) and (3.10.1.2), we get that

$$\boldsymbol{0} = \sum_{\ell \in I} \alpha_\ell \boldsymbol{x}_\ell - \sum_{\ell \in J} \beta_\ell \boldsymbol{x}_\ell = \sum_{\ell \in I \backslash K} \alpha_\ell \boldsymbol{x}_\ell - \sum_{\ell \in J \backslash K} \beta_\ell \boldsymbol{x}_\ell + \sum_{\ell \in K} (\alpha_\ell - \beta_\ell) \boldsymbol{x}_\ell. \tag{3.10.1.3}$$

The affine independence of $\{\boldsymbol{x}_1, \ldots, \boldsymbol{x}_n\}$ yields that all the coefficients of the right-hand side of (3.10.1.3) must be zero. This implies that $\alpha_\ell = 0$ for $\ell \in I \backslash K$, that $\beta_\ell = 0$ for $\ell \in J \backslash K$, and that $\alpha_\ell = \beta_\ell$ for $\ell \in K$. Hence, (3.10.1.1) and (3.10.1.2) become $\boldsymbol{y} = \sum_{\ell \in K} \alpha_\ell \boldsymbol{x}_\ell \in \operatorname{conv}\{\boldsymbol{x}_{k_1}, \ldots, \boldsymbol{x}_{k_t}\}$. □

Theorem 3.10.2 *Every abstract simplicial d-complex can be realised as a simplicial complex in* $\mathbb{R}^{2d+1}$.

Proof Consider the moment curve μ_{2d+1} in $\mathbb{R}^{2d+1}$, namely

$$\mu_{2d+1}(x) = \left(x, \ldots, x^{2d+1}\right)^t \quad \text{for } x \in [a,b],$$

and an abstract simplicial d-complex $\mathcal{K}$ with $\mathcal{V}(\mathcal{K}) = \{x_1, \ldots, x_n\}$. Select n distinct points on μ_{2d+1}, say $\mu_{2d+1}(1), \ldots, \mu_{2d+1}(n)$ and define a simplicial complex $\mathcal{C}$ as follows.

$$\mathcal{C} := \{\operatorname{conv}\{\mu_{2d+1}(i_1), \ldots, \mu_{2d+1}(i_k)\} \mid \{x_{i_1}, \ldots, x_{i_k}\} \in \mathcal{K}\}.$$

This gives a bijection $x_i \to \mu_{2d+1}(i)$ from $\mathcal{V}(\mathcal{K})$ to $\mathcal{V}(\mathcal{C})$. We show that $\mathcal{C}$ is a simplicial complex. Clearly, the faces of each simplex in $\mathcal{C}$ all belong to $\mathcal{C}$. It remains to show that simplices of $\mathcal{C}$ intersect only at faces. Take any two abstract simplices $S_1 := \{x_{i_1}, \ldots, x_{i_r}\}$ and $S_2 := \{x_{j_1}, \ldots, x_{j_s}\}$ in $\mathcal{K}$, and let $S_1 \cap S_2 := \{x_{k_1}, \ldots, x_{k_t}\}$. Then $S_1 \cap S_2 \in \mathcal{K}$, and the set $S_1 \cup S_2$ has cardinality at most $2d + 2$. Since every $2d + 2$ points on μ_{2d+1} are affinely independent in $\mathbb{R}^{2d+1}$ (Proposition 2.9.2), the set

$\{\mu_{2d+1}(i_1), \dots, \mu_{2d+1}(i_r), \mu_{2d+1}(j_1), \dots, \mu_{2d+1}(j_s)\}$ is affinely independent in $\mathbb{R}^{2d+1}$. Let

$$\begin{aligned}\mu_{2d+1}(S_1) &:= \{\mu_{2d+1}(i_1), \dots, \mu_{2d+1}(i_r)\},\\ \mu_{2d+1}(S_2) &:= \{\mu_{2d+1}(j_1), \dots, \mu_{2d+1}(j_s)\},\\ \mu_{2d+1}(S_1 \cap S_2) &:= \{\mu_{2d+1}(k_1), \dots, \mu_{2d+1}(k_t)\}.\end{aligned}$$

Then Lemma 3.10.1 yields that

$$\operatorname{conv} \mu_{2d+1}(S_1) \cap \operatorname{conv} \mu_{2d+1}(S_2) = \operatorname{conv} \mu_{2d+1}(S_1 \cap S_2),$$

which in turn yields that $\operatorname{conv} \mu_{2d+1}(S_1) \cap \operatorname{conv} \mu_{2d+1}(S_2) \in \mathcal{C}$, and so simplices of $\mathcal{C}$ intersect only at faces. □

The case $d = 1$ in Theorem 3.10.2 gives the following.

Corollary 3.10.3 *Every graph has a realisation in $\mathbb{R}^3$ in such a way that every edge is represented by a line segment.*

The geometric realisation of an abstract simplicial complex in Theorem 3.10.2 is unique up to homeomorphism. This means that for combinatorial purposes, there is often no need to distinguish between abstract and geometric simplicial complexes.

We now generalise the fact that not all abstract simplicial 1-complexes, namely graphs, can be realised as a simplicial 1-complex in $\mathbb{R}^2$.

In Section 3.1, we defined a (topological) embedding of a graph in a topological space. Here, we broaden this definition to include embeddings of general topological spaces. Let X and Y be two topological spaces. A (topological) *embedding* of X into Y is a function $\sigma : X \to Y$ that is a homeomorphism from X onto its image $\sigma(X)$; that is, $\sigma(X)$ is a subspace of Y that is a homeomorphic to X. In our cases of interest, namely finite graphs and underlying sets of simplicial complexes, we need to consider only injective continuous functions from X to Y, as their inverses are always continuous.

The *underlying set* of an abstract simplicial complex $\mathcal{K}$, denoted $\operatorname{set} \mathcal{K}$, is the underlying set of a simplicial complex realising $\mathcal{K}$. This underlying set can be considered as a topological space. While the $\operatorname{set} \mathcal{K}$ is not uniquely determined, isomorphic abstract simplicial complexes have homeomorphic underlying sets (Problem 3.11.11).

A topological embedding of an abstract simplicial complex $\mathcal{K}$ into a topological space Y is a topological embedding of $\operatorname{set} \mathcal{K}$ into Y. While a planar realisation of a graph in $\mathbb{R}^2$ that does not require that the edges be represented by line segments, Fáry's theorem (3.1.5) states that this stronger demand can also be met. In other words, if an abstract simplicial 1-complex is (topologically) embeddable in $\mathbb{R}^2$, then it can also be realised as a simplicial 1-complex in $\mathbb{R}^2$. Grünbaum (1969, p. 502) conjectured that this is true for every dimension: if an abstract simplicial d-complex is (topologically) embeddable in $\mathbb{R}^{2d}$, then it can also be realised as a simplicial complex in $\mathbb{R}^{2d}$. Brehm and Sarkaria (1992) disproved the conjecture for every $d \geqslant 2$. Thus, for abstract simplicial d-complexes with $d \geqslant 2$, the notions of topological embeddability and realisations as simplicial complexes are different.

Whereas every abstract simplicial d-complex can be realised as a simplicial complex in $\mathbb{R}^{2d+1}$, Flores (1934) and Van Kampen (1932) showed that the bound $2d + 1$ is best possible by exhibiting abstract simplicial d-complexes that cannot be topologically embedded in $\mathbb{R}^{2d}$. The *abstract d-simplex* is the abstract simplicial complex associated with the d-simplex.

Theorem 3.10.4 (Van Kampen–Flores's theorem)[8] *Let $d \geqslant 1$ and let $\mathcal{K}$ be the d-skeleton of the abstract $(2d + 2)$-simplex. Then $\mathcal{K}$ cannot be topologically embedded in $\mathbb{R}^{2d}$. In other words, for each continuous map $\varphi\colon \operatorname{set}\mathcal{K} \to \mathbb{R}^{2d}$, the images of two disjoint faces of $\mathcal{K}$ intersect.*

Remark 3.10.5 The case $d = 1$ simply states that the graph of the 4-simplex is not embeddable in $\mathbb{R}^2$, namely that K^5 is nonplanar.

Continuing with the analogy with graphs, we have that the graphs K^5 and $K^{3,3}$ are the only minimal graphs nonembeddable in $\mathbb{R}^2$, as any proper subgraph of them is planar and every nonplanar graph contains a subdivision of one of them. This minimality extends to higher dimensions. Grünbaum (1969) and Sarkaria (1991) constructed abstract simplicial d-complexes that cannot be topologically embedded in $\mathbb{R}^{2d}$ and are minimal in the sense that every proper subcomplex can be embedded in $\mathbb{R}^{2d}$. Later on, Schild (1993) presented a class of minimal abstract simplicial complexes that includes all those by Grünbaum (1969) and Sarkaria (1991).

3.11 Problems

3.11.1 Let G be a planar graph, let v be a vertex of G, and let F be a face in some planar realisation of G. Prove that G admits a planar realisation where v is in the boundary of the unbounded face of G and a planar realisation whose unbounded face has the same boundary as F.

[8] A proof is available in Matoušek (2003, ch. 5).

3.11.2 Let $\mathbb{S}^{d-1}$ be the $(d-1)$-dimensional unit sphere of $\mathbb{R}^d$ with centre at the origin, and let P be a d-polytope in $\mathbb{R}^d$. Prove the following.

(i) The radial projection with centre at the origin intersects the boundary of P in at most one point.
(ii) The radial projection with centre at the origin is a homeomorphism between $\mathbb{S}^{d-1}$ and set $\mathcal{B}(P)$.

3.11.3 (Develin, 2004) Prove the following statements.

(i) Every shelling of the d-cube is a line shelling.
(ii) For each $d \geqslant 3$, there is a nonline shelling of the d-crosspolytope.

3.11.4 A numbering of the vertices of a simple d-polytope is *completely unimodal* if every k-face of P, with $k \in [2 \ldots d]$, has a unique local maximum. Prove the following.

(i) A good orientation of $G(P)$ gives rise to a completely unimodal numbering of $G(P)$.
(ii) The completely unimodal numberings of $G(P)$ correspond to the shellings of P^* (Williamson Hoke, 1988).
(iii) A numbering of the vertices of the d-cube is completely unimodal if it has a unique local maximum on each 2-face (Williamson Hoke, 1988).
(iv) There is a numbering of the vertices of the 4-cube that is not completely unimodal, but has a unique local minimum on each k-face, with $k \in [3 \ldots d]$ (Williamson Hoke, 1988).
(v) A numbering of the vertices of a simple d-polytope is completely unimodal if it has a unique local maximum on each 2-face (Joswig et al., 2002).

3.11.5 Prove the following.

(i) A 3-cube is a refinement of a simplicial 3-prim and a simplicial 3-prim is a refinement of a 3-simplex; find concrete refinement homeomorphisms.
(ii) The connected sum of two octahedra along a triangular face is not a refinement of the cube.

3.11.6 (Larman and Mani, 1970) Prove the following.

(i) A 6-connected graph contains a subdivision of $K^{3,3}$.
(ii) The graph of a 4-polytope with at least six vertices contains a subdivision of $K^{3,3}$.
(iii) A centrally symmetric simplicial 4-polytope contains a refinement of K^6.

3.11.7 (Refinements of simplices; Lockeberg, 1977) Let P be a d-polytope. Prove the following.

(i) For $d \geqslant 4$, if P is a simple d-polytope P with $d+4$ facets, then there are two vertices $\boldsymbol{x}$ and $\boldsymbol{y}$ in P such that P cannot be expressed as a refinement of $T(d)$ under which both $\boldsymbol{x}$ and $\boldsymbol{y}$ are principal vertices.
(ii) If P is a d-polytope with $d+2$ facets, then, for any two vertices of P, there is a refinement homeomorphism from set $\mathcal{B}(P)$ to set $\mathcal{B}(T(d))$ under which both vertices are principal vertices.

3.11.8 Prove that the cube is not a topological minor of the Petersen graph.

3.11.9 (Ehrenborg and Hetyei, 1995, lem. 12) Let $\varphi\colon \mathcal{V}(Q(d_1)) \to \mathcal{V}(Q(d_2))$ be an injective function that maps edges of $Q(d_1)$ to edges of $Q(d_2)$. Prove that $\varphi(Q(d_1))$ is a d_1-face of $Q(d_2)$. (We may rewrite the problem as $Q(d_1)$ embeds in $Q(d_2)$ if $G(Q(d_1))$ embeds in $G(Q(d_2))$).

3.11.10 Use Problem 3.11.9 to prove that a cubical complex $\mathcal{C}$ embeds in the cube $Q(d)$ if $G(\mathcal{C})$ embeds in $G(Q(d))$.

3.11.11 In this problem you show that all underlying sets of an abstract simplicial complex are homeomorphic. Let $\mathcal{C}$ be a simplicial complex. For each point $\boldsymbol{x} \in \operatorname{set}\mathcal{C}$, there is a unique simplex F in $\mathcal{C}$ containing $\boldsymbol{x}$ in its relative interior; this simplex is the *support* of $\boldsymbol{x}$.

Let $\mathcal{C}$ and $\mathcal{C}'$ be two simplicial complexes, let $\mathcal{K}$ and $\mathcal{K}'$ be the abstract simplicial complexes associated with $\mathcal{C}$ and $\mathcal{C}'$, respectively, and let $\varphi\colon \mathcal{V}(\mathcal{K}) \to \mathcal{V}(\mathcal{K}')$ be a simplicial map from $\mathcal{K}$ to $\mathcal{K}'$. We regard φ as a map from $\mathcal{V}(\mathcal{K})$ to $\mathcal{V}(\mathcal{K}')$, and associate with it a continuous map ϱ as follows. Each $\boldsymbol{x} \in \operatorname{set}\mathcal{C}$ can be uniquely written as

$$\boldsymbol{x} = \sum_{i=1}^{k+1} \alpha_i \boldsymbol{v}_i,$$

where $\{\boldsymbol{v}_1, \dots, \boldsymbol{v}_{k+1}\}$ is the vertex set of the support of $\boldsymbol{x}$, $\alpha_i \geqslant 0$ ($i = [1 \dots k+1]$), and $\sum_{i=1}^{k+1} \alpha_i = 1$. Then let $\varrho(\boldsymbol{x}) := \sum_{i=1}^{k+1} \alpha_i \varphi(\boldsymbol{v}_i)$.

(i) Prove that ϱ is a continuous map from $\operatorname{set}\mathcal{C}$ to $\operatorname{set}\mathcal{C}'$, that ϱ is injective if φ is injective, and that ϱ is a homeomorphism if φ is an isomorphism.
(ii) Prove that isomorphic abstract simplicial complexes have homeomorphic underlying sets.

3.11.12* (Nonrational polytopes) This problem is about constructing nonrational polytopes.

(i) Construct a nonrational 4-polytope with a minimum number of vertices (Richter-Gebert, 2006, prob. 17.2.1).
(ii) More generally, for each $d \geqslant 4$, construct a nonrational d-polytope with a minimum number of vertices.

All combinatorial types of 4-polytopes with up to nine vertices can be realised with rational coordinates. (Firsching, 2020).

3.11.13* (Lockeberg, 1977) Let P be a simplicial d-polytope and let $\boldsymbol{x}, \boldsymbol{y} \in \mathcal{V}(P)$. Is there a refinement homeomorphism from set $\mathcal{B}(P)$ to set $\mathcal{B}(T(d))$ in which both $\boldsymbol{x}$ and $\boldsymbol{y}$ are principal vertices?

Lockeberg (1977, thm. 6.1) found a nonsimplicial d-polytope P with $d+3$ facets, and two nonadjacent vertices $\boldsymbol{x}, \boldsymbol{y}$ of it for which there is no refinement homeomorphism from set $\mathcal{B}(P)$ to set $\mathcal{B}(T(d))$ in which both $\boldsymbol{x}$ and $\boldsymbol{y}$ are principal vertices.

3.11.14* Let P be a $(2d-2)$-connected d-polytope.

(i) Is the graph of the d-crosspolytope a minor of $G(P)$?
(ii) If P is also simplicial, is the graph of the d-crosspolytope a topological minor of $G(P)$?

The graph of a 4-connected 3-polytope contains the graph of the octahedron as a minor (Maharry, 1999, thm. 5.1), while the graph of a 5-connected 3-polytope contains a subdivision of the graph of the octahedron (Barnette, 1987a).

Simplicial 3-polytopes with minimum degree four are refinements of the octahedron (Batagelj, 1989, thm. 2).

3.11.15* (Grünbaum, 2003, sec. 11.3) Prove that a centrally symmetric simplicial d-polytope contains a refinement of the complete graph on $\lfloor 3d/2 \rfloor$ vertices.

This is true for the crosspolytope (Halin, 1966) and for $d = 4$ (Larman and Mani, 1970).

3.11.16* Let P be a cubical d-polytope. Is P a refinement of the d-cube? In particular, is the graph of the d-cube a topological minor of $G(P)$? Cubical 3-polytopes are refinements of the 3-cube (Brinkmann et al., 2005).

3.11.17* (Larman and Mani, 1970) Does the graph of every d-polytope with at least $2(d-1)$ vertices contains a subdivision of $K^{d-1,d-1}$?

It is trivially true for $d = 2, 3$, and for $d = 4$ it was proved by Larman and Mani (1970).

3.12 Postscript

The polygonal version of Jordan's curve theorem and the existence of planar realisations in $\mathbb{R}^2$ in a way that edges are simple polygonal arcs (Proposition 3.1.1) are very well presented in Mohar and Thomassen (2001, sec. 2.1). We did not need to state Jordan's curve theorem for simple closed curves (Jordan, 1882); a statement and a rigorous proof of it can be also found in Mohar and Thomassen (2001, sec. 2.1).

The results on maximum values of linear functionals over polytopes (Theorems 3.2.1 and 3.2.2, and Corollary 3.2.4) are standard and can be seen in the context of linear programming; see, for instance, Webster (1994, ch. 4). That the removal of the vertices of a face does not disconnect the graph of a polytope (Theorem 3.2.6) is a standard result in polytope theory; a proof can be found in Sallee (1967, sec. 3). Our statement of Theorem 3.2.6 is from Brøndsted (1983, thm. 15.5) and our proof is an adaptation of that in Brøndsted (1983, thm. 15.5).

The proofs that the star and antistar of a vertex are strongly connected $(d-1)$-complexes (Proposition 3.3.2) are adapted from Sallee (1967, cor. 2.11, thm. 3.5).

The presentation of the excess degree of a polytope, including the relevant proofs, is based on the original paper of Pineda-Villavicencio et al. (2018).

The section on acyclic orientations and shellings was inspired by the dualisation of Bruggesser–Mani's shelling theorem (2.12.8) presented in Danaraj and Klee (1974). The derivation of the h-vector of simple polytopes from good orientations dualises the derivation the h-vector of simplicial polytopes described in Section 2.13 and Ziegler (1995, sec. 8.3).

The proof of Kalai (1990) that every 5-polytope has a facial cycle that is a triangle or a quadrangle uses linear programming on inequalities for flag numbers of 5-polytopes. Kalai (1990) also hinted at a proof of the same result for simple polytopes (Theorem 3.6.5); our proof simply expanded on his arguments. The proof that a cycle in the graph of a d-polytope is the symmetric difference of facial cycles (Theorem 3.6.12) via shellings is from Pineda-Villavicencio (2022); it is an adaptation of the proof by Blind and Blind (1994) of the result in Corollary 3.6.14. This method of proving Theorem 3.6.12 was known to experts in polytope theory; for instance, it was sketched in (Joswig, 2002, sec. 6).

Tutte's realisation theorem (3.7.3) is originally from Tutte (1963). Apart from the original proof in Tutte (1963), there are many presentations of this theorem; see, for instance, Lovász (2019, sec. 3.2), Richter-Gebert (2006, ch. 12), Hopcroft and Kahn (1992), Geelen (2012), Spielman (2019, ch. 15), and Erickson (2020) (which is based on Spielman (2019, ch. 15)). Whereas our presentation is mostly based on Spielman (2019, ch. 15), we have been inspired by elements of all these expositions.

The proof of the difficult part of Steinitz's theorem (3.8.1) is inspired by the clear presentations in Richter-Gebert (2006, sec. 13.1), Hopcroft and Kahn (1992), and Lovász (2019, sec. 3.2.2). Mohar and Thomassen (2001, sec. 2.8) and Lovász (2019, ch. 5) give excellent accounts of the circle packings of planar graphs, which also cover the strong version of Steinitz's theorem (3.8.24). An extension of this was proved by Schramm (1992). Sachs (1994) and Felsner and Rote (2019) surveyed, twenty-five years apart, circle packings and their extensions.

Corollary 3.9.6 is from Grünbaum (1965), while Theorem 3.9.5 is from Grünbaum (2003, ex. 11.3). The proof of Theorem 3.9.5 and that of Lemma 3.9.4 follow Lockeberg (1977, pp. 56–57) and Wotzlaw (2009, sec. 2.1). The existence of topological minors or minors in graphs of 3-polytopes has featured undoubtedly more in the literature, especially by considering results for general graphs where the topological minor or minor is a polytopal graph; this was the case of Brinkmann et al. (2005) and Maharry (1999, 2000).

Section 3.10 covers only basic results in embeddability and nonembeddability of simplicial complexes. Our presentation is standard but skipped the proof of Van Kampen–Flores's theorem (3.10.4) and the constructions of minimal abstract simplicial d-complexes that can be embedded in $\mathbb{R}^{2d}$.

4
Connectivity

Graphs of d-polytopes are d-connected, according to Balinski (1961). This chapter also discusses a recent result of Pilaud et al. (2023) on the edge connectivity of simplicial polytopes. We examine the higher connectivity of strongly connected complexes in Section 4.3; recall from Section 3.3 that the star of a vertex and the boundary complex of the polytope are two such complexes.

The second part of the chapter regards linkedness. A graph with at least $2k$ vertices is *k-linked* if, for every set of $2k$ distinct vertices organised in arbitrary k pairs of vertices, there are k vertex-disjoint paths joining the vertices in the pairs. Graphs of d-polytopes are $\lfloor (d+2)/3 \rfloor$-linked (Werner and Wotzlaw, 2011), an improvement on a classic result of Larman and Mani (1970). Moreover, we study a construction of Gallivan (1985) that shows that not all graphs of d-polytopes are $\lfloor d/2 \rfloor$-linked, which answers a question of Larman and Mani (1970). Edge linkedness can be defined similarly by replacing vertex-disjoint paths in the definition of linkedness with edge-disjoint paths.

The d-connectivity of a graph of a d-polytope is a particular case of a more general result of Athanasiadis (2009) on the connectivity of $(r, r+1)$-incidence graphs of a d-polytope. These graphs are defined on the set of r-faces of the polytope, with two r-faces being adjacent if they lie in a common $(r+1)$-face. The chapter ends with a short discussion on the connectivity of incidence graphs.

4.1 Balinski's Theorem and its Consequences

Theorem 3.2.6 yields both that the graph of a polytope is connected and that removing vertices of a proper face of the polytope does not disconnect the graph. Balinski (1961) discovered something stronger.

Theorem 4.1.1 (Balinski, 1961) *The graph of a d-dimensional polytope is d-connected.*

In a graph $G = (V, E)$, we define a set $Z \subseteq V \cup E$ as *separating* two sets $X, Y \subseteq V$ in G if every path from X to Y in G contains an element of Z. Additionally, the set Z is said to *separate* two vertices u and v of G if Z separates the sets $\{u\}, \{v\}$, *and* neither u nor v is in Z; in this case, u and v must be nonadjacent. Furthermore, we say that the set Z *separates* G if it separates any two vertices of G.

Remark 4.1.2 Balinski's theorem is tight. Consider a bipyramid P over a $(d-1)$-simplex T. Let $\boldsymbol{u}$ and $\boldsymbol{v}$ be the apices of P, namely the vertices in $\mathcal{V}(P) \backslash \mathcal{V}(T)$. It clear that $\mathcal{V}(T)$, of cardinality d, separates $G(P)$.

We provide two proofs of the theorem. The first proof yields a geometric structure of certain separators in the graph of the polytope, while the second proof yields that vertices in a d-separator X lie in the link of every other vertex of X. A set of vertices that separates a graph is called a *separator*, and a separator containing exactly r vertices is referred to as an r-*separator*.

First proof of Balinski's theorem (4.1.1) Let P be a d-polytope and let G be its graph. Then G has at least $d+1$ vertices. If G is a complete graph, there is nothing to prove, and so suppose otherwise. Proving that $G(P)$ is d-connected is equivalent to proving that no $d-1$ vertices separate the graph.

Let X be a separator of minimum cardinality of two nonadjacent vertices $\boldsymbol{u}$ and $\boldsymbol{v}$. In a graph with nonadjacent vertices, there is a minimum separator disconnecting two nonadjacent vertices (Corollary C.6.7). Suppose that there is an affine hyperplane H in $\mathbb{R}^d$ passing through X and an extra vertex $\boldsymbol{w}$ of P, and let φ be a linear functional that vanishes on H and is positive at some point in P; this may require a translation of the polytope. We consider two possibilities: either H supports P at a face F or H intersects the interior of P.

In the first possibility, Theorem 3.2.6 ensures that $G - X$ is connected, as $X \subseteq \mathcal{V}(F)$. In the second possibility, we consider three further cases: (i) $\varphi(\boldsymbol{u}) \geqslant 0$ and $\varphi(\boldsymbol{v}) \geqslant 0$, (ii) $\varphi(\boldsymbol{u}) \leqslant 0$ and $\varphi(\boldsymbol{v}) \leqslant 0$, (iii) $\varphi(\boldsymbol{u}) > 0$ and $\varphi(\boldsymbol{v}) < 0$. In Case (i), Lemma 3.2.7 gives a $\boldsymbol{u} - \boldsymbol{v}$ path L whose inner vertices all have φ-value greater than zero, implying that L meets H only at $\boldsymbol{u}$ or $\boldsymbol{v}$, and so it avoids X; the same analysis applies to Case (ii). In Case (iii), we use Lemma 3.2.7 to obtain a $\boldsymbol{u} - \boldsymbol{w}$ path L_{uw} and a $\boldsymbol{w} - \boldsymbol{v}$ path L_{wv} that meet H only at $\boldsymbol{w}$. Combining the paths L_{uw} and L_{wv}, we obtain a $\boldsymbol{u} - \boldsymbol{v}$ path L that avoids X.

The proof works provided that H passes through $X \cup \{\boldsymbol{w}\}$. This is always possible if $\#X \leqslant d-1$, since every affine hyperplane in $\mathbb{R}^d$ passes through any d given points. Hence G is d-connected. □

The first proof of Balinski's theorem 4.1.1 yields a geometric structure of separators in the graph of a polytope.

Corollary 4.1.3 *Let P be a d-polytope in $\mathbb{R}^d$ and let H be a hyperplane in $\mathbb{R}^d$. If X is a proper subset of $H \cap \mathcal{V}(P)$, then removing X does not disconnect $G(P)$. In particular, a separator of $G(P)$ with exactly d vertices must form an affinely independent set in $\mathbb{R}^d$.*

Second proof of Balinski's theorem (4.1.1) We prove a statement that also contains information on minimum separators.

The graph of d-polytope P is d-connected. Additionally, for each $d \geqslant 3$ and each vertex $\boldsymbol{x}$ in a minimum separator X of $G(P)$, the set $X \setminus \{\boldsymbol{x}\}$ is a separator of the link of $\boldsymbol{x}$. (4.1.3.1)

Let P be a d-polytope and let G be its graph. Then G has at least $d+1$ vertices. If G is a complete graph, there is nothing to prove, and so suppose otherwise. For $d = 1, 2$, G is d-connected. And so induct on d, assuming that $d \geqslant 3$ and that the theorem is true for $d-1$. Let X be a separator in G of minimum cardinality, and let $\boldsymbol{y}$ and $\boldsymbol{z}$ be two nonadjacent vertices separated by X (Corollary C.6.7). Then $\boldsymbol{y}, \boldsymbol{z} \notin X$. By Menger's theorem (C.6.1), we may assume that there are $\#X$ independent $\boldsymbol{y}-\boldsymbol{z}$ paths in G, each containing precisely one vertex from X. Let L be one such $\boldsymbol{y}-\boldsymbol{z}$ paths and let $\boldsymbol{x}$ be the vertex in $X \cap \mathcal{V}(L)$; say that $L = \boldsymbol{u}_1 \ldots \boldsymbol{u}_m$ such that $\boldsymbol{y} = \boldsymbol{u}_1$, $\boldsymbol{u}_j = \boldsymbol{x}$, and $\boldsymbol{u}_m = \boldsymbol{z}$.

The graph of the link of $\boldsymbol{x}$ in P is isomorphic to the graph of a $(d-1)$-polytope (Proposition 2.11.12), and by the induction hypothesis it is $(d-1)$-connected. The neighbours of $\boldsymbol{x}$ are all part of $\mathrm{lk}(\boldsymbol{x})$, and so $\boldsymbol{u}_{j-1}, \boldsymbol{u}_{j+1} \in G(\mathrm{lk}(\boldsymbol{x}))$. Again, from Menger's theorem follows the existence of at least $d-1$ independent $\boldsymbol{u}_{j-1} - \boldsymbol{u}_{j+1}$ paths in $G(\mathrm{lk}(\boldsymbol{x}))$. We must have that $X \setminus \{\boldsymbol{x}\}$ separates $\boldsymbol{u}_{j-1}$ from $\boldsymbol{u}_{j+1}$ in $G(\mathrm{lk}(\boldsymbol{x}))$, since X separates $\boldsymbol{y}$ from $\boldsymbol{z}$. Hence $\#(X \setminus \{\boldsymbol{x}\}) \geqslant d-1$, which establishes that G is d-connected.

Finally, let $d \geqslant 3$. As stated above, the set $X \setminus \{\boldsymbol{x}\}$ separates $G(\mathrm{lk}(\boldsymbol{x}))$. The aforementioned path L was arbitrary among the $\#X$ $\boldsymbol{y}-\boldsymbol{z}$ paths separated by X, and each such path contains a unique vertex of X. Hence, for each $\boldsymbol{x} \in X$, the set $X \setminus \{\boldsymbol{x}\}$ separates $G(\mathrm{lk}(\boldsymbol{x}))$. □

As a first corollary of (4.1.3.1), we have the following.

Corollary 4.1.4 *Let P be a simplicial d-polytope in $\mathbb{R}^d$ and let X be a minimum separator of P. Then the subgraph of $G(P)$ induced by X has minimum degree at least $d-1$.*

Proof Since P is simplicial, a vertex of P is adjacent to every vertex in its link. Take a vertex $\boldsymbol{x} \in X$. Since the set $X \setminus \{\boldsymbol{x}\}$ is a separator of the link of $\boldsymbol{x}$ and since, for $d \geqslant 3$, the graph of the link of $\boldsymbol{x}$ in P is isomorphic to the graph of a $(d-1)$-polytope (Proposition 2.11.12), we must have that $\#(\mathcal{V}(\mathrm{lk}(\boldsymbol{x})) \cap (X \setminus \{\boldsymbol{x}\})) \geqslant d-1$, and so $\boldsymbol{x}$ is adjacent to at least $d-1$ vertices in $\mathcal{V}(\mathrm{lk}(\boldsymbol{x})) \cap (X \setminus \{\boldsymbol{x}\})$. This proves that $G[X]$ has minimum degree at least $d-1$. □

A second corollary of (4.1.3.1) yields some properties of d-separators. An *empty* $(d-1)$*-simplex* in a d-polytope P is a set of d vertices of P that does not form a face of P but every proper subset does. An empty $(d-1)$-simplex is also called a *missing* $(d-1)$*-simplex.*

Corollary 4.1.5 *Let P be a d-polytope in $\mathbb{R}^d$ and let X be a d-separator of P. Further suppose that H is a hyperplane in $\mathbb{R}^d$ that contains X and that H^- and H^+ are the closed halfspaces in $\mathbb{R}^d$ defined by H. Then the following hold.*

(i) *For $d \geqslant 3$, each vertex $\boldsymbol{x}$ in X lies in the link of every other vertex of X, and the set $X \setminus \{\boldsymbol{x}\}$ is a $(d-1)$-separator of the link of x.*
(ii) *For $d \geqslant 2$, the graph of P is separated by X into exactly two components.*
(iii) *For $d \geqslant 2$, if P is a simplicial d-polytope, then X forms an empty $(d-1)$-simplex T. Furthermore, P is the connected sum of the d-polytopes $P^- := P \cap H^-$ and $P^+ := P \cap H^+$along the common facet $P \cap H$ of P^+ and P^-.*

Proof Suppose that X is a d-separator of $G(P)$ and that $\boldsymbol{x}$ is a vertex of X. From (4.1.3.1) we obtain that the set $X \setminus \{\boldsymbol{x}\}$, of cardinality $d-1$, separates $G(\mathrm{lk}(\boldsymbol{x}))$, a $(d-1)$-connected graph. It then follows that $X \setminus \{\boldsymbol{x}\} \subseteq \mathcal{V}(\mathrm{lk}(\boldsymbol{x}))$. This completes the proof of (i).

(ii) A 2-separator of a 2-polytope separates its graph into exactly two components. So assume $d \geqslant 3$. By induction, $X \setminus \{\boldsymbol{x}\}$, a $(d-1)$-set, separates $G(\mathrm{lk}(\boldsymbol{x}))$ into exactly two components S^+ and S^-. Each component R of $G(P) - X$ must intersect $\mathrm{lk}(\boldsymbol{x})$, otherwise $G(P)$ would be disconnected by $X \setminus \{\boldsymbol{x}\}$ and R would remain a component, a contradiction. Thus R intersects either S^+ or S^-, implying that $G - X$ has exactly two components.

(iii) For the third part of the corollary, suppose that P is a simplicial d-polytope. A simplicial 2-polytope is a polygon and a 2-separator in it satisfies (iii). So assume that $d \geqslant 3$. We have just shown that every vertex in X is in the link of every other vertex of X. Consequently, the subgraph $G[X]$ of G induced by X is a complete graph, as the set of neighbours of each vertex in P coincides with the vertex set of the link of the vertex.

If $d = 3$, then from $G[X]$ being a complete graph, it follows that it is an empty 2-simplex. And so an inductive argument on d can start. Assume that $d \geqslant 4$. The definition of a link and Proposition 2.11.12 gives that $\mathrm{lk}(\boldsymbol{x})$ is combinatorially isomorphic to the boundary complex of a simplicial $(d-1)$-polytope.

Because $X\setminus\{\boldsymbol{x}\}$ is a $(d-1)$-separator of $G(\mathrm{lk}(\boldsymbol{x}))$, the induction hypothesis on $\mathrm{lk}(\boldsymbol{x})$ yields that every proper subset of $X\setminus\{\boldsymbol{x}\}$ forms a face J of $\mathrm{lk}(\boldsymbol{x})$. And from the definition of $\mathrm{lk}(\boldsymbol{x})$, that face J lies in a facet of P containing $\boldsymbol{x}$, namely a $(d-1)$-simplex containing $\boldsymbol{x}$. As a consequence, if J is a k-face, then the set $\mathrm{conv}(J\cup\{\boldsymbol{x}\})$ is a $(k+1)$-face of P. The vertex $\boldsymbol{x}$ of X was taken arbitrarily, and so the set X is an empty $(d-1)$-simplex.

According to (ii), X separates $G(P)$ into two components S^+ and S^-. Consider a linear functional φ that vanishes on H and that is positive in H^+ (and thus negative in H^-). Let $F := P\cap H$. Then F is a $(d-1)$-polytope whose vertices are the vertices of P in H plus the intersections of H with edges e of P that go from H^+ to H^- (Theorem 2.7.1(iii)). Suppose that such an edge $e := \boldsymbol{uv}$ of P exists, in which case we may assume that $\boldsymbol{u}\in H^+, \boldsymbol{v}\in H^-$, and that $\boldsymbol{u},\boldsymbol{v}\in S^+$ since $\boldsymbol{u},\boldsymbol{v}\notin X$. Consider a vertex $\boldsymbol{w}\in S^-$. If $\boldsymbol{w}\in H^+$, then Lemma 3.2.7 ensures the existence of a $\boldsymbol{u}-\boldsymbol{w}$ path that does not meet X, a contradiction. And if $\boldsymbol{w}\in H^-$, then Lemma 3.2.7 ensures the existence of a $\boldsymbol{v}-\boldsymbol{w}$ path that does not meet X, another contradiction. It follows that the edge e cannot exist, thus implying that

$$\mathcal{V}(F)\subseteq\mathcal{V}(P). \tag{4.1.5.1}$$

By the same reasoning, we may assume that $S^+\subseteq H^+$ and $S^-\subseteq H^-$. In addition, every proper k'-face R' of F is the intersection of H and a k-face R of P, with $k'\leqslant k$ (Theorem 2.7.1(ii)). Since R is a simplex and since $\mathcal{V}(R')\subseteq\mathcal{V}(P)$ by (4.1.5.1), we have that the proper faces of F are all faces of P. It is now clear that P is the connected sum of P^+ and P^- along F. This completes the proof of (iii). □

The proof of the third part of Corollary 4.1.5 yielded another geometric property of separators; namely, if X is a separator of a d-polytope P in $\mathbb{R}^d$ so that X is contained in a hyperplane H of $\mathbb{R}^d$, then $G(P)$ is separated by X into exactly two components S^- and S^+ so that $S^-\subseteq H^-$ and $S^+\subseteq H^+$, where H^- and H^+ are the closed halfspaces defined by H.

We end this section with a generalisation of Balinski's theorem due to Perles and Prabhu (1993).

Theorem 4.1.6 *Let P be a d-polytope and let F be k-face of P with $k\in[-1\ldots d-1]$. Then the subgraph $G(P)-\mathcal{V}(F)$ is $\max\{1,d-k-1\}$-connected.*

Proof If $k=d-1$ the result follows from Theorem 3.2.6. So assume otherwise. Let X be a subset of $\mathcal{V}(P)$ of cardinality at most $d-k-2$, and let $\boldsymbol{w}$ be a vertex in $\mathcal{V}(P)\setminus(\mathcal{V}(F)\cup X)$. We show that there is a hyperplane H in $\mathbb{R}^d$ that passes through $F\cup X\cup\{\boldsymbol{w}\}$; this would imply that $G(P)-\mathcal{V}(F)-X$

is connected by Corollary 4.1.3. The existence of the hyperplane follows from observing that

$$\dim(\operatorname{aff}\{F \cup X \cup \{\boldsymbol{w}\}\}) \leqslant \dim F + \#X + 1 \leqslant k + \#X + 1 \leqslant d - 1.$$

This completes the proof of the theorem. □

Balinski's theorem is recovered from Theorem 4.1.6 by setting $k = -1$ and assuming that the vertex set of the empty face is the empty set. And as in the case of Balinski's theorem, Theorem 4.1.6 is tight.

Remark 4.1.7 Theorem 4.1.6 is tight. Consider again a bipyramid P over a $(d-1)$-simplex T. Let $\boldsymbol{u}$ and $\boldsymbol{v}$ be the apices of P and let F be a k-face of T with $k \in [0 \ldots d-2]$. Then the set $\mathcal{V}(T) - \mathcal{V}(F)$, of cardinality $d - (k+1) = d - k - 1$, separates $\boldsymbol{u}$ from $\boldsymbol{v}$ in $G(P) - \mathcal{V}(F)$. Hence $G(P) - \mathcal{V}(F)$ is not $(d-k)$-connected. For $k = d - 1$, suppose F has dimension $d - 2$ and let $J := \operatorname{conv}(F \cup \{\boldsymbol{v}\})$. Then J is a facet of P. It follows that the subgraph $G(P) - \mathcal{V}(J)$ is an edge, and so it is not 2-connected.

4.2 Edge Connectivity

A consequence of Balinski's theorem (4.1.1) is that the graph of a d-polytope is d-edge-connected; see Section C.4 for the relevant definitions on edge connectivity. Whereas for some polytopes, such as simplicial prisms, the cardinalities of both trivial and nontrivial minimum edge cuts coincide, for other polytopes, such as simplicial polytopes, these cardinalities significantly differ. Recall from Section C.4 that an edge cut is *trivial* if its edges are incident with a single vertex; otherwise, it is *nontrivial*. Moreover, a minimum edge cut is an edge cut with the smallest cardinality.

Theorem 4.2.1 (Pilaud et al., 2023) *For $d \geqslant 3$, every nontrivial minimum edge cut in a simplicial d-polytope has at least $d(d+1)/2$ edges.*

The theorem trivially implies the following.

Corollary 4.2.2 *For $d \geqslant 3$, a simplicial d-polytope with minimum degree δ is $\min\{\delta, d(d+1)/2\}$-edge-connected.*

While the proof of Theorem 4.2.1 relies on the lower bound theorem of Barnette (1971, 1973c) (see also Theorem 8.3.34), the case $d = 3$ has an elementary proof, also given by Pilaud et al. (2023).

Theorem 4.2.3 *In a planar triangulation, every minimum edge cut is trivial.*

Proof Recall that if G is a plane graph with $v(G) \geqslant 3$, then $e(G) \leqslant 3v(G) - 6$, with equality if and only if G is a plane triangulation (Proposition C.7.2).

Assume that G is a plane triangulation. Then the minimum degree of G is at least three (Proposition C.7.2). Consider a partition X and $\overline{X}$ of $\mathcal{V}(G)$, and set $e_{X\overline{X}}$:=$\#\mathcal{E}_G(X,\overline{X})$, $v_X := v(G[X])$, $e_X := e(G[X])$, $v_{\overline{X}} := v(G[\overline{X}])$, and $e_{\overline{X}} := e(G[\overline{X}])$. The set $\mathcal{E}_G(X,\overline{X})$ consists of the edges in G linking a vertex in X to a vertex in $\overline{X}$.

If $v_X = 1$, then the cut $\mathcal{E}(X,\overline{X})$ is trivial, and we are done. If $G[X]$ has two vertices $\boldsymbol{u}$ and $\boldsymbol{v}$ then $e_X \leqslant 1$, implying that

$$e_{X\overline{X}} \geqslant \deg_G \boldsymbol{u} + \deg_G \boldsymbol{v} - 2 \geqslant \deg_G \boldsymbol{u} + 1,$$

which shows that $\mathcal{E}(X,\overline{X})$ is not minimum. Assume now that both $G[X]$ and $G[\overline{X}]$ have at least three vertices. Since they are subgraphs of a plane graph, we have that $e_X \leqslant 3v_X - 6$ and $e_{\overline{X}} \leqslant 3v_{\overline{X}} - 6$. Additionally, from the equalities $v(G) = v_X + v_{\overline{X}}$ and $e(G) = e_X + e_{\overline{X}} + e_{X\overline{X}}$ it follows that

$$e_{X\overline{X}} = e(G) - e_X - e_{\overline{X}} \geqslant (3v(G) - 6) - (3v_X - 6) - (3v_{\overline{X}} - 6) = 6.$$

The graph G has minimum degree five (Proposition C.7.2), and so we have the conclusion of the theorem. □

Theorem 4.2.1 is sharp for $d \geqslant 4$. We show this through an example based on stacked polytopes and cyclic polytopes.

Proposition 4.2.4 (Pilaud et al., 2023) *For each $d \geqslant 4$, there is a simplicial d-polytope with minimum degree at least $\left(d^2 + d\right)/2$ and a nontrivial minimum edge cut with $\left(d^2 + d\right)/2$ edges.*

Proof Consider a family of stacked d-polytopes P_j with $j = d+1, \dots, 2d$ vertices constructed as follows:

(i) P_{d+1} is a d-simplex, with vertices labelled $\boldsymbol{x}_1, \dots, \boldsymbol{x}_{d+1}$. It has $d(d+1)/2$ edges.
(ii) For $j = d+1, \dots, 2d-1$, the polytope P_{j+1} is the connected sum of a d-simplex T_j and the polytope P_j along the facet of P_j containing the vertices $\boldsymbol{x}_{j+1-d}, \boldsymbol{x}_{j+1-d+1}, \dots, \boldsymbol{x}_j$. The vertices $\boldsymbol{x}_1, \dots, \boldsymbol{x}_d$ constitute a facet in P_{j+1}. Let $\boldsymbol{x}_{j+1}$ be the last vertex added, specifically the one in $\mathcal{V}(T_j) \backslash \mathcal{V}(P_j)$. The vertices $\boldsymbol{x}_{j+2-d}, \boldsymbol{x}_{j+2-d+1}, \dots, \boldsymbol{x}_{j+1}$ form a facet of P_{j+1}, since they also span a facet of the simplex T_j. Furthermore, the polytope P_{j+1} has $d(d+1)/2 + d(j-d)$ edges.

It follows that the polytope P_{2d} has $2d$ vertices and $\left(3d^2 - d\right)/2$ edges, and it has two disjoint facets F_0 and F_1 such that F_0 contains the vertices $\boldsymbol{x}_1, \dots, \boldsymbol{x}_d$ and F_1 contains the vertices $\boldsymbol{x}_{d+1}, \dots, \boldsymbol{x}_{2d}$. Let D be the set of the edges in P_{2d} from a vertex in F_0 to a vertex in F_1. Since F_0 and F_1 are $(d-1)$-simplices, the set D is a nontrivial edge cut with $\left(3d^2 - d\right)/2 - d(d-1) = \left(d^2 + d\right)/2$ edges.

Let C be the cyclic d-polytope on $1+(d^2+d)/2$ vertices; then it has degree $(d^2+d)/2$. Let C_0 and C_1 be two copies of C, let J_i be a facet of C_i for $i = 0, 1$, and let P represent the connected sum $C_0 \#_{F_0} P_{2d} \#_{F_1} C_1$ where the first sum goes along the facets J_0 and F_0 and the second sum along the facets J_1 and F_1. The polytope P is the desired simplicial d-polytope. It has minimum degree $(d^2+d)/2$, and the set D in P remains a nontrivial edge cut with $(d^2+d)/2$ edges. The minimality of D is now ensured by Theorem 4.2.1. □

4.3 Higher Connectivity of Strongly Connected Complexes

Proposition 3.3.1 has already established that graphs of strongly connected complexes are connected. But more is true, as discovered by Sallee (1967, sec. 2).

Theorem 4.3.1 (Sallee, 1967) *The graph of a strongly connected $(d-1)$-complex is $(d-1)$-connected.*

Proof Let $\mathcal{C}$ be a strongly connected $(d-1)$-complex and let G be the graph of $\mathcal{C}$. Let $\boldsymbol{u}, \boldsymbol{v}$ be two vertices of G. We show the existence of $d-1$ independent $\boldsymbol{u}-\boldsymbol{v}$ paths in G, which will in turn establish that G is $(d-1)$-connected by Menger's theorem (C.6.1).

Let F_u and F_v be facets of $\mathcal{C}$ containing $\boldsymbol{u}$ and $\boldsymbol{v}$, respectively. Since $\mathcal{C}$ is a strongly connected $(d-1)$-complex, there is a $(d-1, d-2)$-path $L = F_1F_2 \ldots F_n$ where $F_1 = F_u$, $F_n = F_v$, and $F_i \cap F_{i+1}$ is a ridge of $\mathcal{C}$ for each $i \in [1 \ldots n-1]$.

We prove that, for each $j \in [1 \ldots n]$, the graph G_j of $F_1 \cup \cdots \cup F_j$ is $(d-1)$-connected; we do so by induction on the length of a $(d-1, d-2)$-path in $\mathcal{C}$. In the case $j = n$, this will establish that G_n is $(d-1)$-connected, and thus the existence of $d-1$ independent $\boldsymbol{u}-\boldsymbol{v}$ paths, wherefrom we would get that G is $(d-1)$-connected.

For $j = 1$, Balinski's theorem (4.1.1) on the $(d-1)$-face F_1 gives that $G_1 = G(F_1)$ is $(d-1)$-connected, and so the induction on j can start. Suppose that G_j is $(d-1)$-connected. We show the existence of $d-1$ independent paths between every two vertices $\boldsymbol{x}$ and $\boldsymbol{y}$ in G_{j+1}. If $\boldsymbol{x}, \boldsymbol{y} \in G_j$, then the induction hypothesis yields the $d-1$ independent $\boldsymbol{x}-\boldsymbol{y}$ paths in G_j. If instead $\boldsymbol{x}, \boldsymbol{y} \in F_{j+1}$, then Balinski's theorem (4.1.1) on the $(d-1)$-face F_{j+1} gives that $G(F_{j+1})$ is $(d-1)$-connected, and therefore the existence of $d-1$ independent $\boldsymbol{x}-\boldsymbol{y}$ paths in $G(F_{j+1})$. Thus, assume that $\boldsymbol{x} \in G_j \backslash G_{j+1}$ and $\boldsymbol{y} \in G_{j+1} \backslash G_j$.

Since $F_j \cap F_{j+1}$ is a $(d-2)$-face, we have that $v(F_j \cap F_{j+1}) \geqslant d-1$. And because G_j is $(d-1)$-connected, a corollary of Menger's theorem, namely Corollary C.6.3, gives $d-1$ independent $\boldsymbol{x}-(F_j \cap F_{j+1})$ paths $L_1,\ldots,L_{d-1}$ in G_j between $\boldsymbol{x}$ and $F_j \cap F_{j+1}$. Let $Z_j := \{z_1,\ldots,z_{d-1}\}$ such that $z_r := \mathcal{V}(L_r) \cap (F_j \cap F_{j+1})$ for $r \in [1\ldots d-1]$; that is, Z_j is the set of the endvertices in $\mathcal{V}(F_j \cap F_{j+1})$ of the paths $L_1,\ldots,L_{d-1}$. As a consequence, Corollary C.6.3 applied to the polytope $G(F_{j+1})$ gives $d-1$ independent $Z_j - \boldsymbol{y}$ paths $L'_1,\ldots,L'_{d-1}$ in $G(F_{j+1})$ between Z_j and $\boldsymbol{y}$ such that $z_r \in L'_r$. The graph $G_{j+1} = G_j \cup G(F_{j+1})$, and so the concatenation of the paths L_r and L'_r for each $r \in [1\ldots d-1]$ yields $d-1$ disjoint $\boldsymbol{x}-\boldsymbol{y}$ in G_{j+1}, which ensures the $(d-1)$-connectivity of G_{j+1}. The induction is now complete, and thus the theorem follows. □

Combining Proposition 3.3.4 and Theorem 4.3.1, we obtain the following.

Proposition 4.3.2 *Let P be a d-polytope and let $F_1,\ldots,F_s$ be a shelling of P. Then the graph of the complex $\mathcal{C}(F_1 \cup \cdots \cup F_i)$ is $(d-1)$-connected, for each $i \in [1\ldots s]$.*

Proof The complex $\mathcal{C}_i := \mathcal{C}(F_1 \cup \cdots \cup F_i)$ is a strongly connected $(d-1)$-complex, for each $i \in [1\ldots s]$ (Proposition 3.3.4). That $G(\mathcal{C}_i)$ is $(d-1)$-connected now follows from Theorem 4.3.1. □

4.4 Linkedness of Polytopal Graphs

Let G be a graph and X a subset of $2k$ distinct vertices of G. The elements of X are called *terminals*. Let $Y := \{\{s_1,t_1\},\ldots,\{s_k,t_k\}\}$ be an arbitrary labelling and (unordered) pairing of all the vertices in X. We say that Y is *linked* in G if we can find disjoint $s_i - t_i$ paths for $i \in [1\ldots k]$. The set X is *linked* in G if every such pairing of its vertices is linked in G. If G has at least $2k$ vertices and every set of exactly $2k$ vertices is linked in G, we say that G is *k-linked.* If the graph of a polytope is k-linked we say that the polytope is also *k-linked.*

For a set $Y := \{\{s_1,t_1\},\ldots,\{s_k,t_k\}\}$ of pairs of vertices in a graph, a *Y-linkage* $\{L_1,\ldots,L_k\}$ is a set of disjoint paths with the path L_i joining the pair $\{s_i,t_i\}$ for $i \in [1\ldots k]$. The definition of k-linkedness gives the following lemma at once.

Lemma 4.4.1 *Let ℓ and k be positive integers such that $\ell \leqslant k$. Let X be a set of 2ℓ distinct vertices of a k-linked graph G, let Y be a labelling and pairing of the vertices in X, and let Z be a set of $2k-2\ell$ vertices in G such that $X \cap Z = \varnothing$. Then there exists a Y-linkage in G that avoids every vertex in Z.*

Proof Let R be an arbitrary pairing of the vertices of Z and let $S := Y \cup R$. Then $\#S = 2k$. And since G is k-linked, we can find k disjoint paths so that it includes a Y-linkage and an R-linkage. This Y-linkage in G clearly avoids every vertex in Z. □

From an algorithmic point of view, linkedness is closely related to the classical *disjoint paths problem* (Robertson and Seymour, 1995): given a graph G and a set $Y := \{\{s_1, t_1\}, \dots, \{s_k, t_k\}\}$ of k pairs of terminals in G, decide whether or not Y is linked in G. A natural optimisation version of this problem is to find the largest subset of the pairs so that there exist disjoint paths connecting the selected pairs.

From a structural point of view, being k-linked imposes a stronger demand on a graph than just being k-connected. Let G be a graph with at least $2k$ vertices, and let $S := \{s_1, \dots, s_k\}$ and $T := \{t_1, \dots, t_k\}$ be two disjoint k-element sets of vertices in G. If G is k-connected, then the sets S and T can be joined **setwise** by disjoint paths (specifically, k disjoint $S - T$ paths); this is a consequence of Menger's theorem (Corollary C.6.3). However, if G is k-linked then the sets can be joined **pointwise** by disjoint paths.

A k-linked graph needs to be at least $(2k - 1)$-connected, and yet there are $(2k - 1)$-connected graphs that are not k-linked.

Proposition 4.4.2 *A k-linked graph is $(2k - 1)$-connected.*

Proof Let G be a k-linked graph and let Z be a separator of two vertices $\boldsymbol{u}$ and $\boldsymbol{v}$ of G. By way of contradiction, suppose that $\#Z \leqslant 2k - 2$. Let $Y := \{(\boldsymbol{u}, \boldsymbol{v})\}$. Then Lemma 4.4.1 yields a Y-linkage that avoids Z, which contradicts the fact that Z separates $\boldsymbol{u}$ and $\boldsymbol{v}$. This proves the proposition. □

For concrete examples of $(2k - 1)$-connected graphs that are not k-linked, we resort to the classification of 2-linked graphs of Seymour (1980) and Thomassen (1980a), contextualised for 3-polytopes. With the exception of simplicial 3-polytopes, no 3-polytope, despite being 3-connected by Balinski's theorem (4.1.1), is 2-linked. We now give the details.

Let G be a graph and let $X \subseteq \mathcal{V}(G)$. A path in G is X-*valid* if no inner vertex of the path is in X. Moreover, a sequence $a_1, \dots, a_n$ of vertices in a cycle is in *cyclic order* if, while traversing the cycle, the sequence appears in clockwise or counterclockwise order.

Theorem 4.4.3 *Let G be the graph of a 3-polytope and let X be a set of four vertices of G. The set X is linked in G if and only if there is no facet of the polytope containing all the vertices of X.*

Proof Let P be a 3-polytope embedded in $\mathbb{R}^3$, let $G := G(P)$, and let X be an arbitrary set of four vertices in G. We first establish the necessary condition by proving the contrapositive. Let F be a 2-face containing the vertices of X and consider a planar realisation of G in which F is the outer face. Label the vertices of X so that they appear in the cyclic order s_1, s_2, t_1, t_2. Then paths $s_1 - t_1$ and $s_2 - t_2$ in G must inevitably intersect, implying that X is not linked.

Assume that there is no 2-face of P containing all the vertices of X. Let H be a (linear) hyperplane in $\mathbb{R}^3$ that contains s_1, s_2, and t_1. And let φ be a linear function that vanishes on H (this may require a translation of the polytope). Without loss of generality, assume that $\varphi(\boldsymbol{x}) > 0$ for some $\boldsymbol{x} \in P$ and that $\varphi(t_2) \geqslant 0$.

First consider the case that H is a supporting hyperplane of P at a 2-face F. The subgraph $G(F) - \{s_2\}$ is connected by Balinski's theorem (4.1.1), and so there is an X-valid $L_1 := s_1 - t_1$ path on $G(F)$. Then, use Lemma 3.2.7 to find an $L_2 := s_2 - t_2$ path in which each inner vertex has positive φ-value. The paths L_1 and L_2 are clearly disjoint.

Now consider the case that H intersects the interior of P. Then there is a vertex in P with φ-value greater than zero and a vertex with φ-value less than zero. Use Lemma 3.2.7 to find an $s_1 - t_1$ path in which each inner vertex has negative φ-value and an $s_2 - t_2$ path in which each inner vertex has positive φ-value. □

The subsequent corollary follows at once from Theorem 4.4.3.

Corollary 4.4.4 *A 3-polytope is 2-linked if and only if it is simplicial.*

While there are 3-polytopes that are 2-linked, no 3-polytope is 3-linked; planarity is incompatible with 3-linkedness.

Theorem 4.4.5 *No 3-polytope is 3-linked.*

Proof Suppose that P is a 3-polytope that is also 3-linked. Then $\mathcal{V}(P) \geqslant 6$. Let X be a set of six terminals in $G(P)$ and let

$$Y := \{\{s_1, t_1\}, \{s_2, t_2\}, \{s_3, t_3\}\}$$

be a pairing of the vertices of X. A consequence of Corollary 4.4.4 is that P must be simplicial. We place the terminals in X in a configuration incompatible with the planarity of $G(P)$.

Suppose that there is a face F of P that contains the terminals s_1, s_2, s_3, and consider a planar realisation of $G(P)$ in which F is the outer face of the realisation. Let F_{12}, F_{13}, and F_{23} be the other triangular faces containing the

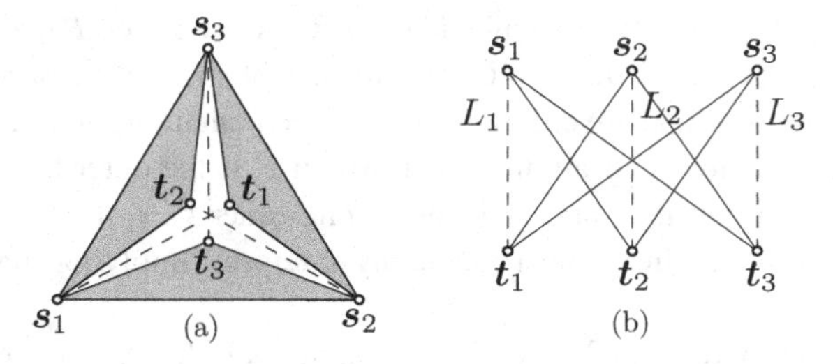

Figure 4.4.1 A planar graph G cannot be 3-linked. (a) A configuration of G with six terminals. (b) A subdivision of $K^{3,3}$ in G; the paths $L_1 := s_1 - t_1, L_2 := s_2 - t_2, L_3 := s_3 - t_3$ are depicted by dashed lines.

edges s_1s_2, s_1s_3, s_2s_3, respectively. Further suppose that

$$t_1 \in \mathcal{V}(F_{23})\backslash\{s_2, s_3\},\ t_2 \in \mathcal{V}(F_{13})\backslash\{s_1, s_3\},\ \text{and}\ t_3 \in \mathcal{V}(F_{12})\backslash\{s_1, s_2\}.$$

Since P is 3-linked, it must be 5-connected (Proposition 4.4.2), and so the minimum degree of P is five (Proposition C.5.1). As a consequence, the vertices t_1, t_2, and t_3 are pairwise distinct. See Fig. 4.4.1(a).

The existence of a Y-linkage $\{L_1 := s_1 - t_1, L_2 := s_2 - t_2, L_3 := s_3 - t_3\}$ in this setting would yield a subdivision of $K^{3,3}$ with the terminals as branch vertices, which would contradict Kuratowski's theorem (C.8.1). Figure 4.4.1(b) depicts this subdivision more clearly. This contradiction completes the proof of the theorem. □

We have shown that not every $(2k - 1)$-connected graph is k-linked. In the case of $2k$-connected graphs, having a k-linked subgraph ensures k-linkedness.

Lemma 4.4.6 *Let G be a $2k$-connected graph. If G contains a k-linked subgraph, then G is k-linked.*

Proof Let X be a set of $2k$ terminals in G, let $Y := \{\{s_1, t_1\}, \ldots, \{s_k, t_k\}\}$ be a pairing of the vertices of X, and let G' be a k-linked subgraph of G. Since the subgraph G' is k-linked, it has at least $2k$-vertices. Thanks to the $2k$-connectivity of G, we can find $2k$ disjoint paths $S_1, \ldots, S_k, T_1, \ldots, T_k$ from X to the subgraph G'; here, $s_i \in S_i$ and $t_i \in T_i$, for each $i \in [1 \ldots k]$.

For each $i \in [1 \ldots k]$, let s_i' be the intersection of the path S_i with G' and let t_i' be the intersection of T_i with G'. Let $Y' := \{\{s_1', t_1'\}, \ldots, \{s_k', t_k'\}\}$. Then, because G' is k-linked, we can find a Y'-linkage $\{L_1', \ldots, L_k'\}$ in G' such that $L_i' := s_i' - t_i'$. A Y-linkage $\{L_1, \ldots, L_k\}$ is now available: for each $i \in [1 \ldots k]$, let $L_i := s_i S_i s_i' L_i' t_i' T_i t_i$. This concludes the proof of the lemma. □

With the help of Lemma 4.4.6, the same reasoning employed in the proof of the sufficient condition of Theorem 4.4.3 settles Theorem 4.4.7.

Theorem 4.4.7 (2-linkedness of 4-polytopes) *A d-polytope is 2-linked, for $d \geqslant 4$.*

Proof We first prove the result for 4-polytopes. Let G be the graph of a 4-polytope P embedded in $\mathbb{R}^4$. Let X be a set of four vertices in G and let $Y := \{\{s_1, s_2\}, \{t_1, t_2\}\}$ be a labelling and pairing of the vertices in X.

Consider a linear function φ that vanishes on a linear hyperplane H in $\mathbb{R}^4$ passing through X. Consider the two cases in which either H is a supporting hyperplane of P at a facet F or H intersects the interior of P.

Suppose H is a supporting hyperplane of P at a facet F. First, find an $L_1 := s_1 - t_1$ path in the subgraph $G(F) - \{s_2, t_2\}$, which is connected by Balinski's theorem (4.1.1). Second, use Lemma 3.2.7 to find an $L_2 := s_2 - t_2$ path that touches F only at $\{s_2, t_2\}$. The Y-linkage $\{L_1, L_2\}$ settles this case.

If, instead, H intersects the interior of P, then there is a vertex in P with φ-value greater than zero and a vertex with φ-value less than zero. Use Lemma 3.2.7 to find an $L_1 := s_1 - t_1$ path in which each inner vertex has a negative φ-value and an $L_2 := s_2 - t_2$ path in which each inner vertex has a positive φ-value. Again, the Y-linkage $\{L_1, L_2\}$ settles this case. Hence, every 4-polytope is 2-linked.

A d-polytope with $d \geqslant 5$ has a 4-face F. Since P is 4-connected by Balinski's theorem (4.1.1) and since $G(F)$ is 2-linked because of what we already proved, Lemma 4.4.6 ensures that P is 2-linked. □

The graph of a d-polytope is d-connected (Balinski's theorem), and there are d-polytopes that are not $(d+1)$-connected, for instance, the simple d-polytopes. If a d-polytope is not $(d+1)$-connected, then it cannot be $(\lfloor (d+1)/2 \rfloor + 1)$-linked as every $(\lfloor (d+1)/2 \rfloor + 1)$-linked graph must be at least $(2(\lfloor (d+1)/2 \rfloor + 1) - 1)$-connected by Proposition 4.4.2; that is, at least $(d+1)$-connected. It follows that classes of d-polytopes with members that are not $(d+1)$-connected can be at most $\lfloor (d+1)/2 \rfloor$-linked. Some such classes of polytopes do achieve this maximum bound.

Theorem 4.4.8 (Linkedness of simplicial polytopes) *For every $d \geqslant 2$, a simplicial d-polytope is $\lfloor (d+1)/2 \rfloor$-linked. Furthermore, this is best possible as there are simplicial d-polytopes that are not $(d+1)$-connected.*

Proof Every 2-polytope is connected, and so it is 1-linked, while every simplicial 3-polytope is 2-linked (Corollary 4.4.4). So we can start an induction on d for the graph of a simplicial d-polytope; assume that $d \geqslant 4$. Let $k := \lfloor (d+1)/2 \rfloor$. Then $d \geqslant 2k-1$.

Let X be a set of $2k$ terminals in G and let $Y := \{\{s_1, t_1\}, \ldots, \{s_k, t_k\}\}$ be a pairing of the terminals in X.

Consider the link $Q := \mathrm{lk}(t_k, \mathcal{B}(P))$ of the vertex t_k in $\mathcal{B}(P)$. From $d \geqslant 2k-1$ and the d-connectivity of G (Balinski's theorem (4.1.1)), we find $2k-1$ disjoint paths $S_1, \ldots, S_k, T_1, \ldots, T_{k-1}$ from the terminals $s_1, \ldots, s_k, t_1, \ldots, t_{k-1}$, respectively, to $G(Q)$. For each $i \in [1 \ldots k]$, let s'_i be the intersection of the path S_i with $G(Q)$, and for each $i \in [1 \ldots k-1]$ let t'_i be the intersection of T_i with $G(Q)$; here, the vertices s'_i and t'_i are the ends of the paths S_i and T_i, respectively; see Fig. 4.4.2(a). In addition, let $Y' := \{\{s'_1, t'_1\}, \ldots, \{s'_k, t_k\}\}$.

Since P is simplicial, every vertex in $G(Q)$ is adjacent to t_k; in particular, s'_k are t_k are adjacent. This implies that the path $L_k := s_k S_k s'_k t_k$ connects s_k and t_k and is disjoint from each of the paths $S_1, \ldots, S_{k-1}, T_1, \ldots, T_{k-1}$.

According to Proposition 2.11.12, Q is combinatorially isomorphic to the boundary complex of a $(d-1)$-polytope, which, in this case, is simplicial as Q is induced by the $(d-2)$-faces of P containing t_k and these faces are simplices. In particular, for $d-1 \geqslant 2$, the graph $G(Q)$ of Q is isomorphic to the graph of a simplicial $(d-1)$-polytope.

We consider the link of s'_k in $\mathcal{B}(Q)$; let $R := \mathrm{lk}(s'_k, \mathcal{B}(Q))$. Resorting to Balinski's theorem (4.1.1) on Q, we get the $(d-1)$-connectivity of $G(Q)$. This gives that $G(Q)$ is at least $(2k-2)$-connected. Thus, there are $2k-2$ disjoint paths $S'_1, \ldots, S'_{k-1}, T'_1, \ldots, T'_{k-1}$ in $G(Q)$ from the terminals $s'_1, \ldots, s'_{k-1}, t'_1, \ldots, t'_{k-1}$ in $G(Q)$, respectively, to $G(R)$. For each $i \in [1 \ldots k-1]$, let s''_i be the intersection of the path S'_i with $G(R)$, and for each $i \in [1 \ldots k-1]$ let t''_i be the intersection of T'_i with $G(R)$; see Fig. 4.4.2(b)–(c). In addition, let $Y'' := \{\{s''_1, t''_1\}, \ldots, \{s''_{k-1}, t''_{k-1}\}\}$.

The complex R is a pure simplicial $(d-2)$-complex that is combinatorially isomorphic to the boundary complex of a simplicial $(d-2)$-polytope (Proposition 2.11.12); in particular, for $d-2 \geqslant 2$, the graph $G(R)$ is isomorphic to the graph of a simplicial $(d-2)$-polytope. By the induction hypothesis on $G(R)$, we get that $G(R)$ is $\lfloor (d-1)/2 \rfloor$-linked or, equivalently, that $G(R)$ is $(k-1)$-linked.

Thanks to the $(k-1)$-linkedness of $G(R)$, we can find $k-1$ disjoint paths $L''_1 := s''_1 - t''_1, \ldots, L''_{k-1} := s''_{k-1} - t''_{k-1}$ in $G(R)$. The idea behind using the link of s'_k in $\mathcal{B}(Q)$ is that the paths $L''_1, \ldots, L''_{k-1}$ do not contain s'_k, and thus do not intersect the path L_k. A Y-linkage $\{L_1, \ldots, L_k\}$ is now available: in addition to the already defined path L_k, for each $i \in [1 \ldots k-1]$ we let

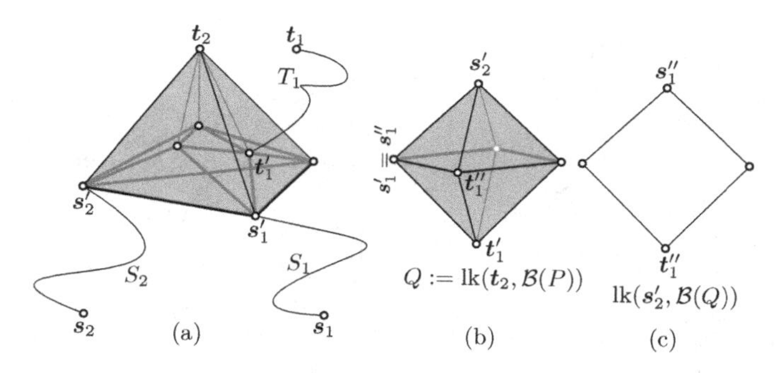

Figure 4.4.2 Linkedness of simplicial polytopes. (a) A vertex t_2 and its link $Q := \mathrm{lk}(t_2, \mathcal{B}(P))$ in a simplicial 4-polytope P; the link $\mathrm{lk}(t_2, \mathcal{B}(P))$ is highlighted in bold. The paths S_1, S_2, and T_1 from s_1, s_2, and t_1 to $G(Q)$, respectively, are depicted. (b) The link Q as a simplicial 3-polytope with the link $\mathrm{lk}(s_2', \mathcal{B}(Q))$ highlighted in bold. (c) The link $\mathrm{lk}(s_2', \mathcal{B}(Q))$.

$L_i := s_i S_i s_i' S_i' s_i'' L_i'' t_i'' T_i' t_i' T_i t_i$. This concludes the proof of the first part of the theorem.

Examples of simplicial d-polytopes that are not $(d+1)$-connected are easy to come by. Let U be a simplicial d-polytope obtained from a simplicial d-polytope W by stacking on a simplex facet F of W. It follows that $\mathcal{V}(U) = \mathcal{V}(W) \cup \{v\}$ where v is the new vertex introduced through the stacking. Removing the d vertices of F from U disconnects v from the vertices in $\mathcal{V}(U) \setminus \mathcal{V}(F)$. Hence $G(U)$ is not $(d+1)$-connected. For further insight, refer to Remark 4.1.2. □

The tight result of Theorem 4.4.8 extends to cubical polytopes.

Theorem 4.4.9 (Linkedness of cubical polytopes, Bui et al., 2024) *For every $d \neq 3$, a cubical d-polytope is $\lfloor (d+1)/2 \rfloor$-linked. Furthermore, this is best possible as there are cubical d-polytopes that are not $(d+1)$-connected, for instance, the d-cube.*

There is a linear function $f(d)$ such that every d-polytope is $f(d)$-linked. This is a consequence of a result of Larman and Mani (1970) where the authors showed that $2k$-connected graphs with subdivisions of K^{3k} are k-linked. We present the slight improvement of Werner and Wotzlaw (2011).

Theorem 4.4.10 (Larman and Mani, 1970; Werner and Wotzlaw, 2011) *Let G be a $2k$-connected graph that contains a subdivision of K^{3k-1} rooted at each vertex of G. Then G is k-linked.*

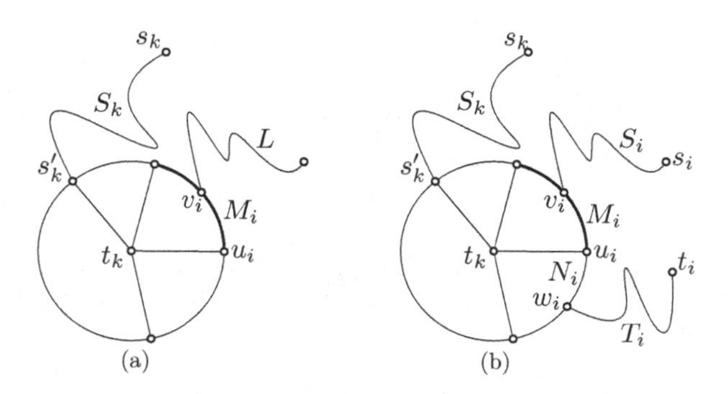

Figure 4.4.3 Auxiliary figure for Theorem 4.4.10: a subdivision K of K^{3k-1} rooted at t_k. The path M_i is highlighted in bold.

Proof Let X be a set of $2k$ terminals in G and let

$$Y := \{\{s_1, t_1\}, \ldots, \{s_k, t_k\}\}$$

be a pairing of the terminals in X. Let K be a subdivision of K^{3k-1} rooted at t_k. Let U be the set of branch vertices of K and let $U := U' \cup \{t_k\}$. Then $U' \subseteq \mathcal{N}_G(t_k)$.

Because G is $2k$-connected, we can find $2k-1$ disjoint paths $S_1, \ldots, S_k$, $T_1, \ldots, T_{k-1}$ in $G - t_k$ from the terminals $s_1, \ldots, s_k, t_1, \ldots, t_{k-1}$, respectively, to U'. We choose these paths in such a way that their total number of edges outside $E(K)$ is as small as possible.

Since K is rooted at t_k, every vertex in U' is adjacent to t_k; in particular, the branch vertex s'_k in $\mathcal{V}(S_k) \cap U'$ and t_k are adjacent. This implies that the path $L_k := s_k S_k s'_k t_k$ connects s_k and t_k and is disjoint from each of the paths $S_1, \ldots, S_{k-1}, T_1, \ldots, T_{k-1}$. See Fig. 4.4.3(a).

Let $\mathcal{L} := \{S_1, \ldots, S_k, T_1, \ldots, T_{k-1}\}$ and let $u_1, \ldots, u_{k-1}$ be the branch vertices of K not contained in any path in $\mathcal{L}$. Furthermore, for each $i \in [1 \ldots k-1]$, let M_i be the path in K from u_i to the branch vertex in $\mathcal{V}(S_i) \cap U'$; here, M_i is a subdivided edge of K between u_i and the vertex in $\mathcal{V}(S_i) \cap U'$. Similarly, let N_i be the path in K from u_i to the branch vertex in $\mathcal{V}(T_i) \cap U'$. We now show that each path in $\mathcal{L}$ other than S_i is disjoint from M_i and each path in $\mathcal{L}$ other than T_i is disjoint from N_i.

Suppose that M_i intersects any path L in $\mathcal{L}$ for the first time at a vertex v_i. Then the subpath $u_i M_i v_i$ of M_i intersects any path in $\mathcal{L}$ precisely at v_i. It follows that L has no more edges outside K than the path $L v_i M_i u_i$; otherwise u_i could have been chosen as a branch vertex instead of the branch vertex in $\mathcal{V}(L) \cap U'$. As a result, the subpath $v_i L$ of L, from v_i to the branch vertex in $\mathcal{V}(L) \cap U'$, completely lies in K, and so $v_i L = v_i M_i$. Hence $L = S_i$; see Fig. 4.4.3(a). It also true that v_i is the vertex of S_i and M_i that is closest to u_i. The same reasoning yields that, for each $k = [1 \ldots k-1]$, the path N_i is disjoint from any path in $\mathcal{L}\backslash\{T_i\}$; let w_i be the vertex of the path N_i and T_i that is closest to u_i. See Fig. 4.4.3(b).

We can now produce a Y-linkage $\{L_1, \ldots, L_k\}$: in addition to the already defined path L_k, for each $i \in [1 \ldots k-1]$ we let

$$L_i := s_i S_i v_i M_i u_i N_i w_i T_i t_i.$$

This concludes the proof of the theorem. □

Theorem 4.4.10 is what is needed to establish a lower bound for the linkedness of every d-polytope.

Corollary 4.4.11 *For each $d \geqslant 2$, a d-polytope is $\lfloor (d+2)/3 \rfloor$-linked.*

Proof Let P be a d-polytope with $d \geqslant 2$. The result is true for $d = 2, 3$ because $G(P)$ is connected, and for $d = 4, 5, 6$ due to Theorem 4.4.7. So assume that $d \geqslant 7$ and let $k := \lfloor (d+2)/3 \rfloor$. Then $d \geqslant 3k - 2$. From Corollary 3.9.8 it follows that $G(P)$ contains a subdivision of K^{d+1} rooted at any vertex, namely a subdivision of K^{3k-1}. Moreover, $G(P)$ is d-connected (Balinski's theorem (4.1.1)) and $d \geqslant 3k - 2 \geqslant 2k$, which implies that we have all the conditions to apply Theorem 4.4.10. Hence $G(P)$ is k-linked. □

By virtue of Corollary 4.4.11, a d-polytope is $\lfloor d/2 \rfloor$-linked, for each $d \leqslant 5$. And by Theorems 4.4.8 and 4.4.9, respectively, so are simplicial and cubical d-polytopes, for each $d \geqslant 2$. So we may wonder whether this is the case for every d-polytope. And so did Larman and Mani (1970, p. 145). This question was answered in the negative by Gallivan (1985). The proof of Gallivan (1985) is a neat application of Gale transforms to linkedness.

Theorem 4.4.12 (Gallivan, 1985) *For every $d \geqslant 2$, there is a d-polytope that is not $\lfloor 2(d+4)/5 \rfloor$-linked.*

Proof A 2-polytope is not 2-linked and there are 3-polytopes that are not 2-linked (Corollary 4.4.4). So assume that $d \geqslant 4$. Let $\ell := \lfloor (d+2)/5 \rfloor$. We produce a sequence $\mathcal{G}$ of $n := d + 1 + \ell$ vectors in $\mathbb{R}^{n-d-1}$ that is the Gale transform of a d-polytope.

Let S be the set of standard vectors in $\mathbb{R}^{\ell}$ plus the set of the reflections of these vectors with respect to the origin:

$$S = \{\boldsymbol{e}_1, \ldots, \boldsymbol{e}_\ell, -\boldsymbol{e}_1, \ldots, -\boldsymbol{e}_\ell\}.$$

Then $\#S = 2\ell$. We construct a Gale transform $\mathcal{G} = (z_1, \ldots, z_n)$ by assigning vectors of multiplicity two to each vector of S and a vector of multiplicity $d + 1 - 3\ell$ to the origin of $\mathbb{R}^\ell$. Then $\mathcal{G}$ has $n = 4\ell + d + 1 - 3\ell = d + 1 + \ell$ vectors.

On the one hand, we have that

$$\sum_{i=1}^{n} z_i = \mathbf{0}_\ell,$$

and on the other hand, the set S positively spans $\mathbb{R}^\ell$ (Theorem 1.10.2). It follows that every open halfspace bounded by a linear hyperplane in $\mathbb{R}^\ell$ contains one element of S and thus at least two vectors of $\mathcal{G}$. Thus, the two conditions of Theorem 2.14.14 are in place, yielding that $\mathcal{G}$ is the Gale transform of a d-polytope P.

Let us now consider P. Let Y' be the set of 2ℓ pairs of terminals in P associated with the 2ℓ elements of S; that is, for each $i \in [1 \ldots 2\ell]$, the pair $\{s_i, t_i\}$ of Y' is formed by the two vertices of P corresponding to one of 2ℓ vectors of S. It follows that, for each $i \in [1 \ldots 2\ell]$, the pair $\{s_i, t_i\} \in Y'$ does not form an edge in P, as $\mathbf{0}_\ell \notin \operatorname{rint}(\operatorname{conv}(\mathcal{G} \setminus \{\mathcal{G}(s_i), \mathcal{G}(t_i)\}))$ (Theorem 2.14.10). If $2\lfloor (d+2)/5 \rfloor < \lfloor 2(d+4)/5 \rfloor$, then let Y'' be a set of further pairs of distinct terminals so that $Y' \cup Y''$ is a set of $\lfloor 2(d+4)/5 \rfloor$ pairs of terminals.

Because the vertices s_i and t_i are not adjacent, every $s_i - t_i$ path in P must pass through one of the $d + 1 + \lfloor (d+2)/5 \rfloor - 2\lfloor 2(d+4)/5 \rfloor$ nonterminals of P corresponding to the origin of the Gale transform. This is impossible as there are $2\lfloor (d+2)/5 \rfloor$ pairs in Y', and

$$d + 1 + \lfloor (d+2)/5 \rfloor - 2\lfloor 2(d+4)/5 \rfloor < 2\lfloor (d+2)/5 \rfloor. \tag{4.4.12.1}$$

This inequality follows from considering the five possible congruence classes of d modulo 5. We show that (4.4.12.1) holds in the case $d = 5p$ for some positive integer p. In this case, (4.4.12.1) becomes $5p+1+p-2(2p+1) < 2p$, which is clearly true. The analysis for the other congruence classes is similar. As a consequence, P is not $\lfloor 2(d+4)/5 \rfloor$-linked. □

As a corollary of Theorem 4.4.12 we get the following.

Corollary 4.4.13 *For $d = 18, 20$ and every $d \geqslant 22$, there is a d-polytope that is not $\lfloor d/2 \rfloor$-linked.*

Proof According to Theorem 4.4.12, for every $d \geqslant 2$ there is a d-polytope that is not $\lfloor 2(d+4)/5 \rfloor$-linked. And for $d = 18, 20$ and $d \geqslant 22$, we have that $\lfloor 2(d+4)/5 \rfloor < \lfloor d/2 \rfloor$. This settles the corollary. □

Some years after Larman and Mani (1970) made their conjecture on the linkedness of polytopes, Thomassen (1980a) conjectured that every $(2k+2)$-connected graph was k-linked. Gallivan's polytopes also disproves this later conjecture.

Corollary 4.4.14 *For $d = 28, 30$ and every $d \geqslant 32$, there is a d-polytope that is not $\lfloor (d-2)/2 \rfloor$-linked.*

Proof Again, for every $d \geqslant 2$, there is a d-polytope that is not $\lfloor 2(d+4)/5 \rfloor$-linked (Theorem 4.4.12). And for $d = 28, 30$ and $d \geqslant 32$, we have that $\lfloor 2(d+4)/5 \rfloor < \lfloor (d-2)/2 \rfloor$. This settles the corollary. □

The d-polytope P presented in Theorem 4.4.12 can be described without resorting to Gale transforms, as Werner and Wotzlaw (2011, sec. 3.2) did. We could also find a realisation of a polytope combinatorially isomorphic to P if we follow the ideas presented in Example 2.14.18.

Remark 4.4.15 (Gallivan's polytopes) Let $\ell := \lfloor (d+2)/5 \rfloor$ and $r := d+1-3\ell$. The d-polytope P presented in Theorem 4.4.12 is an r-fold pyramid over a $(3\ell-1)$-polytope $F_{3\ell-1}$ that can be described as the iteration of ℓ free joins of the 2-cube Q:

$$F_{3\ell-1} = \underbrace{Q * \cdots * Q}_{\ell \text{ factors}}.$$

Then $v(F_{3\ell-1}) = 4\ell$, $\dim F_{3\ell-1} = 3\ell - 1$, and $v(P) = d + 1 + \ell$.

We continue with a result of Thomas and Wollan (2005) that relates large minimum degree with linkedness.

Theorem 4.4.16 (Thomas and Wollan, 2005) *For $k \geqslant 1$, every $2k$-connected graph with minimum degree at least $10k$ is k-linked.*

If we contextualise Theorem 4.4.16 to graphs of polytopes, we get the following.

Corollary 4.4.17 *For $d \geqslant 2$, every d-polytope with minimum degree at least $5d$ is $\lfloor d/2 \rfloor$-linked.*

Finally, we introduce strong linkedness, a property marginally stronger than linkedness. We say that a graph G is *strongly k-linked* if it has at least $2k+1$ vertices and, for every vertex v of G, the subgraph $G - v$ is k-linked. Bui et al. (2021) showed that cubical 4-polytopes are strongly 2-linked and that, for each $d \geqslant 1$, d-dimensional cubes are strongly $\lfloor d/2 \rfloor$-linked.

4.5 Edge Linkedness of Polytopal Graphs

Let G be a graph and let $\{s_1,t_1\},\ldots,\{s_k,t_k\}$ be k pairs of not necessarily distinct vertices in G. As in the case of linkedness, we call the vertices $s_1,\ldots,s_k,t_1,\ldots,t_k$ *terminals* of G. We say that G is *k-edge-linked* if we can find edge-disjoint paths $s_1-t_1,\ldots,s_k-t_k$ in G, for every such k pairs. If the graph of a polytope is k-edge-linked we say that the polytope is also *k-edge-linked*.

If G is k-edge-linked, then, in the case that $s_1=\cdots=s_k$ and $t_1=\cdots=t_k$, we get the existence of k edge-disjoint paths between two vertices of G. Thus, k-edge linkedness implies k-edge connectivity. This is tight as we can find k-edge-connected graphs that are not $(k+1)$-edge-connected but are k-edge-linked; for instance, the graph of a 3-cube is 3-edge-connected, not 4-edge-connected, and 3-edge-linked (Problem 4.7.6). In the other direction, we have the following.

Proposition 4.5.1 *Every* $(2k-1)$*-edge-connected graph is k-edge-linked.*

Proof Let G be a $(2k-1)$-edge-connected graph and let $\{s_1,t_1\},\ldots,\{s_k,t_k\}$ be k pairs of not necessarily distinct vertices in G. Since G is $(2k-1)$-edge-connected, there are $2k-1$ edge-disjoint paths $S_2,\ldots,S_k,T_1,\ldots,T_k$ from s_1 to $s_2,\ldots,s_k,t_1,\ldots,t_k$, respectively. The desired k edge-disjoint paths $L_1,\ldots,L_k$ are as follows:

$$L_i := \begin{cases} s_1T_1t_1, & \\ s_iS_is_1T_it_i, & \text{for } i \in [2\ldots k]. \end{cases}$$

This concludes the proof of the proposition. □

As in the case of connectivity and linkedness, Thomassen (1980a) made a conjecture that relates edge connectivity and edge linkedness.

Conjecture 4.5.2 (Thomassen, 1980a) *For each odd* $k \geqslant 3$*, a graph is k-edge-connected if and only if it is k-edge-linked. And for each even* $k \geqslant 2$*, a* $(k+1)$*-edge-connected graph is k-edge-linked.*

If true, Conjecture 4.5.2 would be best possible since, for even $k \geqslant 2$, there are k-edge-connected graphs that are not k-edge-linked (Thomassen, 1980a, p. 375). A few years later, Huck (1991) came very close to settling Conjecture 4.5.2; this is the best general result so far. We do not present the proof of Huck (1991) because it is too long to be included in this book.

Theorem 4.5.3 (Huck, 1991) *For each odd $k \geqslant 3$, a $(k+1)$-edge-connected graph is k-edge-linked. And for each even $k \geqslant 2$, a $(k+2)$-edge-connected graph is k-edge-linked.*

In the context of polytopes, we have the following.

Corollary 4.5.4 *For each odd $k \geqslant 3$, a $(k+1)$-polytope is k-edge-linked, and for each even $k \geqslant 2$, a $(k+2)$-polytope is k-edge-linked.*

4.6 Connectivity of Incidence Graphs of Polytopes

Several generalisations of Balinski's theorem (4.1.1) have appeared in the literature. Several such generalisations were obtained by Sallee (1967), who investigated the connectivity of incidence graphs of d-polytopes. For integers r, s such that $0 \leqslant r \leqslant s \leqslant d-1$, we define the (r, s) *incidence graph* $G_{r,s}(P)$ of a d-polytope P as the graph on the r-faces of P where two r-faces are adjacent if there is an s-face of P containing them. The graph of P coincides with the $(0, 1)$ incidence graph of P. And the $(d-2, d-1)$ incidence graph of P coincides with the line graph of its dual graph. Balinski's theorem also give the d-connectivity of $G_{d-2,d-1}(P)$.

Proposition 4.6.1 *The $(d-2, d-1)$ incidence graph of a d-polytope is d-connected and $(2d-2)$-edge-connected.*

Proof The $(d-2, d-1)$ incidence graph $G_{d-2,d-1}$ of a d-polytope P is isomorphic to the line graph of its dual graph. And the dual graph of P is isomorphic to the graph of the dual polytope P^*. That is, we have that $L(G(P^*)) = G_{d-2,d-1}(P)$. The graph $G(P^*)$ is d-connected by Balinski's theorem, and so it is d-edge-connected. In this case, Proposition C.5.14 yields that $L(G(P^*))$ is d-connected, and Proposition C.5.15 yields that $L(G(P^*))$ is $(2d-2)$-edge-connected. This settles the theorem. □

Sallee (1967) proved a number of results on different notions of connectivity, including the following. Recall that $\kappa(G)$ gives the connectivity of a graph G.

Theorem 4.6.2 (Sallee, 1967) *Let r, s be integers such that $0 \leqslant r \leqslant s \leqslant d-1$, let G represent the (r, s) incidence graph of a d-polytope, and let*

$$\kappa_{r,s}(d) := \min\{\kappa(G_{r,s}(P)) \mid P \text{ is a } d\text{-polytope}\}.$$

Then

$$(d-1-s)\binom{s}{r} + \binom{s+1}{r+1} \leqslant \kappa(G),$$

$$(d-1-s)\binom{s}{r}+\binom{s+1}{r+1}\leqslant \kappa_{r,s}(d),$$

$$\kappa_{r,s}(d)\leqslant \min\left\{\binom{d}{r+1},\sum_{i=1}^{s-r}\binom{d-r}{i}\binom{r+1}{i}\right\}.$$

For the incidence graph $G_{0,1}$ of a polytope, the graph of the polytope, Theorem 4.6.2 coincides with Balinski's theorem (4.1.1). Additionally, Theorem 4.6.2 yields the d-connectivity of $G_{d-2,d-1}$ as did Proposition 4.6.1. For other values of r and s, the upper and lower bounds of $\kappa_{r,s}(d)$ are different. We remark that the upper bounds of $\kappa_{r,s}(d)$ are both attainable. The bipyramid over the $(d-1)$-simplex gives the upper bound $\binom{d}{r+1}$, while the d-simplex gives the upper bound $\sum_{i=1}^{s-r}\binom{d-r}{i}\binom{r+1}{i}$ (Sallee, 1967, pp. 478, 495).

Athanasiadis (2009) produced tight results on the connectivity of the incidence graphs $G_{r,r+1}$. While Theorem 4.6.2 yields

$$(d-r)(r+1)-r\leqslant \kappa_{r,r+1}(d)\leqslant (d-r)(r+1),$$

Athanasiadis (2009) bettered this as follows.

Theorem 4.6.3 (Athanasiadis, 2009) *Let $0\leqslant r\leqslant d-1$ be an integer and let*

$$\kappa_{r,r+1}(d):=\min\{\kappa(G_{r,r+1}(P))\mid P \text{ is a } d\text{-polytope}\}.$$

Then

$$\kappa_{r,r+1}(d)=\begin{cases} d, & \text{if } r=d-2;\\ (d-r)(r+1), & \text{otherwise.}\end{cases}$$

Wotzlaw (2009) also studied the connectivity of the incidence graphs $G_{r,s}$, improving Theorem 4.6.2 for certain values of r and s.

Theorem 4.6.4 (Wotzlaw, 2009) *Let r,s,d be integers such that $2r+1\leqslant s\leqslant d$, and let $G_{r,s}$ be the incidence graph of a d-polytope P. Then $G_{r,s}(P)$ is $\binom{d}{r+1}$-connected.*

4.7 Problems

4.7.1 (Perles and Prabhu, 1993) Let $F_1,\dots,F_n$ be faces of a d-polytope such that

$$\sum_{i=1}^{n}(1+\dim F_i)=k+1\leqslant d.$$

Prove that $G(P) - \left(\bigcup_1^n \mathcal{V}(F_i)\right)$ is $(d-k-1)$-connected and that this is best possible.

4.7.2 (Monotone Balinski theorem, Holt and Klee, 1999) Let P be a d-polytope in $\mathbb{R}^d$ and let $\boldsymbol{x}, \boldsymbol{y}$ be two vertices of P. Suppose that φ is a linear functional in general position with respect to P such that $\varphi(\boldsymbol{x}) = \min\{\varphi(\boldsymbol{v}) \mid \boldsymbol{v} \in \mathcal{V}(P)\}$ and $\varphi(\boldsymbol{y}) = \max\{\varphi(\boldsymbol{v}) \mid \boldsymbol{v} \in \mathcal{V}(P)\}$. Then there are d independent $\boldsymbol{x}-\boldsymbol{y}$ paths along which the value of φ increases strictly.

4.7.3 (Klee, 1964c) Let P be a d-polytope and let X be a separator of $G(P)$. The number of components $G(P) - X$ is at most

(i) 2, in case $\#X = d$;
(ii) $f_{d-1}(C(\#X, d))$, in case $\#X \geqslant d+1$.

Moreover, this is best possible.

4.7.4 (Edge connectivity of crosspolytopes) Prove that the d-crosspolytope is $(2d-2)$-edge-connected.

4.7.5 (Bui et al., 2021; Mészáros, 2016) Prove that, for every $d \neq 3$, the d-cube is $\lfloor (d+1)/2 \rfloor$-linked.

4.7.6 (Mészáros, 2015, cor. 41) For each $d \geqslant 3$, a d-cube is d-edge-linked.

4.7.7* (Lockeberg, 1977) Let P be a d-polytope and let $d_1, \ldots, d_m$ be natural numbers such that $d_1 + \cdots + d_m = d$. Define a $(k, k-1)$-*path* in P, with $k < d$, as a path $F_1 \ldots F_n$ of k-faces in P such that $F_i \cap F_{i+1}$ is a $(k-1)$-face of P, for each $i \in [1 \ldots n-1]$.

Prove that between any two vertices $\boldsymbol{x}$ and $\boldsymbol{y}$ in $\mathcal{V}(P)$, there is a $(d_1, d_1 - 1)$-path L_1, a $(d_2, d_2 - 1)$-path $L_2, \ldots$, and a $(d_m, d_m - 1)$-path L_m such that $\mathcal{V}(L_i) \cap \mathcal{V}(L_j) = \{\boldsymbol{x}, \boldsymbol{y}\}$ for $i \neq j$.

Balinski's theorem is recovered when $d_1 = d_2 = \cdots = d_d = 1$.

4.7.8* (Critically and minimally connected polytopes) Characterise d-polytopes whose graphs are critically d-connected and those whose graphs are minimally d-connected.

4.7.9* (Critically and minimally edge-connected polytopes) Characterise d-polytopes whose graphs are critically d-edge-connected and those whose graphs are minimally d-edge-connected.

4.7.10* (Werner and Wotzlaw, 2011) Are all 6-polytopes 3-linked?

For each $d \leqslant 5$, a d-polytope is $\lfloor d/2 \rfloor$-linked. Thus, $d = 6$ is the smallest dimension for which we don't know the exact linkedness. Thomas and Wollan (2008) proved that every 6-connected graph with n vertices and at least $5n - 14$ edges is 3-linked. According to the lower bound theorem for simplicial polytopes (8.3.34), this means that 6-polytopes with only triangular faces are 3-linked; that is, 2-simplicial 6-polytopes are 3-linked.

4.7.11* Is a simple d-polytope $\lfloor d/2 \rfloor$-linked?

While not every d-polytope is $\lfloor d/2 \rfloor$-linked, none of the known counterexamples are simple d-polytopes; see also Gallivan (1985).

4.7.12* (Werner and Wotzlaw, 2011) Is there a function $\varphi(d)\colon \mathbb{N} \to \mathbb{N}$ such that every d-polytope with at least $\varphi(d)$ vertices is $\lfloor d/2 \rfloor$-linked?

All the known counterexamples have fewer than $3\lfloor d/2 \rfloor$ vertices.

4.7.13* (Polytopal version of Thomassen, 1980a, conj. 1) Determine whether the following statement is true: For each odd $d \geqslant 3$, a d-polytope is d-edge-linked. Additionally, for each even $d \geqslant 2$, a d-polytope is $(d-1)$-edge-linked.

4.8 Postscript

We presented two proofs of Balinski's theorem (4.1.1). The first proof of Balinski's theorem is essentially the original proof of Balinski (1961) and the second proof is from Pineda-Villavicencio (2021). Beyond Balinski's theorem and its consequences, there are not many other assertions of this kind. Pineda-Villavicencio et al. (2020a) established one such extra result: they proved that, for every $d \geqslant 3$, a cubical d-polytope with minimum degree δ is $\min\{\delta, 2d-2\}$-connected, and that, for every $d \geqslant 4$, its minimum separators of cardinality at most $2d - 3$ are trivial: they consist of all the neighbours of some vertex.

The proof of Sallee's theorem (4.3.1) on the connectivity of strongly connected complexes is inspired by Sallee's original proof. Additional generalisations of Sallee's theorem and Balinski's theorem were obtained by Barnette (1973b, 1982) and Björner and Vorwerk (2015) in the context of pseudomanifolds; a $(d-1)$-*pseudomanifold* is an abstract simplicial $(d-1)$-complex that is also strongly connected and in which every $(d-2)$-face is contained in exactly two $(d-1)$-faces.

The linkedness of simplicial polytopes (Theorem 4.4.8) was established by Larman and Mani (1970, thm. 3); our proof was inspired by theirs. The assertion that every d-polytope is $\lfloor (d+2)/3 \rfloor$-linked (Corollary 4.4.11) is due to Werner and Wotzlaw (2011), who slightly improved the bound of $\lfloor (d+1)/3 \rfloor$ in Larman and Mani (1970, thm. 2). The proof of Werner and Wotzlaw (2011) is based on the proof of Diestel (2017, thm. 3.5.2); our proof follows these two proofs. For general graphs, there is a linear function $f(k)$ such that every $f(k)$-connected graph is k-linked, which follows from Bollobás and Thomason (1996), Kawarabayashi et al. (2006), and Thomas and Wollan (2005). The proof of the existence of a d-polytope that is not $\lfloor 2(d+4)/5 \rfloor$-linked was inspired by the original proof of Gallivan (1985).

For small values of k, namely $k = 3,4$, we can match Thomassen's conjecture (4.5.2). Okamura (1984) proved that every 3-edge-connected graph is 3-edge-linked, while Hirata et al. (1984) proved that every 5-edge-connected graph is 4-edge-linked.

5

Reconstruction

Reconstructing a d-polytope from its k-skeleton ($k \leqslant d - 2$) amounts to determining the face lattice of the polytope from its dimension and skeleton. For each $d \geqslant 4$, there are d-polytopes that have isomorphic $(d-3)$-skeleta and yet are not combinatorially isomorphic. But every d-polytope is reconstructible from its $(d - 2)$-skeleton.

Section 5.2 focusses on reconstructions from 2-skeletons and 1-skeletons. It presents an algorithm that reconstructs a d-polytope with at most $d - 2$ nonsimple vertices from its dimension and 2-skeleton. This result is tight: there are pairs of nonisomorphic d-polytopes with $d - 1$ nonsimple vertices and isomorphic $(d - 3)$-skeleta, for each $d \geqslant 4$. Blind and Mani-Levitska (1987) and, later, Kalai (1988b), showed that a simple polytope can be reconstructed from its dimension and graph. We present a slight generalisation of this result and briefly discuss the theorem of Friedman (2009), which states that the reconstruction of a simple polytope from its dimension and graph can be done in time polynomial in the number of vertices of the polytope.

The chapter ends with variations on the reconstruction problem, including the question of reconstructing a polytope knowing its dimension, k-skeleton, and class of polytopes. This final section also considers a related conjecture of Perles.

5.1 Ambiguity and Reconstruction

Following Grünbaum (2003, sec. 12.1), we say that a d-polytope P and a d'-polytope P' are *combinatorially k-isomorphic* if their k-skeletons for $k \leqslant d$ and $k \leqslant d'$ are isomorphic as posets; if $d = d'$ and $k \geqslant d - 1$ then the polytopes are combinatorially isomorphic.

In Section 2.9 we learnt that, if a d-polytope P shares the k-skeleton for some $k \geqslant \lfloor d/2 \rfloor$ with the d-simplex, then P is the d-simplex (Proposition 2.9.11); this assertion admits a very useful generalisation.

Proposition 5.1.1 (Grünbaum, 2003, thm. 11.1.6) *If P is a d-polytope and $k \geqslant \lfloor d/2 \rfloor$, then the k-skeleton of P doesn't contain any refinement of the k-skeleton of the $(d+1)$-simplex.*

Also in Section 2.9, we learnt that the cyclic d-polytope on $n \geqslant d+1$ vertices, and, more generally, a neighbourly d-polytope on n vertices, shares the k-skeleton of the $(n-1)$-simplex, for each $k < \lfloor d/2 \rfloor$; that is, these three polytopes are combinatorially k-isomorphic. This, however, can not happen when $k \geqslant \lfloor d/2 \rfloor$.

Proposition 5.1.2 *Let P be a d-polytope and P' a d'-polytope with isomorphic k-skeleta for some $k \geqslant \lfloor d/2 \rfloor$. Then $d = d'$.*

Proof Let $\mathcal{B}_k$ be the k-skeleton of P, $\mathcal{B}'_k$ the k-skeleton of P', $\mathcal{K}$ the k-skeleton of the d-simplex, and $\mathcal{K}'$ the k-skeleton of the d'-simplex. According to Corollary 3.9.7, we have that $\mathcal{B}'_k$ contains a refinement of $\mathcal{K}'$. And since $k \geqslant \lfloor d/2 \rfloor$, Proposition 5.1.1 ensures that $\mathcal{B}_k$ doesn't contain a refinement of $\mathcal{K}'$ when $d' > d$. Thus, $\mathcal{B}_k$ and $\mathcal{B}'_k$ cannot be isomorphic in this situation. If instead $d' < d$, then, from $k \geqslant \lfloor d/2 \rfloor \geqslant \lfloor d'/2 \rfloor$, it follows that $\mathcal{B}_k$ contains a refinement of $\mathcal{K}$ but $\mathcal{B}'_k$ doesn't. Hence $d = d'$. □

Grünbaum (2003) showed that the $(d-2)$-skeleton of a d-polytope determines its combinatorial type. The proof of this result is a neat application of Jordan–Lebesgue–Brouwer's separation theorem (B.2.1), a generalisation of the curve theorem of Jordan (1882) to spheres of every dimension.

Theorem 5.1.3 (Grünbaum, 2003, thm. 12.3.1) *Let $d \geqslant 3$. There is a unique d-polytope with a given $(d-2)$-skeleton, up to (combinatorial) isomorphism.*

In the case $d = 3$, the theorem is also a consequence of Steinitz's theorem (3.8.1) and Whitney's theorem (3.2.12). The $(d-2)$-skeleton of a 3-polytope P is its graph G, which is 3-connected and planar by Steinitz's theorem. In this case, Whitney's theorem ensures that the faces of P are uniquely determined by G. Thus, P is the unique 3-polytope with graph G, up to isomorphism.

The classical polytope reconstruction problem can be stated as follows.

Problem 5.1.4 (Reconstruction) *Let $k \geqslant 1$ and $d \geqslant 3$ be given, and let $\mathcal{K}$ be the k-skeleton of a d-polytope. Determine the face lattice of the polytope from k, d, and $\mathcal{K}$.*

We make a couple of observations. First, this problem requires that there is only one d-polytope with the given k-skeleton. Second, by virtue of Remark 2.3.9, determining the face lattice of a polytope from its k-skeleton ($k \geqslant 1$) amounts to determining the facets of the polytope; that is, the facet-vertex incidences of the polytope.

While Theorem 5.1.3 doesn't imply that the reconstruction of the face lattice of a d-polytope can be done from its $(d-2)$-skeleton, there is an algorithm that achieves this in time exponential in the number of vertices of the polytope; the algorithm relies on tools from algebraic topology such as homology groups (Nevo, 2022).

Theorem 5.1.5 (Grünbaum, 2003, thm. 12.3.1) *Every d-polytope can be reconstructed from its $(d-2)$-skeleton.*

Combinatorially $(d-3)$-Isomorphic Polytopes

Theorem 5.1.3 is best possible in the sense that there are d-polytopes with isomorphic $(d-3)$-skeleta and nonisomorphic face lattices. We start with a simple example.

Example 5.1.6 (d-polytopes with isomorphic $(d-3)$-skeleta, excess degree d, and exactly d nonsimple vertices) If we let $B(d)$ be the bipyramid over a $(d-1)$-simplex and let $B := B(d-1)$, then $B(d)$ and the pyramid P over B are combinatorially $(d-3)$-isomorphic and both have excess degree d.

We illustrate the isomorphism between the corresponding $(d-3)$-skeleta. Let $\boldsymbol{w}$ be the apex of the pyramid P, let R be the base of the $(d-1)$-bipyramid B, let $\boldsymbol{u}$ and $\boldsymbol{v}$ be the apices of B, and let $T := \operatorname{conv}(\{\boldsymbol{w}\} \cup R)$. It follows that P has the same $(d-3)$-skeleton as a bipyramid over T with apices $\boldsymbol{u}$ and $\boldsymbol{v}$.

The d-polytopes P_1 and P_2 from Example 5.1.6 can be modified to produce combinatorially $(d-3)$-isomorphic n-vertex d-polytopes for every $n \geqslant d+2$. Let σ be an isomorphism between the $(d-3)$-skeleta of P_1 and P_2. If we stack on simplex facets of P_1 and P_2 whose vertex sets are mapped by σ, we obtain new combinatorially $(d-3)$-isomorphic d-polytopes with an extra vertex. Thus, stacking $n-(d+2)$ times on corresponding simplex facets, we get the desired outcome.

Proposition 5.1.7 *For every $d \geqslant 4$ and every $n \geqslant d + 2$, there is a pair of combinatorially $(d-3)$-isomorphic d-polytopes with nonisomorphic face lattices and n vertices.*

We can also prescribe the number of nonsimple vertices in pairs of combinatorially $(d-3)$-isomorphic d-polytopes. We give a construction with exactly $d-1$ nonsimple vertices; this is best possible as every d-polytope with at most $d-2$ nonsimple vertices can be reconstructed from its 2-skeleton (Theorem 5.2.1). Our construction relies on the inductive construction of polytopes in Theorem 2.10.1.

Proposition 5.1.8 *For every $d \geqslant 4$, there is a pair of combinatorially $(d-3)$-isomorphic d-polytopes with nonisomorphic face lattices and $2d$ vertices, of which exactly $d-1$ vertices are nonsimple.*

Proof The proof of the proposition follows from the following two claims.

Claim 1 For every $d \geqslant 3$, there is a d-polytope Q_d^1 with $2d$ vertices labelled $0, \dots, 2d-1$ in such a way that the set X of vertices with positive even labels contains the $d-1$ nonsimple vertices of Q_d^1 and the $2d$ facets are as follows:

Type A A simplex with vertex set: $\{0\} \cup X$;
Type B $d-1$ facets whose vertex sets take the form: $\{0\} \cup \{2i+1 \mid i = 0, \dots, k-1\} \cup X \backslash \{2k\}$ for $k = 1, \dots, d-1$;
Type C $d-2$ facets whose vertex sets take the form: $\{2d-1\} \cup \{2i+1 \mid i = k-1, \dots, d-2\} \cup X \backslash \{2k\}$ for $k = 1, \dots, d-2$;
Type D A simplex with vertex set: $\{2d-3, 2d-1\} \cup X \backslash \{2(d-1)\}$; and
Type E A simplex with vertex set: $\{2d-1\} \cup X$.

Proof of claim The construction of the polytope Q_d^1 is by induction, with the base case $d = 3$ depicted in Fig. 5.1.1(a). We now construct Q_{d+1}^1 from Q_d^1.

(i) Construct a pyramid P over Q_d^1, label the apex of the pyramid by $2d$, and let $X' := X \cup \{2d\}$.
(ii) Take the convex hull of P with a new vertex $\boldsymbol{v}$ labelled $2d+1$ and positioned on the affine hull of the triangle $\{2d-3, 2d-1, 2d\}$. Arrange the new vertex such that every facet of P not containing the edge $[2d-1, 2d]$ is nonvisible from the point, and *the* facet containing the edge but not the triangle is visible from the point $\boldsymbol{v}$.

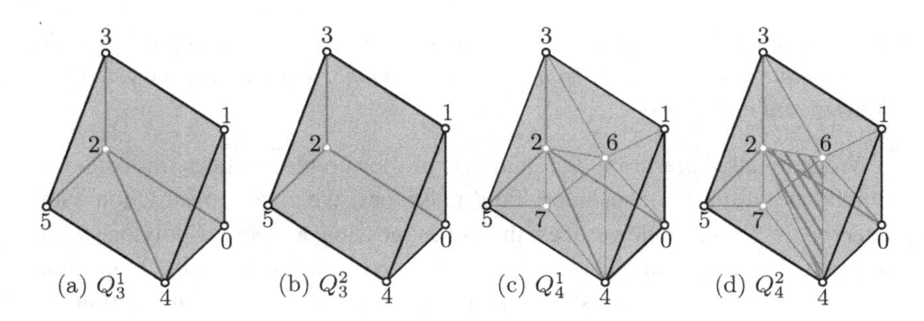

Figure 5.1.1 (a)–(b) The pair of 3-polytopes Q_3^1 and Q_3^2, and (c)–(d) Schlegel diagrams of the pair of 4-polytopes Q_4^1 and Q_4^2, projected on the facet isomorphic to Q_3^1. The missing 2-face of the bipyramid face 02467 in Q_4^2 is highlighted; the corresponding 2-face in Q_4^1 is the intersection of the simplex facets 0246 and 2467 of Q_4^1.

The facets of the polytope Q_{d+1}^1 are as follows. From Theorem 2.10.1(i) it follows that any facet of P not containing the edge $[2d-1, 2d]$ will remain a facet of the polytope. These are our facets of Types A–B.

Type A A simplex with vertex set: $\{0\} \cup X'$;

Type B $d+1-1$ facets whose vertex sets take the form: $\{0\} \cup \{2i+1 \mid i = 0, \ldots, k-1\} \cup X' \backslash \{2k\}$ for $k = 1, \ldots, d+1-1$;

Theorem 2.10.1(ii) gives that any facet F of P containing the triangle $\{2d-3, 2d-1, 2d\}$ (and the edge $[2d-1, 2d]$) is contained in the corresponding facet $F' = \operatorname{conv}(F \cup \{2d+1\})$ in the new polytope. These new facets are of Type C.

Type C $d+1-2$ facets whose vertex sets take the form: $\{2(d+1)-1\} \cup \{2i+1 \mid i = k-1, \ldots, d+1-2\} \cup X' \backslash \{2k\}$ for $k = 1, \ldots, d+1-2$;

Finally, consider the remaining facet F of P, namely the simplex with vertex set $\{2d-1\} \cup X'$, which contains the edge $[2d-1, 2d]$ but not the triangle $\{2d-3, 2d-1, 2d\}$. Theorem 2.10.1(iii) ensures that the convex hull of $v = 2d+1$ and each of the two $(d-1)$-faces of F not containing the edge $[2d-1, 2d]$ will form a facet of Q_{d+1}^1. There are exactly two such facets; the Types D–E.

Type D A simplex with vertex set: $\{2(d+1)-3, 2(d+1)-1\} \cup X' \backslash \{2d\}$; and

Type E A simplex with vertex set: $\{2(d+1)-1\} \cup X'$;

It remains to show that the nonsimple vertices of Q_{d+1}^1 are exactly the elements of X'. By induction, the nonsimple vertices of Q_d^1 are contained in the set

X and so the nonsimple vertices in P are contained in the set $X' = X \cup \{2d\}$. If we take $\operatorname{conv}(P \cup \{\boldsymbol{v}\})$, then the edge $[2d-1, 2d]$ disappears and the edges containing $\boldsymbol{v}$ are created, with one of them being $[\boldsymbol{v}, 2d]$. Thus, it remains to check that $\boldsymbol{v}$ is simple; indeed, $\boldsymbol{v}$ is adjacent to exactly the vertices in $X' \cup \{2d-1\}$, as Theorem 2.10.1(iii) shows. This completes the proof of the claim. □

Claim 2 For every dimension $d \geqslant 4$ there is a polytope Q_d^2 with the same $(d-3)$-skeleton as Q_d^1, whose $2d$ vertices are labelled $0, \ldots, 2d-1$ such that (1) the $d-1$ nonsimple vertices of Q_d^2 have positive even labels, and (2) the $2d-1$ facets are as follows.

Type A′ A bipyramid over a simplex, with vertex set: $\{0, 2d-1\} \cup X$;
Type B′ $d-1$ facets whose vertex sets take the form: $\{0\} \cup \{2i+1 \mid i = 0, \ldots, k-1\} \cup X \backslash \{2k\}$ for $k = 1, \ldots, d-1$;
Type C′ $d-2$ facets whose vertex sets take the form: $\{2d-1\} \cup \{2i+1 \mid i = k-1, \ldots, d-2\} \cup X \backslash \{2k\}$ for $k = 1, \ldots, d-2$; and
Type D′ A simplex with vertex set: $\{2d-3, 2d-1\} \cup X \backslash \{2(d-1)\}$.

In short, the polytope Q_d^2 is created by gluing the simplex facets of Type A and Type E of Q_d^1 along the $(d-2)$-face with vertex set X to create a bipyramid of Q_d^2, the facet of Type A′. The ridge with vertex set X of Q_d^1 then becomes a missing ridge in Q_d^2 (see Fig. 5.1.1).

Proof of claim We construct the d-polytope Q_d^2 by taking the convex hull of Q_d^1 and a new vertex $\boldsymbol{v}^*$.

First, consider the edge $e = [2d-3, 2d-1]$ of Q_d^1 and the unique facet F containing the vertex $2d-1$ but not $2d-3$; recall that the vertex $2d-1$ is simple in Q_d^1. The facet F is a simplex with vertices $\{2d-1\} \cup X$. Place the vertex $\boldsymbol{v}^*$ beyond the facet F along the the ray emanating from the vertex $2d-3$ and containing the edge $[2d-3, 2d-1]$ so that $\boldsymbol{v}^*$ lies in the first hyperplane H encountered that supports some facet $\overline{F}$ of Q_d^1. This ensures that any facet of Q_d^1 different from F, $\overline{F}$, or the $d-1$ facets $F_1^e, \ldots, F_{d-1}^e$ containing the edge e is nonvisible from the vertex $\boldsymbol{v}^*$. To show that the hyperplane H above exists, note that in the construction of Q_d^1 we can place the vertex $\boldsymbol{v} = 2d-1$ arbitrarily close to the vertex $2d-2$, thereby ensuring that the ray emanating from $2d-3$ and containing the edge e intersects a hyperplane that supports a facet containing the vertex $2d-2$ but not the edge e. Such a facet exists; for example, the facet with vertex set $\{0\} \cup X$.The polytope Q_d^2 is the convex hull of Q_d^1 and $\boldsymbol{v}^*$; thus, the vertex set of Q_d^2 is obtained from the vertex set of Q_d^1 by deleting $\boldsymbol{v}$ and adding $\boldsymbol{v}^*$, which we also label as $2d-1$.

Since $v^* \in H$, there is at least one $(d-2)$-face R of $\overline{F}$ that is visible from v^* with respect to $\overline{F}$ in H. This implies that the other facet containing R is visible from v^* with respect to Q_d^1 in $\mathbb{R}^d$. Since there is exactly one facet of Q_d^1 visible from v^* (with respect to Q_d^1), namely F, we must have $R = F \cap \overline{F}$ in Q_d^1 and R is unique.

In addition to not containing the edge $[2d-3, 2d-1]$, the facet $\overline{F}$ also does not contain the vertex $2d-1$. This implies that $\mathcal{V}(R) = X$. In particular, the facet $\overline{F}$ must be the facet of Q_d^1 with vertex set $\{0\} \cup X$.

From Theorem 2.10.1(ii) it follows that the facet $\overline{F}$ of Q_d^1 is replaced by the facet with vertex set $\{0, v^*\} \cup X$, which is a bipyramid over the simplex R. Combinatorially, with the exception of the ridge R, this bipyramid has the same face lattice as the boundaries of the union of the two simplex facets F and $\overline{F}$ (Types A and E) of Q_d^1.

Now consider any face J of Q_d^1 not contained in $\overline{F}$. We have three possibilities:

(i) the face is contained in a facet nonvisible from the vertex v^*,
(ii) the face contains the edge $e = [2d-3, 2d-1]$, and
(iii) the face does not contain the edge and it is not contained in a facet nonvisible from the vertex v^*.

In the first case, by Theorem 2.10.1(i) this face is also a face of Q_d^2. In the second case, the vertex v^* is in the affine hull of the face and, by Theorem 2.10.1(ii), the corresponding face J' in Q_d^2 has the same dimension as J and the form $\operatorname{conv}(J \cup \{v^*\})$. In the third case, the face J must be contained in intersections involving the facet F and some facets in $\{F_1^e, \ldots, F_{d-1}^e\}$. In this case, Theorem 2.10.1(i) assures us that J is not a face of Q_d^2. In summary, the two simplex facets F and $\overline{F}$ in Q_d^1 are replaced by a bipyramid in Q_d^2 with the same $(d-3)$-skeleton as $F \cup \overline{F}$, and every other facet of Q_d^1 falls into the first or second cases: a facet of Q_d^1 falling in the first case remains a facet of Q_d^2 and a facet J of Q_d^1 falling into the second case is replaced by the facet $J' = \operatorname{conv}((J \cup \{v^*\}) \setminus \{v\})$ in Q_d^2.

Note that X remains the set of nonsimple vertices in Q_d^2 as well. This completes the proof of the claim. □

The proof of the proposition is now complete. Refer to Fig. 5.1.1 (c)–(d) for Schlegel diagrams of the polytopes Q_4^1 and Q_4^2, where the projection facet is isomorphic to Q_3^1. □

A `polymake` script (Gawrilow and Joswig, 2000), which implements the ideas presented in Proposition 5.1.8 is available online (Pineda-Villavicencio and Ugon, 2018). Refer to the `polymake` script for information on how to run the program.

Simplicial and Cubical Polytopes

The cyclic d-polytope on $n \geqslant d+1$ vertices is a neighbourly d-polytope, and so it shares the $(\lfloor d/2 \rfloor - 1)$-skeleton of the $(n-1)$-simplex. For each even $d \geqslant 4$, Shemer (1982) constructed many noncyclic neighbourly d-polytopes, which are simplicial (Proposition 2.9.11). Thus, there are nonisomorphic simplicial d-polytopes with the same $(\lfloor d/2 \rfloor - 1)$-skeleton. As a result, the following is best possible.

Theorem 5.1.9 (Perles, 1973; Dancis, 1984) *Every simplicial d-polytope can be reconstructed from its $\lfloor d/2 \rfloor$-skeleton.*

For each $k \geqslant \lfloor d/2 \rfloor$, Dancis (1984) gave a necessary and sufficient topological condition for a union of k-simplices in a simplicial d-polytope to bound a $(k+1)$-simplex, enabling the reconstruction of the face lattice of the polytope. An algorithm exponential in the number of vertices can be derived from Dancis' proof.

The notion of neighbourly polytopes can be naturally extended from simplicial polytopes to other equifacetted polytopes such as cubical polytopes. A cubical d-polytope is *neighbourly* if it has the same $(\lfloor d/2 \rfloor - 1)$-skeleton as some cube; this definition was introduced by Babson et al. (1997). Sanyal and Ziegler (2010, cor. 3.8) constructed many nonisomorphic neighbourly cubical d-polytopes, which means that there are nonisomorphic cubical d-polytopes with the same $(\lfloor d/2 \rfloor - 1)$-skeleton. By building on the work of Dancis (1984), Rowlands (2022) proved an analogous tight result.

Theorem 5.1.10 (Rowlands, 2022) *Every cubical d-polytope can be reconstructed from its $\lfloor d/2 \rfloor$-skeleton.*

As in the case of simplicial polytopes, the reconstruction can be done in time exponential in the number of vertices.

5.2 Reconstruction from Low Skeletons

We consider the reconstruction of polytopes from their dimension and either graphs or 2-skeletons, starting with the latter.

Theorem 5.2.1 *Every d-polytope with at most $d-2$ nonsimple vertices in each facet is reconstructible from its dimension and 2-skeleton. Furthermore, all the facets can be found in linear time in the number of vertices of the polytope.*

For the proof we need the notion of *frames* and a useful observation of Kaibel (2003, prop. 1) that we spell out in Proposition 5.2.2. Let P be a polytope. Define a *k-frame* as a subgraph of $G(P)$ isomorphic to the star $K_{1,k}$ where the vertex of degree k is called the *root* of the frame. If the root of a frame is a simple vertex in P, we say that the frame is *simple*. For a simple vertex $\boldsymbol{v}$ of P contained in a k-face F, we say that the k-frame t_v *defines* F if t_v is the unique k-frame with root $\boldsymbol{v}$ that is contained in F (Theorem 2.8.6).

We next rephrase this observation to suit our needs.

Proposition 5.2.2 *Let $\boldsymbol{uv}$ be an edge of a d-polytope P with $\boldsymbol{u}$ and $\boldsymbol{v}$ being simple vertices in P. Let F be a facet of P containing both $\boldsymbol{u}$ and $\boldsymbol{v}$, with t_u being the frame rooted at $\boldsymbol{u}$ that defines F. Furthermore, suppose that $\boldsymbol{u}'$ is the unique neighbour of $\boldsymbol{u}$ not in t_u and that $\boldsymbol{v}'$ is the neighbour of $\boldsymbol{v}$, other than $\boldsymbol{u}$, that is contained in the 2-face W of P defined by the 2-frame $(\boldsymbol{u},\boldsymbol{u}',\boldsymbol{v})$ with root $\boldsymbol{u}$. Then $\boldsymbol{v}'$ is not in F.*

Proof Let t_v be the $(d-1)$-frame of $\boldsymbol{v}$ defining F. Suppose, by contradiction, that the vertex $\boldsymbol{v}'$ is in F. Since $\boldsymbol{v}$ is in F and every vertex in t_v is in F, the 2-face W, which is also defined by the 2-frame $(\boldsymbol{v},\boldsymbol{u},\boldsymbol{v}')$ with root $\boldsymbol{v}$, would be contained in F, a contradiction because $\boldsymbol{u}' \notin F$. □

Proof of Theorem 5.2.1 We show that a modification of Friedman's algorithm (Friedman, 2009, sec. 7) gives a proof of the theorem. Let P be a d-polytope and assume that the 2-skeleton of P is given. We repeat the following routine until all simple $(d-1)$-frames in $G(P)$ are visited.

(i) Pick a simple vertex $\boldsymbol{u}$ in P (it exists) and select any simple $(d-1)$-frame t_u rooted at $\boldsymbol{u}$. Let $\boldsymbol{u}'$ be the unique vertex adjacent to $\boldsymbol{u}$ that is not in that frame. The frame t_u is contained in a unique facet in P; denote it by F_u.

(ii) Consider any other simple vertex $\boldsymbol{v}$ in the frame t_u with $\boldsymbol{v} \neq \boldsymbol{u}$ (it exists). Then there exists another simple $(d-1)$-frame t_v centred at $\boldsymbol{v}$ in the facet F_u.

(iii) Consider the neighbour $\boldsymbol{v}'$ of $\boldsymbol{v}$, different from $\boldsymbol{u}$, that is present in the 2-face W that contains the 2-frame $(\boldsymbol{u},\boldsymbol{u}',\boldsymbol{v})$ with root $\boldsymbol{u}$; we know all the vertices of W. Then, applying Proposition 5.2.2 to the edge $\boldsymbol{uv}$ gives that the frame t_v is formed by all the vertices adjacent to $\boldsymbol{v}$ other than $\boldsymbol{v}'$.

(iv) Continue this process, always moving along edges incident with simple vertices, and stop when no new simple $(d-1)$-frame can be visited.

When (iv) stops, we obtain the graph of F_u. Indeed, since $G(F_u)$ is $(d-1)$-connected by Balinski's theorem (4.1.1) and there are at most $d-2$ nonsimple vertices in F_u, we can go from any simple vertex to any other simple vertex in F_u through a path formed only by simple vertices. Thus, after step (iv) finishes, we get the vertex set of the facet F_u.

Repeating this process for all simple $(d-1)$-frames will then reveal all the facets of P.

Finally, we show that this process, with some preprocessing, runs in linear time in the number f_0 of vertices of P. The number of simple $(d-1)$-frames is d times the number of simple vertices of P, is thus linear in f_0, and each such frame is visited once. It remains to check that the move from one simple $(d-1)$-frame t_u to the next t_v can be done in constant time (depending on d). Checking if the degree of a neighbour of $\boldsymbol{u}$ (the vertex $\boldsymbol{v}$) is d takes only constant time.

If the neighbour $\boldsymbol{v}$ of $\boldsymbol{u}$ is simple, we need to find $\boldsymbol{v}'$ in step (iii) in constant time. To do so, we preprocess the 2-skeleton to construct the graph G_T (following Friedman [2009]) whose vertices are the 2-frames, and two of them are adjacent if and only if the root of one is a vertex of the other and both belong to the same 2-face. Constructing G_T takes time linear in f_0 since the number of 2-faces in the 2-skeleton of P is linear in f_0, as the next remark shows.

Remark 5.2.3 For $k \in [1, d-1]$, the number of k-faces in a d-polytope with at most $\varphi(d)$ nonsimple vertices is linear in the number n of vertices of the polytope; here, φ is a function in d independent of n.

To see this, note that the number of k-faces in the polytope consists of the k-faces containing a simple vertex, of which there are at most $n\binom{d}{k}$, and the k-faces all whose vertices are nonsimple is at most $\binom{\varphi(d)}{k+1}$. This sum is linear in n.

Once G_T is constructed, finding the vertex $\boldsymbol{v}'$ requires moving along edges of G_T from the 2-frame $(\boldsymbol{u}, \boldsymbol{u}', \boldsymbol{v})$ with root $\boldsymbol{u}$ to the 2-frame $(\boldsymbol{v}, \boldsymbol{u}, \boldsymbol{v}')$ with root $\boldsymbol{v}$, which can be done in constant time. Thus, the complexity result follows. □

Reconstruction from Graphs

We prove that d-polytopes with at most one nonsimple vertex are reconstructible from their graphs.

Theorem 5.2.4 *The face lattice of a polytope with at most one nonsimple vertex can be reconstructed from its dimension and graph.*

By considering a longest directed path in an acyclic orientation of a graph, we get a simple but important remark.

Remark 5.2.5 Let P be a polytope, O an acyclic orientation of $G(P)$, and F a facet of P. Then

(i) the graph $G(F)$ has a sink in O, and
(ii) there is a directed path in $G(F)$ from any vertex in $G(F)$ to some sink in F.

Proof of Theorem 5.2.4 Let P be a d-polytope with at most one nonsimple vertex $\boldsymbol{v}$; if P is a simple polytope, let $\boldsymbol{v}$ be any vertex of it. We partition the facets of P into two sets: the set $\mathcal{F}_v$ of facets containing the vertex $\boldsymbol{v}$ and the set $\mathcal{F}_\varnothing$ of facets not containing $\boldsymbol{v}$.

We say that an acyclic orientation of $G(P)$ is *valid* if it has a unique sink in every facet of P; otherwise the orientation is *invalid*. While every good orientation is valid, we may have valid orientations that are bad; recall that an acyclic orientation of $G(P)$ is good if the graph of every nonempty face of P has a unique sink; otherwise the orientation is bad (Section 3.5).

Let $\mathcal{A}$ be the set of all acyclic orientations of $G(P)$ in which the vertex $\boldsymbol{v}$ is a source. We first show how to distinguish the valid and invalid orientations of $\mathcal{A}$. Let $O \in \mathcal{A}$ and let $h_k(O)$ denote the number of vertices of $G(P)$ with indegree k in O. Define

$$\varphi(O) := h_{d-1}(O) + dh_d(O). \tag{5.2.5.1}$$

Since the vertex $\boldsymbol{v}$ is not a sink in any facet of P, every sink in O is a simple vertex in P. A simple vertex $\boldsymbol{w}$ and every $d-1$ of its neighbours define a unique facet of P (Theorem 2.8.7) and, thus, if $\boldsymbol{w}$ has indegree k in O then it is a sink in $\binom{k}{d-1}$ facets of P. It follows that the function $\varphi(O)$ counts the number of pairs $(F, \boldsymbol{w})$ where F is a facet of P and $\boldsymbol{w}$ is a sink in $G(F)$.

By virtue of Corollary 3.5.4, there is a good orientation of $G(P)$ having $\boldsymbol{v}$ as a source in $G(P)$, which implies that there is a valid orientation in $\mathcal{A}$. Every valid orientation of $\mathcal{A}$ gives the minimum value of φ (from (5.2.5.1)), which is the number f_{d-1} of facets of P. Furthermore, if an orientation $O \in \mathcal{A}$ is invalid, there would be a facet F with at least two sinks, and so F would be present in at least two of the pairs counted by $\varphi(O)$, implying that $\varphi(O) > f_{d-1}$. Thus, an orientation O of $\mathcal{A}$ is valid if and only if $\varphi(O) = f_{d-1}$.

An 2-connected induced subgraph X of $G(P)$ is *feasible* if simple vertices in P have each degree $d-1$ in X and the vertex $\boldsymbol{v}$ (if present) has degree at least $d-1$ in X. The next claim gives a close relation between feasible subgraphs and graphs of facets in P.

Claim 1 Let X be a feasible subgraph of $G(P)$. If the graph $G(F)$ of some facet F of P is contained in X, then $X = G(F)$.

Proof of claim If $\mathcal{V}(F) = \mathcal{V}(X)$, then $G(F) = X$, as $G(F)$ is an induced subgraph of $G(P)$. Otherwise, $\mathcal{V}(F) \subset \mathcal{V}(X)$, in which case any path from a vertex in $\mathcal{V}(X)\backslash\mathcal{V}(F)$ to a vertex in $G(F)$ must pass through a nonsimple vertex, namely through the vertex $\boldsymbol{v}$, since simple vertices of the polytope have the same neighbours in both X and $G(F)$. Consequently, the vertex $\boldsymbol{v}$ would disconnect X, contradicting its 2-connectivity. □

Recall that a set of vertices is *initial* with respect to some orientation if no arc is directed from a vertex not in the set to a vertex in the set (Section 3.5).

Claim 2 A feasible subgraph X of $G(P)$ is the graph of a facet containing $\boldsymbol{v}$ if and only if $\mathcal{V}(X)$ is an initial set with respect to some valid orientation of $\mathcal{A}$.

Proof of claim Let F be a facet containing $\boldsymbol{v}$. It is clear that $G(F)$ is a feasible subgraph of $G(P)$. By Corollary 3.5.4(ii), $\mathcal{V}(F)$ is an initial set with respect to some good orientation that has $\boldsymbol{v}$ as the unique source; this is a valid orientation of $\mathcal{A}$.

We now prove the other direction. Let O be a valid orientation of $\mathcal{A}$ and suppose that X is a feasible subgraph of $G(P)$ containing $\boldsymbol{v}$ and that $\boldsymbol{x}$ is a sink in X with respect to O. Then $\boldsymbol{x}$ has degree d in P and together with its $d-1$ arcs in X defines a unique facet F of P (Theorem 2.8.7). The orientation is valid and so $\boldsymbol{x}$ is the unique sink in F. Since there is a directed path in $G(F)$ from any vertex in $G(F)$ to $\boldsymbol{x}$ (Remark 5.2.5) and since $\mathcal{V}(X)$ is an initial set, we must have that $\mathcal{V}(F) \subseteq \mathcal{V}(X)$. Claim 1 now ensures that $X = G(F)$. □

Running through all the orientations in $\mathcal{A}$, we identify the minimum of φ in (5.2.5.1), which is f_{d-1}. Once this minimum has been detected, we recognise all the valid orientations in $\mathcal{A}$. Then, with the help of Claim 2, we recognise all the graphs of facets in $\mathcal{F}_{\boldsymbol{v}}$; let $f_{d-1}^{\boldsymbol{v}} := \#\mathcal{F}_{\boldsymbol{v}}$.

If P is a simple polytope the vertex $\boldsymbol{v}$ is not special, and so any vertex of P can be taken as $\boldsymbol{v}$. Thus, Claim 2 can be used to recognise all the graphs of facets in this case. As a consequence, we henceforth assume that $\boldsymbol{v}$ has degree more than d in P.

By considering the reverse orientation of each orientation in $\mathcal{A}$, we get the set $\mathcal{B}$ of all acyclic orientations of $G(P)$ in which the vertex $\boldsymbol{v}$ is a sink. We show how to distinguish the valid and invalid orientations of $\mathcal{B}$. Let $O \in \mathcal{B}$ and let $j_k(O)$ denote the number of simple vertices of $G(P)$ with indegree k in O. Define

$$\varrho(O) := j_{d-1}(O) + d j_d(O) + f_{d-1}^{\boldsymbol{v}}. \tag{5.2.5.2}$$

The vertex $\boldsymbol{v}$ is a sink in every facet of P containing it and so there are $f_{d-1}^{\boldsymbol{v}}$ pairs $(F, \boldsymbol{v})$, where F is a facet and $\boldsymbol{v}$ is a sink in $G(F)$. And by virtue of Theorem 2.8.7, a simple vertex $\boldsymbol{w}$ in P and every $d-1$ of its neighbours define a unique facet of P. Thus, if $\boldsymbol{w}$ has indegree k in O, then it is a sink in $\binom{k}{d-1}$ facets of P. It follows that the function $\varrho(O)$ counts the number of pairs $(F, \boldsymbol{z})$ where F is a facet of P and $\boldsymbol{z}$ is a sink in $G(F)$.

By virtue of Proposition 3.5.3 and Corollary 3.5.4, there is a good orientation O' of $G(P)$ having $\boldsymbol{v}$ as a source in $G(P)$ and a unique source in the subgraph of every nonempty face of P. This implies that the reverse orientation of O' is a valid orientation in $\mathcal{B}$. Every valid orientation of $\mathcal{B}$ gives the minimum value of ϱ (see (5.2.5.2)), which is again the number f_{d-1} of facets of P. Furthermore, if an orientation $O \in \mathcal{B}$ is invalid, there would be a facet F with at least two sinks, implying that $\varrho(O) > f_{d-1}$. Thus, an orientation O of $\mathcal{B}$ is valid if and only if $\varrho(O) = f_{d-1}$.

Claim 3 A feasible subgraph X of $G(P)$ not containing $\boldsymbol{v}$ is the graph of a facet not containing $\boldsymbol{v}$ if and only if $\mathcal{V}(X)$ is an initial set with respect to some valid orientation of $\mathcal{B}$.

Proof of claim Let F be a facet not containing $\boldsymbol{v}$. It is clear that $G(F)$ is a feasible subgraph of $G(P)$. By Corollary 3.5.4(i), $\mathcal{V}(F)$ is a final set with respect to some good orientation O that has $\boldsymbol{v}$ as the unique source in $G(P)$ and a unique source in the subgraph of every nonempty face of P; the reverse of this orientation gives a valid orientation of $\mathcal{B}$ where $\mathcal{V}(F)$ is an initial set. Recall that a set of vertices is final with respect to some orientation if no arc is directed from a vertex in the set to a vertex not in the set (Section 3.5).

We prove the other direction. Let O be a valid orientation of $\mathcal{B}$, and suppose that X is a feasible subgraph of $G(P)$ not containing $\boldsymbol{v}$. Then a sink $\boldsymbol{x}$ in X with respect to O has degree d in P. From Theorem 2.8.7 it follows that $\boldsymbol{x}$ together with its $d-1$ arcs in X defines a unique facet F of P. The orientation is valid, and so $\boldsymbol{x}$ is the unique sink in F. Since there is a directed path in $G(F)$ from any vertex in $G(F)$ to $\boldsymbol{x}$ (Remark 5.2.5) and since $\mathcal{V}(X)$ is an initial set, we must have that $\mathcal{V}(F) \subseteq \mathcal{V}(X)$. Claim 1 also ensures that $X = G(F)$ in this case. □

Running through all the valid orientations in $\mathcal{B}$ with the help of Claim 3, we recognise all the graphs of facets in $\mathcal{F}_{\varnothing}$.

Using Claims 2 and 3, we can find all the graphs of all the facets of P, and therefore all the facet-vertex incidences of P. This concludes the proof of the theorem. □

Remark 5.2.6 If P is a polytope with at most one nonsimple vertex, then listing all the $2^{\mathcal{E}}$ orientations of $G(P)$ requires exponential time in the number f_0 of vertices of P; here $2^{\mathcal{E}}$ denotes the power set of $\mathcal{E}$. Thus, the proof of Theorem 5.2.4 gives an algorithm exponential in f_0 for reconstructing the face lattice of P from the dimension and $G(P)$.

Later on, building on a result by Joswig et al. (2002, thm. 1), Friedman (2009) gave a polynomial-time reconstruction algorithm for simple polytopes. A polynomial-time reconstruction algorithm means that one can decide whether a subset of vertices in a polytope is the vertex set of a face in time polynomial in the number of vertices of the polytope. This is the best we can do because a polytope may contain exponentially many faces, and exhibiting them all explicitly may require exponential time.

By duality, Theorem 5.2.4 gives that a simplicial polytope can be reconstructed from its dimension and dual graph. In view of the similarities between simplicial and cubical polytopes depicted in Theorem 5.1.9 and Theorem 5.1.10, one may conjecture that a cubical polytope can also be reconstructed from its dimension and dual graph, and so did Joswig (2000, conj. 3.1). Pineda-Villavicencio and Schröter (2022) provided a counterexample.

Proposition 5.2.7 *Not every cubical polytope is reconstructible from its dimension and dual graph.*

Proof We construct a noncubical 4-polytope P with the same dual graph as the 4-cube $Q(4)$. Consider the 4-polytope R whose eight vertices are grouped into the four pairs:

$$\left((-2,-2,-2,-2)^t,(4,4,4,-2)^t\right);\ \left((-2,-2,-2,4)^t,(4,-2,-2,-2)^t\right);$$
$$\left((-2,-2,4,-2)^t,(4,-5,-2,4)^t\right);\ \left((-2,4,-2,-2)^t,(1,-11,-1,12)^t\right).$$

Each vertex is adjacent to every other vertex except its paired vertex. It follows that the graph of R is isomorphic to the graph of the 4-crosspolytope, which is the dual polytope of $Q(4)$. The polytope R has the following irredundant H-description, arranged in two columns of six facets each:

$$\begin{aligned} -2x_2 - 4x_3 - x_4 &\leqslant 14, & -x_1 - 2x_2 - 15x_3 &\leqslant 36, \\ -3x_1 + 3x_2 + 4x_3 + 4x_4 &\leqslant 8, & -14x_2 - 9x_4 &\leqslant 46, \\ 2x_1 - 2x_2 + 2x_3 - x_4 &\leqslant 10, & -3x_1 - x_2 &\leqslant 8, \\ -x_1 + x_2 + x_3 + x_4 &\leqslant 2, & -x_1 &\leqslant 2, \\ x_1 + 2x_2 - x_3 + 2x_4 &\leqslant 4, & -x_3 &\leqslant 2, \\ 2x_1 + 2x_2 - 2x_3 + x_4 &\leqslant 6, & -x_4 &\leqslant 2. \end{aligned}$$

It can be readily verified that the vertex $\boldsymbol{v} := (-2, -2, -2, -2)^t$ is contained in exactly seven facets: the first facet of the first column and all the facets in the second column.

Consider the dual polytope P of R. The dual graph of P is isomorphic to the dual graph of $Q(4)$. However, the facet of P conjugate to the vertex $\boldsymbol{v}$ has seven vertices, and so P is noncubical. □

We end this section with two other reconstruction results. Polytopes with small excess degrees are a natural generalisation of simple polytopes; recall that the excess degree of a d-polytope P is $2f_1(P) - df_0(P)$. Pineda-Villavicencio et al. (2020b) proved that polytopes with excess degree at most $d - 1$ are reconstructible from their dimension and graphs, and this is best possible as Example 5.1.6 attests.

Let $\boldsymbol{v}_1, \ldots, \boldsymbol{v}_n$ be points in $\mathbb{R}^d$ that span $\mathbb{R}^d$, with no point being a multiple of any other. A *zonotope* in $\mathbb{R}^d$ is a polytope of the form

$$\left\{ \boldsymbol{x} \in \mathbb{R}^d \,\middle|\, \boldsymbol{x} = \sum_{i=1}^{n} \alpha_i \boldsymbol{v}_i,\ |\alpha_1| \leqslant 1, \ldots, |\alpha_n| \leqslant 1 \right\}. \tag{5.2.8}$$

This shows that the point $\boldsymbol{\alpha} = (\alpha_1, \ldots, \alpha_n)^t$ is in the n-cube $Q(n)$ and a zonotope is the image of $Q(n)$ under the affine map $\pi(\boldsymbol{\alpha}) = M\boldsymbol{\alpha} + \boldsymbol{b}$, where $\boldsymbol{\alpha} \in Q(n)$, $\boldsymbol{b}$ is a fixed vector, and M is a $d \times n$ matrix with the points $\boldsymbol{v}_1, \ldots, \boldsymbol{v}_n$ as columns; see also Example 2.1.6 for a representation of the n-cube as a V-polyhedron.

Björner et al. (1990) showed that zonotopes can be reconstructed from their dimension and graphs.

5.3 Variations on the Reconstruction Problem

Proposition 5.2.7 motivates a first variation of the reconstruction problem (5.1.4). While cubical polytopes cannot be reconstructed just from their dimension and dual graphs, they may be if we know that the dual graphs are dual graphs of cubical polytopes.

Problem 5.3.1 (Class reconstruction) *Given a class of polytopes, integers $k \geqslant 1$ and $d \geqslant 3$, and the k-skeleton $\mathcal{K}$ of a d-polytope in the class, determine the face lattice of the polytope from $\mathcal{K}$.*

This problem requires that the class contains exactly one d-polytope with the given k-skeleton.

We say that a polytope within a class of polytopes is *class reconstructible* if some class reconstruction problem can be solved for the polytope. There are partial results in that direction. For example, Joswig (2000, thm. 3.7) proved that capped cubical polytopes are reconstructible from their dimension and dual graphs within the class of capped cubical polytopes. A *capped cubical polytope* is defined as either a cube itself or a polytope formed from a cube through successive cappings. Here, a *capping* refers to the connected sum of a cube and an existing capped cubical polytope along a common facet. Babson et al. (2001) in turn showed that cubical zonotopes are reconstructible from their dimension and dual graphs within the class of cubical zonotopes.

A Conjecture of Perles

Back in 1970, Perles (Haase and Ziegler, 2002, sec. 1) conjectured that, for the graph of a simple d-polytope, every induced connected $(d-1)$-regular subgraph that is also nonseparating is the graph of a facet. While the conjecture holds for several classes of d-polytopes, including for duals of cyclic polytopes, duals of stacked polytopes, and 3-polytopes, it is not true in general (Haase and Ziegler, 2002, sec. 3–6). Haase and Ziegler (2002, sec. 7), however, believed that for simple 4-polytopes, adding planarity and 3-connectivity to the conjecture would make it true. This variation was disproved by Doolittle (2018). Consider the simplicial 4-polytope Q with vertices:

$$(27,-95,120,0)^t,\ (-50,-45,101,-94)^t,\ (-9,-67,126,-35)^t,$$
$$(195,-145,11,125)^t,\ (-40,-10,8,-65)^t,\ (232,-102,-21,198)^t,$$
$$(-63,-25,94,-139)^t,\ (-80,45,-49,-65)^t,\ (-72,4,24,-90)^t,$$
$$(-30,167,-154,92)^t,\ (-43,190,-199,100)^t,\ (-67,80,-61,-26)^t.$$

The dual polytope P of Q is a minimum counterexample to this new variation, with respect first to the number of facets and then the number of vertices. It has f-vector $(38,76,50,12)$. The polytope P contains a connected and planar induced subgraph with 26 vertices that is not the graph of a facet of P.

5.4 Problems

5.4.1 Let P be a d-polytope with fewer than $2d$ vertices, of which at most $d-1$ are nonsimple. Prove P that is a pyramid.

5.4.2 (Free joins of crosspolytopes) Let $k \geqslant 1, d_1 \geqslant k+1, \ldots, d_s \geqslant k+1$, and $d = d_1 + \cdots + d_s$. Prove that the $(d+s-1)$-polytope $I(d_1) * I(d_2) * \cdots I(d_s)$ is combinatorially k-isomorphic to $I(d)$.

5.4.3 (Pfeifle et al., 2012) Prove that a Cartesian product of graphs is the graph of a simple polytope if and only if each factor is also a graph of a simple polytope.

5.4.4* Let P be a d-polytope with at most $d-2$ nonsimple vertices. Can P be reconstructed from its dimension and graph? If a reconstruction is possible, provide an efficient algorithm.

This is true for 4-polytopes, as Doolittle et al. (2018) showed that every polytope with at most two nonsimple vertices can be reconstructed from its dimension and graph.

5.4.5* (Reconstruction of simplicial spheres) A *simplicial sphere* is a simplicial complex $\mathcal{C}$ where set $\mathcal{C}$ is homeomorphic to a unit sphere.

Can every simplicial sphere be determined by its dimension and dual graph? If a reconstruction is possible, provide an efficient algorithm.

This is true for simplicial 2-spheres, a consequence of Steinitz's theorem (3.8.1), but it is still open for simplicial 3-spheres. From the dual version of Theorem 5.2.4 it follows that this is also true for all *polytopal simplicial spheres* (boundary complexes of simplicial polytopes), since polytopes share the dual graph with their boundary complexes. However, there are nonpolytopal simplicial spheres; see, for instance, Barnette (1987b).

5.5 Postscript

Classical topics of reconstruction are covered in Grünbaum (2003, ch. 12), while recent developments are surveyed by Bayer (2018). Section 5.1 is based on the presentation in Grünbaum (2003, sec. 12.1, 12.2). The reconstruction of simplicial polytopes from their $\lfloor d/2 \rfloor$-skeletons is attributed to Perles (Goodman et al., 2017, thm. 19.5.27). Section 5.2 is based on Doolittle et al. (2018); the proof of Theorem 5.2.4 is inspired by the proof of Doolittle et al. (2018, thm. 4.8), which is in turn inspired by Kalai (1988b).

6
Decomposition

The Minkowski decomposability of polytopes is developed via geometric graphs and decomposing functions; if you recall from Chapter 3, a geometric graph is a realisation in $\mathbb{R}^d$ of a graph G where distinct vertices of G correspond to distinct points in $\mathbb{R}^d$, edges in G correspond to line segments, and no three vertices are collinear. One advantage of this approach is its versatility. The decomposability of polytopes reduces to the decomposability of geometric graphs, which are not necessarily polytopal. And the decomposability of geometric graphs often revolves around the existence of suitable subgraphs or useful properties in the graphs. Section 6.3 classifies d-polytopes with at most $2d + 1$ vertices into decomposable and indecomposable.

The chapter concludes with a section on polytopes that admit both decomposable and indecomposable realisations, a phenomenon that occurs exclusively in d-polytopes with at least $2d + 2$ vertices. This emphatically underscores that decomposability is not a combinatorial property. This section is among the few instances in the book where a noncombinatorial property is addressed.

6.1 Minkowski Sums

We recall from Definition 1.6.6 that the Minkowski sum of two sets X and Y in $\mathbb{R}^d$ is the set

$$X + Y := \{\boldsymbol{x} + \boldsymbol{y} \mid \boldsymbol{x} \in X, \boldsymbol{y} \in Y\}.$$

An expression $Z = X + Y$ is said to be a *decomposition* of Z into sets X and Y. In this chapter, we will focus on the Minkowski sum of two polytopes. We provide an alternative definition of a Minkowski sum based on supporting functions. Recall that $h(P,\boldsymbol{x})$ denotes the supporting function of the polytope P (Problem 1.12.16), namely the function given by

$$h(P,\boldsymbol{x}) = \sup\{\boldsymbol{p}\cdot\boldsymbol{x} \mid \boldsymbol{p}\in P\}, \text{ for every } \boldsymbol{x}\in\mathbb{R}^d.$$

An alternative characterisation of the Minkowski sum of two polytopes follows from the definition of a supporting function and Problem 1.12.16.

Lemma 6.1.1 *Let Q and R be polytopes in $\mathbb{R}^d$. Then the polytope $Q+R$ is the polytope in $\mathbb{R}^d$ whose supporting function is given by*

$$h(Q+R,\boldsymbol{x}) = h(Q,\boldsymbol{x}) + h(R,\boldsymbol{x}),$$

for all $\boldsymbol{x}\in\mathbb{R}^d$.

Recall that, for a polytope P in $\mathbb{R}^d$ and a vector $\boldsymbol{a}\in\mathbb{R}^d$, the set of maximisers of the dot product $\boldsymbol{x}\cdot\boldsymbol{a}$ over P is a face of P, denoted by $F(P,\boldsymbol{a})$; see (1.9.7). And each proper face of P is exposed, so it has the form $F(P,\boldsymbol{a})$ for some $\boldsymbol{a}\in\mathbb{R}^d$ (Theorem 2.3.1). The following assertion will play an important role in our deliberations.

Lemma 6.1.2 *Let Q and R be polytopes in $\mathbb{R}^d$ and let $\boldsymbol{a}$ be any nonzero vector in $\mathbb{R}^d$. Then the following relation holds:*

$$F(Q+R,\boldsymbol{a}) = F(Q,\boldsymbol{a}) + F(R,\boldsymbol{a}).$$

Proof Lemma 6.1.1 together with the definition of the sets $F(Q,\boldsymbol{a})$ and $F(R,\boldsymbol{a})$ gives the lemma; see also Problem 1.12.16. □

We characterise the faces of the polytope $Q+R$ thereafter.

Lemma 6.1.3 *Let Q and R be polytopes in $\mathbb{R}^d$ and let $P := Q+R$. A nonempty subset F of P is a face of P if and only if*

(i) *$F = F_Q + F_R$ for a face F_Q of Q and a face F_R of R, and*
(ii) *there exists a vector $\boldsymbol{a}\in\mathbb{R}^d$ such that $F = F(P,\boldsymbol{a})$, $F_Q = F(Q,\boldsymbol{a})$, and $F_R = F(R,\boldsymbol{a})$.*

Furthermore, the decomposition $F = F_Q + F_R$ is unique.

Proof If F is a nonempty face of P, then there exists $\boldsymbol{a}\in\mathbb{R}^d$ such that $F = F(P,\boldsymbol{a})$. From Lemma 6.1.2 it then follows that

$$F(Q+R,\boldsymbol{a}) = F(Q,\boldsymbol{a}) + F(R,\boldsymbol{a}).$$

Letting $F_Q = F(Q,\boldsymbol{a})$ and $F_R = F(R,\boldsymbol{a})$, we obtain the necessity.

Let F be a nonempty subset of P. The existence of the faces $F_Q = F(Q, \boldsymbol{a})$ and $F_R = F(R, \boldsymbol{a})$ for a vector $\boldsymbol{a} \in \mathbb{R}^d$ guarantees that $F = F(P, \boldsymbol{a})$ (Lemma 6.1.2). Hence F is a face of P.

We next prove the uniqueness of the decomposition. Let $F = F(P, \boldsymbol{a})$ for some vector $\boldsymbol{a} \in \mathbb{R}^d$ and let $F = F_Q + F_R$ be any decomposition of F. We first show that $F_Q \subseteq F(Q, \boldsymbol{a})$ and $F_R \subseteq F(R, \boldsymbol{a})$. Take a point $\boldsymbol{v}_Q \in F_Q$ and a point $\boldsymbol{v}_R \in F_R$ and write $\boldsymbol{v} = \boldsymbol{v}_Q + \boldsymbol{v}_R$. Then $\boldsymbol{v} \in F$. The value of $\boldsymbol{v} \cdot \boldsymbol{a}$ over P, which is equal to $h(P, \boldsymbol{a})$, is the sum of $h(Q, \boldsymbol{a})$ and $h(R, \boldsymbol{a})$, according to Problem 1.12.16(vi). Thus $\boldsymbol{v}_Q \in F(Q, \boldsymbol{a})$ and $\boldsymbol{v}_R \in F(R, \boldsymbol{a})$. Since the points $\boldsymbol{v}_Q$ and $\boldsymbol{v}_R$ were taken arbitrarily, we have that

$$F_Q \subseteq F(Q, \boldsymbol{a}) \text{ and } F_R \subseteq F(R, \boldsymbol{a}).$$

We next show that $F(Q, \boldsymbol{a}) = F_Q$ and $F(R, \boldsymbol{a}) = F_R$. Suppose otherwise and let $\boldsymbol{w} \in F(R, \boldsymbol{a}) \setminus F_R$. Then there exists a linear functional $\boldsymbol{b} \cdot \boldsymbol{x}$ such that $\boldsymbol{b} \cdot \boldsymbol{w} > \boldsymbol{b} \cdot \boldsymbol{v}_R$ for all $\boldsymbol{v}_R \in F_R$. Let $\boldsymbol{v}_Q^*$ be any point in $\mathbb{R}^d$ attaining the maximum of $\boldsymbol{b} \cdot \boldsymbol{x}$ over F_Q. Since $\boldsymbol{v}_Q^* \in F_Q$ and since $F_Q \subseteq F(Q, \boldsymbol{a})$, it follows that $\boldsymbol{v}_Q^* \in F(Q, \boldsymbol{a})$. By Lemma 6.1.2, we get that $\boldsymbol{v}_Q^* + \boldsymbol{w} \in F$. This last statement yields the final contradiction as

$$\boldsymbol{b} \cdot \boldsymbol{v}_Q^* + \boldsymbol{b} \cdot \boldsymbol{w} > \boldsymbol{b} \cdot \boldsymbol{v} = \boldsymbol{b}(\boldsymbol{v}_Q + \boldsymbol{v}_R),$$

for every $\boldsymbol{v} \in F$. As a result, $F_Q = F(Q, \boldsymbol{a})$ and $F_R = F(R, \boldsymbol{a})$. □

Because of Lemma 6.1.3 we refer to a decomposition $F = F_Q + F_R$ of a face F of a polytope $Q + R$ as the decomposition of F. Several corollaries are direct consequences of Lemma 6.1.3.

Corollary 6.1.4 *Let Q and R be polytopes in $\mathbb{R}^d$. If $F = F_Q + F_R$ is the decomposition of a face F of $Q + R$, then* $\dim F \geqslant \dim F_Q$ *and* $\dim F \geqslant \dim F_R$.

Corollary 6.1.5 *Let Q and R be polytopes in $\mathbb{R}^d$ and let $P = Q + R$. Let $\boldsymbol{u}$ and $\boldsymbol{v}$ be adjacent vertices of P with decompositions $\boldsymbol{u} = \boldsymbol{u}_Q + \boldsymbol{u}_R$ and $\boldsymbol{v} = \boldsymbol{v}_Q + \boldsymbol{v}_R$. Then*

(i) $[\boldsymbol{u}, \boldsymbol{v}] = [\boldsymbol{u}_Q, \boldsymbol{v}_Q] + [\boldsymbol{u}_R, \boldsymbol{v}_R]$ *for faces* $[\boldsymbol{u}_Q, \boldsymbol{v}_Q]$ *of Q and* $[\boldsymbol{u}_R, \boldsymbol{v}_R]$ *of R;*
(ii) *All one-dimensional faces in* $\{[\boldsymbol{u}, \boldsymbol{v}], [\boldsymbol{u}_Q, \boldsymbol{v}_Q], [\boldsymbol{u}_R, \boldsymbol{v}_R]\}$ *are parallel.*
(iii) $\boldsymbol{v}_Q - \boldsymbol{u}_Q = \beta(\boldsymbol{v} - \boldsymbol{u})$ *and* $\boldsymbol{v}_R - \boldsymbol{u}_R = \gamma(\boldsymbol{v} - \boldsymbol{u})$, *for* $\beta, \gamma \in [0, 1]$.

Proof Let $\boldsymbol{u}$ and $\boldsymbol{v}$ be adjacent vertices of P with decompositions $\boldsymbol{u} = \boldsymbol{u}_Q + \boldsymbol{u}_R$ and $\boldsymbol{v} = \boldsymbol{v}_Q + \boldsymbol{v}_R$.

(i) Let $\boldsymbol{a} \in \mathbb{R}^d$ be such that $[\boldsymbol{u}, \boldsymbol{v}] = F(P, \boldsymbol{a})$ (1.9.7) . Lemma 6.1.3 shows that $[\boldsymbol{u}, \boldsymbol{v}] = F(Q, \boldsymbol{a}) + F(R, \boldsymbol{a})$. The uniqueness of the decompositions of $\boldsymbol{u}$

and $\boldsymbol{v}$ shows that $\boldsymbol{u}_Q, \boldsymbol{v}_Q \in F(Q,\boldsymbol{a})$ and $\boldsymbol{u}_R, \boldsymbol{v}_R \in F(R,\boldsymbol{a})$. Hence $[\boldsymbol{u}_Q, \boldsymbol{v}_Q] \subseteq F(Q,\boldsymbol{a})$ and $[\boldsymbol{u}_R, \boldsymbol{v}_R] \subseteq F(R,\boldsymbol{a})$, and thus

$$[\boldsymbol{u}_Q, \boldsymbol{v}_Q] + [\boldsymbol{u}_R, \boldsymbol{v}_R] \subseteq [\boldsymbol{u}, \boldsymbol{v}].$$

The opposite relation $[\boldsymbol{u}_Q, \boldsymbol{v}_Q] + [\boldsymbol{u}_R, \boldsymbol{v}_R] \supseteq [\boldsymbol{u}, \boldsymbol{v}]$ is easily obtained from the decompositions of $\boldsymbol{u}$ and $\boldsymbol{v}$ and the definition of the Minkowski sum.

(ii) For some $\boldsymbol{a} \in \mathbb{R}^d$, from (i) it follows that

$$[\boldsymbol{u}, \boldsymbol{v}] = F(P,\boldsymbol{a}) = F(Q,\boldsymbol{a}) + F(R,\boldsymbol{a}) = [\boldsymbol{u}_Q, \boldsymbol{v}_Q] + [\boldsymbol{u}_R, \boldsymbol{v}_R]. \tag{6.1.5.1}$$

According to Corollary 6.1.4, the face $[\boldsymbol{u}_Q, \boldsymbol{v}_Q]$ is either an edge or a vertex. And in the case of $[\boldsymbol{u}_Q, \boldsymbol{v}_Q]$ being an edge, (6.1.5.1) yields that it must be parallel to $[\boldsymbol{u}, \boldsymbol{v}]$. Similarly, if $[\boldsymbol{u}_R, \boldsymbol{v}_R]$ is an edge, then it must be parallel to $[\boldsymbol{u}, \boldsymbol{v}]$.

(iii) Let $\boldsymbol{b} \neq \boldsymbol{0}$ such that $\boldsymbol{u} = F(P,\boldsymbol{b})$. If $\boldsymbol{v} = \boldsymbol{u}$, then $\boldsymbol{v}_Q = \boldsymbol{u}_Q$ and $\boldsymbol{v}_R = \boldsymbol{u}_R$ by the uniqueness of the decomposition $\boldsymbol{u} = \boldsymbol{u}_Q + \boldsymbol{u}_R$ (Lemma 6.1.3), which concludes the proof in this case. So suppose $\boldsymbol{v} \neq \boldsymbol{u}$.

From $\boldsymbol{u} = \boldsymbol{u}_Q + \boldsymbol{u}_R$ and Lemma 6.1.3, it follows that $\boldsymbol{u}_Q = F(Q,\boldsymbol{b})$ and $\boldsymbol{u}_R = F(R,\boldsymbol{b})$. Thus the functional $\boldsymbol{b} \cdot \boldsymbol{x}$ for $\boldsymbol{x} \in \mathbb{R}^d$ satisfies

$$\boldsymbol{b} \cdot \boldsymbol{v} < \boldsymbol{b} \cdot \boldsymbol{u} \,, \boldsymbol{b} \cdot \boldsymbol{v}_Q \leqslant \boldsymbol{b} \cdot \boldsymbol{u}_Q \,, \boldsymbol{b} \cdot \boldsymbol{v}_R \leqslant \boldsymbol{b} \cdot \boldsymbol{u}_R. \tag{6.1.5.2}$$

Choose $\boldsymbol{0} \neq \boldsymbol{t} \in \mathbb{R}^d$ so that $\boldsymbol{v} = \boldsymbol{u} + \alpha \boldsymbol{t}$, for $\alpha > 0$. Since $\alpha > 0$, (6.1.5.2) yields that

$$\boldsymbol{b} \cdot \boldsymbol{v} = \boldsymbol{b} \cdot (\boldsymbol{u} + \alpha \boldsymbol{t}) = \boldsymbol{b} \cdot \boldsymbol{u} + \alpha \boldsymbol{b} \cdot \boldsymbol{t} < \boldsymbol{b} \cdot \boldsymbol{u},$$

which implies that $\boldsymbol{b} \cdot \boldsymbol{t} < 0$.

By (ii), all one-dimensional faces in $\{[\boldsymbol{u}, \boldsymbol{v}], [\boldsymbol{u}_Q, \boldsymbol{v}_Q], [\boldsymbol{u}_R, \boldsymbol{v}_R]\}$ are parallel. Thus we can write $\boldsymbol{v}_Q = \boldsymbol{u}_Q + \beta' \boldsymbol{t}$ and $\boldsymbol{v}_R = \boldsymbol{u}_R + \gamma' \boldsymbol{t}$, for $\beta', \gamma' \in \mathbb{R}$. In this case, (6.1.5.2) yields that

$$\boldsymbol{b} \cdot \boldsymbol{v}_Q = \boldsymbol{b} \cdot (\boldsymbol{u}_Q + \beta' \boldsymbol{t}) = \boldsymbol{b} \cdot \boldsymbol{u}_Q + \beta' \boldsymbol{b} \cdot \boldsymbol{t} \leqslant \boldsymbol{b} \cdot \boldsymbol{u}_Q,$$

which together with $\boldsymbol{b} \cdot \boldsymbol{t} < 0$ gives that $\beta' \geqslant 0$. The same analysis gives that $\gamma' \geqslant 0$. As a consequence,

$$\boldsymbol{v}_Q - \boldsymbol{u}_Q = \frac{\beta'}{\alpha}(\boldsymbol{v} - \boldsymbol{u}), \; \boldsymbol{v}_R - \boldsymbol{u}_R = \frac{\gamma'}{\alpha}(\boldsymbol{v} - \boldsymbol{u}),$$

from which, letting $\beta := \beta'/\alpha$ and $\gamma := \gamma'/\alpha$, we obtain that $\beta, \gamma \geqslant 0$. The desired conclusion follows from combining $\beta, \gamma \geqslant 0$ with (i). □

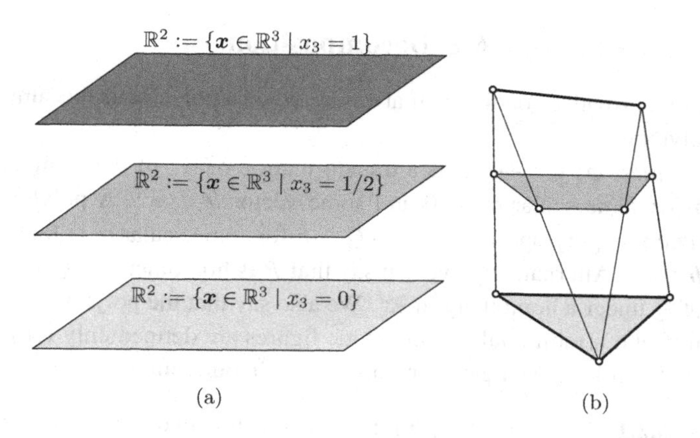

Figure 6.1.1 A Cayley embedding in $\mathbb{R}^3$ of a segment Q and a triangle R. (a) Hyperplanes in $\mathbb{R}^3$. (b) The Cayley polytope with the sum $Q + R$ as a section.

Cayley Embedding of a Sum

The Minkowski sum of two polytopes Q and R in $\mathbb{R}^d$ can be realised as a section of a certain convex hull of the polytopes Q and R. Embed Q in the hyperplane $\{x \in \mathbb{R}^{d+1} | \, x_{d+1} = 1\}$ of $\mathbb{R}^{d+1}$ and R in the hyperplane $\{x \in \mathbb{R}^{d+1} | \, x_{d+1} = 0\}$ of $\mathbb{R}^{d+1}$. The *Cayley polytope* of Q and R is the polytope $\operatorname{conv}(Q \cup R)$. The intersection of the Cayley polytope with a hyperplane $\{x \in \mathbb{R}^{d+1} | \, x_{d+1} = \alpha\}$ for a fixed value of $\alpha \in (0, 1)$, say $\alpha = 1/2$, produces the *weighted Minkowski sum*

$$\alpha Q + (1 - \alpha)R = \{\alpha q + (1 - \alpha)r | \, q \in Q, r \in R\}$$

of Q and R. The weighted sum $\alpha Q + (1 - \alpha)R$ is combinatorially isomorphic to the sum $Q + R$: the weighted sums corresponding to two distinct values of α are clearly combinatorially isomorphic, and the particular case $\alpha = 1/2$ is the sum $Q + R$ scaled by a factor of $1/2$. Consequently, a Minkowski sum is a section of the corresponding Cayley polytope. This procedure is called the *Cayley embedding* of the sum. See Fig. 6.1.1.

A Cayley embedding reduces combinatorial properties of a sum to combinatorial properties of the corresponding Cayley polytope. The k-faces of the sum $Q + R$ correspond bijectively to the $(k + 1)$-faces of the corresponding Cayley polytope that are not contained in either Q or R. This reduction has proven useful in deriving upper and lower bounds for the number of k-faces of the sum $Q + R$ (Adiprasito and Sanyal, 2016; Karavelas and Tzanaki, 2016a,b).

6.2 Decomposition

This section provides conditions that ensure when a polytope is the sum of two other polytopes.

A *homothety* $\varphi\colon \mathbb{R}^d \to \mathbb{R}^d$ is a transformation of $\mathbb{R}^d$ of the form $\varphi(\boldsymbol{x}) := \alpha\boldsymbol{x} + \boldsymbol{b}$ for some scalar $\alpha \geqslant 0$ and some vector $\boldsymbol{b} \in \mathbb{R}^d$. A polytope P is *homothetic* to a polytope Q if $P = \alpha Q + \boldsymbol{b}$ for some scalar $\alpha \geqslant 0$ and some vector $\boldsymbol{b} \in \mathbb{R}^d$. Alternatively, we can say that P is homothetic to Q if P is the image of Q under a homothety in $\mathbb{R}^d$. We also say that the polytopes P and Q are homothetic. Traditionally, homothetic figures are defined only for positive scalars α, but it is convenient to include $\alpha = 0$ in our context.

Remark 6.2.1 A homothety $\varphi(\boldsymbol{v}) := \alpha\boldsymbol{v} + \boldsymbol{b}$ has two parameters, namely $\alpha \geqslant 0$ and $\boldsymbol{b} \in \mathbb{R}^d$, and so is determined by its value at two points.

A (realisation of a) polytope is *decomposable* if it is the sum of two polytopes that are not homothetic to the polytope; otherwise it is *indecomposable*. If a polytope P decomposes as $Q + R$, then it also decomposes as $P = (Q + \boldsymbol{b}) + (R - \boldsymbol{b})$ for each vector $\boldsymbol{b} \in \mathbb{R}^d$; all these decompositions are considered to be the same. If we were to allow homothetic summands, then, for each $\alpha \in [0, 1]$ and each vector $\boldsymbol{b} \in \mathbb{R}^d$, every polytope P in $\mathbb{R}^d$ would decompose as

$$P = (\alpha P + \boldsymbol{b}) + ((1 - \alpha)P - \boldsymbol{b}),$$

making the theory meaningless.

Gale (1954) announced, without proofs, the first results on decomposability, including the assertions that polygons other than triangles are decomposable (Theorem 6.2.16) and that simplices are indecomposable (Proposition 6.2.18). Proofs were provided later by Shephard (1963) and Silverman (1973). We develop the theory of decomposability via geometric graphs and decomposing functions, an approach different from standard sources such as Shephard (1963) and Schneider (2014).

Let $G = (V, E)$ be a (finite) geometric graph. A function $\varphi\colon V \to \mathbb{R}^d$ is a *decomposing function* for the graph if it satisfies

$$\varphi(\boldsymbol{u}) - \varphi(\boldsymbol{w}) = \alpha_e(\boldsymbol{u} - \boldsymbol{w}),$$

for every edge $e = [\boldsymbol{u}, \boldsymbol{w}] \in E$, where the scalar $\alpha_e \geqslant 0$ depends only on the edge e. For the decomposing function φ, we denote by $A(\varphi)$ the set of scalars α_e over all edges $e \in E$. Then $A(\varphi) \subset [0, \infty]$. If $A(\varphi) \subset (0, \infty]$, the decomposing function is said to be *positive*. The identity function id_V is trivially a positive decomposing function. We remark that if a graph has a decomposing function, then it has infinitely many, including positive ones.

Lemma 6.2.2 *If $\varphi\colon V \to \mathbb{R}^d$ is a decomposing function for a graph $G := (V, E)$, then, for every $\beta > 0$, the function $\beta\varphi$ is another decomposition for G and so is the function $\gamma \operatorname{id}_V - \varphi$ for each $\gamma > \max A(\varphi)$.*

Proof Since φ is a decomposition function for G, for every edge $e = [\boldsymbol{u}, \boldsymbol{w}] \in E$, we have that

$$\beta\varphi(\boldsymbol{u}) - \beta\varphi(\boldsymbol{w}) = \beta\alpha_e(\boldsymbol{u} - \boldsymbol{w})$$

and that

$$\begin{aligned}(\gamma \operatorname{id}_V - \varphi)(\boldsymbol{u}) - (\gamma \operatorname{id}_V - \varphi)(\boldsymbol{w}) &= \gamma \boldsymbol{u} - \varphi(\boldsymbol{u}) - (\gamma \boldsymbol{w} - \varphi(\boldsymbol{w}))\\ &= (\gamma - \alpha_e)(\boldsymbol{u} - \boldsymbol{w}),\end{aligned}$$

where $\beta\alpha_e \geqslant 0$ and $\gamma - \alpha_e \geqslant 0$. It follows that $\beta\varphi$ and $\gamma \operatorname{id}_V - \varphi$ are both decomposing functions for G. □

According to Lemma 6.2.2, there was no loss of generality by assuming that the scalars of our decomposing functions are nonnegative. A *decomposing function for a polytope* is a decomposing function for its graph. Abusing notation, we will often write $\varphi(G)$ instead of $\varphi(\mathcal{V}(G))$ and $\varphi(P)$ instead of $\varphi(G(P)) = \{\varphi(\boldsymbol{v}) \mid \boldsymbol{v} \in \mathcal{V}(P)\}$. Example 6.2.4 illustrates geometric graphs and decomposing functions.

Remark 6.2.3 A decomposing function φ is the restriction of a homothety in $\mathbb{R}^d$ if and only if $\#A(\varphi) = 1$.

Example 6.2.4 Consider the graphs in Fig. 6.2.1 and let

$$V := \{\boldsymbol{v}_1, \boldsymbol{v}_2, \boldsymbol{v}_3, \boldsymbol{v}_4, \boldsymbol{v}_5, \boldsymbol{v}_6\}$$

denote their vertex set.

(i) We define two pairs (φ, ϱ) and $(\varphi_\varepsilon, \varrho_\varepsilon)$ of decomposing functions for the graph in Fig. 6.2.1(b) that are not restrictions of homotheties in $\mathbb{R}^3$. Let

$$\varphi(\boldsymbol{v}_1) = \varphi(\boldsymbol{v}_4) = \boldsymbol{v}_1,\ \varphi(\boldsymbol{v}_2) = \varphi(\boldsymbol{v}_5) = \boldsymbol{v}_2,\ \varphi(\boldsymbol{v}_3) = \varphi(\boldsymbol{v}_6) = \boldsymbol{v}_3$$
$$\varrho(\boldsymbol{v}_1) = \varrho(\boldsymbol{v}_2) = \varrho(\boldsymbol{v}_3) = \boldsymbol{v}_2,\ \varrho(\boldsymbol{v}_4) = \varrho(\boldsymbol{v}_5) = \varrho(\boldsymbol{v}_6) = \boldsymbol{v}_5.$$

We then have that $A(\varphi) = A(\varrho) = \{0, 1\}$. Let $\varepsilon \in [0, 1)$ and let

$$\begin{aligned} \varphi_\varepsilon(\boldsymbol{v}_1) &= \boldsymbol{v}_1, & \varrho_\varepsilon(\boldsymbol{v}_1) &= \varrho_\varepsilon(\boldsymbol{v}_2) = \varrho_\varepsilon(\boldsymbol{v}_3) = \boldsymbol{v}_2, \\ \varphi_\varepsilon(\boldsymbol{v}_2) &= \boldsymbol{v}_2, & \varrho_\varepsilon(\boldsymbol{v}_4) &= \varrho_\varepsilon(\boldsymbol{v}_5) = \varrho_\varepsilon(\boldsymbol{v}_6) = (0, 0, 1-\varepsilon)^t . \\ \varphi_\varepsilon(\boldsymbol{v}_3) &= \boldsymbol{v}_3, \\ \varphi_\varepsilon(\boldsymbol{v}_4) &= (0, 1, \varepsilon)^t , \\ \varphi_\varepsilon(\boldsymbol{v}_5) &= (0, 0, \varepsilon)^t , \\ \varphi_\varepsilon(\boldsymbol{v}_6) &= (1, 0, \varepsilon)^t . \end{aligned}$$

We then have that $A(\varphi_\varepsilon) = \{\varepsilon, 1\}$ and $A(\varrho_\varepsilon) = \{0, 1 - \varepsilon\}$. Observe that $\varphi = \varphi_0$ and $\varrho = \varrho_0$.

(ii) In contrast, every decomposing function for the graph in Fig. 6.2.1(a) is the restriction of a homothety in $\mathbb{R}^3$. Let φ be a decomposing function for this graph. We first check the image of the function on the cycle $\boldsymbol{v}_1\boldsymbol{v}_4\boldsymbol{v}_2\boldsymbol{v}_3$. For nonnegative scalars $\alpha_{14}, \alpha_{42}, \alpha_{23}, \alpha_{31}$, we get

$$\begin{aligned} \varphi(\boldsymbol{v}_1) - \varphi(\boldsymbol{v}_4) &= \alpha_{14}(\boldsymbol{v}_1 - \boldsymbol{v}_4), \\ \varphi(\boldsymbol{v}_4) - \varphi(\boldsymbol{v}_2) &= \alpha_{42}(\boldsymbol{v}_4 - \boldsymbol{v}_2), \\ \varphi(\boldsymbol{v}_2) - \varphi(\boldsymbol{v}_3) &= \alpha_{23}(\boldsymbol{v}_2 - \boldsymbol{v}_3), \\ \varphi(\boldsymbol{v}_3) - \varphi(\boldsymbol{v}_1) &= \alpha_{31}(\boldsymbol{v}_3 - \boldsymbol{v}_1). \end{aligned}$$

From which we obtain that

$$\alpha_{14}(\boldsymbol{v}_1 - \boldsymbol{v}_4) + \alpha_{42}(\boldsymbol{v}_4 - \boldsymbol{v}_2) + \alpha_{23}(\boldsymbol{v}_2 - \boldsymbol{v}_3) + \alpha_{31}(\boldsymbol{v}_3 - \boldsymbol{v}_1) = \boldsymbol{0},$$

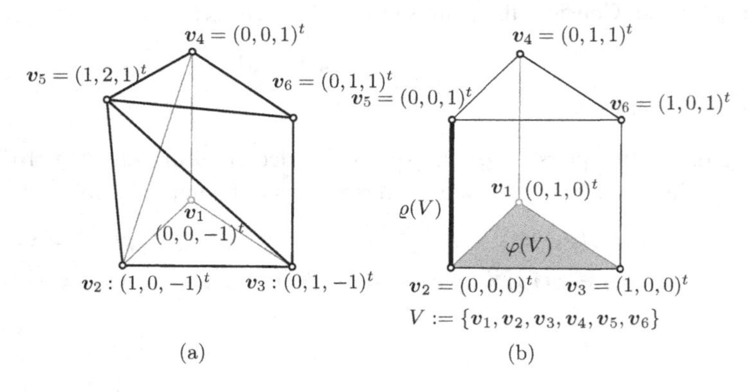

Figure 6.2.1 Decomposing functions of geometric graphs. (a) A geometric graph for which every decomposing function is the restriction of a homothety in $\mathbb{R}^3$. (b) A geometric graph with two decomposing functions φ and ϱ that are not restrictions of homotheties. The image of V under ϱ is $\{\boldsymbol{v}_2, \boldsymbol{v}_5\}$ and the image of V under φ is $\{\boldsymbol{v}_1, \boldsymbol{v}_2, \boldsymbol{v}_3\}$.

or, equivalently, that

$$(\alpha_{14} - \alpha_{31})\boldsymbol{v}_1 + (\alpha_{23} - \alpha_{42})\boldsymbol{v}_2 + (\alpha_{31} - \alpha_{23})\boldsymbol{v}_3 + (\alpha_{42} - \alpha_{14})\boldsymbol{v}_4 = \boldsymbol{0}.$$

On the other hand, the points $\boldsymbol{v}_1, \boldsymbol{v}_2, \boldsymbol{v}_3, \boldsymbol{v}_4$ are affinely independent, and so

$$\alpha_{14} - \alpha_{31} = \alpha_{23} - \alpha_{42} = \alpha_{31} - \alpha_{23} = \alpha_{42} - \alpha_{14} = 0.$$

Consequently, $\alpha_{14} = \alpha_{31} = \alpha_{23} = \alpha_{42} = \alpha$. Each edge of the graph is contained in some triangle. As a result, performing the same analysis on successive triangles for which one knows the scalar of two edges, for instance $\boldsymbol{v}_1\boldsymbol{v}_2\boldsymbol{v}_3$, we get that $A(\varphi) = \{\alpha\}$. That is, the function φ is the restriction of a homothety in $\mathbb{R}^3$ (Problem 6.5.1).

The graph $G = (V, E)$ in Fig. 6.2.1(b) is the graph of a simplicial 3-prism. A simplicial 3-prism is the Minkowski sum of the triangle $\operatorname{conv}\varphi(V)$ and the segment $\operatorname{conv}\varrho(V)$, and of the simplicial 3-prism $\operatorname{conv}\varphi_\varepsilon(V)$ and the segment $\operatorname{conv}\varrho_\varepsilon(V)$ for $\varepsilon \in (0, 1)$. The functions $\varphi, \varrho, \varphi_\varepsilon, \varrho_\varepsilon$ are defined in Example 6.2.4. In this example, the convex hulls of the images of decomposing functions for a polytope are summands of the polytope; this is not a coincidence. It is also not a coincidence that a positive decomposing function gives rise to a summand combinatorially isomorphic to the sum. All these remarks are justified next.

Remark 6.2.5 A decomposition of a polytope is, in general, not unique. A regular hexagon in the plane has at least two distinct decompositions: as the sum of two triangles and as the sum of three segments. A simplicial d-prism admits infinitely many distinct decompositions (Example 6.2.4).

In fact, if a polytope P admits two distinct decompositions, then it admits infinitely many distinct decompositions. The set of all decompositions is convex: if $Q + R$ and $Q' + R'$ are two decompositions of P, so is $(\alpha Q + \beta Q') + (\alpha R + \beta R')$ for $\alpha, \beta \geqslant 0$ and $\alpha + \beta = 1$.

Lemma 6.1.2 provides conditions for the existence of decomposing functions.

Lemma 6.2.6 *Let P and Q be two polytopes in $\mathbb{R}^d$.*

(i) *If $\dim F(P, \boldsymbol{a}) = \dim F(Q, \boldsymbol{a})$ for every $\boldsymbol{a} \in \mathbb{R}^d$, then there exists a one-to-one correspondence φ between the set of k-faces of P and the set of k-faces of Q (for each $k \in [0 \dots d]$) in which the face $F(P, \boldsymbol{a})$ corresponds to the face $F(Q, \boldsymbol{a})$. In particular, φ is a positive decomposing function for P satisfying $\varphi(\mathcal{V}(F(P, \boldsymbol{a}))) = \mathcal{V}(F(Q, \boldsymbol{a}))$.*

(ii) *If* $\dim F(P,\boldsymbol{a}) \geqslant \dim F(Q,\boldsymbol{a})$ *for every* $\boldsymbol{a} \in \mathbb{R}^d$, *then there exists a decomposing function* φ *for* P *satisfying* $\varphi(\mathcal{V}(F(P,\boldsymbol{a}))) = \mathcal{V}(F(Q,\boldsymbol{a}))$.

Proof (i) If $\dim F(P,\boldsymbol{a}) = \dim F(Q,\boldsymbol{a})$ for every $\boldsymbol{a} \in \mathbb{R}^d$, then it is clear that there exists a bijection $\varphi \colon \mathcal{F}(P) \to \mathcal{F}(Q)$ that maps the k-faces of P to the k-faces of Q, and it is also clear that a vertex $\boldsymbol{u} = F(P,\boldsymbol{a})$ with $\boldsymbol{a} \neq \boldsymbol{0}$ is mapped to the vertex $\varphi(\boldsymbol{u}) = F(Q,\boldsymbol{a})$. If an edge e of P joins two vertices $\boldsymbol{u}$ and $\boldsymbol{w}$ of P, then an edge parallel to e must join the vertices $\varphi(\boldsymbol{u})$ and $\varphi(\boldsymbol{w})$ of Q, since $\dim[\boldsymbol{u},\boldsymbol{w}] = \dim[\varphi(\boldsymbol{u}),\varphi(\boldsymbol{w})]$. Hence

$$\varphi(\boldsymbol{u}) - \varphi(\boldsymbol{w}) = \alpha_e(\boldsymbol{u} - \boldsymbol{w}),$$

for some α_e that depends only on $\boldsymbol{u}$ and $\boldsymbol{w}$. Corollary 6.1.5 now gives that $\alpha_e > 0$, showing that φ is a positive decomposing function for P.

In the setting of (ii) we may no longer have a bijection between the k-faces of P and the k-faces of Q ($k \in [0 \ldots d]$). Every vertex $\boldsymbol{u} = F(P,\boldsymbol{a})$ with $\boldsymbol{a} \neq \boldsymbol{0}$ is mapped to the vertex $\varphi(\boldsymbol{u}) = F(Q,\boldsymbol{a})$, but if $\dim F(P,\boldsymbol{a}) > 0$ several vertices in $F(P,\boldsymbol{a})$ may be mapped to the same vertex in $F(Q,\boldsymbol{a})$. The function φ remains surjective.

For every edge $e = F(P,\boldsymbol{a})$ ($\boldsymbol{a} \neq \boldsymbol{0}$) that joins two vertices $\boldsymbol{u}$ and $\boldsymbol{w}$ of P, the face $F(Q,\boldsymbol{a})$ may be either an edge parallel to e that joins the vertices $\varphi(\boldsymbol{u})$ and $\varphi(\boldsymbol{w})$ or a vertex. In either case,

$$\varphi(\boldsymbol{u}) - \varphi(\boldsymbol{w}) = \alpha_e(\boldsymbol{u} - \boldsymbol{w}),$$

for some $\alpha_e \geqslant 0$, where α_e depends only on $\boldsymbol{u}$ and $\boldsymbol{w}$. This shows that φ is a decomposing function for P with $A(\varphi) \subset [0,\infty)$. □

Decomposing functions for a sum encode a wealth of information on the summands. Lemma 6.2.7, Proposition 6.2.8, and Theorem 6.2.9 tell the story.

Lemma 6.2.7 *Let* P *and* Q *be two polytopes in* $\mathbb{R}^d$. *Suppose there exists a decomposing function* φ *for* P *satisfying* $\mathcal{V}(Q) \subseteq \varphi(P)$. *Then it is true that* $\varphi(F(P,\boldsymbol{a})) = \mathcal{V}(F(Q,\boldsymbol{a}))$ *for every* $\boldsymbol{a} \in \mathbb{R}^d$. *In particular,* $\varphi(P) = \mathcal{V}(Q)$.

Proof Let $\boldsymbol{a} \neq \boldsymbol{0}$ be given.

Let $\boldsymbol{u}$ be any vertex of the face $F(P,\boldsymbol{a})$. We first show that $\varphi(\boldsymbol{u}) \in F(Q,\boldsymbol{a})$. Let $\boldsymbol{w}$ be any vertex of P. Since $\boldsymbol{u} \in F(P,\boldsymbol{a})$, there is a nonempty path $L := z_0 \ldots z_\ell$ in $G(P)$ from $z_0 = \boldsymbol{w}$ to $z_\ell = \boldsymbol{u}$ where $\boldsymbol{a} \cdot z_{i-1} \leqslant \boldsymbol{a} \cdot z_i$ for each $i \in [1 \ldots \ell]$ (Corollary 3.2.3). If $\boldsymbol{w} \in F(P,\boldsymbol{a})$ then we can assume that $L \subset F(P,\boldsymbol{a})$, implying that $\boldsymbol{a} \cdot z_{i-1} = \boldsymbol{a} \cdot z_i$ for each $i \in [1 \ldots \ell]$; otherwise we have that $\boldsymbol{a} \cdot z_{i-1} < \boldsymbol{a} \cdot z_i$ for each $i \in [1 \ldots \ell]$.

From φ being a decomposing function for P, we get that

$$\boldsymbol{a} \cdot \varphi(z_\ell) - \boldsymbol{a} \cdot \varphi(z_{\ell-1}) = \alpha(\boldsymbol{a} \cdot z_\ell - \boldsymbol{a} \cdot z_{\ell-1}),$$

and from $\alpha \geqslant 0$ and $\boldsymbol{a} \cdot z_\ell \geqslant \boldsymbol{a} \cdot z_{\ell-1}$, that

$$\boldsymbol{a} \cdot \varphi(z_\ell) - \boldsymbol{a} \cdot \varphi(z_{\ell-1}) \geqslant 0.$$

Applying the same reasoning to the full path L, we obtain that

$$\begin{aligned} \boldsymbol{a} \cdot \varphi(z_\ell) - \boldsymbol{a} \cdot \varphi(z_{\ell-1}) &\geqslant 0, \\ \boldsymbol{a} \cdot \varphi(z_{\ell-1}) - \boldsymbol{a} \cdot \varphi(z_{\ell-2}) &\geqslant 0, \\ &\vdots \\ \boldsymbol{a} \cdot \varphi(z_1) - \boldsymbol{a} \cdot \varphi(z_0) &\geqslant 0. \end{aligned}$$

Summing all these inequalities side by side and doing the appropriate cancellations, we get that

$$\boldsymbol{a} \cdot \varphi(\boldsymbol{u}) - \boldsymbol{a} \cdot \varphi(\boldsymbol{w}) \geqslant 0,$$

which amounts to saying that $\varphi(\boldsymbol{u}) \in \mathcal{V}(F(Q,\boldsymbol{a}))$, as every vertex of Q is the image under φ of some vertex of P. Hence $\varphi(F(P,\boldsymbol{a})) \subseteq \mathcal{V}(F(Q,\boldsymbol{a}))$.

We next show that $\mathcal{V}(F(Q,\boldsymbol{a})) \subseteq \varphi(F(P,\boldsymbol{a}))$. Take any vertex $\boldsymbol{q}$ in $F(Q,\boldsymbol{a})$. Since $\mathcal{V}(Q) \subseteq \varphi(P)$, there exists $\boldsymbol{w}$ in $\mathcal{V}(P)$ such that $\varphi(\boldsymbol{w}) = \boldsymbol{q}$. If $\boldsymbol{w} \in F(P,\boldsymbol{a})$, there is nothing to do, so suppose otherwise. Reasoning as before, we obtain a nonempty path $L := z_0 \dots z_\ell$ in $G(P)$ from $z_0 = \boldsymbol{w}$ to some vertex $z_\ell = \boldsymbol{u} \in F(P,\boldsymbol{a})$, where $\boldsymbol{a} \cdot z_{i-1} < \boldsymbol{a} \cdot z_i$ for $i \in [1 \dots \ell]$. Applying φ to the edge $z_1 z_0$, we get that

$$\boldsymbol{a} \cdot \varphi(z_1) - \boldsymbol{a} \cdot \varphi(z_0) = \alpha(\boldsymbol{a} \cdot z_1 - \boldsymbol{a} \cdot z_0).$$

From $\varphi(z_0) \in F(Q,\boldsymbol{a})$, it follows that $\boldsymbol{a} \cdot \varphi(z_1) - \boldsymbol{a} \cdot \varphi(z_0) \leqslant 0$. And from $\alpha \geqslant 0$ and $\boldsymbol{a} \cdot z_1 > \boldsymbol{a} \cdot z_0$, it follows that $\boldsymbol{a} \cdot \varphi(z_1) - \boldsymbol{a} \cdot \varphi(z_0) \geqslant 0$. Thus $\alpha = 0$, which gives that $\varphi(z_1) = \varphi(z_0)$ and $\varphi(z_1) \in F(Q,\boldsymbol{a})$. By applying this reasoning to the remaining edges of the path L, we see that

$$\varphi(z_0) = \varphi(z_1) = \cdots = \varphi(z_\ell).$$

Hence $\boldsymbol{q} = \varphi(\boldsymbol{u}) \in \varphi(F(P,\boldsymbol{a}))$.

Finally, let $\boldsymbol{a} = \boldsymbol{0}$. We show that $\varphi(F(P,\boldsymbol{a})) = \mathcal{V}(F(Q,\boldsymbol{a}))$, which is the case $\varphi(P) = \mathcal{V}(Q)$. Suppose that there exists a vertex $\boldsymbol{u}$ of P such that $\varphi(\boldsymbol{u})$ is not a vertex of Q. Let $\boldsymbol{b} \neq \boldsymbol{0}$ be such that $\boldsymbol{u} = F(P,\boldsymbol{b})$. Then, from the case where $\boldsymbol{a} \neq \boldsymbol{0}$, we find that $\mathcal{V}(F(Q,\boldsymbol{b})) \subseteq \varphi(F(P,\boldsymbol{b})) = \{\varphi(\boldsymbol{u})\} \nsubseteq \mathcal{V}(F(Q,\boldsymbol{b}))$. This contradiction completes the proof of the lemma. □

The next proposition is in some sense the converse of Lemma 6.2.6(i).

Proposition 6.2.8 *Let P and Q be two polytopes in $\mathbb{R}^d$. Suppose there exists a positive decomposing function φ for P satisfying $\mathcal{V}(Q) \subseteq \varphi(P)$. Then the polytope Q is combinatorially isomorphic to the polytope P.*

Proof We show that φ is a bijection taking $\mathcal{V}(F(P,\boldsymbol{a}))$ and $\mathcal{V}(F(Q,\boldsymbol{a}))$ for every $\boldsymbol{a} \neq \mathbf{0}$. That φ is surjective is established in Lemma 6.2.7. Take $\boldsymbol{u}, \boldsymbol{w} \in \mathcal{V}(P)$ with $\boldsymbol{u} \neq \boldsymbol{w}$. Let $\boldsymbol{a} \neq \mathbf{0}$ such that $\boldsymbol{u} = F(P,\boldsymbol{a})$. There exists a nonempty path $L := z_0 \ldots z_\ell$ in $G(P)$ from $z_0 = \boldsymbol{w}$ to $z_\ell = \boldsymbol{u}$ where $\boldsymbol{a} \cdot z_{i-1} < \boldsymbol{a} \cdot z_i$ for every $i \in [1 \ldots \ell]$ (Corollary 3.2.3). Suppose $\varphi(\boldsymbol{u}) = \varphi(\boldsymbol{w})$. Lemma 6.2.7 ensures that $\{\varphi(\boldsymbol{u})\} = \mathcal{V}(F(Q,\boldsymbol{a}))$. Applying φ to the edge $z_1 z_0$ of L, we get that

$$\boldsymbol{a} \cdot \varphi(z_1) - \boldsymbol{a} \cdot \varphi(z_0) = \alpha(\boldsymbol{a} \cdot z_1 - \boldsymbol{a} \cdot z_0).$$

Since $\varphi(z_0) = \varphi(\boldsymbol{u}) \in F(Q,\boldsymbol{a})$, it follows that $\boldsymbol{a} \cdot \varphi(z_1) - \boldsymbol{a} \cdot \varphi(z_0) \leqslant 0$. But from $\alpha > 0$ and $\boldsymbol{a} \cdot z_1 > \boldsymbol{a} \cdot z_0$, it follows that $\boldsymbol{a} \cdot \varphi(z_1) - \boldsymbol{a} \cdot \varphi(z_0) > 0$. This contradiction yields that $\varphi(\boldsymbol{u}) \neq \varphi(\boldsymbol{w})$ and, thus, that φ is injective.

Second, we show that the geometric graphs $G(P)$ and $G(Q)$ are isomorphic with parallel corresponding edges. For every edge $e := [\boldsymbol{u}, \boldsymbol{w}]$ of P, Lemma 6.2.7 gives that $[\varphi(\boldsymbol{u}), \varphi(\boldsymbol{w})]$ are the vertices of a face of Q. And since φ is a bijective decomposing function, the face $[\varphi(\boldsymbol{u}), \varphi(\boldsymbol{w})]$ must be an edge of Q parallel to e: $\varphi(\boldsymbol{u}) - \varphi(\boldsymbol{w}) = \alpha_{uw}(\boldsymbol{u} - \boldsymbol{w})$. Conversely, let $e := [\varphi(\boldsymbol{u}), \varphi(\boldsymbol{w})]$ be an edge of Q and let $\boldsymbol{b} \neq \mathbf{0}$ such that $e = F(Q,\boldsymbol{b})$. By Lemma 6.2.7, $\varphi(F(P,\boldsymbol{b})) = \mathcal{V}(F(Q,\boldsymbol{b})) = \{\varphi(\boldsymbol{u}), \varphi(\boldsymbol{w})\}$. As a result of φ being a bijective decomposing function, we must have that $\mathcal{V}(F(P,\boldsymbol{b})) = \{\boldsymbol{u}, \boldsymbol{w}\}$ and that $[\boldsymbol{u}, \boldsymbol{w}]$ is parallel to e.

Finally, we extend the isomorphism induced by φ to arbitrary faces. For every face $F(P,\boldsymbol{a})$ of P, Lemma 6.2.7 ensures that $\varphi(\mathcal{V}(F(P,\boldsymbol{a})))$ are the vertices of the face $F(Q,\boldsymbol{a})$ of Q. Because $G(P)$ and $G(Q)$ are isomorphic with corresponding parallel edges, to every edge $e := [\boldsymbol{u}, \boldsymbol{w}]$ of $F(P,\boldsymbol{a})$ there must correspond an edge $[\varphi(\boldsymbol{u}), \varphi(\boldsymbol{w})]$ of $F(Q,\boldsymbol{a})$ parallel to e. As a consequence, $\dim F(P,\boldsymbol{a}) = \dim F(Q,\boldsymbol{a})$. Conversely, let $F(Q,\boldsymbol{b})$ be a face of Q for some $\boldsymbol{b} \neq \mathbf{0}$. By Lemma 6.2.7, $\varphi(F(P,\boldsymbol{b})) = \mathcal{V}(F(Q,\boldsymbol{b}))$. Again, as a result of $G(P)$ and $G(Q)$ being isomorphic with corresponding parallel edges, to every edge $e := [\varphi(\boldsymbol{u}), \varphi(\boldsymbol{w})]$ of $F(Q,\boldsymbol{a})$ there corresponds an edge $[\boldsymbol{u}, \boldsymbol{w}]$ of $F(P,\boldsymbol{a})$ parallel to e. It must follow that $\dim F(P,\boldsymbol{a}) = \dim F(Q,\boldsymbol{a})$. □

Decomposing functions φ that satisfy $A(\varphi) \subseteq [0, 1]$ describe the summands of the corresponding decomposition of the polytope.

Theorem 6.2.9 *Let P and Q be two polytopes in $\mathbb{R}^d$. The polytope Q is a summand of P if and only if there exists a decomposing function φ for P such that $\mathcal{V}(Q) \subseteq \varphi(P)$ and $A(\varphi) \subseteq [0, 1]$.*

Proof Suppose that Q is a summand of P. Then $P = Q + R$ for some polytope R. For each nonzero $\boldsymbol{a} \in \mathbb{R}^d$, the face $F(P, \boldsymbol{a})$ of P decomposes as $F(Q, \boldsymbol{a}) + F(R, \boldsymbol{a})$ by Lemma 6.1.2 and Lemma 6.1.3. This decomposition gives that $\dim F(P, \boldsymbol{a}) \geqslant \dim F(Q, \boldsymbol{a})$, which together with Lemma 6.2.6(ii) ensures the existence of a decomposing function φ for P such that $\varphi(P) = \mathcal{V}(Q)$. We show that $A(\varphi) \subset [0, 1]$. Consider an edge $e := [\boldsymbol{u}, \boldsymbol{w}]$ of P. Then $e = F(P, \boldsymbol{b})$ for some nonzero $\boldsymbol{b} \in \mathbb{R}^d$. Since φ is a decomposing function, we have that

$$\varphi(\boldsymbol{u}) - \varphi(\boldsymbol{w}) = \alpha_e(\boldsymbol{u} - \boldsymbol{w}), \tag{6.2.9.1}$$

with $\alpha_e \geqslant 0$ and $\boldsymbol{u}, \boldsymbol{w} \in \mathcal{V}(P)$. Again, as Q is a summand of P, it follows that

$$F(P, \boldsymbol{b}) = F(Q, \boldsymbol{b}) + F(R, \boldsymbol{b}). \tag{6.2.9.2}$$

Because of Corollary 6.1.5, $F(Q, \boldsymbol{b}) = [\varphi(\boldsymbol{u}), \varphi(\boldsymbol{w})]$ and $F(R, \boldsymbol{b}) = [\boldsymbol{u}_R, \boldsymbol{w}_R]$, where $\boldsymbol{u}$ and $\boldsymbol{w}$ decompose as $\varphi(\boldsymbol{u}) + \boldsymbol{u}_R$ and $\varphi(\boldsymbol{w}) + \boldsymbol{w}_R$, respectively. Hence, we can rewrite (6.2.9.2) as

$$[\boldsymbol{u}, \boldsymbol{w}] = [\varphi(\boldsymbol{u}), \varphi(\boldsymbol{w})] + [\boldsymbol{u}_R, \boldsymbol{w}_R],$$

which, together with Corollary 6.1.5, gives that

$$\|\boldsymbol{u} - \boldsymbol{w}\| \geqslant \|\varphi(\boldsymbol{u}) - \varphi(\boldsymbol{w})\|. \tag{6.2.9.3}$$

Combining (6.2.9.1) with (6.2.9.3), we get that $\alpha_e \in [0, 1]$, as desired.

Suppose that a decomposing function φ for P is given such that $\mathcal{V}(Q) \subseteq \varphi(P)$ and $A(\varphi) \subseteq [0, 1]$. We will construct a polytope R such that $P = Q + R$, which would establish the assertion. Define a new function $\varrho(\boldsymbol{v}) := \boldsymbol{v} - \varphi(\boldsymbol{v})$ for each $\boldsymbol{v} \in \mathcal{V}(P)$. The function ϱ is a decomposing function for P with $A(\varrho) \subset [0, 1]$. We let $R := \operatorname{conv} \varrho(P)$. Then $\mathcal{V}(R) \subseteq \varrho(P)$. We complete the proof by showing for every $\boldsymbol{a} \in \mathbb{R}^d$ that

$$h(P, \boldsymbol{a}) = h(Q, \boldsymbol{a}) + h(R, \boldsymbol{a}), \tag{6.2.9.4}$$

which would imply that $P = Q + R$, according to Lemma 6.1.1.

Let $\boldsymbol{a} \in \mathbb{R}^d$ be given and let $\boldsymbol{v}$ be any vertex of the face $F(P, \boldsymbol{a})$. Then $h(P, \boldsymbol{a}) = \boldsymbol{a} \cdot \boldsymbol{v}$. Both functions φ and ϱ satisfy the premises of Lemma 6.2.7. Consequently, $\varphi(\boldsymbol{v}) \in F(Q, \boldsymbol{a})$, which ensures $h(Q, \boldsymbol{a}) = \boldsymbol{a} \cdot \varphi(\boldsymbol{v})$, and $\varrho(\boldsymbol{v}) \in F(R, \boldsymbol{a})$, which ensures $h(R, \boldsymbol{a}) = \boldsymbol{a} \cdot \varrho(\boldsymbol{v})$. Hence (6.2.9.4) follows, and with it, the theorem. □

The proof of Theorem 6.2.9 yields the following.

Corollary 6.2.10 *Let P and Q be two polytopes in $\mathbb{R}^d$. Let φ be a decomposing function for P such that $\mathcal{V}(Q) \subseteq \varphi(P)$ and $A(\varphi) \subseteq [0,1]$. Define $\varrho(\boldsymbol{v}) := \boldsymbol{v} - \varphi(\boldsymbol{v})$ for every $\boldsymbol{v} \in \mathcal{V}(P)$ and $R := \operatorname{conv} \varrho(P)$. Then $Q = \operatorname{conv} \varphi(P)$ and ϱ is another decomposing function for P with $A(\varrho) \subset [0,1]$ that results in P decomposing as*

$$P = Q + R.$$

By virtue of Corollary 6.2.10, we can say that a decomposition of P is simply a decomposing function φ for P with $A(\varphi) \subseteq [0,1]$, in which case

$$P = \operatorname{conv} \varphi(P) + \operatorname{conv}(\operatorname{id} - \varphi)(P).$$

As a consequence of Lemma 6.2.7, Proposition 6.2.8, and Theorem 6.2.9, we can always find a summand that is combinatorially isomorphic to the sum. Accordingly, the theory of decomposability can be entirely developed using this fact, as was done in Kallay (1982). The next corollary gives the details.

Corollary 6.2.11 *Let P and Q be two polytopes in $\mathbb{R}^d$. Suppose there is a decomposing function φ for P such that $\mathcal{V}(Q) \subseteq \varphi(P)$. Then the following hold.*

(i) *There exists a scalar $\alpha > 0$ such that polytope αQ is a summand of the polytope P, and*
(ii) *in the decomposition $P = \alpha Q + R$ for some polytope R, the polytope αQ or R can be chosen so that it is combinatorially isomorphic to P.*

Proof If φ is a decomposing function for P, then, for every $\alpha > 0$, $\alpha\varphi$ is another decomposing function for P (Lemma 6.2.2). Furthermore, $\mathcal{V}(\alpha Q) \subseteq (\alpha\varphi)(P)$. If necessary, choose $\alpha > 0$ so that $\alpha\varphi$ becomes a decomposing function with $A(\alpha\varphi) \subset [0,1]$. Part (i) now follows from Theorem 6.2.9.

Let $Q' := \alpha Q$ and let $\varphi' := \alpha\varphi$. Then $A(\varphi') \subset [0,1]$, $Q' = \operatorname{conv} \varphi'(P)$ (Lemma 6.2.7), and Q' is a summand of P (Theorem 6.2.9). If $A(\varphi') \subset (0,1]$, then Proposition 6.2.8 yields that Q' is combinatorially isomorphic to P, and we are home with the aforementioned choice of α. Otherwise, we choose a new $\alpha := \frac{1}{2}\alpha$, and introduce two new decomposing functions satisfying $\varphi''(\boldsymbol{v}) := \frac{1}{2}\varphi'(\boldsymbol{v})$ and $\varrho(\boldsymbol{v}) := \boldsymbol{v} - \varphi''(\boldsymbol{v})$ for every $\boldsymbol{v} \in \mathcal{V}(P)$. Then $A(\varphi'') \subset [0,1]$ and $A(\varrho) \subset (0,1]$. Letting $Q'' := \operatorname{conv} \varphi''(P)$ and $R := \operatorname{conv} \varrho(P)$, we obtain that

P decomposes as $P = Q'' + R$ thanks to Corollary 6.2.10. Proposition 6.2.8 finally concludes that R is combinatorially isomorphic to P. □

Decomposability of a polytope is a property of its graph, since decomposing functions are defined only for the graph of the polytope (Theorem 6.2.9). Therefore, it makes sense to define decomposability for geometric graphs as an independent concept. This approach will prove to be very versatile because the graphs under consideration need not be graphs of polytopes.

A geometric graph in $\mathbb{R}^d$ is *decomposable* if there exists a decomposing function for the graph that is not the restriction of a homothety in $\mathbb{R}^d$; if every decomposing function for the graph is the restriction of a homothety in $\mathbb{R}^d$, we say that the graph is *indecomposable*. The graph in Fig. 6.2.1(a) is indecomposable while the graph in Fig. 6.2.1(b) is decomposable.

With the notion of decomposable geometric graphs at hand, Theorem 6.2.9 can be recast as a theorem that focusses on graphs. Theorem 6.2.9 imposes the requirement of a decomposing function φ to have $A(\varphi) \subset [0, 1]$, but this does not inhibit generality. If a general decomposing function exists, so does a decomposing function φ with $A(\varphi) \subset [0, 1]$, as demonstrated in the proof of Corollary 6.2.11.

Theorem 6.2.12 *A polytope is decomposable if and only if its graph is decomposable.*

The whole graph of a polytope is not required to decide the decomposability of the polytope; a certain subgraph suffices. A graph touches a facet if it contains a vertex of the facet, as one would expect.

Theorem 6.2.13 *The graph of a polytope is indecomposable if and only if it contains an indecomposable subgraph that touches every facet of the polytope.*

Proof One direction is trivial: if the graph of a polytope is indecomposable, it itself can be taken as the desired subgraph, in which case Theorem 6.2.12 yields the direction.

Suppose that S is a subgraph satisfying the hypothesis of the theorem. Let $P = Q + R$ be an arbitrary decomposition of P; without loss of generality, assume that $\dim Q = \dim P$ (see Theorem 6.2.9 and Corollary 6.2.11). First, we will show the existence of a decomposing function for P with image $\mathcal{V}(Q)$, and then we will use S to show that Q is homothetic to P.

For the decomposition $P = Q + R$, Corollary 6.1.4 ensures that

$$\dim F(P,\boldsymbol{x}) \geqslant \dim F(Q,\boldsymbol{x}) \text{ for every } \boldsymbol{x} \in \mathbb{R}^d.$$

From this and Lemma 6.2.6(ii) follows the existence of a decomposing function φ for P satisfying

$$\varphi(\mathcal{V}(F(P,\boldsymbol{x}))) = \mathcal{V}(F(Q,\boldsymbol{x})) \text{ for every } \boldsymbol{x} \in \mathbb{R}^d. \tag{6.2.13.1}$$

Now consider an irredundant H-description of P of the form

$$P = \left\{\boldsymbol{x} \in \mathbb{R}^d \,\middle|\, \boldsymbol{a}_i \cdot \boldsymbol{x} \leqslant h(P,\boldsymbol{a}_i),\ i = 1,\ldots,f_{d-1}(P)\right\}, \tag{6.2.13.2}$$

where $P \cap \{\boldsymbol{x} \in \mathbb{R}^d \mid \boldsymbol{a}_i \cdot \boldsymbol{x} = h(P,\boldsymbol{a}_i)\}$ defines the ith facet of P (Theorem 2.3.1). Then $h(P,\boldsymbol{a}_j) \neq 0$ for some $j \in [1 \ldots f_{d-1}(P)]$; otherwise P would be a cone (Proposition 2.1.2), contrary to the fact that P is a polytope. As $h(P,\boldsymbol{a}_j) = h(Q,\boldsymbol{a}_j) + h(R,\boldsymbol{a}_j)$ (Lemma 6.1.1) and $h(P,\boldsymbol{a}_j) \neq 0$, we may assume that $h(Q,\boldsymbol{a}_j) \neq 0$.

At this point we use the fact that S is indecomposable. The restriction of φ to S is a homothety in $\mathbb{R}^d$. That is, there exists a scalar $\alpha \geqslant 0$ and a vector $\boldsymbol{b}$ such that

$$\varphi|_S(\boldsymbol{v}) = \alpha\boldsymbol{v} + \boldsymbol{b} \text{ for each } \boldsymbol{v} \in \mathcal{V}(S). \tag{6.2.13.3}$$

Translating Q and R if necessary, we may assume that $\boldsymbol{b} = \boldsymbol{0}$; the polytope P also decomposes as $(Q + \boldsymbol{c}) + (R - \boldsymbol{c})$ for any vector $\boldsymbol{c} \in \mathbb{R}^d$.

We now show that $Q = \alpha P$. Let F be any facet of P and suppose $F = F(P,\boldsymbol{a})$ for some nonzero $\boldsymbol{a} \in \mathbb{R}^d$. By the hypothesis of the theorem, there exists a vertex $\boldsymbol{v} \in \mathcal{V}(S) \cap \mathcal{V}(F)$. Then $\varphi(\boldsymbol{v}) \in \mathcal{V}(F(Q,\boldsymbol{a}))$ by (6.2.13.1). In this case, we have

$$h(Q,\boldsymbol{a}) = \boldsymbol{a} \cdot \varphi(\boldsymbol{v}) = \boldsymbol{a} \cdot \alpha\boldsymbol{v} = \alpha\boldsymbol{a} \cdot \boldsymbol{v} = \alpha h(P,\boldsymbol{a}) = h(\alpha P,\boldsymbol{a}). \tag{6.2.13.4}$$

The last equality follows from Problem 1.12.16(iii). In particular, the equality (6.2.13.4) holds for the facet $F(P,\boldsymbol{a}_j)$. As a consequence of $h(P,\boldsymbol{a}_j) \neq 0$ and $h(Q,\boldsymbol{a}_j) \neq 0$, we see that $\alpha > 0$ in (6.2.13.4).

Finally, take any facet J of Q. Then $J = F(Q,\boldsymbol{c})$ for some nonzero $\boldsymbol{c} \in \mathbb{R}^d$. The relation $\dim F(P,\boldsymbol{c}) \geqslant \dim F(Q,\boldsymbol{c})$ implies that $F(P,\boldsymbol{c})$ is a facet of P. Combining this with (6.2.13.4), we get that an H-description of Q must be of the form

$$Q = \left\{\boldsymbol{x} \in \mathbb{R}^d \,\middle|\, \boldsymbol{a}_{i_r} \cdot \boldsymbol{x} \leqslant h(\alpha P,\boldsymbol{a}_{i_r}),\ r = 1,\ldots,f_{d-1}(Q)\right\},$$

where $P \cap \{\boldsymbol{x} \in \mathbb{R}^d \mid \boldsymbol{a}_{i_r} \cdot \boldsymbol{x} = h(P,\boldsymbol{a}_{i_r})\}$ is the i_rth facet of P. On the other hand, since (6.2.13.2) is irredundant, all the $f_{d-1}(P)$ inequalities of (6.2.13.2) are required in the description of P. It follows that all the $f_{d-1}(P)$ inequalities of (6.2.13.2) must be present in the H-description of Q, otherwise Q would be unbounded. That is, $Q = \alpha P$, as claimed. □

Theorem 6.2.13 is a very powerful tool. We will obtain most of the known decomposability results as applications of it. Let us start from the beginning, dimension one.

Proposition 6.2.14 *A segment is indecomposable.*

Proof Let $[\boldsymbol{u}, \boldsymbol{w}]$ be a segment and let $G := (\{\boldsymbol{u}, \boldsymbol{w}\}, \{\boldsymbol{u}\boldsymbol{w}\})$ be the corresponding geometric graph. Let φ be a decomposing function for G. Then

$$\varphi(\boldsymbol{u}) - \varphi(\boldsymbol{w}) = \alpha(\boldsymbol{u} - \boldsymbol{w}).$$

Suppose $\varphi(\boldsymbol{u}) = \boldsymbol{p}$ and $\varphi(\boldsymbol{w}) = \boldsymbol{q}$ and define $\boldsymbol{b} := \boldsymbol{p} - \alpha\boldsymbol{u}$. Then φ is the restriction of the homothety $\boldsymbol{x} \mapsto \alpha\boldsymbol{x} + \boldsymbol{b}$. The key point here is that any homothety is determined by its value at two points (Remark 6.2.1). □

The trivial Proposition 6.2.14 in conjunction with Theorem 6.2.13 gives a result of McMullen (1987).

Corollary 6.2.15 *A join of two polytopes is indecomposable.*

Proof Every facet of the join P of two polytopes Q and R contains either Q or R. Hence, if $\boldsymbol{q}$ and $\boldsymbol{r}$ are vertices of Q and R, respectively, then the edge $[\boldsymbol{q}, \boldsymbol{r}]$, which is indecomposable by Proposition 6.2.14, meets every facet of the join, and so P is indecomposable by Theorem 6.2.13. □

The indecomposable and decomposable polygons are also easily determined. This was first proved by Silverman (1973).

Theorem 6.2.16 *A triangle is indecomposable. Any other polygon in the plane is decomposable.*

Proof Consider the geometric graph $G := (\{\boldsymbol{u}, \boldsymbol{v}, \boldsymbol{w}\}, E)$ of a triangle. Then

$$\begin{aligned}
\varphi(\boldsymbol{u}) - \varphi(\boldsymbol{v}) &= \alpha(\boldsymbol{u} - \boldsymbol{v}),\\
\varphi(\boldsymbol{v}) - \varphi(\boldsymbol{w}) &= \beta(\boldsymbol{v} - \boldsymbol{w}),\\
\varphi(\boldsymbol{w}) - \varphi(\boldsymbol{u}) &= \gamma(\boldsymbol{w} - \boldsymbol{u}).
\end{aligned}$$

It suffices to show that $\alpha = \beta = \gamma$ (Problem 6.5.1).

Summing up these equalities side by side, we get that

$$\boldsymbol{0} = \alpha(\boldsymbol{u} - \boldsymbol{v}) + \beta(\boldsymbol{v} - \boldsymbol{w}) + \gamma(\boldsymbol{w} - \boldsymbol{u}) = (\alpha - \gamma)\boldsymbol{u} + (\beta - \alpha)\boldsymbol{v} + (\gamma - \beta)\boldsymbol{w}.$$

This is an affine combination of the affinely independent points $\boldsymbol{u}, \boldsymbol{v}, \boldsymbol{w}$. Hence

$$\alpha - \gamma = \beta - \alpha = \gamma - \beta = 0,$$

wherefrom it follows that $\alpha = \beta = \gamma$.

We next show that any other polygon is decomposable. Consider the graph G of a polygon with at least four vertices $\boldsymbol{a}, \boldsymbol{b}, \boldsymbol{c}, \boldsymbol{d}$. Let ℓ be a line parallel to the edge $[\boldsymbol{b}, \boldsymbol{c}]$ that passes through the vertex $\boldsymbol{a}$. Without loss of generality, either the vertices $\boldsymbol{c}$ and $\boldsymbol{d}$ lie on different sides of ℓ or the vertex $\boldsymbol{d}$ lies in ℓ. Let $\boldsymbol{e}$ be the intersection of ℓ with the edge $[\boldsymbol{c}, \boldsymbol{d}]$. See Fig. 6.2.2.

Define a function φ by $\varphi(\boldsymbol{b}) = \boldsymbol{a}, \varphi(\boldsymbol{c}) = \boldsymbol{e}$, and $\varphi(\boldsymbol{v}) = \boldsymbol{v}$ for every other vertex of G. Then φ is a decomposing function for G with $\#A(\varphi) > 1$ (convince yourself of this). Hence φ is not the restriction of a homothety in $\mathbb{R}^2$ (Problem 6.5.1) and, therefore, G is decomposable. By Theorem 6.2.12, the polygon is decomposable. □

The proof that a triangle is indecomposable also proves the following. Compare with Example 6.2.4(ii).

Proposition 6.2.17 *A cycle whose vertices are affinely independent is indecomposable.*

As a consequence of Proposition 6.2.17 and Theorem 6.2.13, if a polytope contains a cycle that touches every facet and whose vertices are affinely independent, the polytope is indecomposable. This gives the indecomposability of a simplex at once.

Proposition 6.2.18 *A simplex is indecomposable.*

The power of Theorem 6.2.13 becomes evident when applied to indecomposable graphs that do not come from polytopes. We next provide a lemma of Przesławski and Yost (2016) that produces indecomposable geometric graphs from existing indecomposable ones.

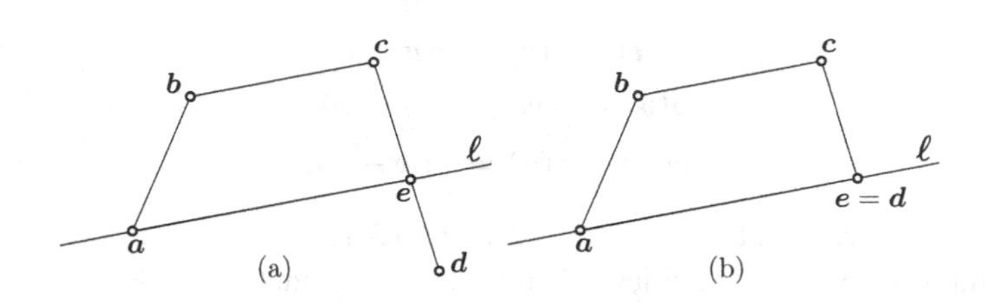

Figure 6.2.2 Decomposable polygons.

Lemma 6.2.19 *Let $G_1 := (V_1, E_1)$ and $G_2 := (V_2, E_2)$ be indecomposable geometric graphs in $\mathbb{R}^d$ that intersect in at least two distinct vertices. Suppose that G is another geometric graph in $\mathbb{R}^d$ whose vertex set V is $V_1 \cup V_2$ and whose edge set E contains the subset $(E_1 \backslash E_1[V_1 \cap V_2]) \cup E_2$. Then G is also indecomposable.*

For a graph $G' := (V', E')$ and a subset X of V', the set $E'[X]$ consists of the edges in E' whose endvertices both lie in X. In Lemma 6.2.19, the edge set E of G is obtained from $E_1 \cup E_2$ by possibly throwing away edges of E_1 whose endvertices lie in the intersection $V_1 \cap V_2$.

Proof Let φ be a decomposing function for G. Since G_2 is indecomposable and is a subgraph of G, the restriction of φ to G_2 is a homothety in $\mathbb{R}^d$. That is, for some scalar α, some vector $\boldsymbol{b}$, and every vertex $\boldsymbol{v}$ in G_2 we have that

$$\varphi|_{G_2}(\boldsymbol{v}) = \alpha \boldsymbol{v} + \boldsymbol{b}.$$

It requires proof to say that the restriction of φ to V_1 is a homothety in $\mathbb{R}^d$, since the subgraph $G[V_1]$ of G may miss edges from G_1. Here is the proof. For every two adjacent vertices $\boldsymbol{u}$ and $\boldsymbol{w}$ of G_1 that are not both contained in $V_1 \cap V_2$ we have that $\varphi(\boldsymbol{u}) - \varphi(\boldsymbol{w}) = \gamma_{uw}(\boldsymbol{u} - \boldsymbol{w})$, and for every two adjacent vertices $\boldsymbol{u}$ and $\boldsymbol{w}$ of G_1 that are both contained in $V_1 \cap V_2$ we have that $\varphi(\boldsymbol{u}) - \varphi(\boldsymbol{w}) = \alpha(\boldsymbol{u} - \boldsymbol{w})$. We have covered all the edges of G_1, and consequently $\varphi|_{V_1}$ is a decomposing function for G_1.

Since G_1 is indecomposable, $\varphi|_{V_1}$ is the restriction of a homothety in $\mathbb{R}^d$. That is, for some scalar β, some vector $\boldsymbol{c}$, and every vertex $\boldsymbol{v}$ in V_1 we have that

$$\varphi|_{V_1}(\boldsymbol{v}) = \beta \boldsymbol{v} + \boldsymbol{c}.$$

Finally, take any two vertices $\boldsymbol{u}, \boldsymbol{w} \in V_1 \cap V_2$. Then

$$\alpha \boldsymbol{u} + \boldsymbol{b} = \beta \boldsymbol{u} + \boldsymbol{c}, \alpha \boldsymbol{w} + \boldsymbol{b} = \beta \boldsymbol{w} + \boldsymbol{c},$$

which forces $\alpha = \beta$ and $\boldsymbol{b} = \boldsymbol{c}$. Hence φ is the restriction of a homothety in $\mathbb{R}^d$. □

Lemma 6.2.19 slightly generalises Kallay (1982, lem. 7 and thm. 8); the following is a useful particular case.

Corollary 6.2.20 *Let G_0 be an indecomposable graph in $\mathbb{R}^d$. And let G be a geometric graph in $\mathbb{R}^d$ obtained from G_0 by adding a new vertex and two edges from this new vertex to two distinct vertices of G_0. Then G is also indecomposable.*

Proof Let $\{v\} = \mathcal{V}(G)\backslash\mathcal{V}(G_0)$, and let $[v, u]$ and $[v, w]$ be the two new edges incident to v in G where $u, w \in \mathcal{V}(G_0)$. Let G_1 be a triangle with vertices u, v, w and let $G_2 := G_0$. Then $\mathcal{V}(G) = \mathcal{V}(G_1) \cup \mathcal{V}(G_2)$ and $\mathcal{E}(G) \subseteq (\mathcal{E}(G_1)\backslash\{[u, w]\}) \cup \mathcal{E}(G_2)$. An application of Lemma 6.2.19 to G, G_1, and G_2 yields that G is indecomposable. □

Theorem 6.2.22 proves the main assertion of McMullen (1987), which in turn generalises a result of Smilansky (1987, thm. 5.1). We give the details hereinafter, but first we need a definition and a lemma.

A family $\mathcal{C}$ of faces in a polytope is *connected* if every pair of faces F and F' of $\mathcal{C}$ is linked by a path $F_1 \ldots F_\ell$ of faces in $\mathcal{C}$ such that $\dim(F_i \cap F_{i+1}) \geqslant 1$, $F_1 = F$, and $F_\ell = F'$; we say that such a path is a *chain of faces*. If $\mathcal{C} := \{F_1, \ldots, F_\ell\}$, then the graph $G(\mathcal{C})$ of $\mathcal{C}$ is $\bigcup_i^\ell G(F_i)$.

Lemma 6.2.21 *Let P be a polytope and let $\mathcal{C}$ be a connected family of faces of P, each of dimension at least one. Then the faces of $\mathcal{C}$ can be linearly ordered as $F_1, \ldots, F_\ell$ so that, for every $i \in [1 \ldots \ell]$,*

(i) *the subgraph $G[F_1, \ldots, F_i]$ of $G(\mathcal{C})$ induced by $\{\mathcal{V}(F_1), \ldots, \mathcal{V}(F_i)\}$ is connected, and*
(ii) *the subgraph $G(F_i) \cap G[F_1, \ldots, F_{i-1}]$ of $G(F_i)$ contains at least one edge.*

Proof A modification of the proof of Proposition C.4.1 gives both results.

Inductively suppose that the subfamily $F_1, \ldots, F_{i-1}$ satisfies the condition. The assertions are clearly true for $i = 1$, so an induction on $i \geqslant 2$ can start. Pick a face F in $\mathcal{C}\backslash\{F_1, \ldots, F_{i-1}\}$. As $\mathcal{C}$ is a connected family, the faces F and F_1 are connected by a chain $L := F_{i_1} \ldots F_{i_r}$ with $F_1 = F_{i_1}$ and $F = F_{i_r}$. Choose F_i as the first face in the chain L that is not in $\{F_1, \ldots, F_{i-1}\}$; that is, for the smallest $j \in [1 \ldots r]$ with F_{i_j} not in $\{F_1, \ldots, F_{i-1}\}$, let $F_i := F_{i_j}$. Successive pairs of faces in L share at least one edge, therefore the result follows from induction. □

Theorem 6.2.22 *A polytope with a connected family of indecomposable faces that touches every facet is indecomposable.*

Proof Consider a connected family $\mathcal{C}$ of indecomposable faces $F_1, \ldots, F_\ell$ that touches every facet of a polytope P. We show that $G(\mathcal{C})$ is indecomposable, which would prove that the polytope itself is indecomposable (Theorem 6.2.13).

By Lemma 6.2.21, we may assume that the subgraph $G[F_1, \ldots, F_i]$ of $G(\mathcal{C})$ induced by $\{\mathcal{V}(F_1), \ldots, \mathcal{V}(F_i)\}$ is connected and the subgraph $G(F_i) \cap G[F_1, \ldots, F_{i-1}]$ of $G(F_i)$ contains at least one edge. This implies that $G[F_1, \ldots, F_i]$ is constructed from $G[F_1, \ldots, F_{i-1}]$ by adding the graph of the new indecomposable face F_i, which intersects some graph in $\{G(F_1), \ldots, G(F_{i-1})\}$ in at least one edge.

The graph $G[F_1] = G(F_1)$ is indecomposable by assumption. By induction on $i \geqslant 2$, suppose that $G[F_1, \ldots, F_{i-1}]$ is indecomposable. Applying Lemma 6.2.19 to the indecomposable graphs $G_1 := G(F_i)$ and $G_2 := G[F_1, \ldots, F_{i-1}]$ yields that the graph $G := G[F_1, \ldots, F_i]$ is indecomposable. Induction now gives that $G(\mathcal{C}) = G[F_1, \ldots, F_\ell]$ is indecomposable, completing the proof of the theorem. □

An immediate corollary of Theorem 6.2.22 is a result of Shephard (1963, thm. 13).

Corollary 6.2.23 *A 2-simplicial polytope is indecomposable.*

Proof A triangle is indecomposable (Theorem 6.2.16). Furthermore, a 2-simplicial polytope has a connected family of triangles that touches every facet of the polytope. Theorem 6.2.22 now gives the corollary. □

A combinatorial indecomposability criterion of a different nature is provided below.

Theorem 6.2.24 *Let P be a d-polytope with fewer than d decomposable facets. Then P is indecomposable.*

Proof Let $F_1, \ldots, F_r$ be the decomposable facets of P. Then $r < d$. Let ψ denote the natural antiisomorphism from $\mathcal{L}(P)$ to $\mathcal{L}(P^*)$.

By Balinski's theorem (4.1.1), the removal of vertices $\psi(F_1), \ldots, \psi(F_r)$ from P^* does not disconnect $G(P^*)$. All the other vertices of $G(P^*)$, which correspond to the indecomposable facets of P, can be linearly ordered in a sequence L so that any successive pair of vertices defines an edge of P (Proposition C.4.1). The existence of L in P^* means that the indecomposable facets in P form a connected family $\mathcal{C}$ of facets: there is a $(d-1, d-2)$-path between any two members of $\mathcal{C}$. This is much more than what is needed to apply Theorem 6.2.22. The family $\mathcal{C}$ touches every facet of P, because there are only $r < d$ decomposable facets and any facet intersects at least d other facets. Hence, Theorem 6.2.22 gives the conclusion. □

Sufficient conditions for decomposability are rare. We list two useful criteria, including one that generalises (Shephard, 1963, thm. 15). Another criterion is provided in Problem 6.5.14.

Proposition 6.2.25 *Let $G_1 := (V_1, E_1)$ and $G_2 := (V_2, E_2)$ be geometric graphs with $V_1 \cap V_2 = \emptyset$. Suppose that $V_1 = \{\boldsymbol{v}_1, \ldots, \boldsymbol{v}_n\}$ and that there* *is a bijection σ from V_1 to V_2 such that the segments $[\boldsymbol{v}_i, \sigma(\boldsymbol{v}_i)]$ are parallel for all $i \in [1 \dots n]$. Then the union G of G_1, G_2, and the edges $[\boldsymbol{v}_i, \sigma(\boldsymbol{v}_i)]$ is decomposable.*

Proof We construct a decomposing function φ for G such that $\#A(\varphi) \geqslant 2$, which by Problem 6.5.1 would show that G is decomposable.

Since $V_1 \cap V_2 = \varnothing$, we have that $\boldsymbol{v}_1 \neq \sigma(\boldsymbol{v}_1)$. Let $\varphi(\boldsymbol{v}_i) := \boldsymbol{v}_1$ and $\varphi(\sigma(\boldsymbol{v}_i)) := \sigma(\boldsymbol{v}_1)$ for each $i \in [1 \dots n]$. This defines φ for all the vertices of G. It follows that φ is a decomposing function for G:

(i) $\varphi(\boldsymbol{v}_i) - \varphi(\boldsymbol{v}_j) = 0(\boldsymbol{v}_i - \boldsymbol{v}_j) = \boldsymbol{0}$ for each edge $[\boldsymbol{v}_i, \boldsymbol{v}_j]$ in G_1,
(ii) $\varphi(\sigma(\boldsymbol{v}_i)) - \varphi(\sigma(\boldsymbol{v}_j)) = 0(\sigma(\boldsymbol{v}_i) - \sigma(\boldsymbol{v}_j)) = \boldsymbol{0}$ for each edge $[\sigma(\boldsymbol{v}_i), \sigma(\boldsymbol{v}_j)]$ in G_2, and
(iii) $\varphi(\boldsymbol{v}_i) - \varphi(\sigma(\boldsymbol{v}_i)) = \alpha_i(\boldsymbol{v}_i - \sigma(\boldsymbol{v}_i)) = \boldsymbol{v}_1 - \sigma(\boldsymbol{v}_1)$ for each edge $[\boldsymbol{v}_i, \sigma(\boldsymbol{v}_i)]$.

The value α_i in (iii) exists for each $i \in [1 \dots n]$ because the segments $[\boldsymbol{v}_i, \sigma(\boldsymbol{v}_i)]$ and $[\boldsymbol{v}_1, \sigma(\boldsymbol{v}_1)]$ are parallel. This gives that $\alpha_1 = 1$, and so $\{0, 1\} \subseteq A(\varphi)$. Consequently, $\#A(\varphi) \geqslant 2$ and the proposition follows. □

Theorem 6.2.26 *Let P be a polytope with a facet in which every vertex has a unique neighbour outside the facet. Suppose that the polytope has at least two vertices outside the facet. Then P is decomposable.*

Proof Let F be a facet of a polytope P that satisfies the conditions of the theorem. Let y be a linear functional that defines a supporting hyperplane of P at F; we may suppose that $y(\boldsymbol{x}) = 1$ for every $\boldsymbol{x} \in F$ and $y(\boldsymbol{x}) < 1$ for every $\boldsymbol{x} \in P \backslash F$.

Let $\mathcal{V}(F) := \{\boldsymbol{u}_1, \dots, \boldsymbol{u}_\ell\}$. For each vertex $\boldsymbol{u}_i$, let $\boldsymbol{w}_i$ be the unique vertex of $P \backslash F$ that is adjacent to $\boldsymbol{u}_i$. Denote by α the maximum number in $\{y(\boldsymbol{w}_1), \dots, y(\boldsymbol{w}_\ell)\}$. Then $\alpha < 1$. We next define a decomposing function φ for P with $\#A(\varphi) \geqslant 2$. For each $i \in [1 \dots \ell]$, let $\boldsymbol{x}_i$ be the unique point on the edge edge $[\boldsymbol{u}_i, \boldsymbol{w}_i]$ that lies in the hyperplane $y(\boldsymbol{x}) = \alpha$. Then

$$\boldsymbol{x}_i = \beta_i \boldsymbol{u}_i + (1 - \beta_i)\boldsymbol{w}_i \text{ for some } \beta_i \in [0, 1).$$

In the definition of the function φ, every vertex $\boldsymbol{u}_i$ of F is mapped to $\boldsymbol{x}_i$, while every other vertex of P is mapped to itself.

Take an edge $e := [\boldsymbol{u}_i, \boldsymbol{u}_j]$ of F. Then $\varphi(\boldsymbol{u}_i) - \varphi(\boldsymbol{u}_j) = \boldsymbol{x}_i - \boldsymbol{x}_j$. There is a 2-face Q of P such that $F \cap Q = \{e\}$ (Problem 2.15.3). From the existence of Q, it follows that the neighbour $\boldsymbol{w}_i$ of $\boldsymbol{u}_i$ and the neighbour $\boldsymbol{w}_j$ of $\boldsymbol{u}_j$ must be both in Q. As a consequence, the line segments $[\boldsymbol{x}_i, \boldsymbol{x}_j]$ and $[\boldsymbol{u}_i, \boldsymbol{u}_j]$ are parallel, that is,

$$\varphi(\boldsymbol{u}_i) - \varphi(\boldsymbol{u}_j) = \alpha_{ij}(\boldsymbol{u}_i - \boldsymbol{u}_j), \text{ for } \alpha_{ij} \geqslant 0.$$

Take an edge $[\boldsymbol{u}_i, \boldsymbol{w}_i]$ with $\boldsymbol{u}_i \in \mathcal{V}(F)$ and $\boldsymbol{w}_i \in \mathcal{V}(P)\backslash\mathcal{V}(F)$. Then

$$\varphi(\boldsymbol{u}_i) - \varphi(\boldsymbol{w}_i) = \boldsymbol{x}_i - \boldsymbol{w}_i = \beta_i \boldsymbol{u}_i + (1 - \beta_i)\boldsymbol{w}_i - \boldsymbol{w}_i = \beta_i(\boldsymbol{u}_i - \boldsymbol{w}_i).$$

Finally, take an edge $[\boldsymbol{v}_i, \boldsymbol{v}_j]$ with $\boldsymbol{v}_i, \boldsymbol{v}_j \in \mathcal{V}(P)\backslash\mathcal{V}(F)$ (the existence of such an edge is crucial for the proof to work). Then

$$\varphi(\boldsymbol{v}_i) - \varphi(\boldsymbol{v}_j) = \boldsymbol{v}_i - \boldsymbol{v}_j.$$

We have covered all the edges of P and therefore shown that φ is decomposing function for P with $\#A(\varphi) > 2$: $1 \in A(\varphi)$ and $\beta_i \in A(\varphi)$ with $\beta_i \neq 1$ for each $i \in [1 \ldots \ell]$. As a result, φ is not the restriction of a homothety in $\mathbb{R}^d$, and thereby P is decomposable by Theorem 6.2.12. $\square$

In Theorem 6.2.26 the two vertices outside the facet cannot be reduced to just one vertex. In that case, the vertices $\boldsymbol{w}_1 = \cdots = \boldsymbol{w}_\ell = \boldsymbol{x}_1 = \cdots = \boldsymbol{x}_\ell = \boldsymbol{w}$, and the proof would break down as $A(\varphi)$ would equal $\{0\}$. Furthermore, our polytope would be a pyramid, and thus indecomposable (Problem 6.5.4).

6.3 Combinatorial Decomposability

Decomposability is not a combinatorial property. There exists, however, a combinatorial version. A polytope is *combinatorially decomposable* if every polytope combinatorially isomorphic to the polytope is decomposable; if every polytope combinatorially isomorphic to the polytope is indecomposable, we say that the polytope is *combinatorially indecomposable*.

We already met combinatorial decomposability in Corollary 6.2.23; Theorem 6.2.24, and Theorem 6.2.26, where sufficient combinatorial criteria are provided for the indecomposability and decomposability of polytopes.

In this section, we classify the d-polytopes with at most $2d$ vertices into decomposable and indecomposable; in fact into combinatorially decomposable and combinatorially indecomposable. The story is told in the next theorem, whose first proof was supplied in Kallay (1979, thm. 7.1). Our argument follows that of Przesławski and Yost (2016, thm. 9).

Theorem 6.3.1 *Let $d \geqslant 2$. Every d-polytope with at most $2d$ vertices is indecomposable unless it is a simplicial d-prism.*

Proof The case $d = 2$ was established in Theorem 6.2.16.

Let P be a decomposable d-polytope with at most $2d$ vertices. If every facet of P is indecomposable, then there would be a connected family of indecomposable faces that touches every facet of the polytope, for instance a shelling of the polytope, and so the polytope would be indecomposable by Theorem 6.2.22. Hence, let F be a decomposable facet of P. Since P is not a pyramid (Problem 6.5.4), there are at least two vertices outside the facet F. Consequently, F has at most $2(d-1)$ vertices, and by induction on d it is a simplicial $(d-1)$-prism. Denote by T_1 and T_2 the two disjoint $(d-2)$-simplices of F, and label their vertices as $\mathcal{V}(T_1) := \{\boldsymbol{u}_1, \ldots, \boldsymbol{u}_{d-1}\}$ and $\mathcal{V}(T_2) := \{\boldsymbol{w}_1, \ldots, \boldsymbol{w}_{d-1}\}$ so that $[\boldsymbol{u}_i, \boldsymbol{w}_i]$ is an edge of F for every $i \in [1 \ldots d-1]$. Denote by $\boldsymbol{x}$ and $\boldsymbol{y}$ the only two vertices of P outside F. Then $[\boldsymbol{x}, \boldsymbol{y}]$ is an edge of P by convexity (Problem 2.15.2).

Suppose that one of these vertices, say $\boldsymbol{x}$, is adjacent to vertices in both simplices, say $[\boldsymbol{x}, \boldsymbol{u}_i]$ and $[\boldsymbol{x}, \boldsymbol{w}_j]$ are both edges of P.

We may assume that $i \neq j$.

This is clear if $d \geqslant 4$, since in this case the vertex $\boldsymbol{x}$ would need to be adjacent to at least two vertices in one simplex in order to have degree at least d in P. Suppose this assumption is not true for $d = 3$. Then $\boldsymbol{x}$ is adjacent to precisely two vertices in F, say $\boldsymbol{u}_1$ and $\boldsymbol{w}_1$. If $\boldsymbol{y}$ were adjacent to $\boldsymbol{u}_1$, then the triangles $\boldsymbol{x}, \boldsymbol{u}_1, \boldsymbol{w}_1$ and $\boldsymbol{x}, \boldsymbol{u}_1, \boldsymbol{y}$ would form a chain that touches every facet of P, and P would therefore be indecomposable (Theorem 6.2.22); the same applies if $\boldsymbol{y}$ were adjacent to $\boldsymbol{w}_1$. It follows that $\boldsymbol{y}$ is adjacent to only $\boldsymbol{x}$, $\boldsymbol{u}_2$, and $\boldsymbol{w}_2$. This implies that P is a simple 3-polytope with the same graph as a simplicial 3-prism. By Theorem 5.2.4, P must be a simplicial 3-prism. Therefore, the assumption that $i \neq j$ is sound.

Relabel the vertices of P so that $i = 1$ and $j = d-1$; that is $[\boldsymbol{x}, \boldsymbol{u}_1]$ and $[\boldsymbol{x}, \boldsymbol{w}_{d-1}]$ are edges of P. Then $C = \boldsymbol{x}\boldsymbol{u}_1\boldsymbol{u}_2 \ldots \boldsymbol{u}_{d-1}\boldsymbol{w}_{d-1}\boldsymbol{x}$ is an affinely independent $(d+1)$-cycle in $\mathbb{R}^d$; it touches every facet of P since P has only $2d$ vertices and a facet has at least d vertices. Proposition 6.2.17 then ensures that P is indecomposable, a contradiction.

As a consequence, each of $\boldsymbol{x}$ and $\boldsymbol{y}$ is adjacent to vertices in only one simplex, say $\boldsymbol{x}$ is adjacent to every vertex in $\mathcal{V}(T_1)$ and $\boldsymbol{y}$ is adjacent to every vertex in $\mathcal{V}(T_2)$. The polytope P is thereby simple with a graph isomorphic to that of a simplicial d-prism. Theorem 5.2.4 now yields that P is a simplicial d-prism, as claimed. □

The d-polytopes with $2d+1$ vertices have been also classified into decomposable and indecomposable (Pineda-Villavicencio et al., 2018). The proof available is long and its study offers no further insight into the concepts of the chapter.

Theorem 6.3.2 *Let $d \geqslant 2$. Every d-polytope with $2d + 1$ vertices is indecomposable unless it is one of four examples.*

Classifying d-polytopes with $2d + 2$ vertices into decomposable and indecomposable seems not possible at least for $d = 3$; there exists a 3-polytope with both a decomposable realisation and an indecomposable realisation. This is the topic of Section 6.4.

6.4 Ambiguity

If you recall, the f-vector of a 3-polytope is the sequence (f_0, f_1, f_2) of the number of proper faces of the polytope. We construct a 3-polytope that has $f_0 \geqslant 8$ vertices and admits a decomposable realisation and an indecomposable realisation. Such polytopes are called *ambiguously decomposable*. Meyer (1969) gave the first known example of ambiguously decomposable polytopes; it was a 3-polytope that has twelve vertices and is combinatorially isomorphic to the cuboctahedron (Problem 6.5.11). Kallay (1982) provided a simpler example, a 3-polytope with ten vertices. Smilansky (1986a) studied ambiguously decomposable 3-polytopes in detail: he showed that they exist if and only if $f_0 \leqslant f_2 \leqslant 2f_0 - 8$.

We start with a construction of an ambiguously decomposable 3-polytope with eight vertices.

Proposition 6.4.1 *There is an ambiguously decomposable 3-polytope with eight vertices.*

Proof Let P be the realisation given in Fig. 6.4.1(a) of a 3-polytope with eight vertices. It has vertex set $V := \{\boldsymbol{v}_1, \boldsymbol{v}_2, \boldsymbol{v}_3, \ldots, \boldsymbol{v}_8\}$ and it is obtained from the 3-cube by perturbing two vertices of two opposite 2-faces. Another realisation P' is given in Fig. 6.4.1(b). It is obtained from P by perturbing the vertices $\boldsymbol{v}_1$ and $\boldsymbol{v}_2$ while keeping all the other vertices. It has vertex set $V' := \{\boldsymbol{v}'_1, \boldsymbol{v}'_2, \boldsymbol{v}_3, \ldots, \boldsymbol{v}_8\}$. Vertex coordinates of both realisations are provided in Fig. 6.4.1.

The realisation P' has the same face lattice as P. Indeed, the quadrilateral $\boldsymbol{v}_1'\boldsymbol{v}_2'\boldsymbol{v}_5\boldsymbol{v}_6$ lies in the plane $2x - 3y + 5z = 5$, the quadrilateral $\boldsymbol{v}_2'\boldsymbol{v}_3\boldsymbol{v}_6\boldsymbol{v}_7$ lies in the plane $y + z = -1$, the quadrilateral $\boldsymbol{v}_3\boldsymbol{v}_4\boldsymbol{v}_7\boldsymbol{v}_8$ has not changed, and the quadrilateral $\boldsymbol{v}_1'\boldsymbol{v}_4\boldsymbol{v}_5\boldsymbol{v}_8$ lies in the plane $y + z = 1$.

The realisation P is decomposable. The segments $[\boldsymbol{v}_1, \boldsymbol{v}_5]$, $[\boldsymbol{v}_2, \boldsymbol{v}_6]$, $[\boldsymbol{v}_3, \boldsymbol{v}_7]$, and $[\boldsymbol{v}_4, \boldsymbol{v}_8]$ lie in parallel lines, and so $G(P)$ is decomposable by Proposition 6.2.25.

The realisation P' is indecomposable. First, the subgraphs $\boldsymbol{v}_5\boldsymbol{v}_6\boldsymbol{v}_7\boldsymbol{v}_8$ and $\boldsymbol{v}_1'\boldsymbol{v}_2'\boldsymbol{v}_3\boldsymbol{v}_4$ are both indecomposable: each is the union of two triangles that intersect at one edge (Lemma 6.2.19). Second, the lines $\operatorname{aff}\{\boldsymbol{v}_1', \boldsymbol{v}_5\}$ and $\operatorname{aff}\{\boldsymbol{v}_3, \boldsymbol{v}_7\}$ are skew: the dimension of $\operatorname{conv}\{\boldsymbol{v}_1', \boldsymbol{v}_5, \boldsymbol{v}_3, \boldsymbol{v}_7\}$ is three; recall that two lines are skew if they do not lie in a plane, or equivalently, if they do not intersect and they are not parallel. Therefore, the union S of these edges and the two aforementioned subgraphs is a indecomposable subgraph by Problem 6.5.9. Since S touches every facet of P', it follows that P' is indecomposable (Theorem 6.2.13). □

To extend Proposition 6.4.1 to 3-polytopes with more than eight vertices, we require a result of Przesławski and Yost (2016).

Lemma 6.4.2 *Let P be a polytope and let F be a simplex facet of it. Suppose a polytope Q is obtained from P by stacking on F. Then P is indecomposable if and only if Q is indecomposable.*

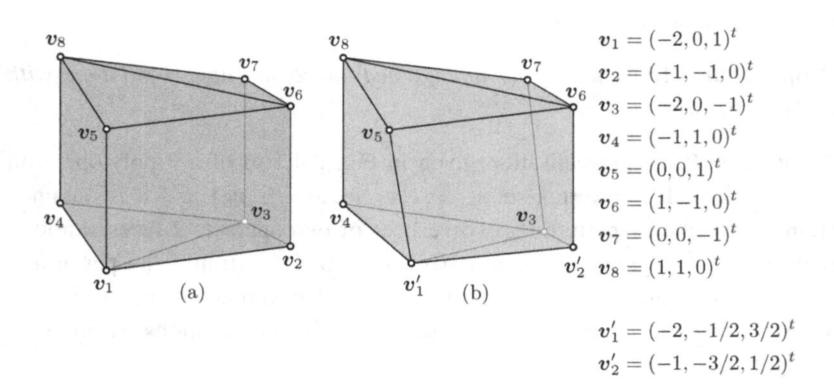

Figure 6.4.1 An ambiguously decomposable polytope. (a) A decomposable realisation of the polytope: the segments $[\boldsymbol{v}_1, \boldsymbol{v}_5]$, $[\boldsymbol{v}_2, \boldsymbol{v}_6]$, $[\boldsymbol{v}_3, \boldsymbol{v}_7]$, and $[\boldsymbol{v}_4, \boldsymbol{v}_8]$ lie in parallel lines. (b) An indecomposable realisation of the polytope: the segments $[\boldsymbol{v}_1', \boldsymbol{v}_5]$ and $[\boldsymbol{v}_3, \boldsymbol{v}_7]$ lie in skew lines.

Proof Let $\boldsymbol{u}$ be the unique vertex of Q that is not in P.

Suppose P is indecomposable. Then $G(P)$ is an indecomposable subgraph of $G(Q)$ that touches every facet of Q. Theorem 6.2.13 now yields the indecomposability of Q.

Suppose P is decomposable. It follows that $\#A(\varphi) > 2$ for some decomposing function φ for P. Since F is indecomposable, $\varphi|_{\mathcal{V}(F)}$ is the restriction of a homothety in $\mathbb{R}^d$, say $\varphi(\boldsymbol{v}) = \alpha\boldsymbol{v} + \boldsymbol{b}$ for some $\alpha \in \mathbb{R}$, some $\boldsymbol{b} \in \mathbb{R}^d$ and every $\boldsymbol{v} \in \mathcal{V}(F)$. Then $\alpha \in A(\varphi)$. Let $\varrho(\boldsymbol{v}) := \varphi(\boldsymbol{v})$ for every $\boldsymbol{v} \in \mathcal{V}(P)$ and $\varrho(\boldsymbol{u}) := \alpha\boldsymbol{u} + \boldsymbol{b}$. Then ϱ is a decomposing function for Q with $A(\varrho) = A(\varphi)$; it holds that $\varrho(\boldsymbol{u}) - \varrho(\boldsymbol{x}) = \alpha(\boldsymbol{u} - \boldsymbol{x})$ for every $\boldsymbol{x} \in \mathcal{V}(F)$. From $\#A(\varrho) > 2$ it follows that Q is decomposable. □

Theorem 6.4.3 *There is an ambiguously decomposable 3-polytope with $f_0 \geqslant 8$ vertices and $f_2 = 2f_0 - 8$ faces.*

Proof Start with the two realisations P and P' given by Proposition 6.4.1, which have eight vertices and eight facets. Stacking on a triangle of P will produce a new decomposable polytope with one more vertex and two more faces (Lemma 6.4.2), including three new triangular faces. Therefore, repeating this stacking operation $f_0 - 9$ more times will yield a decomposable 3-polytope Q with f_0 vertices and $2f_0 - 8$ faces. Mimicking the same $f_0 - 8$ stackings on P' will then produce an indecomposable Q' with f_0 vertices and $2f_0 - 8$ faces. Clearly, the polytopes Q and Q' are combinatorially isomorphic. □

6.5 Problems

6.5.1 Let φ and ϱ be decomposing functions for a geometric graph G in $\mathbb{R}^d$. Prove the following statements.

(i) The function φ is a decomposing function for every subgraph of G.
(ii) The function φ is the restriction of homothety in $\mathbb{R}^d$ if and only if $\#A(\varphi) = 1$.
(iii) The set of decomposing functions φ for G with $A(\varphi) \subset [0, 1]$ forms a convex set. Deduce from this that the set of decompositions for G forms a convex set.

6.5.2 (Deza and Pournin, 2019) Prove that a polytope P has a summand homothetic to a polytope Q if and only if P and $P + Q$ have the same number of vertices.

6.5.3 (Smilansky, 1987) Prove that a polytope with a connected family of triangles that touches every facet is indecomposable.

6.5.4 Prove that a pyramid is indecomposable.

6.5.5 (Shephard, 1963) Prove that a polytope is indecomposable if there exists an edge to which each vertex of the polytope is connected by some chain of indecomposable faces.

6.5.6 (Kallay, 1982) Prove that a polytope P is indecomposable if every two of its vertices belong to some indecomposable subgraph of $G(P)$.

6.5.7 (Shephard, 1963) Let Q and R be two $(d-1)$-polytopes lying in parallel hyperplanes of $\mathbb{R}^d$ $(d \geqslant 3)$ so that no edge of one is parallel to an edge of the other. Prove that $\operatorname{conv}(Q \cup R)$ is indecomposable.

6.5.8 (Kallay, 1982) Let $G_1 := (V_1, E_1), G_2 := (V_2, E_2), G_3 := (V_3, E_3)$ be three indecomposable geometric graphs in $\mathbb{R}^d$ and let $\boldsymbol{v}_{ij} \in V_i \cap V_j$ for $i, j \in [1 \dots 3]$. Prove that $G_1 \cup G_2 \cup G_3$ is indecomposable.

Recall that for geometric graphs $G = (V, E)$ and $G' = (V', E')$, their *union* $G \cup G' := (V \cup V', E \cup E')$ is another geometric graph.

6.5.9 (Kallay, 1982) Let G_1, G_2 be two indecomposable geometric graphs in $\mathbb{R}^d$. Let $\boldsymbol{u}', \boldsymbol{v}'$ be two distinct vertices of G_1 and $\boldsymbol{u}'', \boldsymbol{v}''$ two distinct vertices of G_2. Suppose that G is the union of G_1, G_2 and the two edges $[\boldsymbol{u}', \boldsymbol{u}'']$ and $[\boldsymbol{v}', \boldsymbol{v}'']$. Prove that, if the lines $\operatorname{aff}\{\boldsymbol{u}', \boldsymbol{v}'\}$ and $\operatorname{aff}\{\boldsymbol{u}'', \boldsymbol{v}''\}$ are skew, then G is indecomposable.

6.5.10 (Kallay, 1982) Let P be a d-polytope $(d \geqslant 3)$ and let S be an indecomposable subgraph of $G(P)$ such that the components of $G(P) \setminus S$ are isolated vertices or isolated edges. Prove that P is indecomposable.

6.5.11 (Meyer, 1969) In this problem, we prove that a cuboctahedron is ambiguously decomposable.

(i) Prove that the following realisation of the cuboctahedron is decomposable: $\boldsymbol{v}_1 := (1, -1, 0)^t$, $\boldsymbol{v}_2 := (1, 0, -1)^t$, $\boldsymbol{v}_3 := (1, 0, 0)^t$, $\boldsymbol{v}_4 := (-1, 1, 0)^t$, $\boldsymbol{v}_5 := (0, 1, -1)^t$, $\boldsymbol{v}_6 := (0, 1, 0)^t$, $\boldsymbol{v}_7 := (-1, 0, 1)^t$, $\boldsymbol{v}_8 := (0, -1, 1)^t$, $\boldsymbol{v}_9 := (0, 0, 1)^t$, $\boldsymbol{v}_{10} := (-1, 0, 0)^t$, $\boldsymbol{v}_{11} := (0, -1, 0)^t$, and $\boldsymbol{v}_{12} := (0, 0, -1)^t$.

(ii) Modify the above realisation to obtain one that is indecomposable.

6.5.12 (Decomposability of 3-polytopes, Smilansky, 1987) Suppose that $\frac{1}{2} f_0 + 2 \leqslant f_2 \leqslant 2 f_0 - 4$. Prove the following assertions.

(i) If $f_2 < f_0$ then every 3-polytope with f_0 vertices and f_2 facets is combinatorially decomposable.
(ii) If $f_0 \leqslant f_2 \leqslant 2f_0 - 7$ then there exists a pair of 3-polytopes with f_0 vertices and f_2 facets such that one is combinatorially decomposable and the other is combinatorially indecomposable.
(iii) If $2f_0 - 6 \leqslant f_2 \leqslant 2f_0 - 4$ then every 3-polytope with f_0 vertices and f_2 facets is combinatorially indecomposable.

6.5.13 (Meyer, 1969, sec. 2) Let P be a polytope in $\mathbb{R}^d$ and let

$$\mathcal{S}(P) := \{Q \mid \alpha Q \text{ is a summand of } P \text{ for } \alpha > 0\}.$$

Prove the following.

(i) $\mathcal{S}(P)$ consists entirely of polytopes.
(ii) $\mathcal{S}(P)$ is a cone in the sense that if $Q, R \in \mathcal{S}(P)$ then $\beta Q + \gamma R \in \mathcal{S}(P)$ for $\beta, \gamma \geqslant 0$.
(iii) $Q \in \mathcal{S}(P)$ if and only if Q is a summand of P.
(iv) P is decomposable if and only if $\dim \mathcal{S}(P) > d + 1$.

6.5.14 (Smilansky, 1987) Let P be a d-polytope whose dual is not a free join. Suppose there exists a set S of simple vertices for which the facets of P can be partitioned into disjoint sets $\mathcal{F}_1$ and $\mathcal{F}_2$ and any vertex that belongs to both a facet in $\mathcal{F}_1$ and a facet in $\mathcal{F}_2$ is in S. Then P is decomposable.

6.5.15 (Ambiguous 3-polytopes, Smilansky, 1986a) Consider the following construction. Start with a prism over an n'-gon with $n' \geqslant 4$. Then triangulate the top n'-gon and break the bottom n'-gon into an $(n' - 1)$-gon and a triangle. The resulting 3-polytope Q has $f_0 = 2n'$ vertices and $f_2 = f_0$ faces. To obtain a 3-polytope R with $f_0 = 2n' + 1$ vertices and $f_2 = f_0$, add a new vertex to P so that one quadrilateral face becomes a pentagon. See Fig. 6.5.1. Prove the following assertions.

(i) The polytopes Q and R are decomposable.
(ii) There are indecomposable polytopes Q' and R' combinatorially isomorphic to Q and R, respectively.
(iii) Ambiguously decomposable 3-polytopes exists if and only if $f_0 \leqslant f_2 \leqslant 2f_0 - 8$.

6.5.16 (Decomposable facets, Smilansky, 1986b) Answer the following.

(i) Prove that 3-polytopes without triangular faces are decomposable.

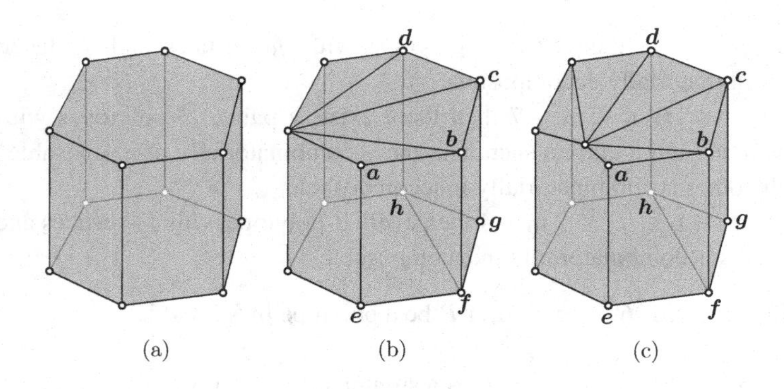

Figure 6.5.1 Auxiliary figure for Problem 6.5.15.

(ii) Construct an indecomposable d-polytope whose facets are all decomposable.

6.5.17 (Diameter of a sum, Deza and Pournin, 2019) Prove that for any two polytopes Q and R the following holds.

(i) $\operatorname{diam} G(Q+R) \geqslant \max\{\operatorname{diam} G(Q), \operatorname{diam} G(Q)\}$.
(ii) $\operatorname{diam} G(Q+R) \leqslant \min\{(\operatorname{diam} G(Q)+1) f_0(Q), (\operatorname{diam} G(R)+1) f_0(R)\}$.

6.5.18 (Decomposability is a projective invariant, Kallay, 1984) Let P be a polytope in $\mathbb{R}^d$ and let ζ be a projective transformation admissible for P. Prove the following.

(i) The sets $\mathcal{S}(P)$ and $\mathcal{S}(\zeta(P))$ of summands of P and $\zeta(P)$, respectively, have the same dimension. (Refer to Problem 6.5.13 for the definition of $\mathcal{S}$.)
(ii) P is decomposable if and only if $\zeta(P)$ is.

6.5.19* (Smilansky, 1987) Let $d \geqslant 4$ and let $\phi_k(d, f)$ denote the maximum number of k-faces in a d-polytope with f facets. The upper bound theorem (McMullen, 1970) states that, for any d-polytope P, it follows that

$$f_0 \leqslant \phi_0(d, f) = \binom{f - \lfloor (d+1)/2 \rfloor}{f-d} + \binom{f - \lfloor (d+2)/2 \rfloor}{f-d}.$$

Equality occurs if and only if P is the dual of a neighbourly polytope.

One would expect that d-polytopes with f facets and number of vertices 'close' to $\phi_0(d, f)$ behave like simple polytopes. Smilansky (1987, sec. 6) made this notion of closeness precise by conjecturing that if

$$f_0 > \phi_0(d-1, f-1) + 1,$$

then a d-polytope is decomposable. It has been verified for $d = 3$ (Problem 6.5.12).

6.5.20* (Smilansky, 1986b) Let $d \geqslant 3$ and let $\zeta(d)$ denote the largest integer k such that any indecomposable d-polytope must have a proper indecomposable k-face.

It is true that $\zeta(3) = 2$ and $1 \leqslant \zeta(d) \leqslant d-2$ (Problem 6.5.16). Find $\zeta(d)$ for every $d \geqslant 4$. In particular,

(i) Are polytopes without triangular faces decomposable?
(ii) Are cubical polytopes decomposable?
(iii) Is there an indecomposable d-polytope where all of its faces of dimensions $\zeta(d)+1, \zeta(d)+2, \ldots, d-1$ are decomposable?
(iv) Is $\zeta(d) = 2$ for every $d \geqslant 3$?

6.5.21* (Ambiguous d-polytopes) Let $a(d)$ denote the minimum number of vertices of an ambiguously decomposable d-polytope. Find $a(d)$. It is true that $a(3) = 8$. Is it the case that $a(d) = 2d + 2$?

6.5.22* (Decomposability of polytopes) Extend Problem 6.5.12 to d-polytopes. Concretely, for a given polytope establish relations between f_0 and f_{d-1} that ensure the decomposability of the polytope. Is it true that a d-polytope with $f_{d-1} < f_0$ is decomposable?

6.6 Postscript

Many results in this chapter extend to general convex sets, as is demonstrated by Schneider (2014, ch. 3).

The presentation in Section 6.1 is in some sense classical and in parts goes along the lines of Fukuda (2004) and Ziegler (1995, secs. 7.1–7.2). Whereas decompositions of polytopes (Sections 6.2 to 6.4) is studied by Grünbaum (2003, ch. 15), our treatment is nonclassical, oriented towards geometric graphs. This is due to Kallay (1982), who reduced the decomposability of a realisation of polytope to that of its geometric graph and, in this way, introduced the decomposability of geometric graphs. Decomposability of geometric graphs has been championed by Yost (2007); Przesławski and Yost (2016, 2008).

Theorem 6.2.9 is essentially Shephard (1963, thm. 4), recast in the language of decomposing functions. Theorem 6.2.13 is from Przesławski and Yost (2008); it combined ideas from Kallay (1982) and (McMullen, 1987, thm. 2). Lemma 6.2.19 is a slight generalisation of (Kallay, 1982, thm. 8). Theorem 6.2.24 is from Pineda-Villavicencio et al. (2018); it seems to be a basic result on decomposability, but we could not find an earlier reference. Theorem 6.2.26 generalises Shephard (1963, thm. 15), where it was assumed that the facet is a simple polytope; this generalisation appeared in Przesławski and Yost (2016, prop. 5).

Proposition 6.2.8 showed that one summand can always be taken to be combinatorially isomorphic to the the sum. It is also true that a summand can be always taken to have fewer vertices than the sum: it arises from a decomposing function φ with $0 \in A(\varphi)$. Obtaining summands with prescribed properties is an interesting line of research, but we are not aware of any work on this direction.

We did not cover algorithmic aspects of Minkowski sums. One such instance is the *Minkowski sum problem*, which is to compute the vertex set of a sum $P = Q + R$ in $\mathbb{R}^d$ given the geometric graphs $G(Q)$ and $G(R)$ of the summands. Fukuda (2004) presented an algorithm polynomial in the number of vertices of P for this problem.

Smilansky (1987) studied the decomposability of a polytope via affine dependences of the vertices of the dual polytope. This approach produced results such as Problem 6.5.14 and new proofs of results from Meyer (1969), McMullen (1987), and Kallay (1984).

Karavelas and Tzanaki (2016b) and, independently, Adiprasito and Sanyal (2016) established tight expressions for the maximum number of k-faces of the Minkowski sum of two d-polytopes as a function in the number of vertices of the summands. This problem is related to the worst-case *combinatorial complexity of Minkowski sums*, their total number of k-faces as a function of the number of k-faces of the summands. Adiprasito and Sanyal (2016) settled the general case: the case of r summands where the summands may have different dimensions. A geometric proof of the case of r summands for $r < d$ was later given by Karavelas and Tzanaki (2016a).

7
Diameter

The *diameter* of a graph G, denoted diam G, is the maximum distance between any two vertices in the graph. The *diameter* of a polyhedron is defined as the diameter of its graph. While the chapter primarily focusses on polytopes, polyhedra also feature in it.

There is a notable connection between diameters of polyhedra and linear programming, and this is partially materialised through the *Hirsch conjectures*, which relate the diameter of a polyhedron to its dimension and number of facets. We first show that the unbounded and monotonic versions of these conjectures are false (Section 7.2). Early on, Klee and Walkup (1967) realised that problems on the diameter of polyhedra could be reduced to problems on the diameter of simple polyhedra; this and other reductions are the focus of Section 7.3. We also present the counterexample of Santos (2012) for the bounded Hirsch conjecture.

Subsequently, we examine lower and upper bounds for the diameter of general polytopes as well as specific polytopes. The final section is devoted to generalisations of polyhedra where diameters may be easier to compute or estimate.

7.1 Linear Programs

Recall from Chapter 2 that the basic problem from linear programming is that of maximising a linear objective function φ subject to a finite set of linear inequalities $A\boldsymbol{x} \leqslant \boldsymbol{b}$ where $A \in \mathbb{R}^{n\times d}$, $\boldsymbol{x} \in \mathbb{R}^d$, and $\boldsymbol{b} \in \mathbb{R}^n$:

$$\text{maximise } \varphi(\boldsymbol{x}) \text{ subject to } A\boldsymbol{x} \leqslant \boldsymbol{b}. \tag{7.1.1}$$

The H-polyhedron defined by the constraints $A\boldsymbol{x} \leqslant \boldsymbol{b}$ is the *feasible region* of the linear programming problem (or linear program for short). Every linear program can be solved in time polynomial in the input size by the ellipsoid method of Khachiyan (1979) or the interior-point method of Karmarkar (1984); the input size is measured as the number of bits required to specify the problem. Despite no proof that the simplex method of Dantzig (1963) can run in polynomial time, it remains one of the most popular solvers of linear programs. The popularity of the simplex method is partly due to the fact that, for problems with n equality constraints, the method typically reaches an optimal solution in at most cn pivot steps, for a small constant c (Todd, 2002). While in practice the ellipsoid method cannot compete with the simplex method, there are implementations of the interior-point method that can.

The maximum of a linear functional φ over a polyhedron P is attained at a vertex of P (Theorem 3.2.1), a fact exploited by the simplex method. The method starts by finding a vertex of P. Then, at each step it moves from a vertex to one of its neighbours with larger φ-value, according to a predefined rule, called a *pivot rule*. The simplex method ends when either it reaches a vertex whose neighbours all have smaller φ-values, in which case this vertex maximises φ, or it determines that the linear program is *infeasible* (the feasible region is empty) or *unbounded* (the linear functional attains arbitrarily large values over P). For almost all major pivot rules, researchers, starting with Klee and Minty (1972), have constructed linear programs for which the method requires an exponential or subexponential number of steps; see also Kalai, 1997; Friedmann, 2011. Despite these results, researchers are still hopeful that there exists a pivot rule ensuring that the simplex method operates in polynomial time.

To understand the performance of the simplex method, we would like to prove that there always exists a 'short' path between any two vertices in the graph G of a d-polyhedron P. By 'short' we mean that the diameter of G is a polynomial function in d and the number of constraints defining P; this is what the polynomial Hirsch conjecture posits. Observe that, even if this polynomial function exists, the simplex method may still not find such a shortest path.

Conjecture 7.1.2 (Polynomial Hirsch conjecture) *The graph of every d-dimensional polyhedron with n facets has a diameter at most polynomial in d and n.*

Hirsch's original conjecture asked whether this diameter was at most $n - d$. This was disproved by Klee and Walkup (1967) for unbounded polyhedra and by Santos (2012) for polytopes. Researchers have also tried to settle a linear version of Conjecture 7.1.2.

Conjecture 7.1.3 (Linear Hirsch conjecture) *The graph of every d-dimensional polyhedron with n facets has a diameter at most linear in d and n.*

We will refer to Hirsch's conjecture, the linear Hirsch conjecture, and the polynomial Hirsch conjecture as the *Hirsch conjectures*. In the study of these conjectures and thus in this chapter, we will consider only pointed polyhedra (Corollary 2.3.4), those with a vertex, and so we won't state this adjective in every instance.

7.2 The Unbounded and Monotonic Hirsch Conjectures

Between 1967 and 1980, several versions of Hirsch's conjecture were disproved, including the versions for unbounded polyhedra (Klee and Walkup, 1967) and for monotone paths on polytopes (Todd, 1980). In these two instances, the relevant counterexamples rely on the *Klee–Walkup polytope*, the unique 4-polytope that has nine facets and diameter five (Firsching, 2020). Figure 7.2.1(c) depicts an irredundant H-description of the polytope, while Fig. 7.2.1(a) shows its graph and its vertex-facet incidence relation. We present some further properties.

Proposition 7.2.1 *Let P be the Klee–Walkup polytope. Following the notation provided in Fig. 7.2.1(a), between the vertices $abcd$ and $efgh$, there is a shortest path of length five in P that is disjoint from the facet w of P.*

Proof Fig. 7.2.1(a) highlighted a path of length five between these vertices. It it not difficult to check that every path from the vertex $abcd$ to the vertex $efgh$ that is disjoint from the facet w has length at least five, and so does every path that uses the facet w: one step is required to enter the facet and four steps to introduce the facets e, f, g, h. □

The Klee–Walkup polytope facilitates a counterexample for the unbounded Hirsch conjecture. Recall that the graph of a polyhedron is formed by its vertices and edges, and so its unbounded 1-faces are not part of the graph.

Theorem 7.2.2 *There is an unbounded 4-polyhedron with eight facets and diameter five.*

Proof Let P be the Klee–Walkup polytope and let F be the facet w of P (Fig. 7.2.1). We use a projective transformation that sends the facet w of P to infinity. As in the scheme of Section 2.5, we pass to $\mathbb{P}(\mathbb{R}^5)$ and embed P in a nonlinear hyperplane H^e. Let H^e_∞ be a linear hyperplane in $\mathbb{R}^5$ parallel to H^e, and within H^e consider an affine 3-space K supporting P at the facet F. We also need both a linear hyperplane H^p_∞ in $\mathbb{R}^5$ intersecting H^e at K and a nonlinear hyperplane H^p parallel to H^p_∞. In this setup, the intersection

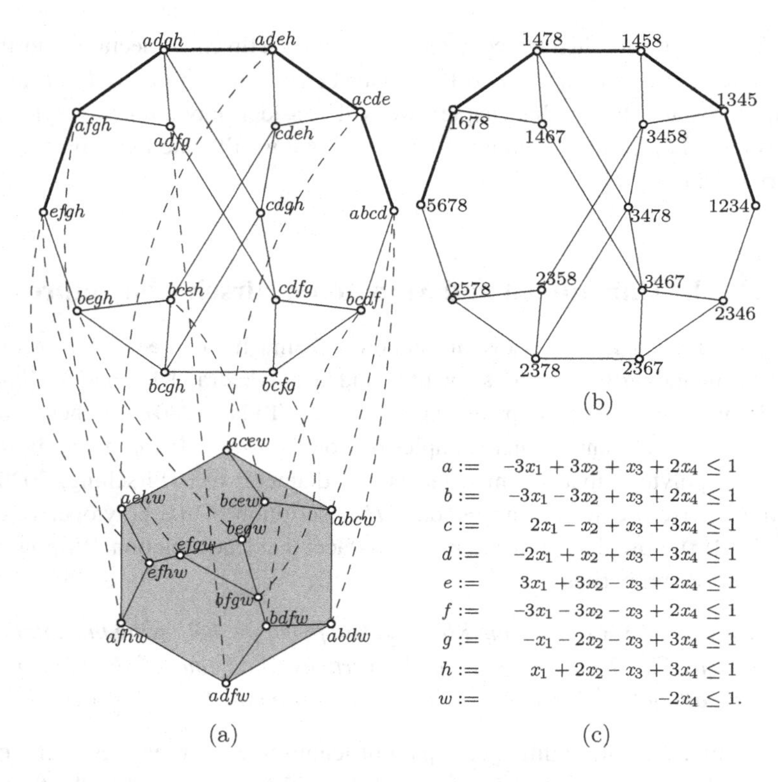

Figure 7.2.1 Klee–Walkup polyhedra. (a) Graph and vertex-facet relation of the Klee–Walkup polytope; a vertex is labelled by the facets containing it. A shortest path between the vertices $abcd$ and $efgh$ is highlighted in bold. (b) Graph and vertex-facet relation of the Klee–Walkup polyhedron; a vertex is labelled by the facets containing it, which are described in (7.2.3). A shortest path between the vertices 1234 and 5678 is highlighted in bold. (c) Irredundant H-description of the Klee–Walkup polytope.

of P and H_∞^p is precisely F. Let ζ be the projective transformation induced by the identity linear map. See Fig. 7.2.2(a) for a sketch of this scheme. The transformation ζ maps P to a polyhedron Q on H^p where the facet F has been sent to infinity. The polytope Q has eight facets, and its graph, depicted in Fig. 7.2.1(b), is a subgraph of the graph of the Klee–Walkup polytope. Moreover, $G(Q)$ has diameter five by Proposition 7.2.1. $\square$

We call the polyhedron of Theorem 7.2.2 the *Klee–Walkup polyhedron*. Klee and Walkup (1967) gave an irredundant H-description of the polyhedron:

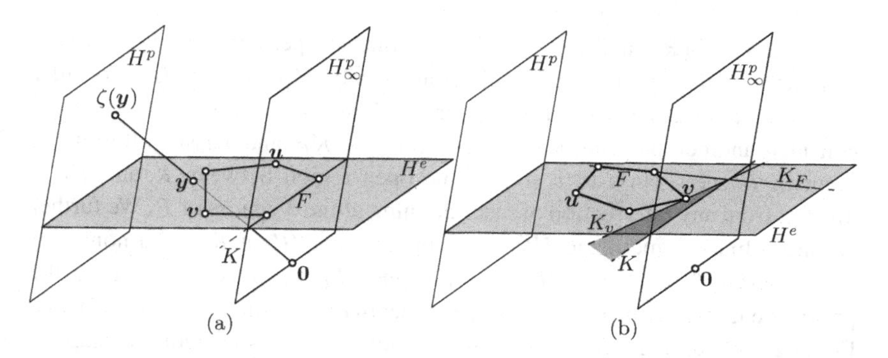

Figure 7.2.2 Projective transformations on the Klee–Walkup polytope P. (a) A scheme for the projective transformation in the proof of Theorem 7.2.2. (b) A scheme for the projective transformation in the proof of Theorem 7.2.4; the open region between the affine spaces K and K_v is highlighted in dark grey.

$$\begin{aligned} &1 := 6x_1 + 3x_2 - x_4 \leqslant 1, && 2 := 3x_1 + 6x_2 - x_3 \leqslant 1,\\ &3 := 35x_1 + 45x_2 - 6x_3 - 3x_4 \leqslant 1, && 4 := 45x_1 + 35x_2 - 3x_3 - 6x_4 \leqslant 1,\\ &5 := -x_1 \leqslant 1, && 6 := -x_2 \leqslant 1,\\ &7 := -x_3 \leqslant 1, && 8 := -x_4 \leqslant 1. \end{aligned} \tag{7.2.3}$$

Figure 7.2.1(b) presents the graph and vertex-facet relation of the Klee–Walkup polyhedron.

In an ideal scenario, the simplex method finds the maximum of a linear functional φ over a polytope by moving to a vertex with larger φ-value at each step, until it reaches a vertex that maximises φ. This scenario motivates the *monotone Hirsch conjecture*: for every d-polytope P in $\mathbb{R}^d$ with at most n facets, every vertex $\boldsymbol{u}$ of P, and every linear functional φ in $\mathbb{R}^d$, there is an φ-increasing path of length at most $n - d$ from $\boldsymbol{u}$ to a vertex of P maximising φ. We present a counterexample. Recall that the distance between any two vertices $\boldsymbol{x}$ and $\boldsymbol{y}$ in a polytope P is defined as the length of a shortest path between the vertices in the graph of the polytope, denoted by $\operatorname{dist}_P(\boldsymbol{x}, \boldsymbol{y})$. Similarly, the distance between any two vertices in a graph is the length of a shortest path connecting them within the graph. (Section C.2).

Theorem 7.2.4 (Todd, 1980) *There is a 4-polytope P in $\mathbb{R}^4$ with eight facets, a vertex $\boldsymbol{u}$ of P, and a linear functional φ in $\mathbb{R}^4$ such that every φ-increasing path from $\boldsymbol{u}$ to a vertex of P maximising φ has length at least five.*

Proof Refer to Fig. 7.2.1. Let P be the Klee–Walkup polytope, F the facet w of P, and let $\boldsymbol{u}$ and $\boldsymbol{v}$ be the vertices $abcd$ and $efgh$ of P, respectively. These vertices are not contained in F and are at distance five (Proposition 7.2.1).

We pass to $\mathbb{P}(\mathbb{R}^5)$ and embed P in a nonlinear hyperplane H^e. Projectively complete H^e by adding a hyperplane at infinity H^e_∞. Within H^e, consider first affine 3-spaces K_v and K_F supporting P at $\boldsymbol{v}$ and F, respectively. Then consider another 3-space K containing $K_v \cap K_F$ and passing through a point in H_e very close to $\boldsymbol{v}$ so that the open region between K and K_v is disjoint from any intersection of facet-defining affine 3-spaces of P. We further require a linear hyperplane H^p_∞ in $\mathbb{R}^5$ intersecting H^e at K and a nonlinear hyperplane H^p parallel to H^p_∞; here, P and H^p_∞ are disjoint. Let ζ be the projective transformation induced by the identity linear map (Section 2.5). See Fig. 7.2.2(b) for a sketch of this configuration. The transformation ζ maps P to a polytope P' on H^p combinatorially isomorphic to P and where the affine 3-spaces $K'_v := \zeta(K_v)$ and $K'_F := \zeta(K_F)$ are parallel. Further, let $\boldsymbol{u}' := \zeta(\boldsymbol{u})$, $\boldsymbol{v}' := \zeta(\boldsymbol{v})$, and $F' := \zeta(F)$. Then the vertex $\boldsymbol{u}'$ lies in H^p between K'_v and K'_F.

We continue the proof on P', assuming that P' is in $\mathbb{R}^d$. We now define a new d-polyhedron Q from P' by removing the inequality associated with the facet F'. The polyhedron Q is bounded because of the relative position of K_v and K when ζ was applied, has new vertices not present in P' (all lying on the side of K'_F opposite to the one containing $\boldsymbol{u}'$ and $\boldsymbol{v}'$), and has eight facets (four containing $\boldsymbol{u}'$ and four containing $\boldsymbol{v}'$). We choose a linear functional φ that is constant on K'_v and K'_F and maximised at $\boldsymbol{v}'$. Then no φ-increasing $\boldsymbol{u}' - \boldsymbol{v}'$ path in Q crosses K'_F, and so every such path in Q is a $\boldsymbol{u}' - \boldsymbol{v}'$ path in P', and hence it has length at least five in P'. □

Repeated Cartesian products of Klee–Walkup polyhedra (Section 2.6 and Problem 7.8.1) yield, for each $r \geqslant 2$, a $4r$-polyhedron Q_r with $8r$ facets and diameter $5r$. Klee and Walkup (1967) and Todd (1980) used this polyhedra Q_r to produce further counterexamples.

Theorem 7.2.5 (Klee and Walkup, 1967; Todd, 1980) *Let $n > d \geqslant 2$ and $r := \min\{\lfloor d/4 \rfloor, \lfloor (n-d)/4 \rfloor\}$.*

(i) *There are d-polyhedra with n facets and diameter $n - d + r$.*
(ii) *There is a d-polytope P in $\mathbb{R}^d$ with n facets, a vertex $\boldsymbol{u}$ of P, and a linear functional φ in $\mathbb{R}^d$ such that every φ-increasing from $\boldsymbol{u}$ to some vertex of P maximising φ has length at least $n - d + r$.*

The proofs of Theorems 7.2.2 and 7.2.5 highlight an important phenomenon in the study of pointed polyhedra: their combinatorics can be reduced to that of polytopes with a distinguished face at infinity; see, for instance, Problem 7.8.2 and Joswig et al. (2001).

In view of Theorems 7.2.2 and 7.2.5 we will henceforth focus on the bounded version of Hirsch's conjecture.

7.3 Reductions

In many ways, simple polytopes behave better than general polytopes; they can, for instance, be reconstructed from their graphs (Theorem 5.2.4). That's why it may be surprising that for disproving Hirsch's conjecture it suffices to consider them. It may also be surprising that, among simple d-polytopes, we need to consider only those with $2d$ facets. These reduction theorems, due to Klee and Walkup (1967), are the focus of this section.

Let us first reduce the Hirsch conjectures to polytopes with exactly n facets.

Proposition 7.3.1 *For every d-polytope P with at most n facets, there is a d-polytope P' with exactly n facets whose diameter is at least that of P. Furthermore, if P is simple then P' can be assumed to be simple.*

Proof Suppose that the number of facets in P is $n - r$ for some $r > 0$. Let $\boldsymbol{u}$ and $\boldsymbol{v}$ be any two vertices of P. We construct another d-polytope P' with exactly n facets by performing r truncations of vertices other than $\boldsymbol{u}$ and $\boldsymbol{v}$; see Section 2.6. After each truncation, a vertex $\boldsymbol{w}$ is replaced by the graph of a new facet. It is plain that $\operatorname{dist}_{P'}(\boldsymbol{u}, \boldsymbol{v}) \geqslant \operatorname{dist}_P(\boldsymbol{u}, \boldsymbol{v})$, and so $\operatorname{diam} P' \geqslant \operatorname{diam} P$.

If P is a simple polytope, then each truncation adds only simple vertices, implying that P' is simple as well. □

The second reduction is about simple polytopes. It is often convenient to consider the dual version of Hirsch's conjecture. For this, we define the *dual diameter* of a polytope P as the diameter of its dual graph. The dual diameter of P coincides with the maximum length of a shortest facet-ridge path in P; the *length* of a facet-ridge path $F_0 F_1 \ldots F_\ell$ is the number ℓ of its facets minus one.

Proposition 7.3.2 (Reduction to simple polytopes) *For every d-polytope P, there is a simple d-polytope whose diameter is at least that of P.*

Proof We prove the proposition in the dual setting. The proof relies on the operation of pushing a vertex into the interior of a polytope (Section 2.10).

Let Q be the dual polytope of P. Suppose that a d-polytope Q' is obtained from Q by pushing a vertex $\boldsymbol{v} \in Q$ to a new vertex $\boldsymbol{v}' \in \operatorname{int} Q$; if you recall, this means that $\mathcal{V}(Q') = (\mathcal{V}(Q) \backslash \{\boldsymbol{v}\}) \cup \{\boldsymbol{v}'\}$ and no hyperplane spanned by the vertices of Q intersects the segment $(\boldsymbol{v}, \boldsymbol{v}']$. According to Theorem 2.10.7, there exists a map φ, from the facets of Q' to the facets of Q, such that every two adjacent facets F_1 and F_2 of Q' are mapped to either the same facet of Q or two adjacent facets $\varphi(F_1)$ and $\varphi(F_2)$ of Q. This immediately implies that the dual diameter of Q' is at least that of Q. Thus, if P' is the dual polytope of Q', then $\operatorname{diam} P \leqslant \operatorname{diam} P'$.

As a consequence, repeatedly pushing the vertices of Q transforms it into a simplicial polytope whose dual diameter is at least that of Q (Corollary 2.10.6). □

Wedges play an important role in the third and fourth reductions; see Section 2.6. It is helpful to think of a wedge W of a polytope P at a facet F of P as a prism Q over P where the facet prism(F) of Q has collapsed into F.

Lemma 7.3.3 (Klee and Walkup, 1967, prop. 1.4) *Let $P \subseteq \mathbb{R}^{d+1}$ be a d-polytope, F a facet of P, and W the wedge of P at F. Then the following hold.*

(i) *Each vertex of P may be considered as a vertex of W. Furthermore, each vertex of F has a unique image in W, while each vertex $\boldsymbol{v} \in P \backslash F$ has a unique neighbour $\boldsymbol{v}'$ in the other base P' of W, in which case $[\boldsymbol{v}, \boldsymbol{v}']$ is a vertical edge of W.*
(ii) *If P is a simple polytope, then so is W.*
(iii) *Each path L in W has a natural projection in P, obtained by projecting each vertex in L onto P.*
(iv) $\operatorname{diam} W \geqslant \operatorname{diam} P$.

Dualising Lemma 7.3.3, we immediately get a corollary on dual wedges.

Corollary 7.3.4 *Let $P \subseteq \mathbb{R}^{d+1}$ be a d-polytope, $\boldsymbol{v}$ a vertex of P, and Q the dual wedge of P at $\boldsymbol{v}$. Then the dual diameter of Q is at least the dual diameter of P.*

The version of Hirsch's conjecture where the number of facets is twice the dimension is called the *d-step conjecture*. It turns out that Hirsch's conjecture is equivalent to the d-step conjecture, although not necessarily dimensionwise. For $n > d \geqslant 2$ let

$$\operatorname{diam}(n,d) := \max\{\operatorname{diam} P : P \text{ is a } d\text{-polytope with at most } n \text{ facets}\}. \tag{7.3.5}$$

The function $\operatorname{diam}(n,d)$ is well defined. It is attained at a simple d-polytope P with exactly n facets (Propositions 7.3.1 and 7.3.2). Each vertex of P is incident with precisely d facets and so it could be labelled with these facets. As a result, P has trivially at most $f_0(P) = \binom{n}{d}$ vertices, and the diameter of P is trivially bounded by a function in n and d. Upper bounds for $\operatorname{diam}(n,d)$ are discussed in Section 7.4 and better upper bounds for $f_0(P)$ are discussed in Section 8.4.

Proposition 7.3.6 (Reduction to $2d$ facets) *For $n > d \geqslant 2$, $\operatorname{diam}(n,d) \leqslant \operatorname{diam}(2(n-d), n-d)$, with equality if $n \leqslant 2d$.*

Proof Let P be a d-polytope with n facets and two vertices that realise $\operatorname{diam}(n,d)$. We first prove the statement for $n \leqslant 2d$ by induction on $r := 2d - n$. The statement holds for $r = 0$, and so suppose it holds for $r - 1 \geqslant 0$. Let $\boldsymbol{u}$ and $\boldsymbol{v}$ be two vertices of P whose distance in P realises $\operatorname{diam}(n,d)$. By Proposition 7.3.2, we may assume that P is simple.

The pair of $\boldsymbol{u}$ and $\boldsymbol{v}$ lies in a common facet F since each vertex is incident with precisely d facets (Section 2.8). Furthermore, each $(d-2)$-face of F is the intersection of F with another facet of P, which ensures that $f_{d-2}(F) \leqslant n-1$. This inequality in conjunction with Proposition 7.3.1 gives that

$$\operatorname{dist}_F(\boldsymbol{u}, \boldsymbol{v}) \leqslant \operatorname{diam}(n-1, d-1). \tag{7.3.6.1}$$

We now combine (7.3.6.1) with $\operatorname{dist}_P(\boldsymbol{u}, \boldsymbol{v}) \leqslant \operatorname{dist}_F(\boldsymbol{u}, \boldsymbol{v})$ to obtain

$$\operatorname{diam}(n,d) \leqslant \operatorname{diam}(n-1, d-1). \tag{7.3.6.2}$$

Because $n - 1 \leqslant 2(d-1)$ and $2(d-1) - (n-1) < 2d - n$, the induction hypothesis guarantees that

$$\begin{aligned} \operatorname{diam}(n-1, d-1) &\leqslant \operatorname{diam}(2(n-1-(d-1)), (n-1-(d-1))) \\ &= \operatorname{diam}(2(n-d), (n-d)), \end{aligned}$$

which together with (7.3.6.2) yields that $\operatorname{diam}(n,d) \leqslant \operatorname{diam}(2(n-d), n-d)$.

We prove the statement for $n \geqslant 2d$ by induction on $r := n - 2d$. If $r = 0$ then the statement holds, and so suppose it holds for $r - 1 \geqslant 0$. We pick a facet F of P and construct the wedge W of P at F. By Proposition 2.6.9, W is a $(d+1)$-polytope with $n+1$ facets, and by Lemma 7.3.3 $\operatorname{diam} W \geqslant \operatorname{diam} P$. Because $n + 1 \geqslant 2(d+1)$ and $n + 1 - 2(d+1) < n - 2d$, the induction hypothesis guarantees that

$$\begin{aligned} \operatorname{diam}(n+1, d+1) &\leqslant \operatorname{diam}(2(n+1-(d+1)), (n+1-(d+1))) \\ &= \operatorname{diam}(2(n-d), (n-d)). \end{aligned}$$

Hence $\operatorname{diam}(n,d) \leqslant \operatorname{diam}(n+1, d+1) \leqslant \operatorname{diam}(2(n-d), n-d)$.

If $n = 2d$ then we have equality. To get equality when $n \leqslant 2d - 1$ we perform a wedge over a $(d-1)$-polytope that has $n - 1$ facets, satisfies

$n - 1 \leqslant 2(d-1)$, and realises $\operatorname{diam}(n-1, d-1)$. This reverses the inequality (7.3.6.2). □

Thanks to Proposition 7.3.1, 7.3.2, and 7.3.6 (and their proofs), to tackle the Hirsch conjectures we may focus only on simple d-polytopes with exactly $2d$ facets. Moreover, we have the following.

Remark 7.3.7 For $n > d \geqslant 2$,

(i) $\operatorname{diam}(n,d) = \operatorname{diam}(n-1, d-1)$ if $n < 2d$;
(ii) the value for $\operatorname{diam}(2d, d)$ is realised as the distance between two vertices $\boldsymbol{u}$ and $\boldsymbol{v}$ in a simple d-polytope with $2d$ facets such that $\boldsymbol{u}$ and $\boldsymbol{v}$ don't lie in a common facet.

The final reduction is specifically for Hirsch's conjecture. Let P be a simple d-polytope and let $\boldsymbol{u}$ and $\boldsymbol{v}$ be two vertices of P. A $\boldsymbol{u}-\boldsymbol{v}$ path is *nonrevisiting* if it does not return to a facet previously visited; otherwise the path is *revisiting*. This definition is equivalent to a remark, which is often more workable.

Remark 7.3.8 Let $L := \boldsymbol{v}_0 \ldots \boldsymbol{v}_\ell$ be a path in a simple polytope. The path L is nonrevisiting if and only if, for $i < j < k$, each facet that includes $\boldsymbol{v}_i$ and $\boldsymbol{v}_k$ also includes $\boldsymbol{v}_j$.

In dealing with Hirsch's original conjecture, one may choose to deal with only nonrevisiting paths.

Conjecture 7.3.9 (Nonrevisiting conjecture) *For every two vertices of a simple polytope, there is a nonrevisiting path between them.*

The nonrevisiting conjecture is equivalent to Hirsch's conjecture, though not necessarily dimension for dimension. The next lemma shows one direction.

Lemma 7.3.10 *The length of a nonrevisiting path in a d-polytope with n facets is at most $n - d$.*

Proof Let $L := \boldsymbol{v}_0 \ldots \boldsymbol{v}_\ell$ be a nonrevisiting path in a d-polytope. Every vertex $\boldsymbol{v}_i$ is contained in at least d facets, and once the path has left one of these facets, it cannot return to it. Suppose that the vertex $\boldsymbol{v}_0$ is contained in the facets $F_1, \ldots, F_d$. Furthermore, for $i \in [1 \ldots \ell]$, there is a facet F_{d+i} that contains $\boldsymbol{v}_i$ but not $\boldsymbol{v}_{i-1}$. Since the path L is nonrevisiting, the facets $F_1, \ldots, F_d, F_{d+1}, \ldots, F_{d+\ell}$ are pairwise distinct. Hence $n \geqslant \ell + d$. □

A variation of the proof of Lemma 7.3.10 shows that, under some conditions, some paths are always nonrevisiting.

Lemma 7.3.11 *Let P be a simple d-polytope with $2d$ facets. If $\boldsymbol{u}$ and $\boldsymbol{v}$ are vertices of P that share no facet, then every $\boldsymbol{u} - \boldsymbol{v}$ path of length d is nonrevisiting.*

Proof Let $L := \boldsymbol{v}_0 \dots \boldsymbol{v}_d$ be a $\boldsymbol{u} - \boldsymbol{v}$ path in P; here, $\boldsymbol{v}_0 = \boldsymbol{u}$ and $\boldsymbol{v}_d = \boldsymbol{v}$. Suppose that the vertex $\boldsymbol{v}_0$ is contained in the facets $F_1, \dots, F_d$, and that, for each $j \in [1 \dots d]$, F_{d+j} is the unique facet that contains $\boldsymbol{v}_j$ but not $\boldsymbol{v}_{j-1}$ (Theorem 2.8.6). For each $i \in [0 \dots d]$, let

$$\mathcal{J}_i := \{F_1, \dots, F_d, F_{d+1}, \dots, F_{d+i}\}.$$

It follows that the collection $\mathcal{J}_i$ contains the d facets containing the vertex $\boldsymbol{v}_i$: this is true for $i = 0$, and if $\mathcal{J}_{i-1}$ contains the d facets containing the vertex $\boldsymbol{v}_{i-1}$ then $\mathcal{J}_{i-1} \cup \{F_{d+i}\} = \mathcal{J}_i$ must contain the facets containing the vertex $\boldsymbol{v}_i$ as $\boldsymbol{v}_{i-1}$ and $\boldsymbol{v}_i$ are adjacent. Thus, $\mathcal{J}_d$ contains all the d facets incident with $\boldsymbol{v}_d$. Since $\boldsymbol{v}_0$ and $\boldsymbol{v}_d$ share no facet, the facets in $\mathcal{J}_d$ are pairwise distinct. Because $\mathcal{J}_i$ contains all the facets visited by the subpath $\boldsymbol{v}_0 \dots \boldsymbol{v}_i$ of L, we find that the path L is nonrevisiting. □

The paths in Lemma 7.3.11 are shortest paths in such a polytope.

Lemma 7.3.12 *For every $d \geqslant 2$ and every $k \in [0 \dots d-1]$, every two vertices that don't lie in a common k-face of a simple d-polytope are at distance at least $k+1$.*

Proof Fix d and proceed by induction on k. Two distinct vertices in a d-polytope are at distance at least one, and so the result follows for $k = 0$. Suppose that $k \geqslant 1$ and that the result is true for $k-1$.

Let P be a simple d-polytope and let $\boldsymbol{v}_0$ and $\boldsymbol{v}_\ell$ be two vertices at distance at most k. Then there is a neighbour $\boldsymbol{v}_{\ell-1}$ of $\boldsymbol{v}_\ell$ at distance at most $k-1$ from $\boldsymbol{v}_0$. By the induction hypothesis, the vertices $\boldsymbol{v}_0$ and $\boldsymbol{v}_{\ell-1}$ lie in a common $(k-1)$-face R. From Theorem 2.8.6 it now follows that $\boldsymbol{v}_\ell$, $\boldsymbol{v}_{\ell-1}$, and the $k-1$ neighbours of $\boldsymbol{v}_{\ell-1}$ in R define a k-face that contains $\boldsymbol{v}_0$ and $\boldsymbol{v}_\ell$. □

Lemma 7.3.13 *Let P be a simple polytope. A path in a face of P is nonrevisiting (in the face) if and only if the path is nonrevisiting in P.*

Proof As every k-face is a facet of a $(k+1)$-face of P, it suffices to prove the result for a facet F of P. Let $\boldsymbol{v}_0$ and $\boldsymbol{v}_\ell$ be two vertices of F and let $L := \boldsymbol{v}_0 \dots \boldsymbol{v}_\ell$ be a nonrevisiting $\boldsymbol{v}_0 - \boldsymbol{v}_\ell$ path in F; that is, no $(d-2)$-face of F is revisited. Aiming for a contradiction, suppose L revisits a facet J of P: for some $i < j < k$ we have that $\boldsymbol{v}_i, \boldsymbol{v}_k \in J$ but $\boldsymbol{v}_j \notin J$ (Remark 7.3.8). Since P is simple, the intersection $J \cap F$ is a $(d-2)$-face R of F (Theorem 2.8.3), which contains $\boldsymbol{v}_i$ and $\boldsymbol{v}_k$ but not $\boldsymbol{v}_j$. This implies that L revisits R within F, a contradiction.

For the other direction, suppose that $\boldsymbol{u}$ and $\boldsymbol{v}$ are vertices of a facet F of P. If a $\boldsymbol{u} - \boldsymbol{v}$ path L in F revisits a $(d-2)$-face R of F, then L revisits the facet J of P satisfying $J \cap F = R$. □

Nonrevisiting paths in a wedge over a polytope P at a facet remain so when projected onto P.

Lemma 7.3.14 *Let P be a d-polytope and W the wedge of P at a facet. If a path is nonrevisiting in W, then so is its projection on P.*

Proof Let $\boldsymbol{u}$ and $\boldsymbol{v}$ be two vertices of P, and let F be a facet of P such that W is the wedge of P at F. Let $L_W := \boldsymbol{u} \ldots \boldsymbol{v}$ be a $\boldsymbol{u} - \boldsymbol{v}$ path in W and let L be the projection of the path L_W onto P (Lemma 7.3.3). We prove the contrapositive.

Suppose that L revisits a facet J of P. If $J = F$, then L_W revisits P or the other base P' of W; if, instead, $J \neq F$ then L_W revisits the d-face of W that projects onto J. □

As in the case of the Hirsch conjectures, we reduce the nonrevisiting conjecture to simple d-polytopes with $2d$ facets.

Proposition 7.3.15 *The following assertions are equivalent.*

(i) *For every two vertices in a simple d-polytope with n facets, there is a nonrevisiting path between them.*
(ii) *For every two vertices in a simple d'-polytope with $2d'$ facets, there is a nonrevisiting path between them.*

Proof Part (i) clearly implies Part (ii). We prove the other direction. We consider two cases according to the relation between n and $2d$, as in the proof of Proposition 7.3.6. Let P be a simple d-polytope with n facets.

Suppose that $n \leqslant 2d$ and proceed by induction on $r := 2d - n$. The case $r = 0$ is true, and so assume the statement holds for $r - 1 \geqslant 0$. Every two vertices $\boldsymbol{u}$ and $\boldsymbol{v}$ are in a facet F of P. Suppose there is no nonrevisiting $\boldsymbol{u} - \boldsymbol{v}$ path in P. By Lemma 7.3.13, there is no nonrevisiting $\boldsymbol{u} - \boldsymbol{v}$ path within F. Thus F violates (i). The facet F has at most $n - 1$ $(d-2)$-faces satisfying $n - 1 \leqslant 2(d-1)$ and $2(d-1) - (n-1) < 2d - n$. By the induction hypothesis, there is a simple d'-polytope with $2d'$ facets that violates (ii).

Suppose that $n \geqslant 2d$, and proceed by induction on $r := n - 2d$ with the case $r = 0$ as the basis. Suppose there are vertices $\boldsymbol{u}$ and $\boldsymbol{v}$ for which there is no nonrevisiting $\boldsymbol{u} - \boldsymbol{v}$ path in P. Since $r \geqslant 1$ and every vertex is incident with precisely d facets of P, there must exist a facet F of P containing neither $\boldsymbol{u}$ nor $\boldsymbol{v}$. We construct the wedge W of P at F. By Lemma 7.3.14, there is no nonrevisiting $\boldsymbol{u}-\boldsymbol{v}$ path in W, where $\boldsymbol{u}$ and $\boldsymbol{v}$ are considered as vertices of W that lie within the facet P. The polytope W has $n+1$ facets satisfying $n+1 \geqslant 2(d+1)$ and $(n+1)-2(d+1) < n-2d$. By the induction hypothesis, there is a simple d'-polytope with $2d'$ facets that violates (ii). □

Proposition 7.3.15 enables the equivalence between the nonrevisiting conjecture and the d-step conjecture.

Proposition 7.3.16 (Klee and Walkup, 1967) *The nonrevisiting conjecture is equivalent to the d-step conjecture.*

Proof If the nonrevisiting conjecture holds, then Lemma 7.3.10 ensures that the d-step conjecture holds. So suppose that the d-step conjecture holds for every $d \geqslant 2$. By Proposition 7.3.15 it suffices to show that the nonrevisiting conjecture holds for simple d-polytopes with $2d$ facets; let P be one such polytope.

Let $\boldsymbol{u}$ and $\boldsymbol{v}$ be two vertices of P sharing exactly r facets of P. If $r=0$, then we let $s:=-1$, $d_s:=d$, $W_s:=P$, and $\boldsymbol{v}_s:=\boldsymbol{v}$. If instead $r \geqslant 1$, then we let W_0 be the intersection of such r facets, $d_0:=\dim W_0$, and $\boldsymbol{v}_0:=\boldsymbol{v}$. Since no facet of W_0 contains both $\boldsymbol{u}$ and $\boldsymbol{v}$, the number of facets of W_0 is $2d_0+s$, where s is the number of facets of W_0 containing neither $\boldsymbol{u}$ nor $\boldsymbol{v}$; we also let $d_s:=d_0+s$.

If $s>0$, then we recursively construct a sequence of wedges $W_1,\ldots,W_s$ as follows. For each $i \in [1\ldots s]$, we pick a facet R_{i-1} of W_{i-1} containing neither $\boldsymbol{u}$ nor $\boldsymbol{v}_{i-1}$, produce the wedge W_i of W_{i-1} at R_{i-1}, and let $\boldsymbol{v}_i$ be the unique neighbour of $\boldsymbol{v}_{i-1}$ in $W_i \setminus W_{i-1}$. The polytope W_i is a simple (d_0+i)-polytope with $2(d_0+i)+s-i$ facets (Lemma 7.3.3). Moreover, no facet of W_i contains both $\boldsymbol{u}$ and $\boldsymbol{v}_i$, and exactly $s-i$ facets of W_i contain neither (why?).

For the cases where $s=0$, -1, and $s>0$, it follows that W_s is a simple d_s-polytope with $2d_s$ facets, and that the vertices $\boldsymbol{u}$ and $\boldsymbol{v}_s$ share no facet of W_s. Then Lemma 7.3.12 shows that $\operatorname{dist}_{W_s}(\boldsymbol{u},\boldsymbol{v}_s) \geqslant d_s$, and the d-step conjecture on W_s shows that $\operatorname{dist}_{W_s}(\boldsymbol{u},\boldsymbol{v}_s) \leqslant d_s$. Thus, $\operatorname{dist}_{W_s}(\boldsymbol{u},\boldsymbol{v}_s)=d_s$, and Lemma 7.3.11 gives that every $\boldsymbol{u}-\boldsymbol{v}_s$ path L_s in W_s is nonrevisiting. If $s=-1$ then $W_s=P$, implying that the path L_s is nonrevisiting in P. Otherwise, for $i \in [1\ldots s]$, the polytope W_{s-i} is a facet of W_{s-i+1}, and by Lemma 7.3.14 the projection of the nonrevisiting $\boldsymbol{u}-\boldsymbol{v}_{s-i+1}$ path L_{s-i+1} in W_{s-i+1} is a nonrevisiting $\boldsymbol{u}-\boldsymbol{v}_{s-i}$ path L_{s-i} in W_{s-i}. As a result, for both cases $s=0$ and $s>0$, we have a nonrevisiting $\boldsymbol{u}-\boldsymbol{v}_0$ path L_0 within the face W_0 of P. By Lemma 7.3.13, this path L_0 is in turn nonrevisiting in P. As required, in every scenario, we have demonstrated a nonrevisiting $\boldsymbol{u}-\boldsymbol{v}$ path in P, which concludes the proof of the proposition. □

7.4 Bounds for Diameters

We start with lower bounds.

Proposition 7.4.1 (Klee, 1964a) *For* $n > d \geqslant 2$,

$$\operatorname{diam}(n,d) \geqslant \left\lfloor \frac{d-1}{d} n \right\rfloor - (d-2).$$

Proof We prove the proposition in the dual setting, for the dual diameter. By Proposition 7.3.2 we can focus on simplicial d-polytopes with n vertices, and by Remark 7.3.7 we can focus on $n \geqslant 2d$.

The d-crosspolytope I has $2d$ vertices and two disjoint facets, F_1 and F_2, connected by a shortest facet-ridge path of length d; see Fig. 7.4.1(a). Thus, the proposition is true for $n = 2d$, and so an induction argument on n can start, for a fixed d.

Consider a simplicial d-polytope P' with n' vertices and two disjoint facets J_1 and J_2 connected by a shortest facet-ridge path of length $\operatorname{diam}(n', d)$. A connected sum of P' and I along the facets J_2 and F_1 yields a simplicial d-polytope P with $n'+d$ vertices and two facets, namely J_1 and F_2, connected by a shortest facet-ridge path of length $\operatorname{diam}(n', d) + d - 1$; performing this sum is always possible (Problem 2.15.9). As a consequence, the dual diameter of P is $\operatorname{diam}(n', d) + d - 1$, and if the proposition holds for $\operatorname{diam}(n', d)$ then it holds for $\operatorname{diam}(n' + d, d)$:

$$\begin{aligned}
\operatorname{diam}(n,d) &\geqslant \operatorname{diam}(n',d) + d - 1 \\
&\geqslant \left\lfloor \frac{d-1}{d} n' \right\rfloor - (d-2) + d - 1 \\
&= \left\lfloor \frac{d-1}{d} (n' + d) \right\rfloor - (d-2) \\
&= \left\lfloor \frac{d-1}{d} n \right\rfloor - (d-2).
\end{aligned}$$

It remains to construct a simplicial d-polytope P' with $n' = 2d + r$ vertices and dual diameter $d + r - 1$, for each $r \leqslant d - 1$. We do so by repeatedly stacking over the crosspolytope. Let $Z_0 := I$ and let $X_0 := F_1$. The first stacking over the facet X_0 of Z_0 produces a polytope Z_1 with a new facet X_1 connected to F_2 by shortest facet-ridge paths of length d. Then, for each $i \geqslant 2$, the ith stacking over the facet X_{i-1} of Z_{i-1} yields a polytope Z_i with a new facet X_i connected to F_2 by a shortest facet-ridge path of length $d + i - 1$. Figure 7.4.1 illustrates the case $i = 2$ and $d = 3$. □

Proposition 7.4.1 facilitates the computation of $\operatorname{diam}(n, 3)$.

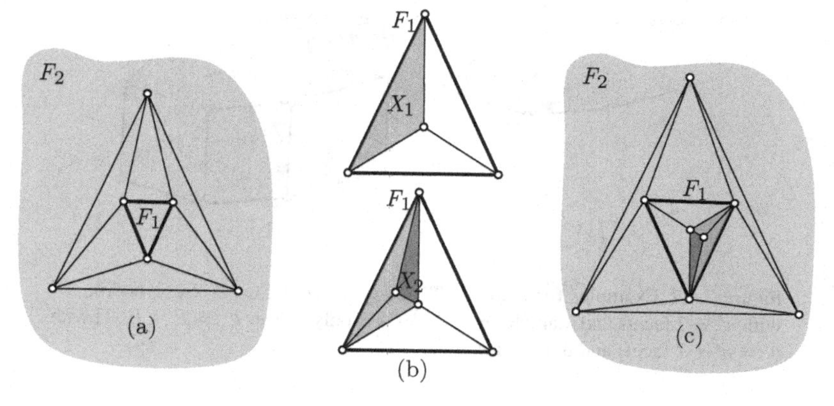

Figure 7.4.1 Auxiliary figure for the proof of Proposition 7.4.1. (a) The graph of the 3-crosspolytope Z_0 with faces F_1 and F_2 highlighted. (b) The face F_1 of Z_0 has been zoomed in. Stacking over F_1 yields a facet X_1, highlighted in grey, in a polytope Z_1. Further, stacking over X_1 yields a facet X_2, highlighted in dark grey, of a polytope Z_2. (c) The 3-polytope Z_2 with $n' = 6 + 2$ vertices and dual diameter $3 + 1$.

Proposition 7.4.2 *For* $n > 3$, $\operatorname{diam}(n,3) = \lfloor 2n/3 \rfloor - 1$.

Proof By Proposition 7.3.2 we can focus on simple 3-polytopes with n faces. Let P represent one such polytope. By Euler's equation (C.7.1), P has $2n - 4$ vertices, as $3f_0(P) = 2f_1(P)$. Let $\boldsymbol{u}$ and $\boldsymbol{v}$ be two vertices realising $\operatorname{diam}(n,3)$. Since P is 3-connected by Balinski's theorem (4.1.1), there are three independent $\boldsymbol{u} - \boldsymbol{v}$ paths. The shortest of these paths must have at most $\lfloor (2n-6)/3 \rfloor + 2 = \lfloor 2n/3 \rfloor$ vertices, and so its length is at most $\lfloor 2n/3 \rfloor - 1$. Hence $\operatorname{diam}(n,3) \leqslant \lfloor 2n/3 \rfloor - 1$. On the other side, Proposition 7.4.1 yields $\operatorname{diam}(n,3) \geqslant \lfloor 2n/3 \rfloor - 1$. □

Upper Bounds

We present the bounds of Todd (2014) and Barnette (1974b). We start with Todd's bound, whose proof is carried in the realm of polyhedra. For $n > d \geqslant 2$ we first define the maximum diameter of d-polyhedra with at most n facets:

$$\operatorname{diam}_u(n,d) := \max\{\operatorname{diam} P : P \text{ a } d\text{-polyhedron with at most } n \text{ facets}\}. \tag{7.4.3}$$

Reasoning as in the definition of $\operatorname{diam}(n,d)$ (7.3.5), we find that the function $\operatorname{diam}_u(n,d)$ is well defined. Two remarks become evident.

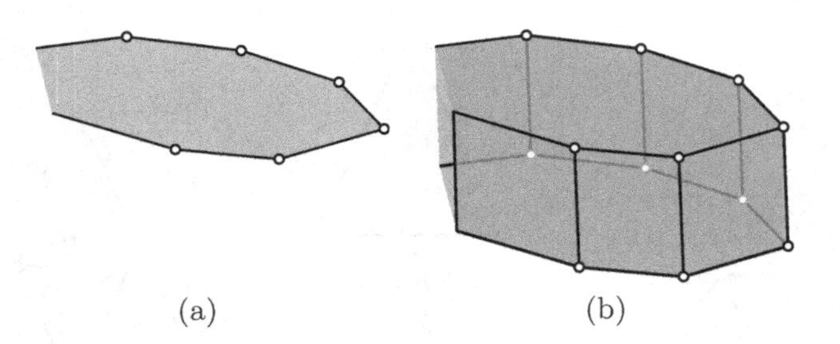

Figure 7.4.2 Examples for $\operatorname{diam}_u(n,2)$ and $\operatorname{diam}_u(n,3)$. (a) A 2-polyhedron P' with $n' = 7$ facets and diameter $n' - 2$. (b) A 3-polyhedron $P := P' \times [0,1]$ with $n := n' + 2$ facets and diameter $n - 3$.

Remark 7.4.4 For $n > d \geqslant 2$, we have the following.

(i) $\operatorname{diam}(n,d) \leqslant \operatorname{diam}_u(n,d)$ because a d-polytope is a bounded d-polyhedron (Theorem 2.2.3).

(ii) Hirsch's conjecture reads as $\operatorname{diam}_u(n,d) \leqslant n - d$.

As in the case of polytopes, we can reduce our proofs to simple polyhedra. This is so because a variation of Theorem 2.10.7, the main ingredient in the proof of Proposition 7.3.2, extends to polyhedra (Klee and Walkup, 1967, thm. 2.6). We next establish the smallest values $\operatorname{diam}_u(n,2)$ and $\operatorname{diam}_u(n,3)$.

Proposition 7.4.5 *For $d = 2,3$ and $n \geqslant d \geqslant 2$, $\operatorname{diam}_u(n,d) = n - d$.*

Proof The proof for $d = 2$ is pictorial (Fig. 7.4.2(a)). If P' is a 2-polyhedron with n' facets and diameter $n' - 2$, then $P := P' \times [0,1]$ is a 3-polyhedron with $n := n' + 2$ facets and diameter $n' - 2 + 1 = n - 3$; see Section 2.6 and Fig. 7.4.2. This gives that $\operatorname{diam}_u(n,3) \geqslant n - 3$. To show that $\operatorname{diam}_u(n,3) \leqslant n - 3$, it suffices to find a nonrevisiting path between every two vertices of a 3-polyhedron (Lemma 7.3.10); this is done carefully in Grünbaum (2003, sec. 16.3). $\square$

The bound of Todd (2014) improves a bound of Kalai and Kleitman (1992), but it requires a lemma of theirs.

Lemma 7.4.6 (Kalai and Kleitman, 1992) *For $n \geqslant 4$ and $\lfloor n/2 \rfloor \geqslant d \geqslant 2$,*

$$\operatorname{diam}_u(n,d) \leqslant \operatorname{diam}_u(n-1,d-1) + 2\operatorname{diam}_u(\lfloor n/2 \rfloor, d) + 2.$$

Proof Let $P \subseteq \mathbb{R}^d$ be a simple d-polyhedron with at most n facets, and let $\boldsymbol{u}$ and $\boldsymbol{v}$ be two vertices of P satisfying $\operatorname{dist}_P(\boldsymbol{u},\boldsymbol{v}) = \operatorname{diam}_u(n,d)$. If $\boldsymbol{u}$ and $\boldsymbol{v}$ lie in a common facet F, which is a simple $(d-1)$-polyhedron with at most

$n-1$ facets, then $\text{dist}_P(\boldsymbol{u},\boldsymbol{v}) \leqslant \text{dist}_F(\boldsymbol{u},\boldsymbol{v}) \leqslant \text{diam}_u(n-1,d-1)$, and so the lemma holds in this case. Thus assume that $\boldsymbol{u}$ and $\boldsymbol{v}$ share no facet.

Let ℓ_v be the largest ℓ so that all paths of length ℓ from $\boldsymbol{v}$ intersect only facets in a set $\mathcal{F}_v$ of cardinality at most $\lfloor n/2 \rfloor$. The number ℓ_v is well defined as $\lfloor n/2 \rfloor \geqslant d$ and all paths of length zero intersect only the d facets containing $\boldsymbol{v}$. Define the number ℓ_u analogously.

Claim 1 $\ell_u \leqslant \text{diam}_u(\lfloor n/2 \rfloor, d)$ and $\ell_v \leqslant \text{diam}_u(\lfloor n/2 \rfloor, d)$.

Proof of claim We prove the claim for ℓ_v. Let P_v be the d-polyhedron obtained from P by considering only the inequalities corresponding to facets in $\mathcal{F}_v$. Then $P \subseteq P_v$. Suppose that $\boldsymbol{w}$ is a vertex of P at distance ℓ_v from $\boldsymbol{v}$ in P. We show that $\text{dist}_{P_v}(\boldsymbol{v},\boldsymbol{w}) = \ell_v$. Let L_v be a shortest path in P_v from $\boldsymbol{v}$ to $\boldsymbol{w}$. If the length of L_v is less than ℓ_v, then L_v uses an edge of P_v not present in P. Let e be the first such edge. Since e is not in P, it intersects a hyperplane that corresponds to a facet F of P not contained in $\mathcal{F}_v$. As a consequence, there is a path from $\boldsymbol{v}$ to F of length less than ℓ_v, which contradicts the definition of $\mathcal{F}_v$ (as F should have been in $\mathcal{F}_v$). Hence

$$\ell_v = \text{dist}_{P_v}(\boldsymbol{v},\boldsymbol{w}) \leqslant \text{diam}_u(\#\mathcal{F}_v, d) \leqslant \text{diam}_u(\lfloor n/2 \rfloor, d).$$

The same proof works for ℓ_u. □

We now consider the set $\mathcal{X}_v$ of facets of P that can be reached from $\boldsymbol{v}$ by a path of length $\ell_v + 1$. By the definition of ℓ_v and $\mathcal{F}_v$, we have that $\mathcal{X}_v \neq \varnothing$. Similarly define $\mathcal{X}_u$. Since $\#\mathcal{X}_v \geqslant \lfloor n/2 \rfloor$ and $\#\mathcal{X}_u \geqslant \lfloor n/2 \rfloor$, there is a facet F of P contained in both $\mathcal{X}_v$ and $\mathcal{X}_u$. This implies that there are vertices $\boldsymbol{w}_v$ and $\boldsymbol{w}_u$ in F, a $\boldsymbol{v} - \boldsymbol{w}_v$ path in P of length at most $\ell_v + 1$, and a $\boldsymbol{u} - \boldsymbol{w}_u$ path in P of length at most $\ell_u + 1$. Because F is a simple $(d-1)$-polyhedron with at most $n-1$ facets and because of Claim 1, we have the desired inequality:

$$\begin{aligned}
\text{diam}_u(n,d) &= \text{dist}_P(\boldsymbol{u},\boldsymbol{v}) \\
&\leqslant \text{dist}_P(\boldsymbol{u},\boldsymbol{w}_u) + \text{dist}_F(\boldsymbol{w}_v,\boldsymbol{w}_u) + \text{dist}_P(\boldsymbol{w}_v,\boldsymbol{v}) \\
&\leqslant \ell_u + 1 + \text{diam}_u(n-1,d-1) + \ell_v + 1 \\
&\leqslant \text{diam}_u(n-1,d-1) + 2\,\text{diam}_v(\lfloor n/2 \rfloor, d) + 2.
\end{aligned}$$

This settles the lemma. □

We use Lemma 7.4.6 to produce upper bounds for some small pairs (n,d).

Lemma 7.4.7 *The following statements hold.*

(i) *For* $d \geqslant 1$ *and* $n = d, d+1$: $\text{diam}_u(n,d) = n - d$.

(ii) *For* $d = 4$: $\operatorname{diam}_u(8,4) \leqslant 6$, $\operatorname{diam}_u(9,4) \leqslant 7$, $\operatorname{diam}_u(10,4) \leqslant 10$, *and* $\operatorname{diam}_u(11,4) \leqslant 11$.
(iii) *For* $d = 5$: $\operatorname{diam}_u(10,5) \leqslant 9$, $\operatorname{diam}_u(11,5) \leqslant 12$, *and* $\operatorname{diam}_u(12,5) \leqslant 15$.
(iv) *For* $d = 6$: $\operatorname{diam}_u(12,6) \leqslant 14$ *and* $\operatorname{diam}_u(13,6) \leqslant 17$.
(v) *For* $d = 7$: $\operatorname{diam}_u(14,7) \leqslant 19$.

Proof Let P be a d-polyhedron with n facets. (i) If $n = d$ then P has exactly one vertex, and so its diameter is zero. If instead $n = d + 1$, then P is a d-simplex and its diameter is one.

(ii)–(iv) These values are obtained from applications of Lemma 7.4.6 and Proposition 7.4.5. □

Theorem 7.4.8 (Todd, 2014) *For* $n \geqslant d \geqslant 1$, $\operatorname{diam}_u(n,d) \leqslant (n-d)^{\log d}$ *where* $\log d := \log_2 d$.

Proof The proof proceeds by induction on $n + d$. The theorem holds for the basis cases $(n,d) = (n,2), (n,3), (d,d), (d+1,d), (8,4), (9,4), (10,4), (11,4), (10,5), (11,5), (12,5), (12,6), (13,6), (14,7)$, as demonstrated by Lemma 7.4.7 and Proposition 7.4.5.

Let P be a simple d-polyhedron with n facets. Suppose the result is true for every pair (n',d') such that $n' + d' < n + d$. In the case $n < 2d$, every two vertices lie in a common facet, which is a simple $(d-1)$-polyhedron with at most $n - 1$ facets. This yields that $\operatorname{diam}_u(n,d) \leqslant \operatorname{diam}_u(n-1,d-1)$. Since $n - 1 + d - 1 < n + d$, the induction hypothesis now settles the case:

$$\operatorname{diam}_u(n,d) \leqslant \operatorname{diam}_u(n-1,d-1) \leqslant (n-d)^{\log(d-1)} < (n-d)^{\log d}.$$

So assume that $n \geqslant 2d$. In view of the basis cases, we can further assume that $d \geqslant 4$ and $n - d \geqslant 8$, in which case we have that $\log(n-d) \geqslant 3$. We also require the identity

$$(n-d)^{\log d} = d^{\log(n-d)},$$

which is a consequence of the equalities

$$\log\left((n-d)^{\log d}\right) = \log\left(d^{\log(n-d)}\right) = \log(n-d) \times \log d.$$

At this stage, the theorem follows from Lemma 7.4.6:

$$\begin{aligned}
\operatorname{diam}_u(n,d) &\leqslant \operatorname{diam}_u(n-1,d-1) + 2\operatorname{diam}_u(\lfloor n/2 \rfloor, d) + 2 \\
&\leqslant (d-1)^{\log(n-d)} + 2d^{\log(n/2-d)} + 2 \\
&\leqslant \left(\frac{d-1}{d}\right)^{\log(n-d)} \times d^{\log(n-d)} + 2d^{\log(n/2-d/2)} + 2
\end{aligned}$$

(as $\log(n-d) \geqslant 3$, we have the following)

$$\leqslant \left(\frac{d-1}{d}\right)^3 \times d^{\log(n-d)} + \frac{2}{d} d^{\log(n-d)} + 2$$

$$= \left[\left(\frac{d-1}{d}\right)^3 + \frac{2}{d}\right] \times d^{\log(n-d)} + 2$$

$$= \left(1 - \frac{3}{d} + \frac{3}{d^2} - \frac{1}{d^3} + \frac{2}{d}\right) \times d^{\log(n-d)} + 2$$

$\left(\text{and from } d^2 \geqslant 4d, \text{ it follows that } \frac{3}{d^2} \leqslant \frac{3}{4d}, \text{ which then yields that}\right)$

$$\leqslant \left(1 - \frac{1}{d} + \frac{3}{4d} - \frac{1}{d^3}\right) \times d^{\log(n-d)} + 2$$

$$= \left(1 - \frac{1}{4d} - \frac{1}{d^3}\right) \times d^{\log(n-d)} + 2$$

$$= d^{\log(n-d)} - \frac{1}{4d} d^{\log(n-d)} - \frac{1}{d^3} d^{\log(n-d)} + 2$$

$\left(\text{since } \frac{1}{4d} d^{\log(n-d)} \geqslant 1 \text{ and } \frac{1}{d^3} d^{\log(n-d)} \geqslant 1, \text{ we get the desired inequality}\right)$

$$\leqslant d^{\log(n-d)}.$$

This completes the proof of the theorem. □

Todd's bound was improved asymptotically to $(n-d)^{\log O(d/\log(d))}$ by Sukegawa (2019).

The bound of Barnette (1974b) is argued by induction on the dimension, starting with the basis case of $d = 3$. Let $e = \boldsymbol{u}\boldsymbol{v}$ be an edge of the graph G of a simple 3-polytope, let $\boldsymbol{u}_1$ and $\boldsymbol{u}_2$ be the other neighbours of $\boldsymbol{u}$, and let $\boldsymbol{v}_1$ and $\boldsymbol{v}_2$ be the other neighbours of $\boldsymbol{v}$. We write $G \dot{-} e$ for the graph obtained from the subgraph $G - \{\boldsymbol{u}, \boldsymbol{v}\}$ by adding the edges $\boldsymbol{u}_1\boldsymbol{u}_2$ and $\boldsymbol{v}_1\boldsymbol{v}_2$; we say that $G \dot{-} e$ is obtained from G by *suppressing the edge* e (Fig. 7.4.3). If, in addition, $G \dot{-} e$ is the graph of a 3-polytope, then we say that the edge e is *suppresible*.

Lemma 7.4.9 *Each face of a simple 3-polytope other than a simplex has a suppressible edge.*

Proof Let G be a planar realisation of the graph of a simple 3-polytope P other than a simplex (Proposition 3.2.11) and let G^* be the standard planar realisation of the dual graph of G. Then G^* is a planar realisation of the graph of the dual polytope P^* (Proposition 3.5.1). Let ψ be an antiisomorphism from $\mathcal{L}(P)$ to $\mathcal{L}(P^*)$.

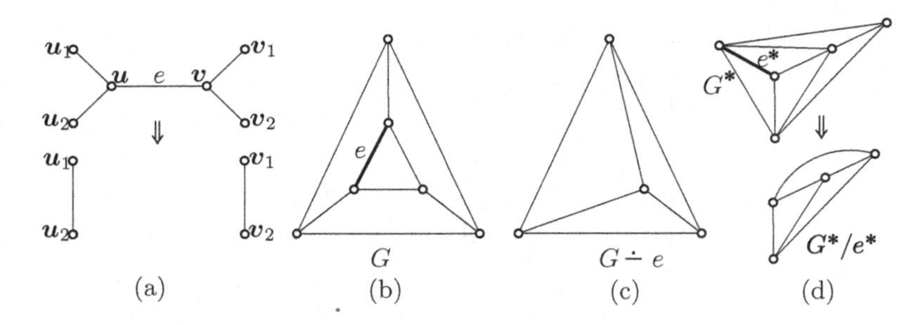

Figure 7.4.3 Edge suppression in a graph. (a) The suppression of an edge e. (b) A 3-connected graph G with an edge e highlighted. (c) The graph $G \dot{-} e$ obtained from G by the suppression of e. (d) The dual graph G^* of G with the edge e^* dual to e highlighted, and the graph G^*/e^* obtained from G^* by contracting e^*.

Suppressing an edge e of G amounts to contracting the edge e^* in G^* dual to e. If the edge e^* of G^* belongs to no nonfacial triangle (Fig. 7.4.3(d)), then G^*/e^* will have no degree-2 vertices and will be a plane triangulation dual to $G \dot{-} e$, and so we will be done. Let F be a face of P, let $\boldsymbol{v} := \psi(F)$, and let $e_1^* := \boldsymbol{v}\boldsymbol{v}_1, \ldots, e_n^* := \boldsymbol{v}\boldsymbol{v}_n$ be the edges of G^* incident with $\boldsymbol{v}$. If no edge e_i^* belongs to a nonfacial triangle, then every edge of F is suppressible and we are done. So suppose some edges incident with $\boldsymbol{v}$ are part of a nonfacial triangle T of G^*. Choose T with the minimum number of neighbours of $\boldsymbol{v}$ in its interior; say e_1^* and e_2^* belong to T and $T = \boldsymbol{v}\boldsymbol{v}_1\boldsymbol{v}_2$. Because G^* is 3-connected, there is at least a neighbour of $\boldsymbol{v}$ in the interior of T, say $\boldsymbol{v}_3$. If the edge e_3^* belonged to a nonfacial triangle T', then, since G^* is a plane graph, the triangle T' would contain fewer neighbours of $\boldsymbol{v}$ in its interior than T, contradicting the minimality of T. The edge e_3 of G dual to e_3^* is a suppressible edge of F. □

The next lemma states the basis case of our induction on d. A face F of a polytope P is at *distance* ℓ from a vertex $\boldsymbol{x}$ of P if ℓ is the minimum distance from $\boldsymbol{x}$ to a vertex in F; this is written $\operatorname{dist}_P(\boldsymbol{x}, F) = \ell$.

Lemma 7.4.10 *Let P be a simple 3-polytope, let $\boldsymbol{v}$ be a vertex of P, and let ℓ be the distance from $\boldsymbol{v}$ to some vertex of P. If there are exactly n facets of P at distance at most $\ell - 1$ from $\boldsymbol{v}$, then $\ell \leqslant \lfloor 2n/3 \rfloor - 1$.*

Proof We consider two cases. In the first case, suppose that all the facets of P are at distance at most $\ell - 1$ from $\boldsymbol{v}$. Then P has n facets, in which case Proposition 7.4.2 yields that $\operatorname{diam}(P) \leqslant \lfloor 2n/3 \rfloor - 1$. Since $\ell \leqslant \operatorname{diam} P$, we are done.

In the second case, suppose that there is a facet F at distance at least ℓ from $\boldsymbol{v}$. According to Lemma 7.4.9, there is a suppressible edge e in F. The edge

e doesn't belong to any path of length at most ℓ from $\boldsymbol{v}$, otherwise F would be at distance at most $\ell - 1$ from $\boldsymbol{v}$. This means that suppressing e produces a simple 3-polytope P' such that no facet of P at distance at most $\ell - 1$ from $\boldsymbol{v}$ becomes a facet of P' at distance at least ℓ from $\boldsymbol{v}$. Thus P' has fewer edges than P and exactly n facets at distance at most $\ell - 1$ from $\boldsymbol{v}$. Let $\boldsymbol{u}$ be a vertex of P at distance ℓ from $\boldsymbol{v}$. We need to find a vertex in P' at distance ℓ from $\boldsymbol{v}$.

If the vertex $\boldsymbol{u}$ is not an end of e, then $\operatorname{dist}_P(\boldsymbol{v}, \boldsymbol{u}) = \operatorname{dist}_{P'}(\boldsymbol{v}, \boldsymbol{u}) = \ell$, and we continue the proof on P'. So instead suppose, without loss of generality, that every such vertex $\boldsymbol{u}$ is an end of e, say $e = \boldsymbol{u}\boldsymbol{w}_1$, that $\boldsymbol{w}_2$ and $\boldsymbol{w}_3$ are the other neighbours of $\boldsymbol{u}$, and that $\boldsymbol{w}_1 \in F$ and $\boldsymbol{w}_2 \in F$. Since $\operatorname{dist}_P(\boldsymbol{v}, F) \geqslant \ell$, we have that $\operatorname{dist}_P(\boldsymbol{v}, \boldsymbol{w}_1) \geqslant \ell$ and $\operatorname{dist}_P(\boldsymbol{v}, \boldsymbol{w}_2) \geqslant \ell$. And since $\operatorname{dist}_P(\boldsymbol{v}, \boldsymbol{u}) = \ell$, we further have that $\boldsymbol{w}_3$ is on a shortest $\boldsymbol{v} - \boldsymbol{u}$ path, implying that $\operatorname{dist}_P(\boldsymbol{v}, \boldsymbol{w}_3) = \ell - 1$. It follows that in P' the vertex $\boldsymbol{w}_2$ is at distance ℓ from $\boldsymbol{v}$ and we can continue the proof on P'. As a consequence, we can continue suppressing edges until there are no facets at distance at least ℓ from $\boldsymbol{v}$, and here the first case applies. □

Theorem 7.4.11 *Let P be a simple d-polytope with $d \geqslant 3$, $\boldsymbol{v}_0$ a vertex of P, and ℓ the distance from $\boldsymbol{v}_0$ to some vertex of P. If there are at most n facets of P at distance at most $\ell - 1$ from $\boldsymbol{v}_0$, then $\ell \leqslant \frac{1}{3}2^{d-2}\left(n - d + \frac{5}{2}\right)$.*

Proof As announced previously, the proof is by induction on d, with the basis case $d = 3$ given in Lemma 7.4.10. So assume that $d \geqslant 4$. Let Y_ℓ be the set of vertices of P at distance at least ℓ from $\boldsymbol{v}_0$.

A facet-ridge path $J_0 J_1 \ldots J_r$ *joins* $\boldsymbol{v}_0$ to another vertex $\boldsymbol{u}$ of P if $\boldsymbol{v}_0 \in J_0$ and $\boldsymbol{u} \in J_r \backslash (J_1 \cup \cdots \cup J_{r-1})$. Let r be the length of a shortest facet-ridge path joining $\boldsymbol{v}_0$ to a vertex in Y_ℓ, and let X_ℓ consist of the vertices in Y_ℓ that are joined to $\boldsymbol{v}_0$ through a facet-ridge of length r. Let $\boldsymbol{u} \in X_\ell$ and let F_0 be the first facet of a shortest facet-ridge path joining $\boldsymbol{v}_0$ to $\boldsymbol{u}$.

(i) Among all shortest facet-ridge paths from $\boldsymbol{v}_0$ to $\boldsymbol{u}$ that starts with F_0, choose one with a second facet F_1 such that $\operatorname{dist}_{F_0}(\boldsymbol{v}_0, F_0 \cap F_1)$ is minimum. Let $\boldsymbol{v}_1$ be a vertex of $F_0 \cap F_1$ realising $\operatorname{dist}_{F_0}(\boldsymbol{v}_0, F_0 \cap F_1)$: $\operatorname{dist}_{F_0}(\boldsymbol{v}_0, \boldsymbol{v}_1) = \operatorname{dist}_{F_0}(\boldsymbol{v}_0, F_0 \cap F_1)$.
(ii) Among all shortest facet-ridge paths from $\boldsymbol{v}_0$ to $\boldsymbol{u}$ that starts with F_0 and F_1, choose one with a third facet F_2 such that $\operatorname{dist}_{F_1}(\boldsymbol{v}_1, F_1 \cap F_2)$ is minimum. Let $\boldsymbol{v}_2$ be a vertex of $F_1 \cap F_2$ realising $\operatorname{dist}_{F_1}(\boldsymbol{v}_1, F_1 \cap F_2)$.
(iii) We continue this way until we get a facet-ridge path $M := F_0 F_1 \ldots F_r$ of length r from $\boldsymbol{v}_0$ to $\boldsymbol{u}$ and corresponding vertices $\boldsymbol{v}_1 \in F_0 \cap F_1, \ldots, \boldsymbol{v}_r \in F_{r-1} \cap F_r$.
(iv) Finally, among all the vertices in $X_\ell \cap \mathcal{V}(F_r)$, let $\boldsymbol{v}_{r+1}$ be one for which $\operatorname{dist}_{F_r}(\boldsymbol{v}_r, \boldsymbol{v}_{r+1})$ is minimum.

Set $\mathcal{F} := \{F_0, \dots, F_r\}$. The construction of M yields a simple claim.

Claim 1 For each $i < r$, every vertex in F_i is at distance at most $\ell - 1$ from $\boldsymbol{v}_0$.

We say that a $(d-2)$-face R of a facet $F_i \in \mathcal{F}$ is *close about* F_i if $\operatorname{dist}_{F_i}(\boldsymbol{v}_i, R) < \operatorname{dist}_{F_i}(\boldsymbol{v}_i, \boldsymbol{v}_{i+1})$. Since $0 = \operatorname{dist}_{F_i}(\boldsymbol{v}_i, F_{i-1} \cap F_i) < \operatorname{dist}_{F_i}(\boldsymbol{v}_i, \boldsymbol{v}_{i+1}) = \operatorname{dist}_{F_i}(\boldsymbol{v}_i, F_i \cap F_{i+1})$, it is clear that $F_{i-1} \cap F_i$ is close about F_i, but $F_i \cap F_{i+1}$ is not. Moreover, if a ridge $F_i \cap F_j$ with $j > i+1$ existed, then r wouldn't be the length of a shortest facet-ridge path joining $\boldsymbol{v}_0$ to $\boldsymbol{u}$. Thus there are r close ridges arising from intersections of facets in $\mathcal{F}$; these are *trivial* close ridges.

Let $\mathcal{F}_c$ be the set of the facets of P not in $\mathcal{F}$ and intersecting some F_i at a close ridge. We examine the elements of $\mathcal{F}_c$.

If a facet $F \in \mathcal{F}_c$ intersects $\mathcal{F}$ on two facets F_i and F_j with $i < j$, then $j = i+1$ or $j = i+2$: if $j > i+2$ then r wouldn't be the length of a shortest facet-ridge path from $\boldsymbol{v}_0$ to $\boldsymbol{u}$. If, instead, such a facet F intersects F_i and F_{i+2} such that $F \cap F_i$ is a close ridge about F_i, then we can replace F_{i+1} with F, which would contradict the selection of F_{i+1} because in this case

$$\operatorname{dist}_{F_i}(\boldsymbol{v}_i, F_i \cap F_{i+1}) = \operatorname{dist}_{F_i}(\boldsymbol{v}_i, \boldsymbol{v}_{i+1}) > \operatorname{dist}_{F_i}(\boldsymbol{v}_i, F_i \cap F).$$

Thus each facet in $\mathcal{F}_c$ must intersect $\mathcal{F}$ in at most two close ridges.

Furthermore, Claim 1 and the selection of $\boldsymbol{v}_{r+1}$ yield another simple claim.

Claim 2 Each facet F in $\mathcal{F}_c$ is at distance at most $\ell - 1$ from $\boldsymbol{v}_0$.

Proof of claim Suppose that $F \cap F_i$ is a ridge close about some F_i with $i \in [0 \dots r]$. If $i < r$ then Claim 1 settles the claim. If instead $i = r$, then there exists a vertex $\boldsymbol{w} \in F \cap F_r$ such that $\operatorname{dist}_{F_r}(\boldsymbol{v}_r, \boldsymbol{w}) < \operatorname{dist}_{F_r}(\boldsymbol{v}_r, \boldsymbol{v}_{r+1})$. If the vertex $\boldsymbol{w}$ were at distance at least ℓ from $\boldsymbol{v}_0$, then its existence would contradict the selection of $\boldsymbol{v}_{r+1}$. □

As a consequence of Claim 2, the facets in $\mathcal{F}_c$ are among the (at most n) facets of P at distance at most $\ell - 1$ from $\boldsymbol{v}_0$. It follows that

$$\#\mathcal{F}_c \leqslant n - \#\mathcal{F} = n - (r+1). \tag{7.4.11.1}$$

Let n_i be the number of $(d-2)$-faces close about F_i. Then

$$\sum_{i=0}^{r} n_i \leqslant 2(n - (r+1)) + r = 2n - r - 2 \leqslant 2n - 2, \tag{7.4.11.2}$$

because the total number of close ridges is at most twice the number of facets in $\mathcal{F}_c$ (see (7.4.11.1)) plus the number of trivial close ridges.

For $i \in [0 \ldots r]$, let L_i be a shortest $\boldsymbol{v}_i - \boldsymbol{v}_{i+1}$ path in F_i and let ℓ_i be its length. It is plain that the concatenation of $L_0, \ldots, L_r$ is a $\boldsymbol{v}_0 - \boldsymbol{v}_{r+1}$ path L in P. Then $\operatorname{dist}(\boldsymbol{v}_0, \boldsymbol{v}_{r+1}) \geqslant \ell$. We bound the length of L. From the definition of a $(d-2)$-face R being close about a facet F_i, it follows that R satisfies $\operatorname{dist}_{F_i}(\boldsymbol{v}_i, R) \leqslant \ell_i - 1$. Then, applying the induction hypothesis for $d-1$ with the parameters $\boldsymbol{v}_i$, ℓ_i, and n_i, we get that the length of L_i is at most $2^{d-3}(n_i - (d-1) + 5/2)/3$. As a result, the length of L satisfies

$$\begin{aligned}\ell \leqslant \sum_{i=0}^{r} \ell_i &\leqslant \frac{1}{3} 2^{d-3} \sum_{i=0}^{r} \left(n_i - (d-1) + \frac{5}{2} \right) \\ &= \frac{1}{3} 2^{d-3} \left(\sum_{i=0}^{r} n_i - (r+1)(d-1) + \frac{5}{2}(r+1) \right). \qquad (7.4.11.3)\end{aligned}$$

If $r \geqslant 1$ then this equation becomes

$$\begin{aligned}\ell &\leqslant \frac{1}{3} 2^{d-3}((2n-2) - 2(d-1) + 5) \qquad \text{(by (7.4.11.2))} \\ &= \frac{1}{3} 2^{d-2} \left(n - d + \frac{5}{2} \right).\end{aligned}$$

If instead $r = 0$, then $\boldsymbol{v}_0$ and $\boldsymbol{v}_{r+1}$ lie in the facet F_0, in which case $\ell_0 = \ell$ and a ridge $F \cap F_0$ is close about F_0 if F is a facet at distance $\ell - 1$ from $\boldsymbol{v}_0$. By assumption, there are at most $n-1$ such facets F, implying that there are at most $n-1$ such ridges. Thus $n_0 = n - 1$, and (7.4.11.3) becomes

$$\begin{aligned}\ell \leqslant \frac{1}{3} 2^{d-3} \left(n - 1 - (d-1) + \frac{5}{2} \right) &= \frac{1}{3} 2^{d-2} \left(\frac{n-d}{2} + \frac{5}{4} \right) \\ &< \frac{1}{3} 2^{d-2} \left(n - d + \frac{5}{2} \right).\end{aligned}$$

This completes the proof of the theorem. □

Corollary 7.4.12 (Barnette, 1974b) *For* $n > d \geqslant 2$, $\operatorname{diam}(n,d) \leqslant \frac{1}{3} 2^{d-2} \left(n - d + \frac{5}{2} \right)$.

Proof Let P be a simple d-polytope with exactly n facets, and let $\boldsymbol{v}$ and $\boldsymbol{u}$ be two vertices of P at distance $\operatorname{diam}(n,d)$. Trivially, there are at most n facets at distance at most $\operatorname{diam} P - 1$ from $\boldsymbol{v}$, and so Theorem 7.4.11 yields the corollary. □

Corollary 7.4.12 improves an earlier bound of Larman (1970).

7.5 The Bounded Hirsch Conjecture

Santos (2012) disproved the bounded Hirsch conjecture by providing a lower bound of $(1+\varepsilon)(n-d)$ for the diameter of d-polytopes with n facets. We present his counterexample, which works for the dual version of the conjecture.

A *d-prismatoid* P is a d-polytope with two disjoint facets F^+ and F^- that contain all the vertices of P. The facets F^+ and F^- are the *bases* of P and may not be unique; a cube is a prismatoid with any pair of opposite facets as bases. The *width* of P is the length ℓ of a shortest facet-ridge path from F^+ to F^-. Thus the dual diameter of P is at least ℓ. The significance of prismatoids is spelled out next.

Proposition 7.5.1 *A d-prismatoid with n vertices and width ℓ gives rise to another $(n-d)$-prismatoid P with $2n-2d$ vertices and width at least $\ell+n-2d$. In the case $\ell > d$, the dual of P is a counterexample to the bounded Hirsch conjecture.*

Proof Let $Q \subseteq \mathbb{R}^d$ be a d-prismatoid with n vertices, width ℓ, and bases Q^+ and Q^-. The proof goes by induction on $s := n-2d$. The case $s=0$ is true, and so assume the proposition holds for $s-1 \geqslant 0$.

Claim 1 The bases of Q (and of any prismatoid) can be assumed to be parallel.

Proof of claim Following the scheme of Section 2.5, we find an admissible projective transformation ζ that sends the intersection of aff Q^+ and aff Q^- to infinity; that is, the linear hyperplane at infinity H_∞^p contains aff $Q^+ \cap$ aff Q^-. The image $\zeta(Q)$ is a prismatoid that is combinatorially isomorphic to Q and has parallel bases. □

Since $s > 0$, one of the bases of Q is not a simplex, say Q^+.

Claim 2 Let $\boldsymbol{q}$ be any vertex of Q^+. Vertices of Q can be slightly perturbed so that every facet incident with $\boldsymbol{q}$ and different from Q^+ is a simplex. Moreover, the resulting polytope is a d-prismatoid with n vertices and dual diameter at least that of Q.

Proof of claim Let H_q be a hyperplane that separates $\boldsymbol{q}$ from the other vertices of Q but is very close to $\boldsymbol{q}$. Let $\boldsymbol{q}' \in$ rint $Q^+ \cap H_q$ so that $\boldsymbol{q}'$ is at Euclidean distance at most ε from $\boldsymbol{q}$ and the only hyperplanes spanned by vertices of Q that intersect the segment $[\boldsymbol{q},\boldsymbol{q}']$ are those containing $\boldsymbol{q}$. Since $\boldsymbol{q}' \in$ rint Q^+, it follows that aff Q^+ is the only hyperplane spanned by a facet of Q that contains $\boldsymbol{q}'$ (Theorem 1.9.10). Let Q' be the d-polytope $\operatorname{conv}((\mathcal{V}(Q)\setminus\{\boldsymbol{q}\}) \cup \{\boldsymbol{q}'\})$. The construction of Q' from Q resembles that of pushing $\boldsymbol{q}$ to $\boldsymbol{q}'$ (Section 2.10), but since q' lies in the boundary of Q and not in its interior as the definition of pushing demands, we will say that the polytope Q' is obtained from Q by *boundary pushing* q to q'.

A slight variation of Theorem 2.10.5 remains valid if the point q' is as we described. In particular, the following apply:

(i) $\mathcal{V}(Q') = (\mathcal{V}(Q)\backslash\{q\}) \cup \{q'\}$ because $q' \notin \operatorname{conv}(\mathcal{V}(Q)\backslash\{q\})$;
(ii) every proper face F of Q that does not contain q is a face of Q' because a hyperplane supporting Q at F is a hyperplane that does not contain q (or q'), and so it is a hyperplane supporting Q' at F;
(iii) the set $J^+ := \operatorname{conv}((Q^+\backslash\{q\}) \cup \{q'\})$ is a facet of Q' because aff Q^+ supports Q' at J^+; and
(iv) every proper face F' of Q' that contains q' and is different from J^+ is a pyramid with apex q'. Indeed, if F' were not a pyramid with apex q', then there would exist a facet $J' \neq J^+$ of Q' containing F' such that aff J' would be spanned by $\mathcal{V}(J') \cap \mathcal{V}(Q)$. This would result in a contradiction to the definition of boundary pushing.

If necessary, we do the same for some vertices of Q^-, pushing them into rint Q^- until we get the desired conclusion for q; this would replace the facet Q^- with a new facet. This settles the first part of the claim.

In fact, repeated boundary pushing of vertices in Q^+ and Q^- can produce a prismatoid where every facet other than its bases is a simplex.

For the second part, we observe, as Santos (2012) did, that the conclusion of Theorem 2.10.7(iii) remains valid under the definition of boundary pushing, yielding the statement on the dual diameter of Q'. □

Thanks to Claim 1, we can assume that Q has parallel bases Q^+ and Q^-, and since Q^+ is not simplex, we find a vertex $q \in Q^+$ that is not an apex of a pyramid over $\operatorname{conv}(Q^+\backslash\{q\})$. We can further assume that q is as described in Claim 2. Embed $Q \times \{0\}$ in the hyperplane $H := \{x \in \mathbb{R}^{d+1} \mid x_{d+1} = 0\}$ of $\mathbb{R}^{d+1}$. Let v be a vertex of Q^-, let X be the dual wedge of Q at v, and let $u := v \times \{-1\}$ and $w := v \times \{1\}$. The facets of X include a dual wedge Y^- of Q^- at v, the pyramid X_u over Q^+ with apex u, and the pyramid X_w over Q^+ with apex w; see Proposition 2.6.10. The faces Q^+ and Y^- of X are disjoint and contain all the $n+1$ vertices of X. We now perturb the vertex q of X.

Claim 3 The vertex q in Q^+ can be perturbed so that X becomes a prismatoid Y with $n+1$ vertices and width $\ell+1$.

Proof of claim We slightly move q in the direction of the segment uw and away from aff Q^+ to a position q' so that Q^+ becomes a d-dimensional pyramid Y^+ over $\operatorname{conv}(\mathcal{V}(Q^+)\setminus\{\boldsymbol{q}\})$ with apex $\boldsymbol{q}'$. In this way, $\operatorname{aff} Y^+ = \operatorname{aff}(Q^+ \cup \{\boldsymbol{q}'\})$ is parallel to $\operatorname{aff} Y^-$. The resulting polytope Y is a d-prismatoid with $n+1$ vertices and parallel bases Y^+ and Y^-.

Because of Claim 2, the only nonsimplex facets of X containing $\boldsymbol{q}$ are X_w and X_v. Since the perturbation of $\boldsymbol{q}$ didn't affect the simplex facets containing it, the face lattices of Y and X coincide except for the $(d-1)$-face Q^+ becoming the d-face Y^+, the facet X_u becoming a complex $\mathcal{C}_u$ formed by the pyramids with apex $\boldsymbol{u}$ over the lower faces of Y^+, and the facet X_w becoming a complex $\mathcal{C}_w$ formed by the pyramids with apex $\boldsymbol{w}$ over the upper faces of Y^+. Recall from Section 2.11 that a face F of a polytope Z in $\mathbb{R}^{d+1}$ is a lower face of Z if $\boldsymbol{x} - \lambda \boldsymbol{e}_{d+1} \notin Z$ for each point $\boldsymbol{x} \in F$ and each $\lambda > 0$; similarly, a face F of Z is an *upper face* if $\boldsymbol{x} + \lambda \boldsymbol{e}_{d+1} \notin Z$ for each point $\boldsymbol{x} \in F$ and each $\lambda > 0$. We prove the assertion on the width of Y.

Since a shortest facet-ridge path in Q from Q^+ to Q^- has length ℓ, Corollary 7.3.4 ensures that, in the prismatoid X, a shortest facet-ridge path L from Y^- to either X_u or X_w has length at least ℓ. Thus, a shortest facet-ridge path in Y from Y^+ to Y^- has length at least $\ell + 1$: it must first go to a facet J of Y in either $\mathcal{C}_u$ or $\mathcal{C}_w$ and then traverse a shortest facet-ridge path in Y from J to Y^-, which has length at least that of L. □

For the prismatoid Y of Claim 3 we have that $n+1-2(d+1) = n-2d-1$. So the induction hypothesis yields a prismatoid P with dimension $n+1-(d+1)$, $2(n+1)-2(d+1)$ vertices, and width at least $\ell + 1 + (n+1) - 2(d+1)$, completing the proof. □

In view of Proposition 7.5.1, we are after d-prismatoids with width greater than d. The 2-prismatoids, the trapezoids in $\mathbb{R}^2$, have width two, and the 3-prismatoids are not suitable either. A prismatoid is *simplicial* if the facets other than the bases are simplices.

Proposition 7.5.2 *The width of a 3-prismatoid is at most three. The simplicial 3-prismatoids are precisely the 3-prismatoids of width three.*

Proof Let P be a 3-prismatoid with bases F^+ and F^-. Since $\mathcal{V}(P) = \mathcal{V}(F^+) \cup \mathcal{V}(F^-)$, if a facet of P other than F^+ and F^- has four vertices then it must intersect both F^+ and F^- at an edge, in which case P has width two. If instead P is simplicial, then every facet F of P is adjacent to either F^+ or F^-, and so F can be accordingly classified as either a $+$ facet or a $-$ facet. Because the boundary complex of P is strongly connected (Proposition 3.3.2), there is a facet-ridge path from F^+ to F^-, which must contain a $+$ facet adjacent to a $-$ facet. Hence P has width three in this case. □

Proofs that 4-prismatoids have width at most four are more complicated (Santos et al., 2012). One such proof connects the notions of the Minkowski sum and the Cayley polytope of a sum (Section 6.1) with those of the normal cone and normal fan (Ziegler, 1995, ch. 7), as well as with the concept of geodesic maps (Santos, 2012). The details can be found in Santos et al. (2012). Dimension five brings the first suitable prismatoids.

Santos (2012) produced a 5-prismatoid with 48 vertices and width six and, thanks to Proposition 7.5.1, this prismatoid gives rise to a 43-prismatoid with 86 vertices and width at least 44. The 43-dimensional polytope is a counterexample to the dual version of the bounded Hirsch conjecture. Matschke et al. (2015) further reduced the dimension of the counterexamples. They constructed a 5-prismatoid P with 25 vertices and width six, which results in a 20-prismatoid with 40 vertices and width at least 21. The vertices of P are given as rows of the two matrices below; the 12 vertices of a base of $P \subseteq \mathbb{R}^5$ lie in the hyperplane $x_5 = 1$ of $\mathbb{R}^5$, while the 13 vertices of the other base of P lie in the hyperplane $x_5 = -1$ of $\mathbb{R}^5$.

$$\begin{pmatrix} 0 & 0 & 20 & -4 & 1 \\ 0 & 0 & -20 & -4 & 1 \\ 0 & 0 & 21 & -7 & 1 \\ 0 & 0 & -21 & -7 & 1 \\ 0 & 0 & 16 & -15 & 1 \\ 0 & 0 & -16 & -15 & 1 \\ 0 & 0 & 0 & 32 & 1 \\ 0 & 0 & 0 & -32 & 1 \\ \frac{3}{50} & -\frac{1}{25} & 0 & -30 & 1 \\ -\frac{3}{50} & -\frac{1}{25} & 0 & 30 & 1 \\ \frac{3}{1000} & \frac{7}{1000} & 0 & -\frac{318}{10} & 1 \\ -\frac{3}{1000} & \frac{7}{1000} & 0 & \frac{318}{10} & 1 \end{pmatrix} \quad \begin{pmatrix} 60 & 0 & 0 & 0 & -1 \\ 8 & -30 & 0 & 0 & -1 \\ 0 & -33 & 0 & 0 & -1 \\ -2 & -32 & 0 & 0 & -1 \\ -55 & 0 & 0 & 0 & -1 \\ -34 & 36 & 0 & 0 & -1 \\ 0 & 76 & 0 & 0 & -1 \\ 44 & 34 & 0 & 0 & -1 \\ -20 & 0 & \frac{1}{5} & -\frac{1}{5} & -1 \\ \frac{2999}{50} & 0 & -\frac{3}{25} & -\frac{1}{5} & -1 \\ \frac{299999}{5000} & 0 & 0 & \frac{1}{100} & -1 \\ -\frac{549}{10} & 0 & \frac{1}{5000} & \frac{1}{800} & -1 \\ -54 & 0 & \frac{1}{500} & -\frac{1}{80} & -1 \end{pmatrix} \tag{7.5.3}$$

Once we have a counterexample to the bounded Hirsch conjecture, we have infinitely many, via Cartesian products, connected sums, and combinations of them. The next proposition is a simple consequence of Problem 7.8.1.

Proposition 7.5.4 (Santos, 2012) *Let P be a d-polytope with n facets and diameter $(1+\varepsilon)(n-d)$ for some $\varepsilon > 0$. Then the k-fold Cartesian product of P is a kd-polytope with kn facets and diameter $(1+\varepsilon)(kn-kd)$.*

If the dimension d is fixed, connected sums facilitate more counterexamples. Let P_1 and P_2 be simple d-polytopes and let $\boldsymbol{v}_1$ and $\boldsymbol{v}_2$ be vertices of P_1 and P_2, respectively. For each i, we truncate the vertex $\boldsymbol{v}_i$ from P_i, obtaining a new simple polytope P_i'. The new facets $F_1' \subseteq P_1'$ and $F_2' \subseteq P_2'$ derived from the truncations are simplices, and so we can perform the connected sum $P := P_1' \# P_2'$ along them (Problem 2.15.9). The d-polytope P is not simple, but the polytopes P_1' and P_2' can be transformed such that each of the d $(d-2)$-faces shared by F_1' and F_2' is the intersection of one facet J_1' of P_1' and one facet J_2' of P_2', and the facets J_1' and J_2' blend together to form a single facet of P. In this way, each edge in P_1 incident with $\boldsymbol{v}_1$ gets blended with an edge in P_2 incident with $\boldsymbol{v}_2$ to form a single edge in P. The resulting polytope P is a simple d-polytope with $f_{d-1}(P_1') + f_{d-1}(P_2') - d$ facets and diameter at least $\operatorname{diam}(P_1) + \operatorname{diam}(P_2) - 1$. This operation is often called the *blending* of P_1 and P_2; see Holt and Klee (1998) for further details. Repeated blending produces infinitely many counterexamples.

Proposition 7.5.5 (Santos, 2012) *If P is a simple d-polytope with n facets and diameter at least $n-d+1$, then, for each $k>0$, there is a simple d-polytope with $k(n-d)+d$ facets and diameter at least $k(n-d)+1$.*

Proof The case $k=1$ is trivially true, and so assume that there is a simple d-polytope P_{k-1} with $(k-1)(n-d)+d$ facets and diameter at least $(k-1)(n-d)+1$. The blending of P_{k-1} and P is the desired simple polytope P_k, and an induction argument on k is complete. □

Matschke et al. (2015) also showed that the counterexamples derived from 5-prismatoids cannot violate the bounded Hirsch conjecture by more than 33%; they prove that no 5-prismatoid with n vertices has width greater than $n/3+3$. They also constructed an infinite family of 5-prismatoids with n vertices and width $\Omega(\sqrt{n})$ for arbitrarily large n. For d-prismatoids with $d \geqslant 6$ and n vertices, we must content ourselves with the general upper bound of Barnette (7.4.12), which yields that their width is no greater than $\frac{1}{3}2^{d-2}\left(n-d+\frac{5}{2}\right)$.

7.6 Diameters of Particular Polytopes

Section 7.4 provided general lower and upper bounds for the diameter of d-polytopes. For certain classes of polytopes, as we saw in Proposition 7.4.2 for $d=3$, better bounds are available. A $(0,k)$ *polytope* in $\mathbb{R}^d$ is a lattice polytope where the vertex coordinates are drawn from the set $[0\ldots k]$. The

(0, 1) polytopes are prominent examples of $(0, k)$ polytopes (Ziegler, 2000), which feature heavily in polyhedral optimisation. Naddef (1989) proved that (0, 1) polytopes have diameter at most their dimension. Kleinschmidt and Onn (1992) extended Naddef's result to every $k \geqslant 1$ by showing that $(0, k)$ polytopes of dimension d have diameter at most kd; we present a proof of this extension, which was suggested by Del Pia and Michini (2016).

We first relate the distance between a vertex and a face to values of a linear functional.

Lemma 7.6.1 *Let P be a lattice polytope in $\mathbb{R}^d$, let $\boldsymbol{u}$ be a vertex of P, and let $\varphi(\boldsymbol{x}) := \boldsymbol{a} \cdot \boldsymbol{x}$ be a linear functional in $\mathbb{R}^d$ with integral vector $\boldsymbol{a}$. If $\alpha := \min\{\varphi(\boldsymbol{x}) \mid \boldsymbol{x} \in P\}$ and $F := \min\{\boldsymbol{x} \in P \mid \varphi(\boldsymbol{x}) = \alpha\}$, then $\operatorname{dist}_P(\boldsymbol{u}, F) \leqslant \varphi(\boldsymbol{u}) - \alpha$.*

Proof Since $\min\{\varphi(\boldsymbol{x}) \mid \boldsymbol{x} \in P\}$ is attained at a vertex of P (Theorem 3.2.1), which has integral coordinates, we find that $\alpha \in \mathbb{Z}$ and that the quantity $\varphi(\boldsymbol{w}) - \alpha$ is a nonnegative integer for each vertex $\boldsymbol{w}$ of P. We proceed by induction on $\varphi(\boldsymbol{u}) - \alpha$. The case $\varphi(\boldsymbol{u}) - \alpha = 0$ implies that $\boldsymbol{u} \in F$, in which case $\operatorname{dist}_P(\boldsymbol{u}, F) = 0$. We assume that $\varphi(\boldsymbol{u}) - \alpha \geqslant 1$. Because $\varphi(\boldsymbol{u})$ is not a minimum over P in this case, there is a neighbour $\boldsymbol{u}'$ of $\boldsymbol{u}$ satisfying $\varphi(\boldsymbol{u}') < \varphi(\boldsymbol{u})$ (Theorem 3.2.2). Since $\varphi(\boldsymbol{u}') - \alpha < \varphi(\boldsymbol{u}) - \alpha$, the induction hypothesis on $\varphi(\boldsymbol{u}') - \alpha$ ensures that $\operatorname{dist}_P(\boldsymbol{u}', F) \leqslant \varphi(\boldsymbol{u}') - \alpha$. It follows that

$$\operatorname{dist}_P(\boldsymbol{u}, F) \leqslant \operatorname{dist}_P(\boldsymbol{u}', F) + \operatorname{dist}_P(\boldsymbol{u}, \boldsymbol{u}') \leqslant \varphi(\boldsymbol{u}') - \alpha + 1 \leqslant \varphi(\boldsymbol{u}) - \alpha,$$

which completes the induction. □

Considering two vertices at once, we get a corollary of Lemma 7.6.1.

Corollary 7.6.2 *Let P be a lattice polytope in $\mathbb{R}^d$, let $\boldsymbol{u}$ and $\boldsymbol{v}$ be two vertices of P, and let $\varphi(\boldsymbol{x}) := \boldsymbol{a} \cdot \boldsymbol{x}$ be a linear functional in $\mathbb{R}^d$ with integral vector $\boldsymbol{a}$. If $\alpha := \min\{\varphi(\boldsymbol{x}) \mid \boldsymbol{x} \in P\}$ and $F := \min\{\boldsymbol{x} \in P \mid \varphi(\boldsymbol{x}) = \alpha\}$, then $\operatorname{dist}_P(\boldsymbol{u}, \boldsymbol{v}) \leqslant \varphi(\boldsymbol{u}) + \varphi(\boldsymbol{v}) - 2\alpha + \operatorname{diam}(F)$.*

Proof From $\operatorname{dist}_P(\boldsymbol{u}, \boldsymbol{v}) \leqslant \operatorname{dist}_P(\boldsymbol{u}, F) + \operatorname{dist}_P(\boldsymbol{v}, F) + \operatorname{diam}(F)$ and Lemma 7.6.1 it follows that $\operatorname{dist}_P(\boldsymbol{u}, \boldsymbol{v}) \leqslant \varphi(\boldsymbol{u}) + \varphi(\boldsymbol{v}) - 2\alpha + \operatorname{diam}(F)$, as desired. □

Theorem 7.6.3 *For $k \geqslant 1$ and $d \geqslant 0$, the diameter of a $(0, k)$ polytope of dimension at most d is at most kd.*

Proof The proof is by induction on $d \geqslant 0$. The zero-dimensional polytopes have diameter zero, and so the case $d = 0$ holds. Let $P \subseteq \mathbb{R}^d$ be a

$(0, k)$ polytope of dimension at most $d \geqslant 1$. We can assume that P is full dimensional; suppose otherwise. Then there is a nonzero vector $\boldsymbol{c} = (c_1, \ldots, c_d)^t \in \mathbb{R}^d$ and a scalar γ such that $\boldsymbol{c} \cdot \boldsymbol{x} = \gamma$ for every $\boldsymbol{x} \in P$. Because $\boldsymbol{c} \neq \boldsymbol{0}$, there is an index i such that $c_i \neq 0$. We consider the projection π of P onto the i-coordinate hyperplane H of $\mathbb{R}^d$, the projection that deletes the i coordinate. This projection is an isomorphism between H and $\mathbb{R}^{d-1}$ that maps $P \subseteq H$ to an affinely isomorphic polytope $\pi(P) \subseteq \mathbb{R}^{d-1}$ (Proposition 2.1.3); in particular, $G(P) = G(\pi(P))$. Moreover, $\pi(P)$ is a $(0, k)$ polytope, as π maps integral vectors to integral vectors. So the proof can be carried out on $\pi(P)$. Repeated orthogonal projections then deliver a d'-dimensional $(0, k)$ polytope P' in $\mathbb{R}^{d'}$ with the same graph as P, and the proof can be carried out on P'. As a consequence, we can assume that P is d-dimensional.

Let $r := \min\{x_1 \mid \boldsymbol{x} = (x_1, \ldots, x_d)^t \in P\}$ and let $h := \max\{x_1 \mid \boldsymbol{x} = (x_1, \ldots, x_d)^t \in P\}$. Since P is full dimensional, we have that $h > r$. From Theorem 3.2.1 it then follows that

$$X_r := \min\{\boldsymbol{x} \in P \mid x_1 = r\} \text{ and } X_h := \min\{\boldsymbol{x} \in P \mid x_1 = h\}$$

are proper faces of P, and so they are $(0, k)$ polytopes of dimension at most $d - 1$. Consider two vertices $\boldsymbol{u} = (u_1, \ldots, u_d)^t$ and $\boldsymbol{v} = (v_1, \ldots, v_d)^t$ of P. If $u_1 + v_1 \leqslant r + h$, then Corollary 7.6.2 on $F = X_r$, $\varphi(\boldsymbol{x}) := \boldsymbol{e}_1 \cdot \boldsymbol{x}$, and $\alpha = r$ yields that

$$\operatorname{dist}_P(\boldsymbol{u}, \boldsymbol{v}) \leqslant u_1 + v_1 - 2r + \operatorname{diam}(X_r) \leqslant h - r + \operatorname{diam}(X_r). \quad (7.6.3.1)$$

If instead $u_1 + v_1 \geqslant r + h$, then Corollary 7.6.2 on $F = X_h$, $\varphi(\boldsymbol{x}) := (-\boldsymbol{e}_1) \cdot \boldsymbol{x}$, and $\alpha = -h$ yields that

$$\operatorname{dist}_P(\boldsymbol{u}, \boldsymbol{v}) \leqslant -(u_1 + v_1) + 2h + \operatorname{diam}(X_h) \leqslant h - r + \operatorname{diam}(X_h). \quad (7.6.3.2)$$

If, for $k \geqslant 1$ and $d \geqslant 1$, we define

$$\operatorname{diam}_\ell^k(d) := \max\{\operatorname{diam} P \mid P \text{ a } (0, k) \text{ polytope of dimension at most } d\},$$

then combining (7.6.3.1) and (7.6.3.2) yields that

$$\operatorname{dist}_P(\boldsymbol{u}, \boldsymbol{v}) \leqslant h - r + \operatorname{diam}_\ell^k(d - 1). \quad (7.6.3.3)$$

The induction hypothesis on $d - 1$ ensures that $\operatorname{diam}_\ell^k(d - 1) \leqslant k(d - 1)$, and P being a $(0, k)$ polytope ensures that $h - r \leqslant k$. Hence, from (7.6.3.3) we get that

$$\operatorname{dist}_P(\boldsymbol{u}, \boldsymbol{v}) \leqslant h - r + k(d - 1) \leqslant k + k(d - 1) = kd,$$

completing the induction argument. □

The case $k = 1$ of Theorem 7.6.3 gives that (0, 1) polytopes of dimension d satisfy Hirsch's conjecture. The bound of Theorem 7.6.3 for $k \geqslant 2$ was later improved by Del Pia and Michini (2016) to $\lfloor (k - 1/2)d \rfloor$, which is tight for $k = 2$. And the bound $\lfloor (k - 1/2)d \rfloor$ was then improved by Deza and Pournin (2018) to $kd - \lceil 2d/3 \rceil - (k - 3)$, which is tight for $k = 3$.

We mention other results on diameters of specific polytopes. Klee and Walkup (1967) established that $\mathrm{diam}(8, 4) = 4$ and $\mathrm{diam}(10, 5) = 5$, while Bremner and Schewe (2011) established that $\mathrm{diam}(12, 6) = 6$. These results together with Proposition 7.3.6 imply the veracity of Hirsch's conjecture for d-polytopes with at most $d + 6$ facets.

Theorem 7.6.4 *The diameter of a d-polytope with $n \leqslant d + 6$ facets is at most $n - d$.*

Duals of cyclic d-polytopes with at most $n \leqslant 2d$ facets have diameter at most $n - d$ (Klee, 1964a), and so they satisfy Hirsch's conjecture. Later, Maksimenko (2009) computed the diameter of such polytopes in the case $n > 2d$.

Theorem 7.6.5 *The dual of cyclic d-polytope with n facets has diameter at most $n - d$ if $n \leqslant 2d$ and diameter at most $n - d - \lceil (n - 2d)/(\lfloor d/2 \rfloor + 1) \rceil$ if $n > 2d$.*

Bonifas et al. (2014) considered H-polytopes given as $A\boldsymbol{x} \leqslant \boldsymbol{b}$ where $A \in \mathbb{Z}^{m \times n}$ (an integer $m \times n$ matrix) and $\boldsymbol{b} \in \mathbb{R}^m$, and they established that that their diameter is at most $O(\Delta^2 n^{3.5} \log(n\Delta))$, where Δ denotes the largest absolute value of a *subdeterminant* of A, the determinant of a square submatrix of A. Yang (2021) produced a similar bound for H-polytopes defined by real $m \times n$ matrices.

Finally, we mention that Adiprasito and Benedetti (2014) proved that flag d-polytopes with n facets have diameter at most $n - d$, which means that they satisfy Hirsch's conjecture; see also Section 3.9.

7.7 Polyhedral Abstractions

To deal with the Hirsch conjectures, researchers have often tried to generalise the conjectures to more general settings, with the aim being to isolate the essential properties that will bound the diameters; this was, for instance, the case in Adler and Dantzig (1974); Kalai (1992); Eisenbrand et al. (2010); Santos (2013). This approach makes sense for at least two reasons: the known proofs of the existing upper bounds use only basic properties of graphs of simple polyhedra, and the previous bounds of Kalai and Kleitman (1992) and Larman (1970) have been rediscovered in combinatorial settings devoid of geometry or topology (Eisenbrand et al., 2010).

The *base abstraction* of Eisenbrand et al. (2010) is a connected graph $G = (V, E)$ where the vertices are labelled by a d-element subset from $\{1, \ldots, n\}$ and the edges are such that a connectivity property holds:

(A1) For vertices u, v there is a path in G from u to v whose intermediate vertices all contain $u \cap v$.

The number d is the *dimension* of the abstraction, while n is its *number of facets*. It is plain that this base abstraction contains graphs of simple polyhedra. This is the case because each vertex of a simple d-polyhedron is incident with precisely d facets, and so it can be labelled with these facets. Furthermore, the connectivity property holds for such polyhedra. Eisenbrand et al. (2010) proved upper bounds $n^{1+\log d} - 1$ and $2^{d-1}n - 1$ for the diameter of a base abstraction of dimension d and n facets, which recovers the bound of Kalai and Kleitman (1992) and a bound slightly worse than that of Larman (1970). Eisenbrand et al. (2010) presented a two-dimensional base abstraction with six facets and diameter five, but Todd's bound (7.4.8) yields $\operatorname{diam}_u(6, 2) \leqslant 4$ and Barnette's bound (7.4.11) is not defined for $d = 2$.

Other abstractions of polyhedra have appeared in the literature. In addition to Condition (A1), Kalai (1992) proposed one on edges, which are labelled by a $(d - 1)$-element subset from $\{1, \ldots, n\}$:

(A2) an edge uv is present in G if and only if $\#(u \cap v) = d - 1$.

For this abstraction, called an *ultraconnected set system*, Kalai (1992) matched the bound of Kalai and Kleitman (1992). Adler and Dantzig (1974) added a third condition to define their *abstract polytopes*:

(A3) a subset of $(d - 1)$ elements from the set $\{1, \ldots, n\}$ is contained either in two vertices of G or in no vertex of G.

Whereas Condition (A2) applies to simple polyhedra, Condition (A3) applies only to simple polytopes. Adler and Dantzig (1974) showed that the diameter of abstract polytopes is bounded by $n - d$ if $n - d \leqslant 5$.

Abstract simplicial complexes offer insight into the combinatorics of simple polyhedra. Consider a simple d-polyhedron P with n facets. Each face of P is the intersection of facets (Theorem 2.3.1), and so it can be associated with a subset from $[1 \ldots n]$. More precisely, we let $\mathcal{K}$ be the abstract simplicial complex whose elements are the subsets of $[1 \ldots n]$ associated with the faces of P. Then $\mathcal{K}$ is a pure abstract simplicial $(d - 1)$-complex with n vertices. Following Santos (2013), we call this complex $\mathcal{K}$ derived from P the *dual face complex* of P. As in the case of polytopes, we define the *dual graph* $G^*(\mathcal{K})$ of the complex $\mathcal{K}$ as the graph whose vertices are the facets of $\mathcal{K}$ and whose edges are the pairs of facets that intersect at a ridge of $\mathcal{K}$. It is easy to see that if P has at least one vertex, then $G^*(\mathcal{K}) = G(P)$.

Dual face complexes of simple polyhedra facilitate the study of the Hirsch conjectures in the framework of abstract simplicial complexes. If the dual graph of an abstract simplicial complex $\mathcal{K}$ is connected, then we say that $\mathcal{K}$ is *strongly connected*; we focus only on strongly connected complexes, and so we will assume this property henceforth. The *dual diameter* of an abstract simplicial complex $\mathcal{K}$ is the diameter of its dual graph. We define the maximum dual diameter over all such complexes:

$$\begin{aligned}\operatorname{diam}_s(n,d) := \max\{&\operatorname{diam}(G^*(\mathcal{K})) \mid \mathcal{K} \text{ a strongly connected, abstract}\\ &\text{simplicial } (d-1)\text{-complex with at most } n \text{ vertices}\}.\end{aligned} \tag{7.7.1}$$

The *complete (d-1)-complex* on n vertices is the abstract simplicial complex whose faces are all the d-element subsets from $[1 \ldots n]$. The dual graph of this complex is the *Johnson graph* $J(n,d)$, whose relevance is highlighted below.

Proposition 7.7.2 *The value of* $\operatorname{diam}_s(n,d)$ *is realised by a longest induced path in* $J(n,d)$.

Proof Let $\mathcal{K}$ be an abstract simplicial $(d-1)$-complex with n vertices and let L be a facet-ridge path in $\mathcal{K}$ realising $\operatorname{diam}_s(n,d)$. The facets in L induce another abstract simplicial $(d-1)$-complex $\mathcal{K}'$ whose dual diameter coincides with that of $\mathcal{K}$, and so $\mathcal{K}'$ also realises $\operatorname{diam}_s(n,d)$. The complex $\mathcal{K}'$ is a subcomplex of the complete $(d-1)$-complex $\mathcal{K}_n$ with n vertices, and therefore $G^*(\mathcal{K}')$ is an induced subgraph of $G^*(\mathcal{K}_n)$, an induced path of $G^*(\mathcal{K}_n)$. □

A corollary of Proposition 7.7.2 is that $\operatorname{diam}_s(n,d)$ is attained by abstract simplicial complexes, called *corridors*, whose dual graphs are paths. This enables a simple upper bound for $\operatorname{diam}_s(n,d)$.

Proposition 7.7.3 (Santos, 2013) *For* $n \geqslant d \geqslant 2$, $\operatorname{diam}_s(n,d) < \frac{1}{d-1}\binom{n}{d-1} - 1$.

Proof Let $\mathcal{K}$ be a $(d-1)$-dimensional corridor on n vertices and $\ell+1$ facets $F_0, \ldots, F_\ell$, where its dual graph is a path $F_0 \ldots F_\ell$ of length ℓ. We estimate the number of ridges in $\mathcal{K}$. Each of the $\ell+1$ facets of $\mathcal{K}$ has d ridges, the ℓ ridges $F_i \cap F_{i+1}$ belong to two facets, and every other ridge belongs to only one facet. Therefore, the number of ridges of $\mathcal{K}$ is $d(\ell+1) - \ell$. On the other hand, $d(\ell+1) - \ell$ is less than the number $\binom{n}{d-1}$ of ridges in the complete $(d-1)$-complex on n vertices. Thus

$$\ell \leqslant \frac{1}{d-1}\binom{n}{d-1} - \frac{d}{d-1} < \frac{1}{d-1}\binom{n}{d-1} - 1.$$

This settles the bound. □

With respect to lower bounds for $\mathrm{diam}_s(n,d)$, Santos (2013) gave a construction that yields $\Omega\left(n^{2d/3}\right) \leqslant \mathrm{diam}_s(n,d)$, and later Criado and Santos (2017) improved upon that.

Proposition 7.7.4 (Criado and Santos, 2017) *For $n \geqslant d \geqslant 2$, $\mathrm{diam}_s(n,d) \geqslant \frac{n^{d-1}}{(d+2)^{d-1}} - 3$.*

A consequence of Propositions 7.7.3 and 7.7.4 is that the polynomial Hirsch conjecture (7.1.2) cannot be proved in the scope of abstract simplicial complexes; other properties of simple polyhedra not captured by these complexes would need to be invoked. While the dual graphs of abstract simplicial complexes satisfy Condition (A2), only those of normal abstract simplicial complexes adhere to Condition (A1). As a result, normal abstract simplicial complexes meet the bounds of Eisenbrand et al. (2010); see also Santos (2013, sec. 3). We define the link of a face in an abstract simplicial complex in the same way as the link of a polytopal complex (Section 2.11). Furthermore, a pure (abstract) simplicial complex is *normal* if the link of every face is strongly connected.

The aforementioned result of Adiprasito and Benedetti (2014) on flag polytopes (Section 7.6) lends further support to the study of normal abstract simplicial complexes; in fact, they showed that flag normal abstract simplicial complexes satisfy Hirsch's original conjecture. Flag abstract simplicial complexes are defined in the same manner as flag simplicial complexes (Section 3.9).

Other polyhedral abstractions came about in the context of the project Polymath 3 of G. Kalai (coordinator) (2010), which aimed to advance the polynomial Hirsch conjecture. Santos (2013, sec. 3) reported on the ideas and results of this project.

7.8 Problems

7.8.1 (Klee and Walkup, 1967) If P_1 is a d_1-polyhedron and P_2 is a d_2-polyhedron, then their Cartesian product $P_1 \times P_2$ is a (d_1+d_2)-polyhedron with $f_{d_1-1}(P_1) + f_{d_2-1}(P_2)$ facets and diameter $\mathrm{diam}(P_1) + \mathrm{diam}(P_2)$.

7.8.2 (Ziegler, 1995, ex. 2.19) Let P be an unbounded pointed polyhedron in $\mathbb{R}^d$. Prove the following assertions.

(i) There is a polytope $P' := \{x \in P \mid a \cdot x \leqslant 1\}$ and a facet $F' := \{x \in P \mid a \cdot x = 1\}$ of P' such that, for $k \geqslant 0$, the k-faces of F' correspond to the unbounded $(k+1)$-faces of P, while the k-faces of P' that are not faces of F' are in one-to-one correspondence with the bounded k-faces of P.

(ii) There is a nonadmissible projective transformation ζ from P to a polytope P'' projectively isomorphic to P. Additionally, ensure that there exists a unique maximal face F'' among the faces of P'' that are not images of faces of P under ζ.

(iii) The facet poset of P is obtained from that of P'' by removing F'' and all its faces.

7.8.3 (Blending of polytopes, Holt and Klee, 1998) Let P_1 and P_2 be simple d-polytopes with n_1 and n_2 facets, respectively. Prove that there is a simple d-polytope, obtained by the 'blending' of P_1 and P_2, that has $n_1 + n_2 - d$ facets and diameter at least $\operatorname{diam}(P_1) + \operatorname{diam}(P_2) - 1$.

7.9 Postscript

The presentation of Sections 7.2 and 7.3 is based on the accounts of Klee and Walkup (1967); Kim and Santos (2010a,b); Santos (2012). For the lower bound theorem in Section 7.4, we consulted Kim and Santos (2010a), while the upper bound theorems are based on the original sources of Todd (2014) and Barnette (1974b). The discussion of Santos' counterexample follows the clear account in Santos (2012). The blending operation of Holt and Klee (1998) is a variation of a construction of Barnette (1969).

Section 7.7 is based on the accounts of Eisenbrand et al. (2010) and Santos (2013). Holmes (2018) considered a generalisation of d-polyhedra with n facets, whose graphs satisfy conditions (A1) and (A2). He established the bound $2^{d-2}(n-2)$ for the diameter of such graphs, thereby improving upon one of the bounds proposed by Eisenbrand et al. (2010).

8

Faces

We investigate numbers of faces of polytopes. We begin with the face numbers of 3-polytopes, including the well-known theorem of Steinitz (1906). The characterisation of f-vectors of d-polytopes ($d \geqslant 4$) is beyond our current means. In view of this, researchers have considered characterisations of the 'projections' of f-vectors, namely the proper subsequences of the f-vector; we review the existing results.

Section 8.2 gives a proof of a theorem of Xue (2021) on the minimum number of faces of d-polytopes with at most $2d$ vertices, answering a conjecture of Grünbaum (2003, sec. 10.2). This is followed by results on the minimum number of faces of d-polytopes with more than $2d$ vertices.

We then discuss the celebrated lower and upper bound theorems for simplicial polytopes, due to Barnette (1973c) and McMullen (1970), respectively. We also explore extensions of these theorems, such as the g-conjecture of McMullen (1971b), now the g-theorem. The proof of the lower bound theorem, based on Kalai (1987), connects rigidity theory with the combinatorics of polytopes.

The chapter ends with a discussion of the flag vector of a polytope, a generalisation of the f-vector. This includes a result of Bayer and Billera (1985) on linear equations for flag vectors, akin to Dehn–Sommerville's equations (2.13.3) for simplicial polytopes.

8.1 Face Numbers

The characterisation of the face numbers of polytopes, which are encoded in the f-vectors of the polytopes, is a fundamental problem in polytope theory.

Problem 8.1.1 (f-vectors) *For $d \geqslant 1$, characterise the f-vectors of d-polytopes.*

The case $d = 1$ is trivial as the f-vector is (2), and so is the case $d = 2$ where the f-vector is (f_0, f_0) for $f_0 \geqslant 3$. The case $d = 3$ was settled by Steinitz (1906).

Theorem 8.1.2 (Steinitz 1906) *The f-vector (f_0, f_1, f_2) of a 3-polytope is uniquely determined by the values f_0 and f_2:*

$$\left\{(f_0, f_1, f_2) \in \mathbb{Z}^3 \middle| f_1 = f_0 + f_2 - 2, 4 \leqslant f_0 \leqslant 2f_2 - 4, 4 \leqslant f_2 \leqslant 2f_0 - 4\right\}.$$

Furthermore, for every such pair (f_0, f_2) there is a 3-polytope with such an f-vector.

It is instructive to use Euler–Poincaré–Schläfli's equation (2.12.17) to write down a proof of the theorem. One such proof can be found in Grünbaum (2003, thm. 10.3.2).

As mentioned before, the characterisation of f-vectors (f_0, f_1, f_2, f_3) of 4-polytopes is currently beyond our grasp. A tractable problem is to consider the characterisations of the 'projections' of the f-vectors, namely the proper subsequences of the f-vector. If three values in $\{f_0, f_1, f_2, f_3\}$ are known, then the fourth can be obtained from Euler–Poincaré–Schläfli's equation (2.12.17):

$$f_0 - f_1 + f_2 - f_3 = 0.$$

Thus, the characterisation of the three-projections of the f-vectors of 4-polytopes is equivalent to the characterisation of the full f-vectors.

All the two-projections of the f-vectors (f_0, f_1, f_2, f_3) of 4-polytopes have been characterised. The two-projections are (f_0, f_1), (f_0, f_2), (f_0, f_3), (f_1, f_2), (f_1, f_3), and (f_2, f_3). But some of these can be obtained from others by duality. If there is a 4-polytope with pair (f_0, f_1), then the dual polytope has pair (f_2, f_3), and so the pair (f_2, f_3) can be obtained from the pair (f_0, f_1) by duality. Similarly, the pair (f_1, f_3) can be obtained from the pair (f_0, f_2) by duality. Grünbaum (2003, sec. 10.4) determined the projections (f_0, f_1) and (f_0, f_3), Barnette and Reay (1973) characterised the pairs (f_0, f_2), and Barnette (1974a) characterised the pairs (f_1, f_2).

Theorem 8.1.3 (Grünbaum 2003, thm. 10.4.1,10.4.2) *The following hold.*

(i) *There is a 4-polytope P satisfying $(f_0, f_1) := (f_0(P), f_1(P))$ if and only if $10 \leqslant 2f_0 \leqslant f_1 \leqslant \binom{f_0}{2}$ and (f_0, f_1) is not one of the pairs $(6, 12), (7, 14), (8, 17), (10, 20)$.*

(ii) *There is a 4-polytope P satisfying $(f_0, f_3) := (f_0(P), f_3(P))$ if and only if $5 \leqslant f_0 \leqslant \frac{1}{2} f_3(f_3 - 3)$ and $5 \leqslant f_3 \leqslant \frac{1}{2} f_0(f_0 - 3)$.*

Theorem 8.1.4 (Barnette and Reay, 1973, thm. 10) *There is a 4-polytope P satisfying $(f_0, f_2) := (f_0(P), f_2(P))$ if and only if*

$$10 \leqslant \frac{1}{2}\left(2f_0 + 3 + \sqrt{8f_0 + 9}\right) \leqslant f_2 \leqslant f_0^2 - 3f_0,$$

$f_2 \neq f_0^2 - 3f_0 - 1$, *and* (f_0, f_2) *is not one of the pairs* $(6, 12), (6, 14), (7, 13), (7, 15), (8, 15), (8, 16), (9, 16), (10, 17), (11, 20), (13, 21)$.

Proof Corollary 2.4.15 ensures that $2f_{d-2} \geqslant df_{d-1}$, which reduces to $2f_2 \geqslant 4f_3$ for 4-polytopes. This together with Euler–Poincaré–Schläfli's equation (2.12.17) yields that

$$f_2 \leqslant 2f_1 - 2f_0, \; f_3 \leqslant f_1 - f_0 \tag{8.1.4.1}$$

hold for each 4-polytope. If we combine the first of these equations with the inequality $f_1 \leqslant \binom{f_0}{2}$, then we obtain that

$$f_2 \leqslant 2f_1 - 2f_0 \leqslant 2\binom{f_0}{2} - 2f_0 = f_0^2 - 3f_0, \tag{8.1.4.2}$$

with equality holding for the neighbourly 4-polytopes.

Solving the inequality $2f_1 - 2f_0 \leqslant f_0^2 - 3f_0$ for the variable f_0 ensures that $f_0 \geqslant 1/2 + \sqrt{1 + 8f_1}/2$, which combined with (8.1.4.1) yields that

$$f_3 \leqslant f_1 - f_0 \leqslant f_1 - \frac{1}{2} - \frac{1}{2}\sqrt{1 + 8f_1}.$$

If we consider this inequality on the dual polytope of P, then this turns into

$$f_0 \leqslant f_2 - \frac{1}{2} - \frac{1}{2}\sqrt{1 + 8f_2}.$$

To solve this inequality for the variable f_2, we multiply it by 8 and solve it for the variable $\sqrt{1 + 8f_2}$ as in

$$8f_0 \leqslant \left(\sqrt{1 + 8f_2}\right)^2 - 5 - 4\sqrt{1 + 8f_2}.$$

The solution is

$$f_2 \geqslant \frac{1}{2}\left(2f_0 + 3 + \sqrt{8f_0 + 9}\right), \tag{8.1.4.3}$$

with equality holding for the duals of the neighbourly 4-polytopes. The inequalities (8.1.4.2) and (8.1.4.3) give a part of the theorem. We now show that $f_2 \neq f_0^2 - 3f_0 - 1$.

Suppose that P is a 4-polytope for which $f_2 = f_0^2 - 3f_0 - 1$. For simplicial 4-polytopes, we have that $f_2 = 2f_3$ (Corollary 2.4.15), in which case f_2 would be even. Since the number $f_0^2 - 3f_0 - 1$ is odd, P is not simplicial. The polytope

P is not neighbourly either as, for even d, every neighbourly d-polytope is simplicial (Proposition 2.9.11). Thus,

$$f_1(P) \leqslant \binom{f_0}{2} - 1. \tag{8.1.4.4}$$

On the other side, from (8.1.4.1) and Euler–Poincaré–Schläfli's equation (2.12.17), it follows that

$$f_3 = f_0 - f_1 + \left(f_0^2 - 3f_0 - 1\right) \leqslant f_1 - f_0,$$

which gives that $f_1 \geqslant \binom{f_0}{2} - 1/2$, or, equivalently, that $f_1 = \binom{f_0}{2}$, contradicting (8.1.4.4). This shows the nonexistence of P.

The nonexistence of a 4-polytope for each of the pairs listed in the theorem, as well as the existence of a 4-polytope P for each suitable pair $(f_0(P), f_2(P))$, are left as exercises. □

As previously mentioned, Barnette (1974a) characterised the pairs (f_1, f_2) such that there is a 4-polytope with f_1 edges and f_2 2-faces; some mistakes were corrected in Höppner and Ziegler (2000).

Theorem 8.1.5 (Barnette, 1974a, Höppner and Ziegler, 2000) *There is a 4-polytope P satisfying $(f_1, f_2) := (f_1(P), f_2(P))$ if and only if*

$$10 \leqslant \frac{1}{2}f_1 + \left\lceil \frac{1}{2} + \sqrt{f_1 + \frac{9}{4}} \right\rceil + 1 \leqslant f_2,\ f_2 \neq \frac{1}{2}f_1 + \sqrt{f_1 + \frac{13}{4}} + 2,$$

$$10 \leqslant \frac{1}{2}f_2 + \left\lceil \frac{1}{2} + \sqrt{f_2 + \frac{9}{4}} \right\rceil + 1 \leqslant f_1,\ f_1 \neq \frac{1}{2}f_2 + \sqrt{f_2 + \frac{13}{4}} + 2,$$

and (f_1, f_2) is not one of the pairs (12, 12), (13, 14), (14, 13), (14, 14), (15, 15), (15, 16), (16, 15), (16, 17), (16, 18), (17, 16), (17, 20), (18, 16), (18, 18), (19, 21), (20, 17), (20, 23), (20, 24), (21, 19), (21, 26), (23, 20), (24, 20), (26, 21).

Beyond dimension four, we are aware only of the projection (f_0, f_1) of f-vectors of 5-polytopes, independently found by Pineda-Villavicencio et al. (2018) and Kusunoki and Murai (2019).

Theorem 8.1.6 (Pineda-Villavicencio et al., 2018, Kusunoki and Murai, 2019) *There is a 5-polytope P satisfying $(f_0, f_1) := (f_0(P), f_1(P))$ if and only if $15 \leqslant \frac{5}{2}f_0 \leqslant f_1 \leqslant \binom{f_0}{2}$, $f_1 \neq \left\lfloor \frac{5}{2}f_0 + 1 \right\rfloor$, and (f_0, f_1) is not one of the pairs* (8, 20), (9, 25), (13, 35).

8.2 Polytopes with a Small Number of Vertices

Grünbaum (2003, sec. 10.2) defined the function

$$\theta_k(d+s,d) := \binom{d+1}{k+1} + \binom{d}{k+1} - \binom{d+1-s}{k+1}, \text{ for } 1 \leqslant s \leqslant d. \quad (8.2.1)$$

He conjectured that $\theta_k(d+s,d)$ gives the minimum number of k-faces of a d-polytope with $d+s$ vertices, and proved the conjecture for $s = 2,3,4$. Xue (2021) then completed the proof, including a characterisation of the unique minimisers for $k \in [1 \ldots d-2]$. Earlier, Pineda-Villavicencio et al. (2019) had settled the conjecture for some values of k.

When $k \in [1 \ldots d-2]$, each minimiser of (8.2.1) is a $(d-s)$-fold pyramid over a simplicial s-prism for $s \in [1 \ldots d]$; we call such a polytope the *$(s,d-s)$-triplex* and denote it by $M(s,d-s)$. In particular, $M(1,d-1)$ is a d-simplex and $M(d,0)$ is a simplicial d-prism. Triplices were introduced in Pineda-Villavicencio et al. (2019, sec. 3). The name alludes to the fact that a triplex is the convex hull of three simplices. It is easy to prove by induction on d for all $s \leqslant d$ that $f_k(M(s,d-s)) = \theta_k(d+s,d)$. We are now ready to state the main theorem of the section.

Theorem 8.2.2 (Lower bound theorem for d-polytopes with at most $2d$ vertices, Xue, 2021) *Let $d \geqslant 2$ and $1 \leqslant s \leqslant d$. If P is a d-polytope with $d+s$ vertices, then*

$$f_k(P) \geqslant \theta_k(d+s,d), \text{ for all } k \in [1 \ldots d-1].$$

Also, if $f_k(P) = \theta_k(d+s,d)$ for some $k \in [1 \ldots d-2]$, then P is the $(s,d-s)$-triplex.

The proof of Theorem 8.2.2 relies on estimating the number of k-faces outside a given a facet. This is done next.

Proposition 8.2.3 *Let $d \geqslant 2$, and let P be a d-polytope. In addition, suppose that $S := (\boldsymbol{v}_1, \ldots, \boldsymbol{v}_r)$ is a sequence of distinct vertices of P where $1 \leqslant r \leqslant d+1$. Then the following hold.*

(i) *There is a sequence $F_1, \ldots, F_r$ of faces of P such that F_i has dimension $d-i+1$, contains $\boldsymbol{v}_i$, but doesn't contain any $\boldsymbol{v}_j$ with $j < i$.*
(ii) *The number of k-faces of P that contains at least one of the vertices in S is at least*

$$\sum_{i=1}^{r} f_{k-1}(F_i/\boldsymbol{v}_i).$$

Proof (i) We prove the result by induction on $r \geqslant 1$. If $r = 1$, then we let $F_1 := P$, in which case the number of k-faces in F_1 containing $\boldsymbol{v}_1$ equates to the number of $(k-1)$-faces in the vertex figure $F_1/\boldsymbol{v}_1$ of F_1 at $\boldsymbol{v}_1$ (Theorem 2.7.2). Suppose that $r \geqslant 2$. By the induction hypothesis, for the subsequence $(\boldsymbol{v}_1, \ldots, \boldsymbol{v}_{r-2}, \boldsymbol{v}_{r-1})$ there exists a sequence $F_1, \ldots, F_{r-2}, F_{r-1}$ of faces of P satisfying the proposition statement. Likewise, for the subsequence $(\boldsymbol{v}_1, \ldots, \boldsymbol{v}_{r-2}, \boldsymbol{v}_r)$ there exists a sequence $J_1, \ldots, J_{r-2}, J$ of faces of P satisfying the proposition statement. In particular, J has dimension $d-(r-1)+1$, contains $\boldsymbol{v}_r$, but doesn't contain any vertex in $S\backslash\{\boldsymbol{v}_{r-1}, \boldsymbol{v}_r\}$. Since $d-r+2 \geqslant 1$, regardless of whether $\boldsymbol{v}_{r-1}$ is present or not in J, J has at least one facet F_r that contains $\boldsymbol{v}_r$ but doesn't contain $\boldsymbol{v}_{r-1}$. Since F_r doesn't contain any vertex in $S\backslash\{\boldsymbol{v}_r\}$, the sequence $F_1, \ldots, F_{r-1}, F_r$ satisfies (i).

(ii) Because of the nature of the sequence $F_1, \ldots, F_{r-1}, F_r$, the k-faces in P containing at least one vertex in S include the k-faces in F_i containing $\boldsymbol{v}_i$, whose number coincides with the number of $(k-1)$-faces in the vertex figure $F_i/\boldsymbol{v}_i$ of F_i at $\boldsymbol{v}_i$. This settles this part. □

We give two corollaries of Proposition 8.2.3, but first another characterisation of simplices.

Proposition 8.2.4 *A d-polytope is a pyramid over $d-1$ or more of its facets if and only if it is a simplex.*

Proof A d-simplex is a pyramid over each of its $d+1$ facets; we prove the other direction. The assertion is true for $d=2$, and so an inductive argument on $d \geqslant 3$ can start. Suppose that P is a d-dimensional pyramid over $d-1$ or more of its facets. Pick one such facet F. It follows that P is an r-fold pyramid with $r \geqslant d-1$ (Problem 2.15.6), implying that F is a pyramid over $d-2$ or more of its facets. By the induction hypothesis, F is a simplex, causing P to be a simplex. □

Corollary 8.2.5 *For each $d \geqslant 2$ and each $k \in [0 \ldots d]$, if P is a simplex, then $f_k(P) = \binom{d+1}{k+1}$, otherwise*

$$f_k(P) \geqslant \binom{d+1}{k+1} + \binom{d-1}{k} = \theta_k(d+2, d).$$

Proof Suppose that P is not a simplex. Then it is not a pyramid over some facet F of it (Proposition 8.2.4). The corollary is true for $d=2$, and so by an inductive argument on d, assume that it is true for every dimension in $[2 \ldots d-1]$. Let $\boldsymbol{v}_1$ and $\boldsymbol{v}_2$ be two vertices outside F. The k-faces of P include the set of k-faces in F and the set of k-faces containing $\boldsymbol{v}_1$ or $\boldsymbol{v}_2$.

According to Proposition 8.2.3, there are two faces F_1, F_2 of P satisfying the proposition statement. In particular, $\dim F_i = d - i + 1$, and the number of k-faces of P that contains $\boldsymbol{v}_1$ or $\boldsymbol{v}_2$ is at least $f_{k-1}(F_1/\boldsymbol{v}_1) + f_{k-1}(F_2/\boldsymbol{v}_2)$. By the induction hypothesis, $f_k(F) \geqslant \binom{d}{k+1}$, and $f_{k-1}(F_i/\boldsymbol{v}_i) \geqslant \binom{d-i+1}{k}$, as F, $F_1/\boldsymbol{v}_1$, or $F_2/\boldsymbol{v}_2$ may be a simplex. Thus

$$f_k(P) \geqslant \binom{d}{k+1} + \binom{d}{k} + \binom{d-1}{k} = \binom{d+1}{k+1} + \binom{d-1}{k}.$$

This proves the corollary. □

Corollary 8.2.6 *Let $d \geqslant 2$, and let P be a d-polytope. In addition, suppose that $S := \{\boldsymbol{v}_1, \ldots, \boldsymbol{v}_r\}$ are distinct vertices of P where $1 \leqslant r \leqslant d+1$. The number of k-faces containing some vertex in S is at least*

$$\sum_{i=1}^{r} \binom{d-i+1}{k}.$$

In case of equality, for any ordering $\boldsymbol{v}_{\ell 1}, \ldots, \boldsymbol{v}_{\ell r}$ of the vertices of S, we must have that the number of k-faces containing $\boldsymbol{v}_{\ell i}$ and not containing any vertex $\boldsymbol{v}_{\ell j}$ with $j < i$ is precisely $\binom{d-i+1}{k}$. In particular, every vertex in S is simple.

Proof Consider the ordering $\boldsymbol{v}_{\ell 1}, \ldots, \boldsymbol{v}_{\ell r}$ of the vertices in S. Proposition 8.2.3 then ensures that the number of k-faces containing some vertex in S is at least $\alpha := \sum_{i=1}^{r} f_{k-1}(F_i/\boldsymbol{v}_{\ell i})$ for faces $F_1, \ldots, F_r$, each satisfying $\dim F_i = d-i+1$. Furthermore, for each $i \in [1 \ldots r]$, the face F_i contains $\boldsymbol{v}_{\ell i}$ but does not contain any $\boldsymbol{v}_{\ell j}$ with $j < i$. The minimum of α is achieved when each vertex figure $F_i/\boldsymbol{v}_{\ell i}$ is a $(d-i)$-simplex (Corollary 8.2.5), in which case we get $\alpha = \sum_{i=1}^{r} \binom{d-i+1}{k}$, the hypothesis of the corollary. Thus the number of k-faces containing $\boldsymbol{v}_{\ell i}$ and not containing any vertex $\boldsymbol{v}_{\ell j}$ with $j < i$ is precisely $\binom{d-i+1}{k}$, as desired. In particular, $F_1/\boldsymbol{v}_{\ell 1}$ is a simplex, and $\boldsymbol{v}_{\ell 1}$ is simple. As we have the freedom to choose $\boldsymbol{v}_{\ell 1}$, we conclude that every vertex in S is simple, which proves the corollary. □

We also need some identities of binomial coefficients, whose proofs are based on repeated applications of the dentity $\binom{n}{m} = \binom{n-1}{m-1} + \binom{n-1}{m}$.

Lemma 8.2.7 *For all integers d, k, a, b with $a > b$, the following hold.*

(i) $\theta_k(d+a,d) - \theta_k(d+b,d) = \binom{d+1-b}{k+1} - \binom{d+1-a}{k+1} = \sum_{i=1}^{a-b} \binom{d-i+1-b}{k}$.

(ii) $\theta_k(d+a,d-1) + \binom{d}{k} + \binom{d-1}{k} = \theta_k(d+a+2,d)$.

If you recall from Proposition 2.6.18, when two polytopes Q and Q' have the origin in their relative interiors we find that $(Q \oplus Q')^* = Q^* \times (Q')^*$. The dual version of Theorem 2.14.19 is the following; see, for instance, McMullen (1971a, sec. 3).

Proposition 8.2.8 (McMullen, 1971a, sec. 3) *Let P be a d-polytope with $d+2$ facets, where $d \geqslant 2$. Then there exist $a > 0$, $b > 0$, and $t \geqslant 0$ such that $d = a+b+t$, and P is a t-fold pyramid over $T(a) \times T(b)$. In particular,*

(i) *the number of vertices of P is $(a+1)(b+1)+t$, and*
(ii) *the facets of P are either t-fold pyramids over $T(a-1) \times T(b)$, t-fold pyramids over $T(a) \times T(b-1)$, or $(t-1)$-fold pyramids over $T(a) \times T(b)$.*

We are now ready to prove Theorem 8.2.2.

Proof of the lower bound theorem for polytopes with at most 2d vertices
The case $k = d-1$ is simple: the d-simplex has $d+1$ facets, and every other d-polytope with $d+s$ vertices has at least $d+2 = \theta_{d-1}(d+s,d)$ facets by Corollary 8.2.5. So henceforth assume that $1 \leqslant k \leqslant d-2$.

We will say that a d-polytope with $d+s$ vertices has parameters (d,s), and we will prove the theorem by induction on $s \geqslant 1$, for all $d \geqslant 2$. In the case of $s=1$, the polytope is a simplex, and so the result is true for all $d \geqslant 2$. For $s=2$, we know that the d-polytopes with $d+2$ vertices are t-fold pyramids over $T(a) \oplus T(b)$ where $a,b \geqslant 1$, $t \geqslant 0$, and $a+b+t=d$ (Theorem 2.14.19). From this, we get that their number of k-faces is

$$\binom{d+2}{d-k+1} - \binom{t+a+1}{d-k+1} - \binom{d-a+1}{d-k+1} + \binom{t+1}{d-k+1}.$$

When $a=1$, $b=1$, and $t=d-2$, we get the $(2,d-2)$-triplex $M(2,d-2)$, which has $\theta_k(d+2,d)$ k-faces. By Corollary 8.2.5, $f_k(P) \geqslant \theta_k(d+2,d)$, and thus $M(2,d-2)$ is a minimiser of (8.2.1) when $s=2$; we leave to the reader to prove that it is the unique minimiser; see also Problem 8.7.1. Therefore, the theorem holds for $s=2$ and every $d \geqslant 2$. Suppose that P has parameters (d,s) with $d \geqslant 3$ and $3 \leqslant s \leqslant d$ and that the theorem is true for every pair (d',s') satisfying $s' < s$ and $d' \leqslant d$.

If P were a pyramid over each of its facets, then P would be a simplex (Proposition 8.2.4). Thus, there is a facet F of P that has $d+s-r$ vertices with $2 \leqslant r \leqslant s$. The facet F has parameters $(d-1, s-r+1)$ with $s-r+1 < s$, and so the induction hypothesis on s ensures that

$$f_k(F) \geqslant \theta_k(d-1+s-r+1, d-1) = \theta_k(d+s-r, d-1), \qquad (8.2.8.1)$$

with equality only if F is a $(s-r+1, d-s+r-2)$-triplex.

Let $S := \{\boldsymbol{v}_1, \dots, \boldsymbol{v}_r\}$ be the vertices of P outside F, and consider the ordering $\boldsymbol{v}_1, \dots, \boldsymbol{v}_r$. The k-faces of P can be partitioned into the set of k-faces in F and the set of k-faces containing at least one vertex in S. Proposition 8.2.3 bounds the number of k-faces in the latter set via a sequence of faces $F_1, \dots, F_r$ such that F_i has dimension $d - i + 1$, contains $\boldsymbol{v}_i$, but doesn't contain any $\boldsymbol{v}_j$ with $j < i$. Corollary 8.2.6 and (8.2.8.1) now give that

$$f_k(P) \geqslant f_k(F) + \sum_{i=1}^{r} f_{k-1}(F_i/\boldsymbol{v}_i) \geqslant \theta_k(d+s-r, d-1) + \sum_{i=1}^{r} \binom{d-i+1}{k}. \tag{8.2.8.2}$$

We analyse this inequality.

Claim 1 $\theta_k(d+s-r, d-1) + \sum_{i=1}^{r} \binom{d+1-i}{k} \geqslant \theta_k(d+s, d)$, with equality only if $r = 2$ or $r = s$.

Proof of claim The proof relies on the identities in Lemma 8.2.7. We let $\beta := \theta_k(d+s-r, d-1) + \sum_{i=1}^{r} \binom{d+1-i}{k}$. Then

$$\begin{aligned}\beta &= \theta_k(d+s-r, d-1) + \binom{d}{k} + \binom{d-1}{k} + \sum_{i=3}^{r} \binom{d+1-i}{k} \\ &= \theta_k(d+s-r+2, d) + \sum_{i=3}^{r} \binom{d+1-i}{k}.\end{aligned}$$

The last equality uses the relation $\theta_k(d+s-r+2, d) = \theta_k(d+s-r, d-1) + \binom{d}{k} + \binom{d-1}{k}$. Now, the identity $\theta_k(d+s-r+2, d) = \theta_k(d+s, d) - \sum_{j=1}^{r-2} \binom{d-j+1-(s-r+2)}{k}$ yields that

$$\begin{aligned}\beta &= \left(\theta_k(d+s, d) - \sum_{j=1}^{r-2} \binom{d-j-1-(s-r)}{k}\right) + \sum_{i=3}^{r} \binom{d+1-i}{k} \\ &= \theta_k(d+s, d) + \sum_{j=1}^{r-2} \left(-\binom{d-j-1-(s-r)}{k} + \binom{d-j-1}{k}\right) \\ &\geqslant \theta_k(d+s, d).\end{aligned}$$

If $\sum_{j=1}^{r-2} \left(-\binom{d-j-1-(s-r)}{k} + \binom{d-j-1}{k}\right) = 0$ then $r = 2$ or $3 \leqslant r = s$. This proves the claim. □

Claim 1 and (8.2.8.2) ensure that $f_k(P) \geqslant \theta_k(d+s, d)$ for every $s \geqslant 3$. It remains to characterise the equality case. If some facet of P has $d + s - r$

vertices with $3 \leqslant r \leqslant s-1$, then $f_k(P) > \theta_k(d+s,d)$. As a result, we have the following.

Claim 2 If $f_k(P) = \theta_k(d+s,d)$ for some $k \in [1 \dots d-2]$, then each facet of P has either $d+s-1$, $d+s-2$, or d vertices.

Another consequence of Claim 1 and (8.2.8.2) is that if $f_k(P) = \theta_k(d+s,d)$ then the number of k-faces of P containing at least one vertex in S is precisely $\sum_{i=1}^{r} \binom{d-i+1}{k}$. In this case, Corollary 8.2.6 yields the following.

Claim 3 The number of k-faces in P containing the vertex $\boldsymbol{v}_i$ and not containing any vertex $\boldsymbol{v}_j$ with $j < i$ is precisely $\binom{d-i+1}{k}$. In particular, every vertex in S is simple.

Claim 4 If $f_k(P) = \theta_k(d+s,d)$ for some $k \in [1 \dots d-2]$, then P has $d+2$ facets.

Proof of claim The facets of P can be partitioned as follows: the facet F, and for each $i \in [1 \dots r]$ the set X_i of the facets containing the vertex $\boldsymbol{v}_i$ but no vertex $\boldsymbol{v}_j$ with $j < i$. Since $\boldsymbol{v}_1$ is a simple vertex (Claim 3), $\#X_1 = d$. We show that $\#X_2 = 1$, and that $\#X_\ell = 0$ for $\ell \in [3 \dots r]$.

By way of contradiction, suppose there are two facets J_2 and J_2' in X_2. Then there must exist a k-face R of J_2' that is not a face of J_2. Since there are already $\binom{d-1}{k}$ k-faces in J_2 containing $\boldsymbol{v}_2$ (and not $\boldsymbol{v}_1$), the number of k-faces in P containing $\boldsymbol{v}_2$ but not $\boldsymbol{v}_1$ is greater than $\binom{d-1}{k}$, contradicting Claim 3. Thus $\#X_2 = 1$.

The same idea proves that $\#X_3 = 0$. If there were a facet J_3 containing $\boldsymbol{v}_3$ but not $\boldsymbol{v}_1$ or $\boldsymbol{v}_2$, then the number of k-faces in P containing $\boldsymbol{v}_3$ but not $\boldsymbol{v}_1$ or $\boldsymbol{v}_2$ would be at least the number of k-faces in J_3 containing $\boldsymbol{v}_3$, namely $\binom{d-1}{k}$, which again contradicts Claim 3. Since we have the freedom to choose a different ordering of the vertices in S, any vertex vertex in $S \backslash \{\boldsymbol{v}_1, \boldsymbol{v}_2\}$ can be picked as $\boldsymbol{v}_3$. Thus $\#X_\ell = 0$ for $\ell \in [3 \dots r]$, yielding that P has $d+2$ facets. □

We now conclude the equality case. Since P has $d+2$ facets, it is a t-fold pyramid over $T(a) \times T(b)$ with $a+b+t = d$ (Proposition 8.2.8); here, $Q := T(a) \times T(b)$ is a simple $(a+b)$-polytope with $(a+1)(b+1)$ vertices. Because each facet of P has either $a+b+t+s-1$, $a+b+t+s-2$, or $a+b+t$ vertices (Claim 2), and because Q is not a pyramid, it follows that each facet of Q has either $a+b+s-2$ or $a+b$ vertices. We consider a facet J of Q with $a+b+s-2$ vertices, which must exist because Q is not simplicial (the simplex is the only simple and simplicial polytope (Theorem 2.8.8)). From Proposition 8.2.8 it follows that either $J = T(a-1) \times T(b)$ or $J = T(a) \times$

$T(b-1)$. If $J = T(a-1) \times T(b)$, then $f_0(J) = a(b+1)$ and considering the number of vertices of J and Q, we form the system of equations:

$$a+b+t=d,\ a(b+1)=a+b+s-2,\ (a+1)(b+1)=a+b+s,$$

whose solution is $b = 1$, $a = s-1$, and $t = d-s$. This implies that P is $(d-s)$-fold pyramid over $T(s-1) \times T(1)$. That is, P is a $(s,d-s)$-triplex. In the case that $J = T(a) \times T(b-1)$, we equally get that P is a $(s,d-s)$-triplex. This completes the proof of the theorem. □

Pineda-Villavicencio and Yost (2022) proved a lower bound theorem for d-polytopes with $2d+1$ vertices; this problem was also studied by Xue (2022). Each minimiser is either a d-polytope with $d+2$ facets or a d-polytope obtained from truncating a simple vertex from the triplex $M(2,d-2)$. The latter is called the *d-pentasm.*

Theorem 8.2.9 (Pineda-Villavicencio and Yost, 2022) *Let $d \geqslant 6$. Suppose that $Pm(d)$ is the d-pentasm, and that P is a d-polytope with $2d+1$ vertices satisfying $f_k(P) \leqslant f_k(Pm(d))$ for some $k \in [1 \dots d-2]$.*

If P is not a pyramid and at least one vertex in P is not simple, then P has $d+3$ facets and is a d-pentasm. Otherwise, P has $d+2$ faces and so it is a t-fold pyramid over $T(a) \times T(b)$ for some $a,b > 0$ and $t \geqslant 0$ satisfying $d = a+b+t$.

For d-polytopes with at least $2d+2$ vertices, we are aware only of a result of Pineda-Villavicencio et al. (2022) on the minimum number of edges of d-polytopes with $2d+2$ vertices: for $d \geqslant 8$, such d-polytopes have at least d^2+2d-3 edges. Problem 8.7.11 proposes a lower bound conjecture for d-polytopes with at most $3d-1$ vertices.

8.3 Rigidity, Stress, and the Lower Bound Theorem

In this section, we return to the notion of a framework of a graph, which we encountered in Section 3.7 in the context of Tutte's realisation theorem (3.7.3). While in Section 3.7, the motivation came from thinking of the vertices of a certain facial cycle C of a framework G as being nailed to the plane and the edges with one endvertex in $\mathcal{V}(G)\backslash\mathcal{V}(C)$ as being rubber bands; here, the motivation comes from thinking of the vertices of G as joints and its edges as rigid bars. Our ultimate goal is to present a version of the proof in Kalai (1987) of the lower bound theorem of Barnette (1971, 1973c).

Let X be a graph with a vertex set $\mathcal{V}(X) = \{1,\dots,n\}$. We assign points or positions $\boldsymbol{p}_1,\dots,\boldsymbol{p}_n$ in $\mathbb{R}^d$ to the vertices of X to form a graph G in $\mathbb{R}^d$ whose edges are precisely the pairs $\boldsymbol{p}_i\boldsymbol{p}_j$ such that ij is an edge of X. We join adjacent vertices of G by line segments. The graph G, or more precisely the pair $(X, \boldsymbol{p})$ where $\boldsymbol{p} := (\boldsymbol{p}_1, \ldots, \boldsymbol{p}_n)$, is said to be a *framework* of X in $\mathbb{R}^d$. The framework is *d-dimensional* or *full-dimensional* if its positions don't lie in any affine hyperplane of $\mathbb{R}^d$. We will be concerned only with full-dimensional frameworks, and so we will often drop the adjective. We may write $\boldsymbol{p}(G)$ instead $\boldsymbol{p}$ to avoid ambiguity and we may sometimes treat the sequence $\boldsymbol{p}$ as a set. If $\boldsymbol{p}(G)$ gives the vertices of a d-polytope P in $\mathbb{R}^d$, then we will say that the framework G is *d-polytopal*, or simply *polytopal*; thus, the realisation of the graph of P can be regarded as a d-polytopal framework.

By mimicking concepts and operations on graphs (Section C.1), we can define their counterparts for frameworks. Suppose that $G := (X, \boldsymbol{p})$ is a framework. If Y is a subgraph of X, then the restriction $\boldsymbol{p}|_Y$ of $\boldsymbol{p}$ to $\mathcal{V}(Y)$ defines the *subframework* $(Y, \boldsymbol{p}|_Y)$ of G; similarly, we can define *induced subframeworks* and *spanning subframeworks*. With regard to operations on frameworks, for instance, if $D \subseteq [\mathcal{V}(G)]^2$, then $G - D$ denotes the spanning subframework of G whose edges are in $\mathcal{E}(G)\backslash D$, and $G + D$ denotes the framework on the vertex set $\mathcal{V}(G)$ whose edges are in $\mathcal{E}(G) \cup D$.

Two frameworks $(X, \boldsymbol{p})$ and $(X, \boldsymbol{q})$ in $\mathbb{R}^d$ of a graph X with n vertices are *congruent* if the corresponding distances between any two vertices of X are preserved: $\|\boldsymbol{p}_i - \boldsymbol{p}_j\| = \|\boldsymbol{q}_i - \boldsymbol{q}_j\|$ for every $1 \leqslant i < j \leqslant n$, or, equivalently, if there is an isometry of $\mathbb{R}^d$ that maps $\boldsymbol{p}_i$ to $\boldsymbol{q}_i$. An *isometry* of $\mathbb{R}^d$ is a bijective function from $\mathbb{R}^d$ to $\mathbb{R}^d$ that preserves the distance between any pair of points in $\mathbb{R}^d$. In other words, the distance between any two points in $\mathbb{R}^d$ remains unchanged before and after the transformation. Common examples of isometries in $\mathbb{R}^d$ are translations and rotations.

Suppose that each vertex $\boldsymbol{p}_i$ of a framework $G := (X, \boldsymbol{p})$ with n vertices follows a trajectory traced by a differentiable function $\boldsymbol{m}_i \colon [0,1] \to \mathbb{R}^d$. At time $t = 0$, we have that $\boldsymbol{p}_i = \boldsymbol{m}_i(0)$ for each $i \in [1 \ldots n]$. The path $\boldsymbol{m}(t) := (\boldsymbol{m}_1(t), \ldots, \boldsymbol{m}_n(t))$ is a *motion* of G if the length of each edge $\boldsymbol{p}_i\boldsymbol{p}_j$ of G remains constant throughout the trajectory:

$$\|\boldsymbol{m}_i(t) - \boldsymbol{m}_j(t)\| = \|\boldsymbol{p}_i - \boldsymbol{p}_j\|, \text{ for every } ij \in \mathcal{E}(X) \text{ and every } t \in [0,1]. \tag{8.3.1}$$

At time $t = 1$, we have a new framework $(X, \boldsymbol{q})$ where $\boldsymbol{q}_i = \boldsymbol{m}_i(1)$ for each $i \in [1 \ldots n]$. In this case, we say that the path $\boldsymbol{m}(t)$ moved $(X, \boldsymbol{p})$ to $(X, \boldsymbol{q})$. In the definition of a motion, we could have equally required that each function $\boldsymbol{m}_i(t)$ be continuous or analytic, without affecting the theory (Connelly and Guest, 2022, thm. 2.4.1). The motion is *rigid* in $\mathbb{R}^d$ if all the $\binom{n}{2}$ distances between the vertices of G are preserved during the trajectory:

$$\|\boldsymbol{m}_i(t) - \boldsymbol{m}_j(t)\| = \|\boldsymbol{p}_i - \boldsymbol{p}_j\|, \text{ for every } 1 \leqslant i < j \leqslant n \text{ and every } t \in [0,1]. \tag{8.3.2}$$

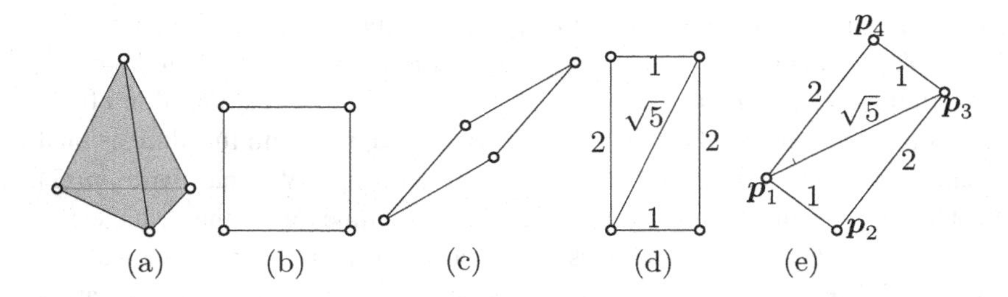

Figure 8.3.1 Rigid and nonrigid frameworks. (a) A rigid framework. (b)–(c) A nonrigid motion of a quadrangle, showing that the quadrangle is nonrigid. (d)–(e) A rigid framework at time $t = 0$ and at time $t = t_1$, respectively.

Otherwise, the motion is *nonrigid* or *flexible* in $\mathbb{R}^d$. Every isometry in $\mathbb{R}^d$ is a rigid motion of G.

A framework is *rigid* in $\mathbb{R}^d$ if all its motions are rigid; otherwise, the framework is *nonrigid* or *flexible* in $\mathbb{R}^d$. Equivalently, a framework is rigid if its rigid motions are only the restrictions of isometries of $\mathbb{R}^d$ (Asimow and Roth, 1978).

Example 8.3.3 We consider the framework G in Fig. 8.3.1(d)–(e) and differentiable functions $\boldsymbol{m}_i\colon [0,1] \to \mathbb{R}^2$ $(i \in [1\ldots 4])$ that form a motion of G from time $t = 0$ to time $t = t_1$. We show that the motion is rigid. For this, we let $\boldsymbol{p}_i := \boldsymbol{m}_i(t_1)$ with $\boldsymbol{p}_i = (x_i, y_i)^t$ $(i \in [1\ldots 4])$. The definition of a motion means that, at time $t = t_1$, we have the system

$$(x_1 - x_2)^2 + (y_1 - y_2)^2 = 1,\ (x_2 - x_3)^2 + (y_2 - y_3)^2 = 4,$$
$$(x_3 - x_4)^2 + (y_3 - y_4)^2 = 1,\ (x_4 - x_1)^2 + (y_4 - y_1)^2 = 4,$$
$$(x_1 - x_3)^2 + (y_1 - y_3)^2 = 5.$$

We remove two variables with the change $r_i := x_i - x_1$ and $s_i := y_i - y_1$, for $i = 2, 3, 4$:

$$r_2^2 + s_2^2 = 1,\ (r_2 - r_3)^2 + (s_2 - s_3)^2 = 4,\ (r_3 - r_4)^2 + (s_3 - s_4)^2 = 1,$$
$$r_4^2 + s_4^2 = 4,\ r_3^2 + s_3^2 = 5.$$

We remove two further variables with the change $u_i := r_2 r_i + s_2 s_i$ and $v_i := -s_2 r_i + r_2 s_i$, for $i = 2, 3, 4$:

$$u_2 = 1,\ v_2 = 0,\ (u_2 - u_3)^2 + (v_2 - v_3)^2 = 4,$$
$$(u_3 - u_4)^2 + (v_3 - v_4)^2 = 1,\ u_4^2 + v_4^2 = 4,\ u_3^2 + v_3^2 = 5.$$

There are four sets of solutions for this system:

$$u_3 = 1, u_4 = 8/5, v_3 = 2, v_4 = 6/5; u_3 = 1, u_4 = 0, v_3 = 2, v_4 = 2;$$
$$u_3 = 1, u_4 = 8/5, v_3 = -2, v_4 = -6/5; u_3 = 1, u_4 = 0, v_3 = -2, v_4 = -2.$$

Under the aforementioned changes of variables, the distance $\|\boldsymbol{p}_2 - \boldsymbol{p}_4\|$ becomes $(1 - u_4)^2 + v_4^2$. Thus, for the pairs $(u_4, v_4) = (8/5, \pm 6/5), (0, \pm 2)$, we get that $\|\boldsymbol{p}_2 - \boldsymbol{p}_4\| = \sqrt{5}$ or $\sqrt{9/5}$ throughout the trajectory. At time $t = 0$, we have that $\|\boldsymbol{m}_2(0) - \boldsymbol{m}_4(0)\| = \sqrt{5}$, and because the distance function is continuous, at any time we must have that $\|\boldsymbol{m}_2(t) - \boldsymbol{m}_4(t)\| = \sqrt{5}$. Consequently, $\|\boldsymbol{p}_2 - \boldsymbol{p}_4\| = \sqrt{5}$, which implies that the motion is rigid. Since the motion was arbitrary, the framework is also rigid.

Example 8.3.3 shows that the framework in Fig. 8.3.1(d)–(e) is rigid in $\mathbb{R}^2$. But the same framework is flexible in $\mathbb{R}^3$. Thus, being rigid or being flexible depends on the ambient space. A framework that is rigid in $\mathbb{R}^d$ and remains rigid when embedded isometrically in every $\mathbb{R}^{d'}$ with $d' \geqslant d$ is called *universally rigid*; we will not return to this concept, but the interested reader may consult Goodman et al. (2017, ch. 63). Example 8.3.3 also shows that deciding the rigidity of even simple frameworks, like the one in Fig. 8.3.1(d)–(e), is very demanding. In fact, deciding whether a given framework is rigid is NP-hard, even for two-dimensional frameworks (Abbott, 2008). This discussion motivates the introduction of a more approachable notion, the notion of infinitesimal rigidity.

As before, we suppose that each vertex $\boldsymbol{p}_i$ of a framework $G := (X, \boldsymbol{p})$ in $\mathbb{R}^d$ with n vertices follows a trajectory traced by a differentiable function $\boldsymbol{m}_i : [0, 1] \to \mathbb{R}^d$. At time $t = 0$, we have that $\boldsymbol{p}_i = \boldsymbol{m}_i(0)$ for each $i \in [1 \ldots n]$. Let $\boldsymbol{m}(t) := (\boldsymbol{m}_1(t), \ldots, \boldsymbol{m}_n(t))$ be the relevant motion. The definition of a motion gives, for every $ij \in \mathcal{E}(X)$ and every $t \in [0, 1]$, that

$$(\boldsymbol{m}_i(t) - \boldsymbol{m}_j(t)) \cdot (\boldsymbol{m}_i(t) - \boldsymbol{m}_j(t)) = \|\boldsymbol{p}_i - \boldsymbol{p}_j\|^2,$$
$$\frac{d}{dt}\big((\boldsymbol{m}_i(t) - \boldsymbol{m}_j(t)) \cdot (\boldsymbol{m}_i(t) - \boldsymbol{m}_j(t))\big) = 0,$$
$$2(\boldsymbol{m}_i(t) - \boldsymbol{m}_j(t)) \cdot (\boldsymbol{m}_i'(t) - \boldsymbol{m}_j'(t)) = 0.$$

Here, if $\boldsymbol{m}_i(t) = (x_1(t), \ldots, x_n(t))$ then $\boldsymbol{m}_i'(t) = (x_1'(t), \ldots, x_n'(t))$. At time $t = 0$, we have that $(\boldsymbol{p}_i - \boldsymbol{p}_j) \cdot (\boldsymbol{m}_i'(0) - \boldsymbol{m}_j'(0)) = 0$, which can be seen as the infinitesimal version of (8.3.1). In view of this, we say that an assignment $\boldsymbol{p}_i \mapsto \boldsymbol{v}_i$ of vectors to the points of G is an *infinitesimal motion* of G if

$$(\boldsymbol{p}_i - \boldsymbol{p}_j) \cdot (\boldsymbol{v}_i - \boldsymbol{v}_j) = 0, \text{ for every } ij \in \mathcal{E}(X). \tag{8.3.4}$$

It follows that, if $\boldsymbol{m}(t)$ is a motion of G then $\boldsymbol{m}'(0)$ is an infinitesimal motion of G. An infinitesimal motion is *infinitesimally rigid* in $\mathbb{R}^d$ if the condition in (8.3.4) holds for all the $\binom{n}{2}$ pairs i, j in $\mathcal{V}(X)$; that is, for every $1 \leqslant i < j \leqslant n$. Otherwise, the infinitesimal motion is *infinitesimally nonrigid* or *infinitesimally flexible* in $\mathbb{R}^d$. We will also say that an infinitesimal rigid motion of the framework G is an *infinitesimal isometry* of G.

By analogy with the notion of rigidity, a framework is *infinitesimally rigid* in $\mathbb{R}^d$ if all its infinitesimal motions are infinitesimally rigid; otherwise, the framework is *infinitesimally nonrigid* or *infinitesimally flexible* in $\mathbb{R}^d$.

A realisation of a d-polytope P in $\mathbb{R}^d$ is *rigid* or *infinitesimally rigid* in $\mathbb{R}^d$ if the framework corresponding to the graph of P is rigid or infinitesimally rigid in $\mathbb{R}^d$, respectively. The same applies to the notions of being flexible and infinitesimally flexible.

The notions of rigidity and infinitesimal rigidity are different, as the next (classical) example attests.

Example 8.3.5 Consider the framework $(X, \boldsymbol{p})$ in Fig. 8.3.2(b), where $\boldsymbol{p} := (\boldsymbol{p}_1, \ldots, \boldsymbol{p}_5)$. We leave to the reader the proof that the framework $(X, \boldsymbol{p})$ is rigid. It may be helpful to notice that $\|\boldsymbol{p}_1 - \boldsymbol{p}_3\|$ remains constant in any motion of $(X, \boldsymbol{p})$; see Example 8.3.3.

The framework $(X, \boldsymbol{p})$ is, however, infinitesimally nonrigid. Let $\boldsymbol{v}_i := \boldsymbol{0}$, for $i \in [1 \ldots 4]$, and let $\boldsymbol{v}_5$ be a nonzero vector perpendicular to the segment $\boldsymbol{p}_1\boldsymbol{p}_3$. Then $(\boldsymbol{p}_1 - \boldsymbol{p}_5) \cdot \boldsymbol{v}_5 = (\boldsymbol{p}_3 - \boldsymbol{p}_5) \cdot \boldsymbol{v}_5 = 0$. The assignment $(\boldsymbol{v}_1, \ldots, \boldsymbol{v}_5)$ is an infinitesimal motion of $(X, \boldsymbol{p})$:

$$(\boldsymbol{p}_1 - \boldsymbol{p}_2) \cdot (\boldsymbol{v}_1 - \boldsymbol{v}_2) = (\boldsymbol{p}_1 - \boldsymbol{p}_4) \cdot (\boldsymbol{v}_1 - \boldsymbol{v}_4) = (\boldsymbol{p}_2 - \boldsymbol{p}_3) \cdot (\boldsymbol{v}_2 - \boldsymbol{v}_3) = 0,$$
$$(\boldsymbol{p}_2 - \boldsymbol{p}_4) \cdot (\boldsymbol{v}_2 - \boldsymbol{v}_4) = (\boldsymbol{p}_3 - \boldsymbol{p}_4) \cdot (\boldsymbol{v}_3 - \boldsymbol{v}_4) = 0,$$
$$(\boldsymbol{p}_1 - \boldsymbol{p}_5) \cdot (\boldsymbol{v}_1 - \boldsymbol{v}_5) = -(\boldsymbol{p}_1 - \boldsymbol{p}_5) \cdot \boldsymbol{v}_5 = 0,$$
$$(\boldsymbol{p}_3 - \boldsymbol{p}_5) \cdot (\boldsymbol{v}_3 - \boldsymbol{v}_5) = -(\boldsymbol{p}_3 - \boldsymbol{p}_5) \cdot \boldsymbol{v}_5 = 0.$$

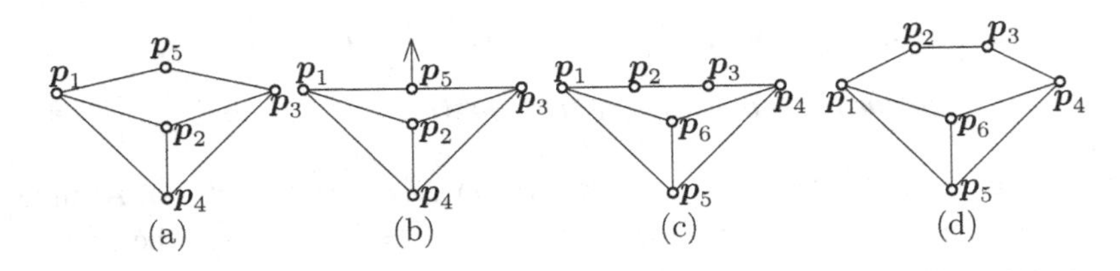

Figure 8.3.2 Infinitesimal rigid and nonrigid frameworks. (a) A rigid and infinitesimal rigid framework in $\mathbb{R}^2$. (b) A rigid and infinitesimal flexible framework in $\mathbb{R}^2$. (c) A rigid and infinitesimal flexible framework in $\mathbb{R}^2$. (d) A flexible and infinitesimal flexible framework in $\mathbb{R}^2$.

Furthermore, $(\boldsymbol{p}_2 - \boldsymbol{p}_5) \cdot (\boldsymbol{v}_2 - \boldsymbol{v}_5) = (\boldsymbol{p}_2 - \boldsymbol{p}_5) \cdot \boldsymbol{v}_5 \neq 0$, since $\boldsymbol{v}_5$ is not perpendicular to the segment $\boldsymbol{p}_2\boldsymbol{p}_5$. Hence, the motion is nonrigid, and so is $(X, \boldsymbol{p})$.

We develop tools from linear algebra to decide the infinitesimal rigidity of a framework. The set $\mathcal{R}$ of infinitesimal motions in $\mathbb{R}^d$ of a framework is a linear space, called the *motion space* of the framework. To see this, consider a framework $G := (X, \boldsymbol{p})$ in $\mathbb{R}^d$ of a graph X with n vertices. We define the *rigidity matrix* $R(G)$ of G as a matrix whose rows are indexed by the edges of G, whose columns are grouped into n sets of d consecutive columns, and where the row of the edge ij has the form

$$\begin{array}{cccccccccc} & 1 & 2 & \cdots & i & \cdots & j & \cdots & n \\ ij & \boldsymbol{0}_d^t & \boldsymbol{0}_d^t & \cdots & (\boldsymbol{p}_i - \boldsymbol{p}_j)^t & \cdots & (\boldsymbol{p}_j - \boldsymbol{p}_i)^t & \cdots & \boldsymbol{0}_d^t. \end{array} \tag{8.3.6}$$

The consecutive d rows corresponding to the vertex ℓ is the ℓth *block* of the row: for the row ij of $R(G)$, the ith block is $(\boldsymbol{p}_i - \boldsymbol{p}_j)^t$, and the first block is $\boldsymbol{0}_d^t$ if $i \neq 1$. We now write an infinitesimal motion $(\boldsymbol{v}_1, \boldsymbol{v}_2, \ldots, \boldsymbol{v}_n)$ of G as the row vector $\boldsymbol{v}^t := \boldsymbol{v}_1^t \boldsymbol{v}_2^t \ldots \boldsymbol{v}_n^t$. Then

$$R(G)\boldsymbol{v} = \boldsymbol{0}, \text{ for } \boldsymbol{v} \in \mathbb{R}^{dn}. \tag{8.3.7}$$

The space $\mathcal{I}(G)$ of infinitesimal isometries of G is a subspace of $\mathcal{R}(G)$ that is obtained from $\mathcal{R}(G)$ by adding the condition (8.3.4) for each pair of nonadjacent vertices of G. Thus, this discussion and the nullity–rank theorem (see Problem 1.12.5) give the next proposition.

Proposition 8.3.8 *For a framework G in $\mathbb{R}^d$ with n vertices, the following hold.*

(i) *The motion space $\mathcal{R}(G)$ of G is the nullspace of the rigidity matrix $R(G)$ of G, and the set $\mathcal{I}(G)$ of infinitesimal isometries is a linear subspace of $\mathcal{R}(G)$.*
(ii) $\operatorname{rank} R(G) = dn - \dim(\operatorname{null}(R(G))) = dn - \dim \mathcal{R}(G)$.

We illustrate the construction of rigidity matrices.

Example 8.3.9 Suppose the vertices of the 2-simplex in $\mathbb{R}^2$ are $\boldsymbol{p}_i = (x_i, y_i)^t$ ($i \in [1 \ldots 3]$). The rigidity matrix has three rows and 2×3 columns:

$$\begin{array}{c} \\ 12 \\ 13 \\ 23 \end{array} \begin{array}{c} \begin{array}{ccc} \quad 1 \quad & \quad 2 \quad & \quad 3 \quad \end{array} \\ \left(\begin{array}{cc|cc|cc} x_1 - x_2 & y_1 - y_2 & x_2 - x_1 & y_2 - y_1 & 0 & 0 \\ x_1 - x_3 & y_1 - y_3 & 0 & 0 & x_3 - x_1 & y_3 - y_1 \\ 0 & 0 & x_2 - x_3 & y_2 - y_3 & x_3 - x_2 & y_3 - y_2 \end{array}\right). \end{array}$$

Suppose the vertices of the 3-simplex in $\mathbb{R}^3$ are $\boldsymbol{p}_i$ ($i \in [1 \ldots 4]$); see Fig. 8.3.1(a). The rigidity matrix has six rows and 3×4 columns:

$$\begin{array}{c} \\ 12 \\ 13 \\ 14 \\ 23 \\ 24 \\ 34 \end{array} \begin{array}{c} \begin{array}{cccc} 1 & 2 & 3 & 4 \end{array} \\ \left(\begin{array}{c|c|c|c} (\boldsymbol{p}_1 - \boldsymbol{p}_2)^t & (\boldsymbol{p}_2 - \boldsymbol{p}_1)^t & \boldsymbol{0}_3^t & \boldsymbol{0}_3^t \\ (\boldsymbol{p}_1 - \boldsymbol{p}_3)^t & \boldsymbol{0}_3^t & (\boldsymbol{p}_3 - \boldsymbol{p}_1)^t & \boldsymbol{0}_3^t \\ (\boldsymbol{p}_1 - \boldsymbol{p}_4)^t & \boldsymbol{0}_3^t & \boldsymbol{0}_3^t & (\boldsymbol{p}_4 - \boldsymbol{p}_1)^t \\ \boldsymbol{0}_3^t & (\boldsymbol{p}_2 - \boldsymbol{p}_3)^t & (\boldsymbol{p}_3 - \boldsymbol{p}_2)^t & \boldsymbol{0}_3^t \\ \boldsymbol{0}_3^t & (\boldsymbol{p}_2 - \boldsymbol{p}_4)^t & \boldsymbol{0}_3^t & (\boldsymbol{p}_4 - \boldsymbol{p}_2)^t \\ \boldsymbol{0}_3^t & \boldsymbol{0}_3^t & (\boldsymbol{p}_3 - \boldsymbol{p}_4)^t & (\boldsymbol{p}_4 - \boldsymbol{p}_3)^t \end{array}\right) \end{array}.$$

For the framework in Fig. 8.3.2(b), suppose that $\boldsymbol{p}_i = (x_i, y_i)^t$ ($i \in [1 \ldots 5]$) lie in $\mathbb{R}^2$ and that $\boldsymbol{p}_5 = (\boldsymbol{p}_1 + \boldsymbol{p}_3)/2$. The rigidity matrix has seven rows and 2×5 columns:

$$\begin{array}{c} \\ 12 \\ 14 \\ 15 \\ 23 \\ 24 \\ 34 \\ 35 \end{array} \begin{array}{c} \begin{array}{ccccc} 1 & 2 & 3 & 4 & 5 \end{array} \\ \left(\begin{array}{c|c|c|c|c} (\boldsymbol{p}_1 - \boldsymbol{p}_2)^t & (\boldsymbol{p}_2 - \boldsymbol{p}_1)^t & \boldsymbol{0}_2^t & \boldsymbol{0}_2^t & \boldsymbol{0}_2^t \\ (\boldsymbol{p}_1 - \boldsymbol{p}_4)^t & \boldsymbol{0}_2^t & \boldsymbol{0}_2^t & (\boldsymbol{p}_4 - \boldsymbol{p}_1)^t & \boldsymbol{0}_2^t \\ \frac{1}{2}(\boldsymbol{p}_1 - \boldsymbol{p}_3)^t & \boldsymbol{0}_2^t & \boldsymbol{0}_2^t & \boldsymbol{0}_2^t & \frac{1}{2}(\boldsymbol{p}_3 - \boldsymbol{p}_1)^t \\ \boldsymbol{0}_2^t & (\boldsymbol{p}_2 - \boldsymbol{p}_3)^t & (\boldsymbol{p}_3 - \boldsymbol{p}_2)^t & \boldsymbol{0}_2^t & \boldsymbol{0}_2^t \\ \boldsymbol{0}_2^t & (\boldsymbol{p}_2 - \boldsymbol{p}_4)^t & \boldsymbol{0}_2^t & (\boldsymbol{p}_4 - \boldsymbol{p}_2)^t & \boldsymbol{0}_2^t \\ \boldsymbol{0}_2^t & \boldsymbol{0}_2^t & (\boldsymbol{p}_3 - \boldsymbol{p}_4)^t & (\boldsymbol{p}_4 - \boldsymbol{p}_3)^t & \boldsymbol{0}_2^t \\ \boldsymbol{0}_2^t & \boldsymbol{0}_2^t & \frac{1}{2}(\boldsymbol{p}_3 - \boldsymbol{p}_1)^t & \boldsymbol{0}_2^t & \frac{1}{2}(\boldsymbol{p}_1 - \boldsymbol{p}_3)^t \end{array}\right) \end{array}.$$

We next show how to compute the dimension of the motion space of the d-simplex.

Proposition 8.3.10 *The dimension of the motion space of the d-simplex in $\mathbb{R}^d$ is $\binom{d+1}{2}$.*

Proof Let T be the d-simplex with vertices $\boldsymbol{p}_1, \ldots, \boldsymbol{p}_{d+1}$. We consider the rigidity matrix $R(T)$ of the d-simplex in $\mathbb{R}^d$; Example 8.3.9 shows the rigidity matrix for the 2-simplex and the 3-simplex.

The rigidity matrix of T has $d(d+1)$ columns and $\binom{d+1}{2}$ rows. We consider each row $\boldsymbol{r}_{i,j}^t$ of $R(T)$, where $1 \leqslant i < j \leqslant d+1$, as a vector of the form described in (8.3.6). Then the ℓth block of $\boldsymbol{r}_{\ell,i}^t$ or $\boldsymbol{r}_{i,\ell}^t$, which corresponds to the vertex ℓ, is the row vector $(\boldsymbol{p}_\ell - \boldsymbol{p}_i)^t$. In the linear combination

$$\sum_{1 \leqslant i < j \leqslant d+1} \alpha_{i,j} \boldsymbol{r}_{i,j}^t = \boldsymbol{0}_{d(d+1)},$$

the sums of the ℓth blocks in the left-hand side has the form

$$\alpha_{1,\ell}(\boldsymbol{p}_\ell - \boldsymbol{p}_1)^t + \cdots + \alpha_{\ell-1,\ell}(\boldsymbol{p}_\ell - \boldsymbol{p}_{\ell-1})^t$$
$$+ \alpha_{\ell,\ell+1}(\boldsymbol{p}_\ell - \boldsymbol{p}_{\ell+1})^t + \cdots + \alpha_{\ell,d+1}(\boldsymbol{p}_\ell - \boldsymbol{p}_{d+1})^t = \boldsymbol{0}_d^t.$$

And since the vertices of T are affinely independent, the d vectors

$$(\boldsymbol{p}_\ell - \boldsymbol{p}_1), \ldots, (\boldsymbol{p}_\ell - \boldsymbol{p}_{\ell-1}), (\boldsymbol{p}_\ell - \boldsymbol{p}_{\ell+1}), \ldots, (\boldsymbol{p}_\ell - \boldsymbol{p}_{d+1})$$

are linearly independent in $\mathbb{R}^d$. From this, we conclude that

$$\alpha_{1,\ell} = \cdots = \alpha_{\ell-1,\ell} = \alpha_{\ell,\ell+1} = \cdots = \alpha_{\ell,d+1} = 0.$$

By running over the $d+1$ blocks, we finally find that $\alpha_{i,j} = 0$ for $1 \leqslant i < j \leqslant d+1$, yielding the linear independence of the rows of $R(T)$. Thus, $\operatorname{rank} R(T) = \binom{d+1}{2}$, and the proposition now follows from the nullity–rank theorem (Problem 1.12.5). □

An *infinitesimal isometry* of $\mathbb{R}^d$ is a bijection $\sigma\colon \mathbb{R}^d \to \mathbb{R}^d$ satisfying $(\boldsymbol{x} - \boldsymbol{y}) \cdot (\sigma(\boldsymbol{x}) - \sigma(\boldsymbol{y})) = 0$ for every pair $\boldsymbol{x}, \boldsymbol{y} \in \mathbb{R}^d$. Every infinitesimal isometry of $\mathbb{R}^d$ can be expressed in matrix form as $\boldsymbol{x} \mapsto M\boldsymbol{x} + \boldsymbol{b}$, where $M \in \mathbb{R}^{d\times d}$ is a skew symmetric $d \times d$ matrix and $\boldsymbol{b}$ is a constant vector; a matrix A is *skew symmetric* if $-A^t = A$. The following is plain.

Remark 8.3.11 The restriction of an infinitesimal isometry of $\mathbb{R}^d$ to any framework G in $\mathbb{R}^d$ is an infinitesimal isometry of G; that is, $\mathcal{I}(\mathbb{R}^d) \subseteq \mathcal{I}(G)$.

We compute the dimension of $\mathcal{I}(\mathbb{R}^d)$; we start by listing a set of linearly independent vectors. An *infinitesimal translation* of $\mathbb{R}^d$ assigns the same vector, say $\boldsymbol{z}$, to every $\boldsymbol{x} \in \mathbb{R}^d$: $\tau_{\boldsymbol{z}}(\boldsymbol{x}) = \boldsymbol{z}$. In the case that $\boldsymbol{z} = \boldsymbol{e}_i$, the ith standard vector of $\mathbb{R}^d$, we simply write τ_i. We next consider infinitesimal isometries that are rotations with an axis of rotation spanned by any $d-2$ coordinate axes; these rotations are examples of infinitesimal rotations in $\mathbb{R}^d$. Concretely, for $1 \leqslant i < j \leqslant d$ and every $\boldsymbol{x} = (x_1, \ldots, x_2)^t \in \mathbb{R}^d$, we let $\rho_{i,j}$ be defined as $\boldsymbol{x} \mapsto x_i\boldsymbol{e}_j - x_j\boldsymbol{e}_i$.

Proposition 8.3.12 *The infinitesimal translations τ_ℓ $(1 \leqslant \ell \leqslant d)$ and infinitesimal rotations $\rho_{i,j}$ $(1 \leqslant i < j \leqslant d)$ of $\mathbb{R}^d$ are linearly independent vectors of $\mathcal{I}(\mathbb{R}^d)$.*

Proof Consider a linear combination of these functions

$$\sum_{\ell=1}^{d} \alpha_\ell \tau_\ell(\boldsymbol{x}) + \sum_{1\leqslant i<j\leqslant d} \beta_{i,j}\rho_{i,j}(\boldsymbol{x}) = \boldsymbol{0}_d, \text{ for every } \boldsymbol{x} \in \mathbb{R}^d. \qquad (8.3.12.1)$$

As customary, we specify distinct values for $\boldsymbol{x}$ to force all the scalars to be zero. We start with $\boldsymbol{x} = \boldsymbol{0}_d$. Then, every infinitesimal rotation $\rho_{i,j}$ becomes the zero vector, and so (8.3.12.1) becomes

$$\sum_{\ell=1}^{d} \alpha_\ell \boldsymbol{e}_\ell = \boldsymbol{0}_d.$$

The linear independence of $\boldsymbol{e}_1, \ldots, \boldsymbol{e}_d$ ensures that $\alpha_1 = \cdots = \alpha_d = 0$. The new version of (8.3.12.1) is as follows:

$$\sum_{1 \leqslant i < j \leqslant d} \beta_{i,j} \rho_{i,j}(\boldsymbol{x}) = \boldsymbol{0}_d, \text{ for every } \boldsymbol{x} \in \mathbb{R}^d. \tag{8.3.12.2}$$

We take $\boldsymbol{x} = \boldsymbol{e}_\ell$, for $\ell \in [1 \ldots d]$. From the definition of $\rho_{i,j}$, we get that

$$\rho_{i,j}(\boldsymbol{e}_\ell) = \begin{cases} \boldsymbol{0}_d, & \text{if } i \neq \ell \text{ and } j \neq \ell; \\ \boldsymbol{e}_j, & \text{if } i = \ell \text{ (and } j > \ell); \\ -\boldsymbol{e}_i, & \text{if } j = \ell \text{ (and } i < \ell). \end{cases}$$

Thus, evaluating (8.3.12.2) at $\boldsymbol{e}_\ell$, we find that

$$(-\beta_{1,\ell}, -\beta_{2,\ell}, \ldots, -\beta_{\ell-1,\ell}, 0, \beta_{\ell,\ell+1}, \ldots, \beta_{\ell,d}) = \boldsymbol{0}_d^t,$$

implying that

$$-\beta_{1,\ell} = -\beta_{2,\ell} = \cdots = -\beta_{\ell-1,\ell} = \beta_{\ell,\ell+1} = \ldots = \beta_{\ell,d} = 0.$$

As a result, evaluating (8.3.12.2) at each of the standard vectors in $\mathbb{R}^d$, we get that $\beta_{i,j} = 0$ for $1 \leqslant i < j \leqslant d$. This settles the proposition. □

We show that the vectors from Proposition 8.3.12 form a basis of $\mathcal{I}(\mathbb{R}^d)$.

Lemma 8.3.13 *Let $X \subseteq \mathbb{R}^d$ contain a set T of $d+1$ affinely independent points and let Y be a d-subset of T. If $\boldsymbol{m}$ and $\tilde{\boldsymbol{m}}$ are two infinitesimal isometries of X that coincide on each point of Y, then $\boldsymbol{m} = \tilde{\boldsymbol{m}}$.*

Proof Assume that $Y = \{\boldsymbol{p}_1, \ldots, \boldsymbol{p}_d\}$ and that $\boldsymbol{x} \in X$. Let $\varphi(\boldsymbol{z}) := \boldsymbol{m}(\boldsymbol{z}) - \tilde{\boldsymbol{m}}(\boldsymbol{z})$ for every $\boldsymbol{z} \in X$. Then φ is an infinitesimal isometry of X satisfying $\varphi(\boldsymbol{p}_i) = \boldsymbol{0}$ for each $i \in [1 \ldots d]$ (convince yourself of this). Since φ is an infinitesimal isometry of X, we have that $(\boldsymbol{x} - \boldsymbol{p}_i) \cdot (\varphi(\boldsymbol{x}) - \varphi(\boldsymbol{p}_i)) = \boldsymbol{0}$, wherefrom it follows that

$$(\boldsymbol{x} - \boldsymbol{p}_i) \cdot \varphi(\boldsymbol{x}) = \boldsymbol{0}, \text{ for each } i \in [1 \ldots d]. \tag{8.3.13.1}$$

If $\boldsymbol{x} \notin \operatorname{aff} Y$, then the d vectors $\boldsymbol{x} - \boldsymbol{p}_i$ in (8.3.13.1) are linearly independent in $\mathbb{R}^d$. If we form a $d \times d$ matrix A with the vectors $(\boldsymbol{x} - \boldsymbol{p}_i)^t$ as rows, then (8.3.13.1) is equivalent to the equation $A\varphi(\boldsymbol{x}) = \boldsymbol{0}$, which yields that $\varphi(\boldsymbol{x}) = \boldsymbol{0}$ (as A is nonsingular); in particular, $\varphi = \boldsymbol{0}$, for each point in T. Suppose instead that $\boldsymbol{x} \in \operatorname{aff} Y$. Then there exists a hyperplane in $\mathbb{R}^d$ that is spanned by a set Z of d vertices in T and doesn't contain $\boldsymbol{x}$. Using Z instead of Y in the previous analysis, we also find that $\varphi(\boldsymbol{x}) = \boldsymbol{0}$. As a result, in any case, we have that $\boldsymbol{m}(\boldsymbol{x}) = \tilde{\boldsymbol{m}}(\boldsymbol{x})$. □

Theorem 8.3.14 *Let G be a framework in $\mathbb{R}^d$ whose vertices contain the vertices of a d-simplex. Then the following hold.*

(i) *The space of infinitesimal isometries of $\mathbb{R}^d$ has dimension $\binom{d+1}{2}$.*
(ii) *The space of infinitesimal isometries of the d-simplex has dimension $\binom{d+1}{2}$.*
(iii) *The space of infinitesimal isometries of G has dimension $\binom{d+1}{2}$.*
(iv) *Each infinitesimal isometry of G is the restriction of an infinitesimal isometry of $\mathbb{R}^d$.*

Proof Since aff $\boldsymbol{p}(G) = \mathbb{R}^d$, there is a $(d+1)$-set T of affinely independent points in $\boldsymbol{p}(G)$.

(i)–(iii) By Proposition 8.3.10, $\dim(\mathcal{I}(T)) \leqslant \binom{d+1}{2}$. The restriction of each infinitesimal isometry of G to T is an infinitesimal isometry of T, and by Lemma 8.3.13, for each infinitesimal isometry of T there is at most one infinitesimal isometry of G. Thus $\dim(\mathcal{I}(G)) \leqslant \binom{d+1}{2}$. The same analysis yields that $\dim\left(\mathcal{I}(\mathbb{R}^d)\right) \leqslant \binom{d+1}{2}$, which combined with Proposition 8.3.12, yields that $\dim\left(\mathcal{I}\left(\mathbb{R}^d\right)\right) = \binom{d+1}{2}$. By Remark 8.3.11, $\mathcal{I}\left(\mathbb{R}^d\right) \subseteq \mathcal{I}(T)$ and $\mathcal{I}(\mathbb{R}^d) \subseteq \mathcal{I}(G)$, which implies that $\dim(\mathcal{I}(T)) = \dim(\mathcal{I}(G)) = \binom{d+1}{2}$.

(iv) Let φ be an infinitesimal isometry of G. Then the restriction of φ to T is an infinitesimal isometry of T, and by (ii), the restriction $\varphi|_T$ is the restriction of an infinitesimal isometry ϱ of $\mathbb{R}^d$ to T. It follows that the restriction $\varrho|_G$ of ϱ to G coincides with φ on T. Lemma 8.3.13 now ensures that $\varrho|_G = \varphi$, and φ is the restriction of ϱ to G. □

With some effort, one could explicitly show that every infinitesimal isometry $\boldsymbol{m}$ of a d-simplex uniquely extends to an infinitesimal isometry of $\mathbb{R}^d$. Indeed, if we take $\boldsymbol{x}, \boldsymbol{x}' \in \mathbb{R}^d$ and let $\boldsymbol{p}_1, \ldots, \boldsymbol{p}_{d+1}$ represent the vertices of the simplex, then there exist scalars α_i and β_i ($i \in [1 \ldots d+1]$) such that

$$\boldsymbol{x} = \sum_{i=1}^{d+1} \alpha_i \boldsymbol{p}_i,\ \boldsymbol{x}' = \sum_{i=1}^{d+1} \beta_i \boldsymbol{p}_i, \text{where} \sum_{i=1}^{d+1} \alpha_i = \sum_{i=1}^{d+1} \beta_i = 1.$$

From this, we extend the infinitesimal isometry $\boldsymbol{m}$ as

$$\boldsymbol{m}(\boldsymbol{x}) := \sum_{i=1}^{d+1} \alpha_i \boldsymbol{m}(\boldsymbol{p}_i),\ \boldsymbol{m}(\boldsymbol{x}') := \sum_{i=1}^{d+1} \beta_i \boldsymbol{m}(\boldsymbol{p}_i). \tag{8.3.15}$$

We then find that $(\boldsymbol{x} - \boldsymbol{x}') \cdot (\boldsymbol{m}(\boldsymbol{x}) - \boldsymbol{m}(\boldsymbol{x}')) = 0$.

We list several corollaries of Theorem 8.3.14.

Corollary 8.3.16 *The following hold.*

(i) *The d-simplex is infinitesimally rigid in $\mathbb{R}^d$.*
(ii) *A full-dimensional framework in $\mathbb{R}^d$ is infinitesimally rigid if and only if its motion space has dimension $\binom{d+1}{2}$.*

Proof (i) From Proposition 8.3.10 we find that $\dim \mathcal{R}(T(d)) = \binom{d+1}{2}$, and from Theorem 8.3.14 we find that $\dim \mathcal{I}(T(d)) = \binom{d+1}{2}$.

(ii) According to Theorem 8.3.14, if G is a framework in $\mathbb{R}^d$ then $\dim \mathcal{I}(G) = \binom{d+1}{2}$. As a result, the equality $\dim \mathcal{R}(G) = \binom{d+1}{2}$ ensures that $\mathcal{R}(G) = \mathcal{I}(G)$, or equivalently that G is infinitesimally rigid. □

We return to the frameworks G_a and G_b in Fig. 8.3.2(a)–(b), respectively. It can be shown that $R(G_a)$ has rank seven, wherefrom we get that $\dim \mathcal{R}(G_a) = 3$ (Proposition 8.3.8). Thus Corollary 8.3.16 yields that G_a is infinitesimally rigid. For the framework G_b, Example 8.3.9 constructed its rigidity matrix $R(G_b)$, which has rank six. This gives that $\dim \mathcal{R}(G_b) = 4 > \dim \mathcal{I}(G_b) = 3$ (Corollary 8.3.16), confirming that G_b is infinitesimally nonrigid, as we saw in Example 8.3.5.

Corollary 8.3.17 *Let G be a (full-dimensional) framework in $\mathbb{R}^d$ with n vertices. If $e(G) < dn - \binom{d+1}{2}$, then G is infinitesimally nonrigid.*

Proof The rigidity matrix $R(G)$ of G has $e(G)$ rows, and so at most $e(G)$ linearly independent rows. Thus rank $R(G) \leqslant e(G)$. By Proposition 8.3.8 and the assumption, we get that

$$\dim \mathcal{R}(G) = dn - \operatorname{rank} R(G) \geqslant dn - e(G) > \binom{d+1}{2}.$$

Corollary 8.3.16 now ensures that G is infinitesimally nonrigid. □

As a corollary of Theorem 8.3.14, we also get a useful operation on frameworks.

Lemma 8.3.18 (Glueing lemma, Asimow and Roth, 1979, thm. 2) *If G and $\tilde{G}$ are two infinitesimally rigid frameworks in $\mathbb{R}^d$ that share d affinely independent vertices, then their union is also infinitesimally rigid.*

Proof Consider an infinitesimal motion $\boldsymbol{m}$ of $G \cup \tilde{G}$. Then $\boldsymbol{m}$ is an infinitesimal motion of both G and $\tilde{G}$; in fact, an infinitesimal isometry of them as they are infinitesimally rigid. By Theorem 8.3.14, there is a unique infinitesimal isometry φ of $\mathbb{R}^d$ whose restriction to G coincides with $\boldsymbol{m}$ and there is a unique infinitesimal isometry $\tilde{\varphi}$ of $\mathbb{R}^d$ whose restriction to $\tilde{G}$ coincides with $\boldsymbol{m}$. The infinitesimal isometries φ and $\tilde{\varphi}$ coincide on d affinely independent vertices, in which case Lemma 8.3.13 ensures that $\varphi = \tilde{\varphi}$. It follows that $\boldsymbol{m} = \varphi$. □

Our exposition has focussed on infinitesimal rigidity, which is more manageable than rigidity – as Examples 8.3.3 and 8.3.5 anticipated, and Corollary 8.3.16(ii) confirmed. Example 8.3.5 also suggests that infinitesimal rigidity is a stronger condition than rigidity, but we still don't know whether a framework can be both nonrigid and infinitesimally rigid. Gluck (1975) settled this issue.

Theorem 8.3.19 (Gluck, 1975) [1] *If a d-dimensional framework in $\mathbb{R}^d$ is infinitesimally rigid, then it is rigid.*

Stresses

As we have seen, the (right) nullspace of the rigidity matrix of a framework $G := (X, \boldsymbol{p})$ in $\mathbb{R}^d$ is the motion space of the framework (Proposition 8.3.8). We are now concerned with the left nullspace of $R(G)$, the nullspace of the transpose of $R(G)$. Suppose that a 'force' or *stress* s_{ij} has been placed on each edge $\boldsymbol{p}_i\boldsymbol{p}_j$ of G so that the following equilibrium holds for each vertex $\boldsymbol{p}_\ell$ of G:

$$\sum_{j\ell \in \mathcal{E}(X)} s_{j\ell}(\boldsymbol{p}_j - \boldsymbol{p}_\ell) = \boldsymbol{0}_d; \tag{8.3.20}$$

compare this equilibrium condition with that of (3.7.1). This assignment $\boldsymbol{s}$ of edge stresses is called a *stress* of the framework G. Since we deal with undirected graphs, we assume the symmetry condition $s_{j\ell} = s_{\ell j}$. If we write the stress $\boldsymbol{s}$ of G as a row vector with $e(G)$ entries, then (8.3.20) is equivalent to

$$\boldsymbol{s}^t R(G) = \boldsymbol{0}^t, \text{ for } \boldsymbol{s} \in R^{e(G)}. \tag{8.3.21}$$

For instance, the vector $\boldsymbol{s}^t$ for the 2-simplex has the form (s_{12}, s_{13}, s_{23}); Example 8.3.9 presents the rigidity matrix of this simplex. Thus, the set $\mathcal{T}(G)$ of stresses of a framework is a linear space, called the *stress space* of the framework.

Proposition 8.3.22 *For a framework G in $\mathbb{R}^d$ with n vertices, the following hold.*

[1] A proof is available in Lovász (2019, lem. 15.5).

(i) *The stress space* $\mathcal{T}(G)$ *of* G *is the left nullspace of the rigidity matrix* $R(G)$.
(ii) rank $R(G) = e(G) - \dim\left(\text{null}(R(G)^t)\right) = e(G) - \dim \mathcal{T}(G)$.

We say that a framework is *stressed* if it admits a nonzero stress; that is, if $\dim \mathcal{T}(G) > 0$. Otherwise the framework is *stress-free*. Putting together the information on Propositions 8.3.8 and 8.3.22 and Corollary 8.3.16, we get a test for infinitesimal rigidity, via stresses.

Proposition 8.3.23 *A framework* G *in* $\mathbb{R}^d$ *is infinitesimally rigid if and only if* $\dim \mathcal{T}(G) = e(G) - dv(G) + \binom{d+1}{2}$.

We remark that, while the converse of Theorem 8.3.19 is false in general (Example 8.3.5), it remains true for stress-free frameworks.

Proposition 8.3.24 [2] *If a framework is both stress-free and infinitesimally nonrigid, then it is nonrigid.*

We end this part with results on the affine and projective invariance of the infinitesimal rigidity and stresses of frameworks. Suppose that G is a d-dimensional framework embedded in a hyperplane H of $\mathbb{R}^{d+1}$ and $\boldsymbol{p}_0$ a point $\mathbb{R}^{d+1} \setminus H$. The $(d+1)$-dimensional framework in $\mathbb{R}^{d+1}$ with vertex set $\mathcal{V}(G) \cup \{\boldsymbol{p}_0\}$ and edge set $\mathcal{E}(G) \cup \{\boldsymbol{p}_0\boldsymbol{p} \mid \boldsymbol{p} \in \mathcal{V}(G)\}$, denoted by $G * \boldsymbol{p}_0$, is the *(framework) cone* over G with apex $\boldsymbol{p}_0$; if G is the graph of a d-polytope P in $\mathbb{R}^d$ then $G * \boldsymbol{p}_0$ is the graph of the pyramid $P * \boldsymbol{p}_0$.

Lemma 8.3.25 (Cone lemma) *Let* G *be a framework in* $\mathbb{R}^d$, *and let* $\boldsymbol{p}_0$ *is a point in* $\mathbb{R}^{d+1}$ *so that* $G * \boldsymbol{p}_0$ *is the framework cone over* G. *Then* $\dim \mathcal{T}(G) = \dim \mathcal{T}(G * \boldsymbol{p}_0)$ *and* $\dim \mathcal{R}(G) + d + 1 = \dim \mathcal{R}(G * \boldsymbol{p}_0)$.

In particular, the framework G *is infinitesimally rigid in* $\mathbb{R}^d$ *if and only if* $G * \boldsymbol{p}_0$ *is infinitesimally rigid in* $\mathbb{R}^{d+1}$.

Proof Suppose that $G := (X, \boldsymbol{p})$ lies in a hyperplane $H := \{\boldsymbol{x} \mid \boldsymbol{a} \cdot \boldsymbol{x} = b\}$ in $\mathbb{R}^{d+1}$, and let $\mathcal{V}(G) := \{\boldsymbol{p}_1, \ldots, \boldsymbol{p}_n\}$. We show a bijection between $\mathcal{T}(G)$ and $\mathcal{T}(G * \boldsymbol{p})$. If s_d is a stress on G, then, assigning a zero stress to each edge incident with $\boldsymbol{p}_0$ and leaving unchanged the stress on each edge of G, we obtain a stress s_{d+1} on $G * \boldsymbol{p}_0$. Suppose now that s_{d+1} is a stress on $G * \boldsymbol{p}_0$. Pick a vertex $\boldsymbol{p}_\ell$ of $G * \boldsymbol{p}_0$ other than $\boldsymbol{p}_0$. The equilibrium condition on $\boldsymbol{p}_\ell$ implies the following:

[2] A proof is available in Lovász (2019, lem. 15.7).

$$0 = \boldsymbol{a} \cdot \boldsymbol{0}_{d+1} = \boldsymbol{a} \cdot \left(\sum_{j\ell \in \mathcal{E}(X)} s_{j\ell}(\boldsymbol{p}_j - \boldsymbol{p}_\ell) + s_{\ell 0}(\boldsymbol{p}_0 - \boldsymbol{p}_\ell) \right)$$
$$= \sum_{j\ell \in \mathcal{E}(X)} s_{j\ell}\boldsymbol{a} \cdot (\boldsymbol{p}_j - \boldsymbol{p}_\ell) + s_{\ell 0}\boldsymbol{a} \cdot (\boldsymbol{p}_0 - \boldsymbol{p}_\ell).$$

Since G lies in H, we find that $\sum_{j\ell \in \mathcal{E}(X)} s_{j\ell}\boldsymbol{a} \cdot (\boldsymbol{p}_j - \boldsymbol{p}_\ell) = 0$, and so the last equality becomes

$$s_{\ell 0}\boldsymbol{a} \cdot (\boldsymbol{p}_0 - \boldsymbol{p}_\ell) = 0.$$

Because the point $\boldsymbol{p}_0$ is not in H, this equality ensures that $s_{\ell 0} = 0$. Running over all vertices of $G * \boldsymbol{p}_0$ other than $\boldsymbol{p}_0$, we find that $s_{10} = \cdots = s_{n0} = 0$. Thus $\boldsymbol{s}_{d+1}$ induces a stress $\boldsymbol{s}_d$ on G.

As a consequence, $\dim \mathcal{T}(G) = \dim \mathcal{T}(G * \boldsymbol{p}_0)$. The relation between $\dim \mathcal{R}(G)$ and $\dim \mathcal{R}(G * \boldsymbol{p}_0)$ follows at once from combining the equality $\dim \mathcal{T}(G) = \dim \mathcal{T}(G * \boldsymbol{p}_0)$ with Proposition 8.3.8 and Proposition 8.3.22.

Suppose that G is infinitesimally rigid in $\mathbb{R}^d$. Then, using Proposition 8.3.23 together with $v(G * \boldsymbol{p}_0) = v(G) + 1$ and $e(G * \boldsymbol{p}_0) = e(G) + v(G)$, we find that

$$\dim \mathcal{T}(G * \boldsymbol{p}_0) = \dim \mathcal{T}(G) = e(G) - dv(G) + \binom{d+1}{2}$$
$$= e(G * \boldsymbol{p}_0) - v(G) - d(v(G * \boldsymbol{p}_0) - 1) + \binom{d+1}{2}$$
$$= e(G * \boldsymbol{p}_0) - (d+1)v(G * \boldsymbol{p}_0) + \binom{d+2}{2}.$$

Now resorting to Proposition 8.3.23 again, we get the infinitesimal rigidity of $G * \boldsymbol{p}_0$ in $\mathbb{R}^{d+1}$. A similar computation shows that if $G * \boldsymbol{p}_0$ is infinitesimally rigid in $\mathbb{R}^{d+1}$ then G is infinitesimally rigid in $\mathbb{R}^d$. □

Proposition 8.3.26[3] *Let G be a framework in $\mathbb{R}^d$, let $\varphi\colon \mathbb{R}^d \to \mathbb{R}^d$, and let $\varphi(G)$ be the application of φ to $\boldsymbol{p}(G)$.*

(i) *If $\boldsymbol{s}$ is a stress of G and φ is a nonsingular affine transformation in $\mathbb{R}^d$, then $\boldsymbol{s}$ is a stress of the framework $\varphi(G)$. Thus,* $\dim \mathcal{T}(G) = \dim \mathcal{T}(\varphi(G))$.
(ii) *If φ is a nonsingular affine transformation of $\mathbb{R}^d$, then* $\dim \mathcal{R}(G) = \dim \mathcal{R}(\varphi(G))$.
(iii) *If φ is a projective transformation in $\mathbb{R}^d$ that doesn't map any point in $\boldsymbol{p}$ to a point at infinity, then* $\dim \mathcal{T}(G) = \dim \mathcal{T}(\varphi(G))$ *and* $\dim \mathcal{R}(G) = \dim \mathcal{R}(\varphi(G))$.

Proof We prove (i). Suppose that $G = (X, \boldsymbol{p})$ with $\boldsymbol{p} := (\boldsymbol{p}_1, \ldots, \boldsymbol{p}_n)$, that $\boldsymbol{s}$ is a stress of G, and that $\varphi(\boldsymbol{x})$ is defined by $A\boldsymbol{x} + \boldsymbol{b}$ where A is a nonsingular

[3] A proof is available in Izmestiev (2009).

$d \times d$ matrix and $\boldsymbol{b} \in \mathbb{R}^d$. For each vertex $\boldsymbol{p}_\ell$ of G, the equilibrium condition on G yields that

$$\sum_{j\ell \in \mathcal{E}(X)} s_{j\ell}(A\boldsymbol{p}_j + \boldsymbol{b} - (A\boldsymbol{p}_\ell + \boldsymbol{b})) = A \sum_{j\ell \in \mathcal{E}(X)} s_{j\ell}(\boldsymbol{p}_j - \boldsymbol{p}_\ell) = A\boldsymbol{0}_d = \boldsymbol{0}_d,$$

which is an equilibrium condition for the vertex $\varphi(\boldsymbol{p}_\ell)$ of $\varphi(G)$. Using the inverse matrix A^{-1}, we can see that if $\boldsymbol{s}$ is a stress of $\varphi(G)$ then it is also a stress of G. This settles this part.

(ii)–(iii) We leave these parts to the reader. The Cone lemma (8.3.25) together with the scheme described in Section 2.5 facilitates proofs of the assertions on projective transformations. □

Infinitesimal Rigidity of Simplicial Polytopes

According to Corollary 8.3.17 (and Proposition C.7.2), simplicial 3-polytopes are the only 3-polytopes that can possibly be infinitesimally rigid. Dehn (1916) settled this case. While Roth (1981, sec. 6) offers a very clear proof of this result, which is based on stresses, for the sake of completeness, we offer a slight variation of his proof.

Theorem 8.3.27 *The framework in* $\mathbb{R}^3$ *of a 3-polytope is stress-free.*

Proof By contradiction, suppose that the framework $G := (X, \boldsymbol{p})$ in $\mathbb{R}^3$ of a 3-polytope P is stressed and let $\boldsymbol{s}$ be a nonzero stress of G. Let E' be the set of edges of G with nonzero stress and let $G' := (Y, \boldsymbol{p}|_Y)$ be the subframework of G whose edges are those in E' and whose vertices are those incident with at least one edge in E'. Then E' has at least two elements because every point in $\boldsymbol{p}$ is distinct and each such point satisfies the equilibrium condition (8.3.20).

The framework G' (or more precisely Y) is a plane graph since G is embedded in the boundary of P, which is homeomorphic to the 2-sphere (Theorem 3.1.8). We now focus on the plane subframework G' (and forget the rest of G for the moment); we label the edges of G' with the signs of their stresses. If possible, we further add edges to G' with positive sign labels to make it a 3-connected plane graph (Proposition C.7.3).

The edges incident with a vertex $\boldsymbol{p}_i$ of G' can be ordered cyclically in a clockwise direction with respect to a vector perpendicular to a hyperplane supporting P at $\boldsymbol{p}_i$ and pointing outward (outside P). We count the number $\varphi(\boldsymbol{p}_i)$ of sign changes at $\boldsymbol{p}_i$ as we traverse the edges incident with it in this cyclic order. See Fig. 8.3.3(a)–(b). Moreover, we let $\varphi(G') := \sum_{\boldsymbol{p}_i \in \mathcal{V}(G')} \varphi(\boldsymbol{p}_i)$; this gives the total number of sign changes in G'.

Claim 1 There is a vertex $\boldsymbol{p}_\ell$ in G' with either zero or two sign changes.

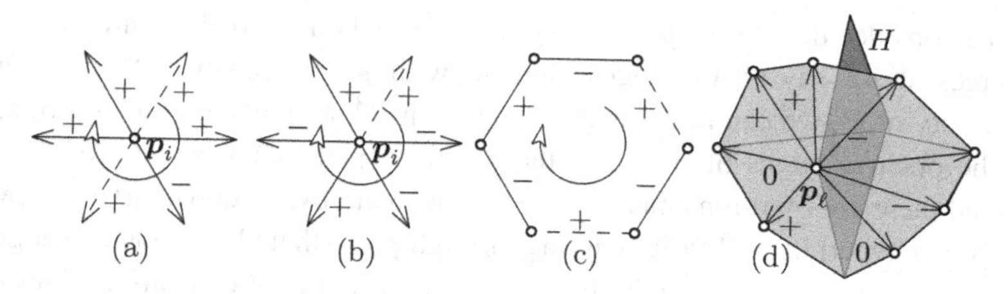

Figure 8.3.3 Auxiliary figure for Theorem 8.3.27. Sign changes at vertices. (a) A vertex $\boldsymbol{p}_i$ with two sign changes; edges added to make the framework G' 3-connected are depicted by dashed lines. (b) A vertex $\boldsymbol{p}_i$ with four sign changes; edges added to make G' 3-connected are depicted by dashed lines. (c) A facial cycle of length six with six sign changes in its boundary; edges added to make G' 3-connected are depicted by dashed lines. (d) A hyperplane H passing through a vertex $\boldsymbol{p}_\ell$ and the interior of a 3-polytope so that all the positive edges and negative edges at $\boldsymbol{p}_\ell$ lie in different sides of H.

Proof of claim The number of sign changes at each vertex in G' is even, and so suppose that this number is at least four for every vertex in G'. Then

$$\varphi(G') \geqslant 4v(G'). \tag{8.3.27.1}$$

We let f be the number of faces in G' and let f^k be the number of faces of G' with k edges. Because G' is 2-connected, every face of G' is bounded by a facial cycle, and so each of its edges is shared by two faces. Thus, the next equalities are plain.

$$2e(G') = \sum_{k\geqslant 3} kf^k,\ f(G') = \sum_{k\geqslant 3} f^k. \tag{8.3.27.2}$$

We now compute $\varphi(G')$ by traversing the facial cycle of each face of G'. The number of sign changes around a facial cycle of length $2r$ or $2r+1$ is at most $2r$; see Fig. 8.3.3(c). This gives another inequality for $\varphi(G')$:

$$\varphi(G') \leqslant 2f^3 + 4f^4 + 4f^5 + \cdots + 2rf^{2r} + 2rf^{2r+1} + \cdots.$$

Using this inequality, (8.3.27.2), and Euler's formula $v(G')-e(G')+f(G') = 2$, we find that

$$\begin{aligned}\varphi(G') \leqslant 2f^3 + 4f^4 + 4f^5 + \cdots + 2rf^{2r} + 2rf^{2r+1} + \cdots &\leqslant \sum_{k\geqslant 3}(2k-4)f^k\\ = 2\sum_{k\geqslant 3} kf^k - 4\sum_{k\geqslant 3} f^k = 4e(G') - 4f(G') &= 4v(G') - 8.\end{aligned}$$

This contradicts (8.3.27.1) and settles the claim. □

We remark that the proof of Claim 1 yields two vertices in G' with either zero or two sign changes. But we need only one such vertex.

We consider the vertex $\boldsymbol{p}_\ell$ of G' granted by Claim 1. If there are no sign changes at $\boldsymbol{p}_\ell$, say all the edges incident with $\boldsymbol{p}_\ell$ are positive, then by the convexity of P, we can find a hyperplane H in $\mathbb{R}^3$ supporting P at $\boldsymbol{p}_\ell$ so that all the positive edges of $\boldsymbol{p}_\ell$ lie in the positive open halfspace H^+ of H. If instead there are two sign changes at $\boldsymbol{p}_\ell$, then, again by the convexity of P, we can find a hyperplane H in $\mathbb{R}^3$ passing through $\boldsymbol{p}_\ell$ so that all the positive edges of $\boldsymbol{p}_\ell$ lie in the positive open halfspace H^+ of H and all the negative edges of $\boldsymbol{p}_\ell$ lie in the negative open halfspace H^- of H.

We now return to the framework G and remove all the edges added to G' to make it 3-connected. The edges incident with the point $\boldsymbol{p}_\ell$ in G and not currently present in G' all have a zero stress. Thus, the hyperplane H still separates the edges of $\boldsymbol{p}_\ell$ with a positive stress from the edges with a negative stress; there may be edges with a zero stress in both sides H^+ and H^- (see Fig. 8.3.3(d)). As a consequence, in any case, if we let $H := \{\boldsymbol{x} \in \mathbb{R}^3 \mid \boldsymbol{a} \cdot \boldsymbol{x} = b\}$, then for every edge $\boldsymbol{p}_j\boldsymbol{p}_\ell$ of G with nonzero stress, we have that

$$\boldsymbol{a} \cdot (\boldsymbol{p}_j - \boldsymbol{p}_\ell) \begin{cases} > 0, & \text{if } s_{j\ell} > 0; \\ < 0, & \text{if } s_{j\ell} < 0. \end{cases}$$

It then follows that $\boldsymbol{a} \cdot s_{j\ell}(\boldsymbol{p}_j - \boldsymbol{p}_\ell) > 0$ for any such edge $\boldsymbol{p}_j\boldsymbol{p}_\ell$. The equilibrium condition on $\boldsymbol{p}_\ell$ (see (8.3.20)) now yields that

$$0 = \boldsymbol{a} \cdot \boldsymbol{0}_3 = \boldsymbol{a} \cdot \left(\sum_{j\ell \in \mathcal{E}(X)} s_{j\ell}(\boldsymbol{p}_j - \boldsymbol{p}_\ell) \right) = \sum_{j\ell \in \mathcal{E}(X)} s_{j\ell}\boldsymbol{a} \cdot (\boldsymbol{p}_j - \boldsymbol{p}_\ell) > 0.$$

This contradiction shows that G cannot have a nonzero stress. □

Dehn's theorem on infinitesimal rigidity is now at hand.

Corollary 8.3.28 (Dehn, 1916) *A framework in $\mathbb{R}^3$ of a 3-polytope is infinitesimally rigid if and only if the polytope is simplicial.*

Proof Combining Proposition 8.3.23 and Theorem 8.3.27, we find that a framework G of a 3-polytope P is infinitesimally rigid if and only if $0 = \dim \mathcal{T}(G) = e(G) - 3v(G) + 6$, which is equivalent to P being simplicial (Proposition C.7.2). □

Convexity played an important role in the proof of Theorem 8.3.27 and, thus, in the veracity of Corollary 8.3.28. There are full-dimensional frameworks in $\mathbb{R}^3$ whose graph is isomorphic to the graph of a simplicial 3-polytope and yet are not infinitesimally rigid. The octahedra of Bricard (1897) epitomise this phenomenon; Gallet et al. (2021) presents a modern treatment of the combinatorics of these nonconvex objects.

Whiteley 1984 extended the work of Dehn (1916) to simplicial d-polytopes.

Theorem 8.3.29 (Whiteley, 1984) *For $d \geqslant 3$, the framework in $\mathbb{R}^d$ of a simplicial d-polytope is infinitesimally rigid in $\mathbb{R}^d$.*

Proof Let G be the framework of a simplicial d-polytope P in $\mathbb{R}^d$. We show the statement by induction on $d \geqslant 3$, with the basis case $d = 3$ granted by Theorem 8.3.27.

Claim 1 The framework in $\mathbb{R}^d$ of the star of a vertex in P is infinitesimally rigid in $\mathbb{R}^d$.

Proof of claim Let $\boldsymbol{p}_0$ be a vertex of G, let X_{st} be the framework of the star of $\boldsymbol{p}_0$ in the boundary complex $\mathcal{B}(P)$ of P, and let X be the graph of the link of $\boldsymbol{p}_0$ in $\mathcal{B}(P)$. We prove that

$$\dim \mathcal{T}(X_{st}) = e(X_{st}) - dv(X_{st}) + \binom{d+1}{2},$$

which would yield that X_{st} is infinitesimally rigid in $\mathbb{R}^d$ by Proposition 8.3.23.

To facilitate the computation, we make three assumptions: (1) $\boldsymbol{p}_0 = \boldsymbol{0}_d$, (2) the hyperplane H defining the vertex figure $P/\boldsymbol{p}_0$ of P at $\boldsymbol{p}_0$ is $\{(x_1, \ldots, x_d)^t \mid x_d = 1\}$, and (3) the positions of the vertices in X have the form $\boldsymbol{p}_i := (\boldsymbol{q}_i, \alpha_i)^t$ where $\alpha_i > 1$ and $i \in [1 \ldots n]$. The assumptions are sound because a stress is preserved under rescaling of the framework or under translations and rotations of $\mathbb{R}^d$; see (8.3.20).

We also let X_{vf} be the framework of $P/\boldsymbol{p}_0$. For $i \in [1 \ldots n]$, the position $\boldsymbol{r}_i$ of each vertex in X_{vf} has the form $(\boldsymbol{q}_i/\alpha_i, 1)^t$. We now show that $\dim \mathcal{T}(X_{st}) = \dim \mathcal{T}(X_{vf})$.

Since P is simplicial, the graph X is isomorphic to the frameworks $X_{st} - \boldsymbol{p}_0$ and X_{vf} (when regarded as abstract graphs). Consider a stress $\boldsymbol{s}$ of X_{st} and a vertex $\boldsymbol{p}_\ell \in X_{st}$ other than $\boldsymbol{p}_0$. For each $\ell \in [1 \ldots n]$, the equilibrium condition (8.3.20) on $\boldsymbol{p}_\ell$ gives the following:

$$\sum_{j\ell \in \mathcal{E}(X)} s_{j\ell}\left(\begin{pmatrix}\boldsymbol{q}_j\\ \alpha_j\end{pmatrix} - \begin{pmatrix}\boldsymbol{q}_\ell\\ \alpha_\ell\end{pmatrix}\right) + s_{\ell 0}\left(\boldsymbol{0} - \begin{pmatrix}\boldsymbol{q}_\ell\\ \alpha_\ell\end{pmatrix}\right) = \boldsymbol{0}_d. \tag{8.3.29.1}$$

From this we get the equality

$$\sum_{j\ell \in \mathcal{E}(X)} s_{j\ell}(\alpha_j - \alpha_\ell) = \alpha_\ell s_{\ell 0}. \tag{8.3.29.2}$$

We next show that

$$s'_{j\ell} := \alpha_\ell \alpha_j s_{j\ell}, \text{ for every } j\ell \in \mathcal{E}(X) \tag{8.3.29.3}$$

defines a stress of X_{vf}. Indeed, for each $\ell \in [1 \ldots n]$ we have that

$$\begin{aligned}
\sum_{j\ell \in \mathcal{E}(X)} s'_{j\ell}(\boldsymbol{r}_j - \boldsymbol{r}_\ell) &= \sum_{j\ell \in \mathcal{E}(X)} \alpha_\ell \alpha_j s_{j\ell}\left(\begin{pmatrix} \boldsymbol{q}_j/\alpha_j \\ 1 \end{pmatrix} - \begin{pmatrix} \boldsymbol{q}_\ell/\alpha_\ell \\ 1 \end{pmatrix}\right) \\
&= \sum_{j\ell \in \mathcal{E}(X)} s_{j\ell}\left(\alpha_\ell \begin{pmatrix} \boldsymbol{q}_j \\ \alpha_j \end{pmatrix} - \alpha_j \begin{pmatrix} \boldsymbol{q}_\ell \\ \alpha_\ell \end{pmatrix}\right) \\
&= \sum_{j\ell \in \mathcal{E}(X)} s_{j\ell}\left(\alpha_\ell \begin{pmatrix} \boldsymbol{q}_j \\ \alpha_j \end{pmatrix} - \alpha_\ell \begin{pmatrix} \boldsymbol{q}_\ell \\ \alpha_\ell \end{pmatrix}\right) \\
&\quad + \sum_{j\ell \in \mathcal{E}(X)} s_{j\ell}\left(\alpha_\ell \begin{pmatrix} \boldsymbol{q}_\ell \\ \alpha_\ell \end{pmatrix} - \alpha_j \begin{pmatrix} \boldsymbol{q}_\ell \\ \alpha_\ell \end{pmatrix}\right) \\
&= \sum_{j\ell \in \mathcal{E}(X)} s_{j\ell}\left(\alpha_\ell \begin{pmatrix} \boldsymbol{q}_j \\ \alpha_j \end{pmatrix} - \alpha_\ell \begin{pmatrix} \boldsymbol{q}_\ell \\ \alpha_\ell \end{pmatrix}\right) \\
&\quad + \sum_{j\ell \in \mathcal{E}(X)} s_{j\ell}(\alpha_\ell - \alpha_j) \begin{pmatrix} \boldsymbol{q}_\ell \\ \alpha_\ell \end{pmatrix},
\end{aligned}$$

and using (8.3.29.2) and (8.3.29.1) in this equation, we get that

$$\begin{aligned}
\sum_{j\ell \in \mathcal{E}(X)} s'_{j\ell}(\boldsymbol{r}_j - \boldsymbol{r}_\ell) &= \alpha_\ell \sum_{j\ell \in \mathcal{E}(X)} s_{j\ell}\left(\begin{pmatrix} \boldsymbol{q}_j \\ \alpha_j \end{pmatrix} - \begin{pmatrix} \boldsymbol{q}_\ell \\ \alpha_\ell \end{pmatrix}\right) - s_{\ell 0}\alpha_\ell \begin{pmatrix} \boldsymbol{q}_\ell \\ \alpha_\ell \end{pmatrix} \\
&= \alpha_\ell \left(\sum_{j\ell \in \mathcal{E}(X)} s_{j\ell}\left(\begin{pmatrix} \boldsymbol{q}_j \\ \alpha_j \end{pmatrix} - \begin{pmatrix} \boldsymbol{q}_\ell \\ \alpha_\ell \end{pmatrix}\right) - s_{\ell 0} \begin{pmatrix} \boldsymbol{q}_\ell \\ \alpha_\ell \end{pmatrix} \right) \\
&= \alpha_\ell \cdot \boldsymbol{0}_d = \boldsymbol{0}_d.
\end{aligned}$$

This shows that the values $s'_{j\ell}$ give a stress of X_{vf}. Conversely, if we are given values $s'_{j\ell}$ that define a stress of X_{vf}, each $s_{j\ell}$ (with $\ell, j \neq 0$) is defined according to (8.3.29.3), and $s_{\ell 0}$ (for each $\ell \in [1 \ldots n]$) is subsequently defined as per (8.3.29.2), then the values $s_{j\ell}$ define a stress of X_{st}. The computation simply reverses the steps in the previous one:

$$\begin{aligned}
&\alpha_\ell \left(\sum_{j\ell \in \mathcal{E}(X)} s_{j\ell}\left(\begin{pmatrix} \boldsymbol{q}_j \\ \alpha_j \end{pmatrix} - \begin{pmatrix} \boldsymbol{q}_\ell \\ \alpha_\ell \end{pmatrix}\right) - s_{\ell 0} \begin{pmatrix} \boldsymbol{q}_\ell \\ \alpha_\ell \end{pmatrix} \right) \\
&\qquad = \sum_{j\ell \in \mathcal{E}(X)} \alpha_\ell s_{j\ell}\left(\begin{pmatrix} \boldsymbol{q}_j \\ \alpha_j \end{pmatrix} - \begin{pmatrix} \boldsymbol{q}_\ell \\ \alpha_\ell \end{pmatrix}\right) - \alpha_\ell s_{\ell 0} \begin{pmatrix} \boldsymbol{q}_\ell \\ \alpha_\ell \end{pmatrix}
\end{aligned}$$

(we now use (8.3.29.2) and (8.3.29.3))

$$
\begin{aligned}
&= \sum_{j\ell\in\mathcal{E}(X)} \alpha_\ell s_{j\ell}\left(\begin{pmatrix}\boldsymbol{q}_j\\ \alpha_j\end{pmatrix}-\begin{pmatrix}\boldsymbol{q}_\ell\\ \alpha_\ell\end{pmatrix}\right)-\left(\sum_{j\ell\in\mathcal{E}(X)} s_{j\ell}(\alpha_j-\alpha_\ell)\right)\begin{pmatrix}\boldsymbol{q}_\ell\\ \alpha_\ell\end{pmatrix}\\
&= \sum_{j\ell\in\mathcal{E}(X)} s_{j\ell}\left(\alpha_\ell\begin{pmatrix}\boldsymbol{q}_j\\ \alpha_j\end{pmatrix}-\alpha_j\begin{pmatrix}\boldsymbol{q}_\ell\\ \alpha_\ell\end{pmatrix}\right)\\
&= \sum_{j\ell\in\mathcal{E}(X)} \frac{s'_{j\ell}}{\alpha_\ell\alpha_j}\left(\alpha_\ell\begin{pmatrix}\boldsymbol{q}_j\\ \alpha_j\end{pmatrix}-\alpha_j\begin{pmatrix}\boldsymbol{q}_\ell\\ \alpha_\ell\end{pmatrix}\right)\\
&= \sum_{j\ell\in\mathcal{E}(X)} s'_{j\ell}\left(\begin{pmatrix}\boldsymbol{q}_j/\alpha_j\\ 1\end{pmatrix}-\begin{pmatrix}\boldsymbol{q}_\ell/\alpha_\ell\\ 1\end{pmatrix}\right)\\
&= \mathbf{0}_d.
\end{aligned}
$$

Since $\alpha_\ell > 0$, we get that (8.3.29.1) holds for each $\ell \in [1\dots n]$, which corresponds to the equilibrium condition for the vertex $(\boldsymbol{q}_\ell,\alpha_\ell)^t$ of X_{st}. The task of verifying that the vertex $\boldsymbol{p}_0$ is also in equilibrium under these $s_{j\ell}$ values is left as an exercise for the reader. Consequently, the values $s_{j\ell}$ constitute a stress of X_{st}, thus completing the proof that $\dim\mathcal{T}(X_{st}) = \dim\mathcal{T}(X_{vf})$.

The vertex figure $P/\boldsymbol{p}_0$ is a simplicial $(d-1)$-polytope (Theorem 2.7.1), and so by the induction hypothesis on $d-1$, its framework X_{vf} is infinitesimally rigid in $\mathbb{R}^{d-1}$. Therefore, Proposition 8.3.23 gives that

$$
\begin{aligned}
\dim\mathcal{T}(X_{st}) = \dim\mathcal{T}(X_{vf}) &= e(X_{vf})-(d-1)v(X_{vf})+\binom{d}{2}\\
&= e(X_{st})-dv(X_{st})+d+\binom{d}{2}\\
&= e(X_{st})-dv(X_{st})+\binom{d+1}{2}.
\end{aligned}
$$

Resorting again to Proposition 8.3.23 we get that X_{st} is infinitesimally rigid in $\mathbb{R}^d$. There was nothing special about the star of $\boldsymbol{p}_0$, and so the star of every vertex in P is infinitesimally rigid in $\mathbb{R}^d$. □

The framework G is connected, and so we can find a path between any two vertices of G. Let $\boldsymbol{p}_u$ and $\boldsymbol{p}_v$ be two adjacent vertices of G. Then the frameworks X_u and X_v of their respective stars are infinitesimally rigid in $\mathbb{R}^d$ by Claim 1. Since the vertices of every facet of P containing the edge $\boldsymbol{p}_u\boldsymbol{p}_v$ lie in both X_u and X_v of P, the intersection $X_u \cap X_v$ contains d affinely independent vertices in $\mathbb{R}^d$. Thus, the glueing lemma (8.3.18) yields that the framework $X_u \cup X_v$ is also infinitesimally rigid in $\mathbb{R}^d$. Hence, we can show that G is infinitesimally rigid in $\mathbb{R}^d$ by building it from stars of adjacent vertices and applications of the glueing lemma. □

In fact, Whiteley (1984, sec. 8) proved his theorem on infinitesimal rigidity for a generalisation of frameworks of 2-simplicial d-polytopes. Suppose that a framework G in $\mathbb{R}^d$ is the graph of a d-polytope P and that, for each 2-face of P, we have added edges to G to triangulate the 2-face. The resulting framework is said to be *2-simplicial d-polytopal*. Realisations of graphs of 2-simplicial d-polytopes in $\mathbb{R}^d$ are examples of 2-simplicial d-polytopal frameworks in $\mathbb{R}^d$.

Theorem 8.3.30 (Whiteley, 1984) *A 2-simplicial d-polytopal framework in $\mathbb{R}^d$ is infinitesimally rigid.*

Combinatorial Rigidity

Figure 8.3.2 shows that rigidity and infinitesimal rigidity are not combinatorial properties; for example, Figure 8.3.2(a) is infinitesimally rigid, while Fig. 8.3.2(b) is not. And Fig. 8.3.2(c) and (d) are rigid or flexible according to whether or not the path $\boldsymbol{p}_1\boldsymbol{p}_2\boldsymbol{p}_3\boldsymbol{p}_4$ is pulled taut. There exists, however, a combinatorial version.

Suppose that $G := (X, \boldsymbol{p})$ is a full-dimensional framework in $\mathbb{R}^d$; this means that the positions of G don't lie in any affine hyperplane of $\mathbb{R}^d$. We can further say that the framework G is in *general position* if the set $\boldsymbol{p}(G)$ is in general position; that is, every subset of $\boldsymbol{p}(G)$ with at most $d+1$ points is affinely independent. Full-dimensionality and being in general position are two examples of nondegeneracy conditions imposed on the framework. We define a further such condition. Suppose that $\boldsymbol{p} = (\boldsymbol{p}_1, \ldots, \boldsymbol{p}_n)$ and write $\boldsymbol{p}$ as a $d \times n$ vector in $\mathbb{R}^{dn}$ where $\boldsymbol{p}^t = \boldsymbol{p}_1^t \cdots \boldsymbol{p}_n^t$, say $\boldsymbol{p}^t = (x_{11}, x_{12}, \ldots, x_{1d}, x_{2d} \ldots, x_{nd})^t$. We say that G (or more precisely the configuration $\boldsymbol{p}$) is *generic* if, for every nonzero polynomial $\varphi(z_{11}, z_{12}, \ldots, z_{1d}, z_{2d} \ldots, z_{nd})$ with integer coefficients, $\varphi(\boldsymbol{p}^t) \neq 0$. Being generic is a far more restrictive assumption than the aforementioned two conditions. From the definition, it follows that the set of generic frameworks (with the positions as points in $\mathbb{R}^{dn}$) is a dense subset of $\mathbb{R}^{dn}$: for every point $\boldsymbol{p} \in \mathbb{R}^{dn}$, we can find another point $\boldsymbol{q} \in \mathbb{R}^{dn}$ that is generic and is arbitrarily close to $\boldsymbol{p}$. The combinatorial aspects of generic frameworks is highlighted by the next theorem.

Theorem 8.3.31 *If a framework $(X, \boldsymbol{p})$ is a full-dimensional infinitesimally rigid framework in $\mathbb{R}^d$, then every generic framework $(X, \boldsymbol{q})$ in $\mathbb{R}^d$ is infinitesimally rigid. Furthermore, if $(X, \boldsymbol{p})$ is stress-free in $\mathbb{R}^d$, then so is every generic framework $(X, \boldsymbol{q})$ in $\mathbb{R}^d$.*

Proof Suppose that the framework $G := (X, \boldsymbol{p})$ has n vertices and that $(X, \boldsymbol{q})$ is a generic framework. Then $R(G)$ has dn columns. By Proposition 8.3.8

and Corollary 8.3.16, G is infinitesimally rigid if and only if $\operatorname{rank} R(G) = dn - \binom{d+1}{2}$; say $r := \operatorname{rank} R(G)$. It is a standard result from linear algebra that if $\operatorname{rank} R(G) = r$ then the determinant $\varphi(z_{11}, z_{12}, \ldots, z_{1d}, z_{2d}, \ldots, z_{nd})$ of some $r \times r$ submatrix of $R(G)$ is nonzero. This determinant is a nonzero polynomial and so $\varphi(\boldsymbol{q}^t) \neq 0$ (since $\boldsymbol{q}$ is generic). It follows that

$$\operatorname{rank} R(X, \boldsymbol{q}) \geqslant \operatorname{rank} R(X, \boldsymbol{p}) = dn - \binom{d+1}{2}.$$

On the other hand, Proposition 8.3.8 and Theorem 8.3.14 give that $\operatorname{rank} R(X, \boldsymbol{q}) \leqslant dn - \binom{d+1}{2} = \operatorname{rank} R(X, \boldsymbol{p})$. Thus, $(X, \boldsymbol{q})$ is infinitesimally rigid in $\mathbb{R}^d$.

The part about the stress is proved analogously. □

For generic frameworks, rigidity and infinitesimal rigidity are equivalent.

Theorem 8.3.32[4] *A generic framework is infinitesimally rigid if and only if it is rigid.*

Theorem 8.3.31 tells us that two generic frameworks with the same graph are both either infinitesimally rigid or infinitesimally nonrigid. And so the rigidity and infinitesimal rigidity of such frameworks are combinatorial properties, properties of the underlying graph. We say that a graph X is *generically rigid* in $\mathbb{R}^d$ if it has a generic framework in $\mathbb{R}^d$ that is infinitesimally rigid or, equivalently, if there is an infinitesimally rigid framework having X as the graph (Theorem 8.3.31). The graph X in Fig. 8.3.2(c)–(d) is not generically rigid, since no framework derived from it is infinitesimally rigid, while the graph in Fig. 8.3.2(a)–(b) is generically rigid by Theorem 8.3.31. We similarly define *generically stress-free* graphs in $\mathbb{R}^d$.

Characterisations of generically rigid and generically stress-free graphs in $\mathbb{R}^d$ is an active research area (Lovász, 2019, sec. 15.3). While the situation in $\mathbb{R}^2$ is well understood, for higher dimensions not much is known. We will stop here but encourage the reader to pursue this topic further in Connelly and Guest (2022, ch. 7) and Lovász (2019, sec. 15.3).

The Lower Bound Theorem

The lower bound theorem of Barnette (1971, 1973c) gives the minimum number of k-faces of a simplicial d-polytope with n vertices. The theorem is a major result on the combinatorics of polytopes. We prove the first part of the theorem in this section.

[4] A proof is available in Lovász (2019, thm. 15.21).

For the parameters $d \geqslant 2$ and $n \geqslant d+1$, we define the function

$$\eta_k(n,d) := \begin{cases} \binom{d}{k}n - \binom{d+1}{k+1}k, & \text{for } k \in [1\dots d-2]; \\ (d-1)n - (d+1)(d-2), & \text{for } k = d-1. \end{cases} \tag{8.3.33}$$

While not all stacked d-polytopes with n vertices are combinatorial isomorphic, they all have the same f-vector, whose values are given by (8.3.33). These values can be readily computed by a recurrence equation that starts with a simplex and quantifies the contributions of each stacking. We next state the theorem.

Theorem 8.3.34 (Lower bound theorem, Barnette, 1971, 1973c) *Let $d \geqslant 2$ and let P be a simplicial d-polytope with n vertices. Then, for each $k \in [1\dots d-1]$, we have that*

(i) $f_k(P) \geqslant \eta_k(n,d)$.
(ii) *Moreover, if equality is attained for $d \geqslant 4$ and some $1 \leqslant k \leqslant d-1$, then P is a stacked d-polytope.*

The lower bound theorem holds with equality for $d = 2$, and by Proposition C.7.2 and Euler's equation (C.7.1), a simplicial 3-polytope with n vertices has $3n-6$ edges and $2n-4$ faces, implying that the theorem also holds with equality for all simplicial 3-polytopes. Walkup (1970) proved Part (i) of the theorem for $d = 4,5$, while Barnette (1971, 1973c) extended it to all $d \geqslant 4$. The case of equality, namely Part(ii), was settled by Barnette (1971) for $k = d-1$ and for all $k \in [1\dots d-1]$ by Billera and Lee (1981).

The proof of lower bound theorem relies on a reduction to the case of edges that was first proved by Perles (Barnette, 1973c). The reduction is also sketched in McMullen and Walkup (1971), and so it is called the *MPW reduction*.

Theorem 8.3.35 (MPW reduction) *Let P be a simplicial d-polytope with n vertices. If $f_1(P) \geqslant dn - \binom{d+1}{2}$, then, for each $k \in [1\dots d-1]$, we have that $f_k(P) \geqslant \eta_k(n,d)$.*

Proof The lower bound theorem is true for $d = 2,3$, and so we can start an induction on $d \geqslant 4$. We first prove that $f_k(P) \geqslant \eta_k(n,d)$, for each $k \in [2\dots d-2]$. Suppose the theorem is true for the edges of simplicial d-polytopes and the k-faces of simplicial $(d-1)$-polytopes satisfying $1 \leqslant k \leqslant d-3$.

Let $\boldsymbol{u}_1,\dots,\boldsymbol{u}_n$ be the vertices of P and let $n_i := \deg_P(\boldsymbol{u}_i)$ for each $i \in [1\dots n]$. We consider the vertex figure $P/\boldsymbol{u}_i$ of P at each vertex $\boldsymbol{u}_i$. The vertex figure $P/\boldsymbol{u}_i$ is a simplicial $(d-1)$-polytope on n_i vertices (Section 2.7). By induction, $f_{k-1}(P/\boldsymbol{u}_i)$ is at least $\eta_{k-1}(n_i,d-1)$, for each $k \in [2\dots d-2]$, which implies that the vertex $\boldsymbol{u}_i$ is contained in at least $\eta_{k-1}(n_i, d-1)$ k-faces of P (Theorem 2.7.2). We count the number of incidences between vertices and k-faces in P in two different ways. From the vertices' point of view, this number of incidences is at least $\sum_{i=1}^{n} \eta_{k-1}(n_i, d-1)$. We bound this sum:

$$\begin{aligned}\sum_{i=1}^{n} \eta_{k-1}(n_i, d-1) &= \sum_{i=1}^{n}\left(\binom{d-1}{k-1}n_i - \binom{d}{k}(k-1)\right)\\ &= -n\binom{d}{k}(k-1) + \binom{d-1}{k-1}\sum_{i=1}^{n} n_i\\ &= -n\binom{d}{k}(k-1) + \binom{d-1}{k-1}2f_1(P).\end{aligned}$$

Since $f_1(P) \geqslant dn - 2\binom{d+1}{2}$, we get that

$$\begin{aligned}\sum_{i=1}^{n} \eta_{k-1}(n_i, d-1) &\geqslant -n\binom{d}{k}(k-1) + \binom{d-1}{k-1}2\left(dn - \binom{d+1}{2}\right)\\ &= -n\binom{d}{k}(k-1) + \frac{d}{k}\binom{d-1}{k-1}(2kn - k(d+1))\\ &= -n\binom{d}{k}(k-1) + \binom{d}{k}2kn - \binom{d}{k}\frac{d+1}{k+1}k(k+1)\\ &= (k+1)\binom{d}{k}n - k(k+1)\binom{d+1}{k+1}. \qquad (8.3.35.1)\end{aligned}$$

In the last two steps, we used the identity $\frac{a}{b}\binom{a-1}{b-1} = \binom{a}{b}$.

On the other side, the number of incidences between vertices and k-faces of P is exactly $(k+1)f_k(P)$, as each k-face is incident with exactly $k+1$ vertices. Combining this with (8.3.35.1), we get that

$$\begin{aligned}(k+1)f_k(P) &\geqslant (k+1)\binom{d}{k}n - k(k+1)\binom{d+1}{k+1}\\ f_k(P) &\geqslant \binom{d}{k}n - k\binom{d+1}{k+1}, \quad \text{for each } k \in [2 \ldots d-2].\end{aligned}$$

The statement for $k = d-1$ can be obtained similarly. □

Remark 8.3.36 Because of the MPW reduction and (2.13.12), Part (i) of the lower bound theorem can be stated in terms of the h-vector of a simplicial polytope (2.13.12): $h_1 \leqslant h_2$.

We are now ready to present a version of the proof of the lower bound theorem by Kalai (1987), using the rigidity arguments developed before.

Proof of the lower bound theorem (8.3.34) By the MPW reduction, it suffices to prove that the simplicial polytope P satisfies the theorem for the edges. Whiteley's theorem on infinitesimal rigidity (8.3.29) ensures that the graph G of a realisation of P in $\mathbb{R}^d$ is infinitesimally rigid in $\mathbb{R}^d$. In this case, Proposition 8.3.23 implies that the dimension of the stress space $\mathcal{T}$ of G equals $h_2(P)-h_1(P) = f_1(P)-dn+\binom{d+1}{2}$ (Remark 8.3.36). The desired inequality follows from the fact that $\dim \mathcal{T}(G) \geqslant 0$.

For Part (ii), we direct the reader to Blind and Blind (1999). □

The lower bound theorem of Barnette (1971, 1973c) has been generalised to many settings. For instance, Kalai (1987) extended it to $(d-1)$-pseudomanifolds; see also Tay (1995). McMullen and Walkup (1971) conjectured that the lower bound theorem extends to other entries in the h-vector of a simplicial polytope P (see Remark 8.3.36):

$$1 \leqslant h_0(P) \leqslant h_1(P) \leqslant h_2(P) \leqslant \cdots \leqslant h_{\lfloor d/2 \rfloor}(P). \tag{8.3.37}$$

And they conjectured conditions for the equality case. This extension, now known as the *generalised lower bound theorem*, was settled by Murai and Nevo (2013).

8.4 The Upper Bound Theorem

The upper bound theorem, conjectured by Motzkin (1957) and proved by McMullen (1970), gives the maximum number of k-faces of a d-polytope with n vertices. Like the lower bound theorem of Barnette (1971, 1973c), this is a major result on the combinatorics of polytopes.

Repeatedly pulling vertices of a d-polytope P with n vertices transforms the polytope into a simplicial d-polytope P' with n vertices and at least as many k-faces as P. And if some i-face of P is not a simplex then $f_k(P) < f_k(P')$ for each $i-1 \leqslant k \leqslant d-1$ (Theorem 2.10.4). Thus, we can focus on simplicial d-polytopes to prove the upper bound theorem.

A simplicial d-polytope with n vertices has at most $\binom{n}{k}$ $(k-1)$-faces, and so a simplicial neighbourly d-polytope on n vertices attains the maximum possible values for the numbers $f_0, \ldots, f_{\lfloor d/2 \rfloor -1}$ (Proposition 2.9.11). The cyclic d-polytope $C(n,d)$ on n-vertices is one such simplicial polytope. Motzkin (1957) conjectured that $C(n,d)$ also attains the maximum possible values for the numbers $f_{\lfloor d/2 \rfloor}, \ldots, f_{d-1}$.

For a simplicial d-polytope Q, using (2.13.12), we can compute the numbers $h_0, \ldots, h_{\lfloor d/2 \rfloor}$ from the numbers $f_0, \ldots, f_{\lfloor d/2 \rfloor -1}$. And using the h-vector version of Dehn–Sommerville's equations for simplicial polytopes (2.13.13), we can compute the remaining numbers $h_{\lfloor d/2\rfloor+1}, \dots, h_d$. Moreover, from the h-vector of Q we can compute its f-vector (2.13.11). That is, the numbers $f_0, \dots, f_{\lfloor d/2\rfloor-1}$ determine the complete f-vector of Q and, likewise, the numbers $h_0, \dots, h_{\lfloor d/2\rfloor}$ determine the complete f-vector of Q. In fact, each number in $\{f_1, \dots, f_{d-1}\}$ is a nonnegative combination of numbers in $\{h_0, \dots, h_{\lfloor d/2\rfloor}\}$; see also (2.13.11). Thus, we can focus on bounds for the numbers $h_0, \dots, h_{\lfloor d/2\rfloor}$ and state the upper bound theorem as follows.

Theorem 8.4.1 (Upper bound theorem, McMullen, 1970) *Let P be a d-polytope with n vertices. Then*

$$f_k(P) \leqslant f_k(C(n,d)), \textit{ for each } 1 \leqslant k \leqslant d-1,$$

or, equivalently,

$$h_k(P) \leqslant h_k(C(n,d)), \textit{ for each } 0 \leqslant k \leqslant \lfloor d/2\rfloor.$$

Moreover, if equality is attained in all these inequalities involving h-numbers, then P is simplicial and neighbourly.

We now compute the numbers $h_0, \dots, h_d$ of the cyclic d-polytope $C(n,d)$ on n vertices. For each $1 \leqslant k \leqslant \lfloor d/2\rfloor$, since $f_{k-1}(C(n,d)) = \binom{n}{k}$, (2.13.12) becomes

$$\begin{aligned} h_k(C(n,d)) &= \sum_{i=0}^{k}(-1)^{k-i}\binom{d-i}{k-i} f_{i-1} \\ &= \sum_{i=0}^{k}(-1)^{k-i}\binom{d-i}{k-i}\binom{n}{i}. \end{aligned}$$

Using the binomial identity $\binom{r}{t} = (-1)^t\binom{t-r-1}{t}$ first and then Vandermonde's identity $\sum_t \binom{r}{t}\binom{s}{k-t} = \binom{r+s}{k}$ (see, for instance, Graham et al. (1994, table 174)), for $0 \leqslant k \leqslant \lfloor d/2\rfloor$ we get that

$$h_k(C(n,d)) = \sum_{i=0}^{k}\binom{k-d-1}{k-i}\binom{n}{i} = \binom{n-d-1+k}{k}. \qquad (8.4.2)$$

The other entries of the h-vector of $C(n,d)$ can be obtained using the equality $h_k = h_{d-k}$ from (2.13.13).

Proof of the upper bound theorem (8.4.1) As mentioned previously, repeatedly pulling vertices of a polytope makes the polytope simplicial. Consequently, we can assume that P is a simplicial d-polytope. We run the proof on the dual polytope Q of P. Then Q is a simple d-polytope. We adapt the statement for Q.

Claim 1 $h_k(P) = h_k(Q)$, for each $k \in [0 \dots d]$.

Proof of claim We first use Dehn–Sommerville's equations for simplicial polytopes (2.13.13), and then the expressions for $h_{d-k}(P)$ in (2.13.12) and for $h_k(Q)$ in (3.5.12), respectively:

$$\begin{aligned} h_k(P) = h_{d-k}(P) &= \sum_{i=0}^{d-k} (-1)^{d-k-i} \binom{d-i}{d-k-i} f_{i-1}(P) \\ &= \sum_{i=0}^{d-k} (-1)^{d-k-i} \binom{d-i}{d-k-i} f_{d-1-(i-1)}(Q) \\ &= h_k(Q). \end{aligned}$$

We have also used the relation between the conjugate faces of P and Q (Theorem 2.4.12). □

In view of Claim 1, it suffices to prove that

$$h_k(Q) \leqslant \binom{n-d-1+k}{k}, \text{ for } 1 \leqslant k \leqslant \lfloor d/2 \rfloor. \tag{8.4.2.1}$$

If you recall from Section 3.5, the number $h_k(Q)$ can be thought of as the number of vertices with indegree k in a good orientation O_φ of $G(Q)$ induced by a linear functional φ in general position; in this orientation, we orient the edge $\boldsymbol{x}_i\boldsymbol{x}_j$ of $G(Q)$ from $\boldsymbol{x}_j$ to $\boldsymbol{x}_i$ if $\varphi(\boldsymbol{x}_i) > \varphi(\boldsymbol{x}_j)$. Let F be any facet of Q. Then F is a simple polytope, and the restriction $O|_F$ of the orientation O of $G(Q)$ to $G(F)$ is a good orientation of $G(F)$ (Lemma 3.5.6). Thus, we can define the number $h_k(F)$ as the number of vertices in F with indegree k in $O|_F$.

We proceed by counting and bounding the number of incidences $(F, \boldsymbol{w})$ between the facets F of Q and vertices $\boldsymbol{w}$ with indegree k in the orientations $O|_F$ of the graphs $G(F)$.

Claim 2 For $k \in [0 \ldots d]$,

$$\sum_{F \in \mathcal{F}_{d-1}(Q)} h_k(F) = (k+1)h_{k+1}(Q) + (d-k)h_k(Q).$$

Proof of claim We count the incidences $(F, \boldsymbol{w})$ in two different ways. The left-hand side of the equation clearly counts these incidences by summing over the facets of Q. For the right-hand side, consider a vertex $\boldsymbol{w}$ of Q. Suppose that $\boldsymbol{w}$ has indegree $k+1$ in O, and let these edges be $e_1, \ldots e_{k+1}$. Since Q is a simple polytope, there is a unique facet F_i of Q containing the $d-1$ edges of $\boldsymbol{w}$ other than e_i (Theorem 2.8.7). In each of these $k+1$ facets the vertex $\boldsymbol{w}$ has indegree k with respect to $O|_F$. If instead the vertex $\boldsymbol{w}$ has indegree k in

O, then for each of the $d-k$ edges $e'_1, \ldots, e'_{d-k}$ having $\boldsymbol{w}$ as a tail, there is a unique facet F_i of Q containing the $d-1$ edges of $\boldsymbol{w}$ other than e'_i. In each of these $d-k$ facets, the vertex $\boldsymbol{w}$ has indegree k with respect to $O|_F$. For other possible indegrees of $\boldsymbol{w}$, the indegree of $\boldsymbol{w}$ in each facet containing it is different from k. This analysis gives the right-hand side of the equation. □

Claim 3 For each facet F of Q, and for each $k \in [0 \ldots d]$, $h_k(F) \leqslant h_k(Q)$.

Proof of claim Pick a facet F of Q. Since the numbers $h_k(F)$ and $h_k(Q)$ are independent of any good orientation, we have the freedom to pick the orientation. According to Corollary 3.5.4, there is a good orientation O' induced by a linear functional φ in general position where $\mathcal{V}(F)$ is an initial set; recall that an initial set with respect to O' is a set of vertices such that no arc is directed from a vertex not in the set to a vertex in the set. In this orientation, every vertex with indegree k in $O'|_F$ is a vertex with indegree k in O'. So the inequality follows. □

If we sum over the n facets of Q, Claim 3 yields the following.

Claim 4 For $k \in [0 \ldots d]$, $\sum_{F \in \mathcal{F}_{d-1}(F)} h_k(F) \leqslant n h_k(Q)$.

The inequality (8.4.2.1) now follows from Claim 2, Claim 4, and an easy induction on k.

If we have an equality in (8.4.2.1) for all $0 \leqslant k \leqslant \lfloor d/2 \rfloor$, then we must also have equality in Claim 3 for each facet of the simple d-polytope Q and each $0 \leqslant k \leqslant \lfloor d/2 \rfloor - 1$. If Q is the dual of a simplicial neighbourly d-polytope then all these equalities hold, and vice versa. For this part, the key observation is that any set of k facets of Q with $2k \leqslant d$ shares a $(d-k)$-face. We leave the details as an exercise. □

8.5 Characterisation of f-Vectors of Simplicial Polytopes

The lower bound theorem (8.3.34), the upper bound theorem (8.3.34), and the generalised lower bound theorem (8.3.37) provide inequalities for the f-vectors of simplicial polytopes, but they certainly provide no characterisation of such f-vectors among the d-tuples of positive integers. Such a characterisation, conjectured by McMullen (1971b), was known as the *g-conjecture*. The g-conjecture gave conditions for a d-tuple of positive integers to be the h-vector of a simplicial d-polytope. We need some definitions to state the theorem.

For any positive integers r and k, there are unique positive integers $n_k, \ldots, n_j$ satisfying $n_k \geqslant n_{k-1} \geqslant \cdots \geqslant n_j \geqslant j \geqslant 1$ and

$$r = \binom{n_k}{k} + \binom{n_{k-1}}{k-1} + \cdots + \binom{n_j}{j}. \tag{8.5.1}$$

Following Grünbaum (2003, sec. 10.1), we call this expression the kth *canonical representation* of r. To construct the representation, we choose the largest number n_k satisfying $r \geqslant \binom{n_k}{k}$, then the largest number n_{k-1} satisfying $r - \binom{n_k}{k} \geqslant \binom{n_{k-1}}{k-1}$, and so on. With the kth canonical representation of r at hand, we define the kth *pseudopower* of r as

$$r^{\langle k \rangle} = \binom{n_k+1}{k+1} + \binom{n_{k-1}+1}{k} + \cdots + \binom{n_j+1}{j+1}. \tag{8.5.2}$$

Here, we adopt the convention that $0^{\langle k \rangle} = 0$, for every k.

From the definition, it is clear that, for a fixed k, the kth pseudopower is increasing:

$$\text{if } r < s \text{ then } r^{\langle k \rangle} < s^{\langle k \rangle}. \tag{8.5.3}$$

The g-conjecture of McMullen (1971b), now the g-theorem, reads as follows.

Theorem 8.5.4 (g-theorem, h-vector version) *A sequence $(h_0, h_1, \ldots, h_d)$ of positive integers is the h-vector of a simplicial d-polytope if and only if*

(i) $h_k = h_{d-k}$, *for* $0 \leqslant k \leqslant \lfloor d/2 \rfloor$;
(ii) $h_k \leqslant h_{k+1}$, *for* $0 \leqslant k \leqslant \lfloor d/2 \rfloor - 1$; *and*
(iii) $h_0 = 1$ *and* $h_{k+1} - h_k \leqslant (h_k - h_{k-1})^{\langle k \rangle}$, *for* $1 \leqslant k \leqslant \lfloor d/2 \rfloor - 1$.

McMullen (1971b) settled the conjecture for simplicial d-polytopes with at most $d + 3$ vertices. For all simplicial polytopes, Stanley (1980) proved the necessity of these conditions, while Billera and Lee (1981) proved their sufficiency.

The g-theorem encompasses all the main results on face numbers of simplicial polytopes. Part (i) is Dehn–Sommerville's equations for simplicial polytopes (2.13.13), and Part(ii) is the generalised lower bound theorem for simplicial polytopes (8.3.37), which in turn extended the lower bound theorem of Barnette (1971, 1973c) (Remark 8.3.36). With some work, we find that Part (iii) implies the upper bound theorem of McMullen (1970).

Proposition 8.5.5 *Let $(h_0, h_1, \ldots, h_d)$ be the h-vector of a simplicial d-polytope P with n vertices. If $h_0 = 1$ and $h_{k+1} - h_k \leqslant (h_k - h_{k-1})^{\langle k \rangle}$, for $1 \leqslant k \leqslant \lfloor d/2 \rfloor - 1$, then $h_k(P) \leqslant \binom{n-d-1+k}{k}$.*

Proof The proof is based on an inductive argument on k. Using (2.13.12), we find that $h_1 - h_0 = n - d - 1$, and so

$$(h_1 - h_0)^{\langle 1\rangle} = \binom{n-d}{2}, \; h_2 \leqslant \binom{n-d}{2} + h_1 = \binom{n-d+1}{2}.$$

This is our basis case, $k = 1$. Suppose that $k \geqslant 2$ and that

$$(h_{k-1} - h_{k-2})^{\langle k-1\rangle} \leqslant \binom{n-d-2+k}{k}, \; h_k \leqslant \binom{n-d-1+k}{k}. \tag{8.5.5.1}$$

Since the kth pseudopower is an increasing function (8.5.3), the assumption $h_k - h_{k-1} \leqslant (h_{k-1} - h_{k-2})^{\langle k-1\rangle}$ together with the induction hypothesis gives that

$$(h_k - h_{k-1})^{\langle k\rangle} \leqslant \left((h_{k-1} - h_{k-2})^{\langle k-1\rangle}\right)^{\langle k\rangle} \leqslant \binom{n-d-2+k}{k}^{\langle k\rangle}. \tag{8.5.5.2}$$

The kth canonical representation of $\binom{n-d-2+k}{k}$ is $\binom{n-d-2+k}{k}$ (see (8.5.1)), and thus its kth pseudopower is $\binom{n-d-1+k}{k+1}$. As a consequence, from (8.5.5.2) and the assumption $h_{k+1} - h_k \leqslant (h_k - h_{k-1})^{\langle k\rangle}$, we find that

$$h_{k+1} - h_k \leqslant (h_k - h_{k-1})^{\langle k\rangle} \leqslant \binom{n-d-1+k}{k+1},$$

wherefrom the induction hypothesis yields that

$$\begin{aligned} h_{k+1} &\leqslant \binom{n-d-1+k}{k+1} + h_k \\ &\leqslant \binom{n-d-1+k}{k+1} + \binom{n-d-1+k}{k} = \binom{n-d+k}{k+1}. \end{aligned}$$

This concludes the proof of the proposition. □

From Proposition 8.5.5 it is now plain that Part (iii) of the g-theorem implies the upper bound theorem.

The g-conjecture and g-theorem are named after the *g-vector* of a simplicial d-polytope P, which is defined as the sequence $\left(g_0, \ldots, g_{\lfloor d/2\rfloor}\right)$, where

$$g_k(P) := \begin{cases} h_0(P) = 1, & \text{for } k = 0; \\ h_k(P) - h_{k-1}(P), & \text{for } 1 \leqslant k \leqslant \lfloor d/2\rfloor. \end{cases} \tag{8.5.6}$$

8.6 Flag Vectors

The flag vector of a d-polytope is a useful generalisation of the f-vector of the polytope. For a flag

$$F_1 \subset F_2 \subset \cdots \subset F_\ell$$

of faces of a d-polytope P satisfying $\dim F_i = d_i$, we let

$$S := \{d_1, \ldots, d_\ell\} \subseteq \{0, \ldots, d-1\}$$

and call the flag an *S-flag*. On all subsets $S \subseteq \{0, \ldots, d-1\}$ we impose an order, first by increasing cardinality, and then lexicographically for subsets of the same cardinality: for $X := \{x_1, \ldots, x_r\} \subseteq \{0, \ldots, d-1\}$ and $Y := \{y_1, \ldots, y_s\} \subseteq \{0, \ldots, d-1\}$, we say that

$$X < Y \text{ if and only if either } r < s \text{ or } r = s,\ x_a < y_a \text{ for some } a \in [1 \ldots r], \text{ and } x_b = y_b \text{ for each } b \in [1 \ldots a). \tag{8.6.1}$$

The number of S-flags in P is a *flag number* of the polytope P and is denoted by $f_S(P)$. The sequence $\left(f_{\{0\}}(P), \ldots, f_{\{d-1\}}(P), \ldots, f_{\{0,\ldots,d-1\}}(P)\right)$ of 2^d flag numbers is the *flag vector* of P. If there is no ambiguity, we write f_0 rather than $f_{\{0\}}$ or f_{01} rather than $f_{\{0,1\}}$.

Some flag numbers are easily computed for each d-polytope P. Each edge contains two vertices, and so it appears precisely in two $\{0,1\}$-flags. Thus $f_{01}(P) = 2f_1(P)$. Applying this result to the dual polytope, we get that $f_{\{d-2,d-1\}}(P) = 2f_{d-2}(P)$. Additionally, a 2-face has the same number of vertices as edges, and so $f_{02}(P) = f_{12}(P)$. We summarise these observations.

Remark 8.6.2 For a d-polytope P, we have that

$$f_{01}(P) = 2f_1(P),\ f_{\{d-2,d-1\}}(P) = 2f_{d-2}(P),\ f_{02}(P) = f_{12}(P).$$

We explore the flag vector $(f_0, f_1, f_2, f_{01}, f_{02}, f_{12}, f_{012})$ of 3-polytopes. Remark 8.6.2 for a 3-polytope P yields that

$$f_{01}(P) = f_{02}(P) = f_{12}(P) = 2f_1(P).$$

Fix a flag X of faces $F_0 \subset F_2$ where $\dim F_i = i$ $(i = 0, 2)$, and consider all the edges F_1 such that $F_0 \subset F_1 \subset F_2$. Then this flag X appears in precisely two $\{0,1,2\}$-flags. Thus, $2f_{02}(P) = f_{0,1,2}(P)$, which gives that $f_{0,1,2}(P) = 4f_1(P)$. We summarise all these assertions:

$$f_{01}(P) = f_{02}(P) = f_{12}(P) = 2f_1(P),\ 2f_{02}(P) = f_{0,1,2}(P) = 4f_1(P). \tag{8.6.3}$$

Hence the flag vector of a 3-polytope depends on its f-vector, and so it does not contain any additional information.

The face figure F_j/F_i of two faces F_i, F_j of a d-polytope P is useful when computing flag numbers; recall from Section 2.7 that F_j/F_i is the set of faces F satisfying $F_i \subseteq F \subseteq F_j$, for $-1 \leqslant i < j \leqslant d$. Fix a 0-face F_0 and a 3-face F_3 of a d-polytope P, and consider the fixed face figure F_3/F_0. According to Theorem 2.7.4, the face figure F_3/F_0 is combinatorially isomorphic to a 2-polytope Q. Euler–Poincaré–Schläfli's equation (2.12.17) on Q ensures that

$$
\begin{aligned}
-f_{-1}(Q) + f_0(Q) - f_1(Q) &= -1,\\
-f_0(F_3/F_0) + f_1(F_3/F_0) - f_2(F_3/F_0) &= -1.
\end{aligned}
$$

Since $f_{-1}(Q) = f_0(F_3/F_0) = 1$, the above equality becomes

$$f_1(F_3/F_0) - f_2(F_3/F_0) = 0.$$

If we now let $S := \{0, 3\}$ and sum this equality over all S-flags of P, then we have that

$$
\begin{aligned}
\sum_{S\text{-flags}} \sum_{j=1}^{2} (-1)^{j-1} f_j(F_3/F_0) &= 0 \times f_{03}(P),\\
\sum_{j=1}^{2} (-1)^{j-1} \sum_{S\text{-flags}} f_j(F_3/F_0) &= 0 \times f_{03}(P).
\end{aligned}
$$

The expression $\sum_{S\text{-flags}} f_j(F_3/F_0)$ coincides with $f_{S \cup \{j\}}(P)$ for each $j = 1, 2$, and so

$$
\begin{aligned}
\sum_{j=1}^{2} (-1)^{j-1} f_{S \cup \{j\}}(P) &= 0 \times f_{03}(P),\\
f_{\{0,1,3\}}(P) - f_{\{0,2,3\}}(P) &= 0.
\end{aligned}
\tag{8.6.4}
$$

That is, we have used the face figure F_3/F_0 of a d-polytype P to prove that $f_{\{0,1,3\}}(P) - f_{\{0,2,3\}}(P) = 0$. This idea was exploited by Bayer and Billera (1985) to produce linear equations for flag vectors, similar to Dehn–Sommerville's equations (2.13.3) for simplicial polytopes.

Theorem 8.6.5 *Let P be a d-polytope. Let $S \subseteq \{0, \ldots, d-1\}$, and let $i, k \in S \cup \{-1, d\}$ such that $i \leqslant k-2$ and S contains no integer j with $i < j < k$. Then*

$$\sum_{j=i+1}^{k-1} (-1)^{j-i-1} f_{S \cup \{j\}}(P) = \left(1 - (-1)^{k-i-1}\right) f_S(P).$$

Remark 8.6.6 Euler–Poincaré–Schläfli's equation (2.12.17) is recovered when $S = \varnothing$, $i = -1$, and $k = d$.

Proof Let $S := \{d_1, \dots, d_\ell\} \subseteq \{0, \dots, d-1\}$, and let α be a fixed S-flag in P, say

$$\alpha = F^\alpha_{d_1} \subset F^\alpha_{d_2} \subset \cdots \subset F^\alpha_{d_j} \subset F^\alpha_{d_{j+1}} \subset \cdots \subset F^\alpha_{d_\ell}.$$

Further suppose that $\dim F^\alpha_{d_r} = d_r$ for $r \in [1 \dots \ell]$ and that $d_j = i$ and $d_{j+1} = k$. For the sake of simplicity, let $F^\alpha_i := F^\alpha_{d_j}$ and $F^\alpha_k := F^\alpha_{d_{j+1}}$.

We now consider all the faces F^α of P such that

$$F^\alpha_i \subseteq F^\alpha \subseteq F^\alpha_k.$$

This set of faces gives the face figure F^α_k/F^α_i, which is combinatorially isomorphic to a $(k-1-i)$-polytope Q (Theorem 2.7.4). Euler–Poincaré–Schläfli's equation (2.12.17) on Q ensures that

$$-f_{-1}(Q) + f_0(Q) + \cdots + (-1)^{k-2-i} f_{k-2-i}(Q) = -(-1)^{k-1-i}.$$

Additionally, to each aforementioned face F^α of P there corresponds a face in Q of dimension $\dim(F^\alpha) - 1 - i$. Thus, the previous equality becomes

$$-f_i(F^\alpha_k/F^\alpha_i) + \cdots + (-1)^{k-2-i} f_{k-1}(F^\alpha_k/F^\alpha_i) = -(-1)^{k-1-i}.$$

Because $f_i(F^\alpha_k/F^\alpha_i) = 1$, we can rewrite this equality as

$$\sum_{j=i+1}^{k-1} (-1)^{j-i-1} f_j(F^\alpha_k/F^\alpha_i) = 1 - (-1)^{k-1-i}. \tag{8.6.6.1}$$

We let $\mathcal{S}$ be the set of all S-flags of P and sum (8.6.6.1) over all flags β in $\mathcal{S}$:

$$\sum_{\beta \in \mathcal{S}} \sum_{j=i+1}^{k-1} (-1)^{j-i-1} f_j\left(F^\beta_k/F^\beta_i\right) = \left(1 - (-1)^{k-1-i}\right) f_S(P),$$

which amounts to

$$\sum_{j=i+1}^{k-1} (-1)^{j-i-1} \sum_{\beta \in \mathcal{S}} f_j\left(F^\beta_k/F^\beta_i\right) = \left(1 - (-1)^{k-1-i}\right) f_S(P). \tag{8.6.6.2}$$

For each $j \in [i+1 \dots k-1]$, the sum $\sum_{\beta \in \mathcal{S}} f_j\left(F^\beta_k/F^\beta_i\right)$ counts the number of $(S \cup \{j\})$-flags of P, namely it gives $f_{S\cup\{j\}}(P)$. Hence (8.6.6.2) becomes

$$\sum_{j=i+1}^{k-1} (-1)^{j-i-1} f_{S\cup\{j\}}(P) = \left(1 - (-1)^{k-1-i}\right) f_S(P),$$

which proves the theorem. □

Let $S \subseteq \{0,\dots,d-1\}$, and let Π_d be the set of subsets of $\{0,\dots,d-2\}$ with no consecutive integers. It turns out that the number f_S of S-flags can be expressed as a linear combination of flag numbers f_T with $T \in \Pi_d$.

Theorem 8.6.7 *Let $d \geqslant 1$ and let $S \subseteq \{0,\dots,d-1\}$. The flag number f_S of a d-polytope P can be expressed as a linear combination of flag numbers $f_T(P)$ with $T \in \Pi_d$.*

Proof Let P be a d-polytope. If $S \in \Pi_d$ then there is nothing to prove. So suppose that $S \notin \Pi_d$. Then, for some $k \in [1\dots d]$, we have that $\{k-1,k\} \subseteq S \cup \{d\}$. Let

$$T := S\backslash\{k-1\} \text{ and } i := \max\{j \in S \cup \{-1\} \mid j < k-1\}.$$

Then the set T contains no elements between i and k. An application of Theorem 8.6.5 to the set T, and integers i and k yields that

$$\sum_{j=i+1}^{k-2} (-1)^{j-i-1} f_{T\cup\{j\}}(P) + (-1)^{k-1-i-1} f_{T\cup\{k-1\}}(P)$$
$$= f_T(P)\left(1-(-1)^{k-i-1}\right).$$

If we multiply both sides of this equality by $(-1)^{k+i}$ and solve for $f_{T\cup\{k-1\}}(P)$, we obtain that

$$f_S(P) = f_{T\cup\{k-1\}}(P) = \sum_{j=i+1}^{k-2} (-1)^{k+j} f_{T\cup\{j\}}(P) + f_T(P)\left(1+(-1)^{k+i}\right). \tag{8.6.7.1}$$

All the elements in $X := \{T, T\cup\{i+1\},\dots,T\cup\{k-2\}\}$ are smaller than S in the order (8.6.1). Repeating this reduction for any element in $X\backslash\Pi_d$, we transform (8.6.7.1) into the desired linear expression of $f_S(P)$. □

It is not difficult to prove that the cardinality of Π_d is the dth Fibonacci number c_d.

Proposition 8.6.8 *For $d \geqslant 1$, the cardinality of Π_d is the dth Fibonacci number c_d.*

Proof It is clear that $\Pi_1 = \{\varnothing\}$ and $\Pi_2 = \{\varnothing,\{0\}\}$, and so

$$\#\Pi_1 = 1 \text{ and } \#\Pi_2 = 2. \tag{8.6.8.1}$$

So assume that $d \geqslant 3$. Let $T \in \Pi_d$. Then either $d-2 \notin T$ or $d-2 \in T$. In the former case we have that $T \in \Pi_{d-1}$, and in the latter case we have that $T\backslash\{d-2\} \in \Pi_{d-2}$ since $d-3 \notin T$. It follows that $\#\Pi_d = \#\Pi_{d-1} + \#\Pi_{d-2}$, which together with (8.6.8.1) yields the recurrence equation for the Fibonacci numbers. The lemma is proved. □

Every index S of a flag of a d-polytope is a subset of $\{0,\ldots,d-1\}$, and every flag number f_S can be expressed in terms of elements of Π_d (Theorem 8.6.7). Since there are precisely c_d elements in Π_d (Proposition 8.6.8), it follows that the affine dimension of the set of flag vectors is at most $c_d - 1$. This number is actually the affine dimension of this set of flag vectors. Problem 8.7.7 asks for families of c_d d-polytopes whose flag vectors are affinely independent in $\mathbb{R}^{c_d}$.

In the same way that the flag vector of a 3-polytope is determined by its f-vector, the flag vectors of simplicial polytopes and cubical polytopes are determined by their f-vectors.

Theorem 8.6.9 (Flag numbers of simplicial polytopes) *Let P be a simplicial d-polytope. Then the flag vector of P is determined by its f-vector. Specifically, for a subset $S := \{d_1,\ldots,d_s\} \subseteq \{0,\ldots,d-1\}$ with $s \geqslant 2$ and $0 \leqslant d_1 < \cdots < d_s \leqslant d-1$, it holds that*

$$\begin{aligned} f_S(P) &= f_{d_s}(P) \times f_{S\backslash\{d_s\}}(T(d_s)) \\ &= f_{d_s}(P) \times \prod_{j=1}^{s-1} f_{d_j}(T(d_{j+1})), \end{aligned}$$

where $T(k)$ denotes the k-simplex.

Proof Let $S := \{d_1, d_2\}$ with $0 \leqslant d_1 < d_2 \leqslant d-1$. The flag number $f_S(P)$ counts the incidences between the d_1-faces and d_2-faces of P, both of which are simplices. It follows that

$$f_S(P) = f_{d_2}(P) \times f_{d_1}(T(d_2)) = f_{d_2}(P) \times f_{S\backslash\{d_2\}}(T(d_2)).$$

We generalise this idea. Let $S := \{d_1,\ldots,d_s\}$ with $0 \leqslant d_1 < \cdots < d_s \leqslant d-1$. Then the flag number f_S counts the number of $(S\backslash\{d_s\})$-flags in a d_s-simplex of P. Thus

$$f_S(P) = f_{d_s}(P) \times f_{S\backslash\{d_s\}}(T(d_s)),$$

which inductively gives that

$$f_S(P) = f_{d_s}(P) \times \prod_{j=1}^{s-1} f_{d_j}(T(d_{j+1})),$$

concluding in this way the proof of the theorem. □

The same analysis in the proof of Theorem 8.6.9 gives an analogous result for cubical polytopes (Problem 8.7.6).

By repeatedly pulling vertices of a d-polytope P, we arrive at a simplicial d-polytope with the same number of vertices as P and whose f-vector is not smaller componentwise than that of P (Theorem 2.10.3). A modification of this approach establishes the existence of a simplicial d-polytope with the same number of vertices as P and whose flag vector is not smaller componentwise than that of P. By virtue of the upper bound theorem (8.4.1), the cyclic d-polytope $C(n,d)$ on n vertices maximises the number of k-faces among the d-polytopes with n vertices, and by virtue of Theorem 8.6.9, $C(n,d)$ simultaneously maximises the flag numbers f_S for each $S \subseteq [0 \dots d-1]$, among the d-polytopes with n vertices.

We end this part with a theorem of Kalai (1987), which is a direct consequence of Theorem 8.3.30.

Theorem 8.6.10 (Kalai, 1987) *Let P be d-polytope in $\mathbb{R}^d$. Then*

$$f_{02}(P) - 3f_2(P) + f_1(P) - df_0(P) + \binom{d+1}{2} \geqslant 0.$$

Proof Let G' be a 2-simplicial d-polytopal framework in $\mathbb{R}^d$ obtained from a realisation of the graph G of P in $\mathbb{R}^d$ (Section 8.3). To obtain G' from G, we added $\ell - 3$ edges to each 2-face F of P with ℓ edges in order to triangulate the face. As a result, if we let f_2^ℓ be the number of 2-faces of P with ℓ edges, then

$$e(G') = e(G) + \sum_{\ell \geqslant 3} (\ell - 3) f_2^\ell = f_1(P) + f_{02}(P) - 3f_2(P); \qquad (8.6.10.1)$$

here, we used that $f_{02}(P) = \ell f_2^\ell$ and $f_2(P) = \sum_{\ell \geqslant 3} f_2^\ell$.

Since G' is infinitesimally rigid (Theorem 8.3.30), Proposition 8.3.23 ensures that $e(G') - dv(G') + \binom{d+1}{2} \geqslant 0$. Thus, (8.6.10.1) becomes

$$f_1(P) + f_{02}(P) - 3f_2(P) \geqslant df_0(P) - \binom{d+1}{2};$$

here, we used the equality $f_0(P) = v(G')$. □

4-polytopes

We devote the last part of the section to studying the flag vector

$$(f_0, f_1, f_2, f_3, f_{01}, f_{02}, f_{03}, f_{12}, f_{13}, f_{23}, f_{012}, f_{013}, f_{023}, f_{123}, f_{0123})$$

of a 4-polytope P. Euler–Poincaré–Schläfli's equation (2.12.17) gives that

$$f_0(P) - f_1(P) + f_2(P) = f_3(P).$$

To apply Theorem 8.6.5, we require a triple (S, i, k) as in $S := \{0\}$, $i := 0$, and $k := 4$, which in this case yields that

$$\sum_{j=1}^{3} (-1)^{j-0-1} f_{0j}(P) = \left(1 - (-1)^{4-0-1}\right) f_0(P),$$

$$f_{01}(P) - f_{02}(P) + f_{03}(P) = 2f_0(P).$$

From the triples $(S := \{0,3\}, i := 0, k := 3)$, $(S := \{1\}, i := 1, k := 4)$, $(S := \{1,2\}, i := 2, k := 4)$, $(S := \{0,2\}, i := 0, k := 2)$, $(S := \{0,2\}, i := 2, k := 4)$, and $(S := \{0,1,2\}, i := 2, k := 4)$, we respectively get the following relations.

$$\begin{aligned} f_{013}(P) &= f_{023}(P), \\ f_{12}(P) &= f_{13}(P), \\ f_{123}(P) &= 2f_{12}(P), \\ f_{012}(P) &= 2f_{02}(P), \\ f_{023}(P) &= 2f_{02}(P), \\ f_{0123}(P) &= 2f_{012}(P). \end{aligned}$$

For every d-polytope, we have that $f_{01} = 2f_1$, $f_{23} = 2f_2$, and $f_{02} = f_{12}$ (Remark 8.6.2). We summarise this analysis.

Remark 8.6.11 The characterisation of flag vectors of 4-polytopes reduces to the characterisation of the tuple (f_0, f_1, f_2, f_{02}).

(i) $f_3 = f_0 - f_1 + f_2$,
(ii) $f_{01} = 2f_1, f_{03} = 2f_0 - f_{01} + f_{02} = 2f_0 - 2f_1 + f_{02}, f_{12} = f_{02}, f_{13} = f_{12} = f_{02}, f_{23} = 2f_2$,
(iii) $f_{012} = f_{013} = f_{023} = f_{123} = 2f_{12} = 2f_{02}$,
(iv) $f_{0123} = 2f_{012} = 4f_{12} = 4f_{02}$.

Alternatively, the characterisation of flag vectors of 4-polytopes reduces to the characterisation of the tuple (f_0, f_1, f_3, f_{03}) (Problem 8.7.9). Bayer (1987) gave linear inequalities on the flag vectors of 4-polytopes.

Theorem 8.6.12 (Bayer, 1987) *Suppose that P is a 4-polytope and suppose that $(f_0, f_1, f_2, f_{02}) := (f_0(P), f_1(P), f_2(P), f_{02}(P))$. Then*

(i) $f_{02} - 3f_2 \geqslant 0$,

(ii) $f_{02} - 3f_1 \geqslant 0$,
(iii) $f_{02} - 3f_2 + f_1 - 4f_0 + 10 \geqslant 0$,
(iv) $6f_1 - 6f_0 - f_{02} \geqslant 0$,
(v) $f_0 - 5 \geqslant 0$,
(vi) $f_2 - f_1 + f_0 - 5 \geqslant 0$.

Proof (i) Each 2-face has at least three vertices, so $f_{02} - 3f_2 \geqslant 0$.

(ii) Since each edge is the intersection of at least three facets of P (Theorem 2.4.16), it follows that $f_{13} \geqslant 3f_1$, which together with $f_{02} = f_{13}$ (Remark 8.6.11) implies the result.

(iii) This is Kalai's inequality (Theorem 8.6.10) for $d = 4$.

(iv) For each 3-polytope F, we have that $2f_1(F) \geqslant 3f_0(F)$. Summing this inequality over all the facets F of P we get that

$$2f_{13}(P) = \sum_{F \in \mathcal{F}_3(P)} 2f_1(F) \geqslant \sum_{F \in \mathcal{F}_3(P)} 3f_0(F) = 3f_{03}(P).$$

Combining this equation with the equalities $f_{13} = f_{02}$ and $f_{03} = 2f_0 - 2f_1 + f_{02}$ (Remark 8.6.11), we get that

$$2f_{13} = 2f_{02} \geqslant 3f_{03}(P) \geqslant 3(2f_0 - 2f_1 + f_{02}),$$
$$2f_{02} \geqslant 6f_0 - 6f_1 + 3f_{02}.$$

(v)–(vi) The inequality (v) is trivial, and so is (vi) by duality: $5 \leqslant f_3 = f_2 - f_1 + f_0$. □

The inequalities in Theorem 8.6.12 can be readily translated into inequalities about face numbers.

Theorem 8.6.13 *Let P be a 4-polytope and let*

$$(f_0, f_1, f_2) := (f_0(P), f_1(P), f_2(P)).$$

Then

(i) $2f_1 - 2f_0 - f_2 \geqslant 0$,
(ii) $f_1 - 2f_0 \geqslant 0$,
(iii) $-3f_2 + 7f_1 - 10f_0 + 10 \geqslant 0$,
(iv) $f_0 - 5 \geqslant 0$,
(v) $f_2 - f_1 + f_0 - 5 \geqslant 0$.

Bayer (1987) also proved some nonlinear inequalities.

Theorem 8.6.14 (Bayer, 1987) *Suppose that P is a 4-polytope and suppose that $(f_0, f_1, f_2, f_{02}) := (f_0(P), f_1(P), f_2(P), f_{02}(P))$. Then*

(i) $2(f_{02} - 3f_2) + f_1 \leqslant \binom{f_0}{2}$,
(ii) $2(f_{02} - 3f_1) + f_2 \leqslant \binom{f_2 - f_1 + f_0}{2}$,
(iii) $f_{02} - 4f_2 + 3f_1 - 2f_0 \leqslant \binom{f_0}{2}$,
(iv) $f_{02} - f_2 - 2f_1 - 2f_0 \leqslant \binom{f_2 - f_1 + f_0}{2}$.

Proof (i) The number $\binom{f_0}{2}$ of pairs of vertices in P can be partitioned into three subsets: (1) the set of pairs determining an edge, (2) the set of pairs on some 2-face but not on an edge, and (3) the set of pairs not on any 2-face of P. The cardinality of the first set is f_1. We estimate the cardinality of the second set. Consider a 2-face F with n vertices. Then the number of nonedge pairs of vertices on F is $\binom{n}{2} - n = \frac{1}{2}n(n-3) \geqslant 2(n-3)$, for $n \geqslant 3$. Summing this last term over all the 2-faces of P, we get that

$$\sum_{F \in \mathcal{F}_2(P)} 2(f_0(F) - 3) = 2f_{02} - 6f_2,$$

which is a lower bound for the cardinality of the second set. Hence

$$f_1 + 2f_{02} - 6f_2 \leqslant \binom{f_0}{2}.$$

(ii) Applying (i) to the dual polytope of P we get (ii).
(iii)–(iv) We leave the proof of these as an exercise. □

Ling (2007) obtained other nonlinear inequalities for flag numbers of 4-polytopes.

8.7 Problems

8.7.1 (Xue, 2022) Let P and P' be two d-polytopes, with respective f-vectors $f(P)$ and $f(P')$. Suppose that $f(P) \leqslant f(P')$; that is, $f_i(P) \leqslant f_i(P')$ for each $i \in [0 \dots d-1]$. If the f-vectors in a set $X := \{P_1, P_2, P_3 \dots\}$ of polytopes are all distinct and satisfy $f(P_1) \leqslant f(P_2) \leqslant f(P_3) \leqslant \cdots$, then X is said to be *completely ordered by f-vectors*.

Prove that, for a fixed s, the d-polytopes with $d+s$ vertices and $d+2$ facets are completely ordered by f-vectors.

8.7.2 (Whiteley, 1996) A *vertex d-addition* to a framework G in $\mathbb{R}^d$ with d vertices $\boldsymbol{p}_1, \dots, \boldsymbol{p}_d$ involves adding to G a new vertex $\boldsymbol{p}_0$ and d new edges $\boldsymbol{p}_0\boldsymbol{p}_1, \dots, \boldsymbol{p}_0\boldsymbol{p}_d$. And a framework G' is an *edge d-split* of a framework G

with a given edge $\boldsymbol{p}_1\boldsymbol{p}_2$ and $d-1$ other vertices $\boldsymbol{p}_3, \ldots, \boldsymbol{p}_{d+1}$ if G' is obtained from G by adding a new vertex $\boldsymbol{p}_0$, removing the edge $\boldsymbol{p}_1\boldsymbol{p}_2$, and adding $d+1$ new edges $\boldsymbol{p}_0\boldsymbol{p}_1, \ldots, \boldsymbol{p}_0\boldsymbol{p}_{d+1}$. Prove the following statement:

If G is a framework in $\mathbb{R}^d$ and G' is obtained from G through a vertex d-addition such that the vertices $\boldsymbol{p}_0, \boldsymbol{p}_1, \ldots, \boldsymbol{p}_d$ are in general position in $\mathbb{R}^d$, or if G' is derived from G by an edge d-split such that the vertices $\boldsymbol{p}_1, \ldots, \boldsymbol{p}_{d+1}$ are in general position in $\mathbb{R}^d$, then rank $R(G') = R(G) + d$.

8.7.3 (Replacement lemma, Kalai, 1987) Let G be a full-dimensional framework in $\mathbb{R}^d$, let G' be an induced subframework of G that is infinitesimally rigid in $\mathbb{R}^d$, and let G'' be another framework in $\mathbb{R}^d$ on the vertex set $\mathcal{V}(G')$. If both G'' and the framework $(G - \mathcal{E}(G')) + \mathcal{E}(G'')$ are infinitesimally rigid in $\mathbb{R}^d$, then G is infinitesimally rigid in $\mathbb{R}^d$. That is, the edges of G' in G have been replaced by those of G''. See Appendix C for the description of these graph operations.

8.7.4 (Facet-forming and nonfacets) A d-polytope P is a *nonfacet* if there is no $(d+1)$-polytope all of whose facets are combinatorially equivalent to P; otherwise it is *facet-forming*.

(i) Prove that a 2-polytope with at most five vertices is facet-forming, and nonfacet otherwise.
(ii) Prove that every d-polytope with at most $d+2$ vertices is facet-forming (Perles and Shephard, 1967).
(iii) Prove that, for $d \geqslant 4$, the d-crosspolytope is nonfacet.
(iv) Find a simple 3-polytope that is nonfacet.
(v) Find a simple 4-polytope that is nonfacet (Barnette, 1969).
(vi) Prove that the 24-cell, the 120-cell, and the 600-cell are nonfacets (Kalai, 1990).

8.7.5 (Equifacetted polytopes, Blind and Blind, 2003) Let F be the pyramid over a $(d-2)$-cube, and let P be d-polytope whose facets are all combinatorially isomorphic to F. Prove that, for $d \geqslant 5$, the minimum number of vertices of P is $2^{d-1}+2$, in which case P is the bipyramid over a $(d-1)$-cube.

8.7.6 (Flag numbers of cubical polytopes) Let P be a cubical d-polytope. Then the flag vector of P is determined by its f-vector. Specifically, for a subset $S := \{d_1, \ldots, d_s\} \subseteq \{0, \ldots, d-1\}$ with $s \geqslant 2$ and $d_1 < \cdots < d_s < d$, it holds that

$$f_S(P) = f_{d_s}(P) \times \prod_{j=1}^{s-1} f_{d_j}(Q(d_{j+1})),$$

where $Q(d_{j+1})$ denotes the d_{j+1}-cube.

8.7.7 (Bayer and Billera, 1985; Kalai, 1988a) Find a family of c_d d-polytopes whose flag vectors are affinely independent in $\mathbb{R}^{c_d}$.

8.7.8 Prove that, for a fixed number n of vertices, the flag numbers f_S for each $S \subseteq [0 \dots d-1]$ of a d-polytope are simultaneously maximised by a simplicial d-polytope. Conclude from this that the cyclic d-polytope on n vertices simultaneously maximises the flag numbers f_S among the d-polytopes with n vertices.

8.7.9 Prove that the characterisation of flag vectors of 4-polytopes can be reduced to the characterisation of the tuple (f_0, f_1, f_3, f_{03}). Specifically, we have that

(i) $f_2 = -f_0 + f_1 + f_3$,
(ii) $f_{01} = 2f_1, f_{02} = -2f_0 + 2f_1 + f_{03}, f_{12} = -2f_0 + 2f_1 + f_{03}, f_{13} = -2f_0 + 2f_1 + f_{03}, f_{23} = 2f_2$,
(iii) $f_{012} = f_{013} = f_{023} = f_{123} = -4f_0 + 4f_1 + 2f_{03}$,
(iv) $f_{0123} = 2f_{012} = 4f_{02} = -8f_0 + 8f_1 + 4f_{03}$.

8.7.10 (Ling, 2007) Let P be a 4-polytope and suppose that

$$(f_0, f_1, f_2, f_{02}) := (f_0(P), f_1(P), f_2(P), f_{02}(P)).$$

Prove the following inequalities.

(i) For each integer ℓ,

$$(\ell - 1)f_{02} - \binom{\ell+1}{2} f_2 + f_1 \leqslant \binom{f_0}{2}.$$

Equality holds only if $\ell = 2, 3$, every two vertices of P lies in a common 2-face, and every 3-face of P is either the 3-simplex, the pyramid over a square, or the simplicial 3-prism.

(ii) For each integer ℓ,

$$2(\ell - 1)f_{02} - 2\binom{\ell+1}{2} f_2 + \left(\ell^2 - 3\ell + 4\right) f_1 - \ell(\ell - 3)f_0 \leqslant 4\binom{f_0}{2}.$$

Equality holds only if every 3-face of P has ℓ or $\ell + 1$ vertices, every two vertices of P lies in a common 2-face, and every 3-face of P is either the pyramid over a $(\ell - 1)$-gon, or the pyramid over a ℓ-gon, or in the case of $\ell = 3$, also the simplicial 3-prism.

8.7.11* (A lower bound conjecture for polytopes with at most $3d-1$ vertices) Let $d \geqslant 3$ and $1 \leqslant s \leqslant d-1$. Truncating a simple vertex from the $(s+1, d-s-1)$-triplex yields a new d-polytope $J(s,d)$ with $2d+s$ vertices and $d+3$ facets.

Let $c(s)$ be a constant that depends on s, let $d \geqslant c(s)$, and let P be a d-polytope with $2d+s$ vertices.

(i) If P has at least $d+3$ facets, then $f_k(P) \geqslant f_k(J(s,d))$ for each $k \in [1 \ldots d-2]$.
(ii) If P has $d+2$ facets, then it is a t-fold pyramid over $T(a) \times T(b)$ for some $a, b > 0$ and $t \geqslant 0$ satisfying $d = a+b+t$.

The conjecture is true for $s = 1$ and each $k \in [1 \ldots d-2]$ (Pineda-Villavicencio and Yost, 2022), and for $s = 2$ and $k = 1$ (Pineda-Villavicencio et al., 2022).

8.7.12* (Cubical lower bound conjecture, Jockusch, 1993) Let $d \geqslant 2$ and let P be a cubical d-polytope with n vertices. Then we have that

$$f_k(P) \geqslant \begin{cases} \left(\binom{d}{k} + \binom{d-1}{k-1}\right)\frac{n}{2^k} - \binom{d-1}{k-1}2^{d-k}, & \text{for } k \in [1 \ldots d-2]; \\ (2d-2)\frac{n}{2^{d-1}} - (2d-4), & \text{for } k = d-1. \end{cases}$$

Capped cubical d-polytopes are minimisers. If you recall, a capped cubical polytope is either a cube itself or a polytope derived from a cube through successive cappings. A capping is defined as the connected sum of a cube and an existing capped cubical polytope along a shared facet.

If the conjecture is true for edges, then it is true for all faces.

8.7.13 (Cubical MPW reduction, Jockusch, 1993) Let P be a cubical d-polytope with n vertices. If the cubical lower bound conjecture is true for $k = 1$, then it is true for each $k \in [1 \ldots d-1]$.

8.7.14* (Kalai's cubical upper bound conjecture, Babson et al., 1997, conj. 4.2) Let $d \geqslant 2$ and let P be a cubical d-polytope with 2^n vertices. Suppose that Q is a neighbourly cubical d-polytope with 2^n vertices. Then we have that

$$f_k(P) \leqslant f_k(Q), \text{ for each } 1 \leqslant k \leqslant d-1.$$

Joswig and Ziegler (2000) proved the special case $n = d+1$. Initially, Kalai stated the conjecture for cubical $(d-1)$-spheres, but Joswig and Ziegler (2000) found a counterexample for $d = 4$ and $n = 6$.

8.7.15* (Nonfacets, Perles and Shephard, 1967) Answer the following.

(i) Is the icosahedron nonfacet?

(ii) For each $d \geqslant 3$, what is the minimum number of vertices of a nonfacet d-polytope?
(iii) For each $d \geqslant 3$, does there exist a finite number $n(d)$ such that every simplicial d-polytope with at least $n(d)$ vertices is nonfacet? Is $n(d) = d + 4$?
(iv) For every $d \geqslant 5$, does there exist a simple d-polytope that is nonfacet?

8.7.16* (A lower bound conjecture) Let F be the pyramid over a $(d-2)$-cube, and let P be d-polytope whose facets are all combinatorially isomorphic to F. Bipyramids over cubical $(d-1)$-polytopes are examples of such polytopes. Suppose that P has n vertices.

Is is true that, for $d \geqslant 5$, the minimum number of k-faces is realised when P is a bipyramid over a capped cubical $(d-1)$-polytope?

For background on the problem, consult Blind and Blind (2003).

8.8 Postscript

The proof of the lower bound theorem for d-polytopes with at most $2d$ vertices (Theorem 8.2.2) was based on the original proof of Xue (2021). Proposition 8.2.4 and Corollary 8.2.5 first appeared in Klee (1964b).

Our presentation of rigidity and infinitesimal rigidity has been mostly utilitarian and directed towards the proof by Kalai (1987) of the lower bound theorem of Barnette (1971, 1973c). We have only scratched the surface of this rich topic. For further exploration, we recommend Lovász (2019, ch. 14,15), Graver et al. (1993), and Connelly and Guest (2022), and also the expository sources of Roth (1981) and Graver (2001).

Example 8.3.3 has been adapted from an example in Graver (2001), while Example 8.3.5 is a classical example in this area, which can be found, for instance, in Whiteley (1992, p. 4). The proof of Proposition 8.3.12 is a standard application of how to prove that functions are linearly independent as vectors; a similar proof was presented in Graver et al. (1993, thm. 2.3.2). Lemma 8.3.13 is a standard result; our proof is inspired by that of Graver et al. (1993, lem. 2.3.2). Many proofs demonstrating that the graph of a 3-polytope in $\mathbb{R}^3$ carries no nonzero stress (Theorem 8.3.27) have appeared in the literature; see, for instance, Lovász (2019, thm. 14.5), Pak (2010), Roth (1981, sec 6), and Lee (2013). Our proof has benefited from all these sources. The presentation of Whiteley's theorem on infinitesimal rigidity (8.3.29) and the proof of the lower bound theorem were based on the notes of Lee (2013). The proof of the MPW reduction is essentially that in Barnette (1973c). Two other accessible proofs of the lower bound theorem can be found in Brøndsted (1983, sec. 19) and Blind and Blind (1999); the former is based on the original proof of Barnette (1973c), while the latter is based on shellings.

The original proof of the upper bound theorem in McMullen (1970) is based on shellings and the interpretation of h-numbers given in Section 2.13, and so is the proof in Ziegler (1995, sec. 8.4). McMullen (1970) also sketched a proof in the dual setting; complete proofs in this setting were provided by Brøndsted (1982) and Brøndsted (1983, sec. 18). Our proof was inspired by these two proofs. In addition, Mulmuley (1993, thm. 7.2.6) proved the inequality part of the theorem for simple polytopes. The proof of McMullen (1970) may be extended to *shellable simplicial spheres*, simplicial spheres that admit a shelling; shellable simplicial spheres need not be polytopal; see, for instance, Lutz (2008). Alon and Kalai (1985) supplied another proof that works for shellable simplicial spheres; Ewald (1996, ch. II, sec. 7) presents this proof as well. There are, however, simplicial spheres that are not shellable (Lutz, 2004). The case of all simplicial spheres was finally settled by Stanley (1975); see also Stanley (1996, ch. 2, sec. 3).

For further extensions of the lower bound and upper bound theorems, please consult Goodman et al. (2017, ch. 17).

McMullen (1993) provided another proof of the necessity of the conditions in the g-theorem. Fleming and Karu (2010) further simplified the poof in McMullen (1993). Stanley (1980), among others, asked whether there is an analogous g-theorem for simplicial spheres. Adiprasito (2019) was the first who proved this extension and Papadakis and Petrotou (2020) were the second. The ideas of these two papers were combined in Adiprasito et al. (2021) to produced a third proof of the g-theorem for simplicial spheres.

Nevo et al. (2019) extended both the lower bound theorem (8.3.34) and the upper bound theorem (8.3.34) to *almost simplicial polytopes*, polytopes with at most one nonsimplex facet F. Since every ridge of the facet F is shared with another facet, the facet F is necessarily simplicial and every ridge of the polytope is a simplex.

The discussion in Section 8.6 is based on the presentation in Bayer and Billera (1984, sec. 7). The proofs of Theorems 8.6.12 and 8.6.14 follow the original ones from Bayer (1987).

Appendix A

Open Problems

This appendix gathers all the open problems that we posed throughout the book. Every problem has in brackets a reference to the original label in the book. We endeavour to keep updates on the problems on the author's website for the book:

www.guillermo.com.au/graphs-of-polytopes/

A.1 Polytopes

A.1.1 (Problem 2.15.20; Ziegler, 1995, ex. 8.9) Let $F_1, \ldots, F_s$ be a shelling of a polytope. For each $i \in [1, \ldots, s]$, order the $(d-2)$-faces of F_i as they appear in the list $F_1 \cap F_i$, $F_2 \cap F_i$, $F_{i-1} \cap F_i$, $F_{i+1} \cap F_i, \ldots, F_s \cap F_i$. If, for each $i \in [1, \ldots, s]$, this ordering of the $(d-2)$-faces of F_i is a shelling of F_i, then we say that the shelling is *perfect*.

(i) Does every polytope have a perfect shelling?
(ii) Does every simple polytope have a perfect shelling?
(iii) Does every cubical polytope have a perfect shelling?

Ziegler attributes the first question to Gil Kalai. It is known that simplicial polytopes, duals of cyclic polytopes, d-cubes, and 3-polytopes all have perfect shellings.

A.2 Polytopal Graphs

A.2.1 (Problem 3.11.12) This problem is about constructing nonrational polytopes.

(i). Construct a nonrational 4-polytope with a minimum number of vertices (Richter-Gebert, 2006, Prob. 17.2.1).
(ii). More generally, for each $d \geqslant 4$, construct a nonrational d-polytope with a minimum number of vertices.

All combinatorial types of 4-polytopes with up to nine vertices can be realised with rational coordinates. (Firsching, 2020).

A.2.2 (Problem 3.11.13; Lockeberg, 1977) Let P be a simplicial d-polytope, and let $\boldsymbol{x}, \boldsymbol{y} \in \mathcal{V}(P)$. Is there a refinement homeomorphism from set $\mathcal{B}(P)$ to set $\mathcal{B}(T(d))$ in which both $\boldsymbol{x}$ and $\boldsymbol{y}$ are principal vertices?

Lockeberg (1977, thm. 6.1) found a nonsimplicial d-polytope P with $d+3$ facets, and two nonadjacent vertices $\boldsymbol{x}, \boldsymbol{y}$ of it for which there is no refinement homeomorphism from set $\mathcal{B}(P)$ to set $\mathcal{B}(T(d))$ in which both $\boldsymbol{x}$ and $\boldsymbol{y}$ are principal vertices.

A.2.3 (Problem 3.11.14) Let P be a $(2d-2)$-connected d-polytope.

(i) Is the graph of the d-crosspolytope a minor of $G(P)$?
(ii) If P is also simplicial, is the graph of the d-crosspolytope a topological minor of $G(P)$?

The graph of a 4-connected 3-polytope contains the graph of the octahedron as minor (Maharry, 1999, thm. 5.1), while the graph of a 5-connected 3-polytope contains a subdivision of the graph of the octahedron (Barnette, 1987a).

Simplicial 3-polytopes with minimum degree four are refinements of the octahedron (Batagelj, 1989, thm. 2).

A.2.4 (Problem 3.11.15; Grünbaum, 2003, Sec. 11.3) Prove that a centrally symmetric simplicial d-polytope contains a refinement of the complete graph on $\lfloor 3d/2 \rfloor$ vertices.

This is true for the crosspolytope (Halin, 1966) and for $d = 4$ (Larman and Mani, 1970).

A.2.5 (Problem 3.11.16) Let P be a cubical d-polytope. Is P a refinement of the d-cube? In particular, is the graph of the d-cube a topological minor of $G(P)$?

Cubical 3-polytopes are refinements of the cube (Brinkmann et al., 2005).

A.3 Connectivity

A.3.1 (Problem 4.7.7; Lockeberg, 1977) Let P be a d-polytope and let $d_1, \dots, d_m$ be natural numbers such that $d_1 + \cdots + d_m = d$. Define a $(k, k-1)$-*path* in P, with $k < d$, as a path $F_1 \dots F_n$ of k-faces in P such that $F_i \cap F_{i+1}$ is a $(k-1)$-face of P, for $i \in [1, \dots, n-1]$.

Prove that between any two vertices $\boldsymbol{x}$ and $\boldsymbol{y}$ in $\mathcal{V}(P)$, there is a $(d_1, d_1 - 1)$-path L_1, a $(d_2, d_2 - 1)$-path $L_2, \dots$, and a $(d_m, d_m - 1)$-path L_m such that $\mathcal{V}(L_i) \cap \mathcal{V}(L_j) = \{\boldsymbol{x}, \boldsymbol{y}\}$ for $i \neq j$.

Balinski's theorem (4.1.1) is recovered when $d_1 = d_2 = \cdots = d_d = 1$.

A.3.2 (Problem 4.7.8) Characterise d-polytopes whose graphs are critically d-connected and those whose graphs are minimally d-connected.

A.3.3 (Problem 4.7.9) Characterise d-polytopes whose graphs are critically d-edge-connected and those whose graphs are minimally d-edge-connected.

A.3.4 (Problem 4.7.10; Werner and Wotzlaw, 2011) Are all 6-polytopes 3-linked?

For each $d \leqslant 5$, a d-polytope is $\lfloor d/2 \rfloor$-linked. Thus, $d = 6$ is the smallest dimension for which we don't know the exact linkedness. Thomas and Wollan (2008) proved that every 6-connected graph with n vertices and at least $5n - 14$ edges is 3-linked. According to the lower bound theorem for simplicial polytopes (8.3.34), this means that 6-polytopes with only triangular faces are 3-linked; that is, 2-simplicial 6-polytopes are 3-linked.

A.3.5 (Problem 4.7.11) Is a simple d-polytope $\lfloor d/2 \rfloor$-linked?

While not every d-polytope is $\lfloor d/2 \rfloor$-linked, none of the known counterexamples are simple d-polytopes; see also Gallivan (1985).

A.3.6 (Problem 4.7.12; Werner and Wotzlaw, 2011) Is there a function $\varphi(d) \colon \mathbb{N} \to \mathbb{N}$ such that every d-polytope with at least $\varphi(d)$ vertices is $\lfloor d/2 \rfloor$-linked?

All the known counterexamples have fewer than $3\lfloor d/2 \rfloor$ vertices.

A.3.7 (Problem 4.7.13; Polytopal version of Thomassen, 1980a, conj. 1) Determine whether the following statement is true: For each odd $d \geqslant 3$, a d-polytope is d-edge-linked. Additionally, for each even $d \geqslant 2$, a d-polytope is $(d-1)$-edge-linked.

A.4 Reconstruction

A.4.1 (Problem 5.4.4) Let P be a d-polytope with at most $d-2$ nonsimple vertices. Can P be reconstructed from its dimension and graph? If a reconstruction is possible, provide an efficient algorithm.

This is true for 4-polytopes, as Doolittle et al. (2018) showed that every polytope with at most two nonsimple vertices can be reconstructed from its dimension and graph.

A.4.2 (Problem 5.4.5) Can every simplicial sphere be determined by its dimension and dual graph. If a reconstruction is possible, provide an efficient algorithm.

This is true for simplicial 2-spheres, a consequence of Steinitz's theorem (3.8.1), but it is still open for simplicial 3-spheres. From the dual version of Theorem 5.2.4 it follows that this is also true for all polytopal simplicial spheres (boundary complexes of simplicial polytopes), since polytopes share the dual graph with their boundary complexes. However, there are nonpolytopal simplicial spheres; see, for instance, Barnette (1987b).

A.5 Decomposition

A.5.1 (Problem 6.5.19; Smilansky, 1987) Let $d \geqslant 4$ and let $\phi_k(d,f)$ denote the maximum number of k-faces in a d-polytope with f facets. The upper bound theorem (McMullen, 1970) states that, for any d-polytope P, it follows that

$$f_0 \leqslant \phi_0(d,f) = \binom{f - \lfloor (d+1)/2 \rfloor}{f-d} + \binom{f - \lfloor (d+2)/2 \rfloor}{f-d}.$$

Equality occurs if and only if P is the dual of a neighbourly polytope.

One would expect that d-polytopes with f facets and number of vertices 'close' to $\phi_0(d,f)$ behave like simple polytopes. Smilansky (1987, sec. 6) made this notion of closeness precise by conjecturing that if

$$f_0 > \phi_0(d-1, f-1) + 1,$$

then a d-polytope is decomposable. It has been verified for $d = 3$ (Problem 6.5.12).

A.5.2 (Problem 6.5.20; Smilansky, 1986b) Let $d \geqslant 3$ and let $\zeta(d)$ denote the largest integer k such that any indecomposable d-polytope must have a proper indecomposable k-face.

It is true that $\zeta(3) = 2$ and $1 \leqslant \zeta(d) \leqslant d - 2$ (Problem 6.5.16). Find $\zeta(d)$ for every $d \geqslant 4$. In particular,

(i) Are polytopes without triangular faces decomposable?
(ii) Are cubical polytopes decomposable?
(iii) Is there an indecomposable d-polytope where all of its faces of dimensions $\zeta(d) + 1, \zeta(d) + 2, \ldots, d - 1$ are decomposable?
(iv) Is $\zeta(d) = 2$ for every $d \geqslant 3$?

A.5.3 (Problem 6.5.21) Let $a(d)$ denote the minimum number of vertices of an ambiguously decomposable d-polytope. Find $a(d)$. It is true that $a(3) = 8$. Is it the case that $a(d) = 2d + 2$?

A.5.4 (Problem 6.5.22) Extend Problem 6.5.12 to d-polytopes. Concretely, for a given polytope establish relations between f_0 and f_{d-1} that ensure the decomposability of the polytope. Is it true that a d-polytope with $f_{d-1} < f_0$ is decomposable?

A.6 Diameter

A.6.1 (Polynomial Hirsch conjecture (7.1.2)) The graph of every d-dimensional polyhedron with n facets has a diameter at most polynomial in d and n.

A.6.2 (Linear Hirsch conjecture (7.1.3)) The graph of every d-dimensional polyhedron with n facets has a diameter at most linear in d and n.

A.7 Faces

A.7.1 (Problem 8.1.1) For $d \geqslant 1$, characterise the f-vectors of d-polytopes.
It is done for $d \leqslant 3$ and is likely unfeasible for $d \geqslant 4$.

A.7.2 (Problem 8.7.11) Let $d \geqslant 3$ and $1 \leqslant s \leqslant d - 1$. Truncating a simple vertex from the $(s + 1, d - s - 1)$-triplex yields a new d-polytope $J(s,d)$ with $2d + s$ vertices and $d + 3$ facets.

Let $c(s)$ be a constant that depends on s, let $d \geqslant c(s)$, and let P be a d-polytope with $2d + s$ vertices.

(i) If P has at least $d + 3$ facets, then $f_k(P) \geqslant f_k(J(s,d))$ for each $k \in [1 \ldots d - 2]$.
(ii) If P has $d+2$ facets, then it is a t-fold pyramid over $T(a) \times T(b)$ for some $a, b > 0$ and $t \geqslant 0$ satisfying $d = a + b + t$.

The conjecture is true for $s = 1$ and each $k \in [1 \dots d-2]$ (Pineda-Villavicencio and Yost, 2022), and for $s = 2$ and $k = 1$ (Pineda-Villavicencio et al., 2022).

A.7.3 (Problem 8.7.12; Jockusch, 1993) Let $d \geqslant 2$ and let P be a cubical d-polytope with n vertices. Then we have that

$$f_k(P) \geqslant \begin{cases} \left(\binom{d}{k} + \binom{d-1}{k-1}\right)\frac{n}{2^k} - \binom{d-1}{k-1}2^{d-k}, & \text{for } k \in [1 \dots d-2]; \\ (2d-2)\frac{n}{2^{d-1}} - (2d-4), & \text{for } k = d-1. \end{cases}$$

Capped cubical d-polytopes are minimisers. If you recall, a capped cubical polytope is either a cube itself or a polytope derived from a cube through successive cappings. A capping is defined as the connected sum of a cube and an existing capped cubical polytope along a shared facet.

If the conjecture is true for edges, then it is true for all faces.

A.7.4 (Problem 8.7.14; Babson et al., 1997, Conj. 4.2) Let $d \geqslant 2$ and let P be a cubical d-polytope with 2^n vertices. Suppose that Q is a neighbourly cubical d-polytope with 2^n vertices. Then we have that

$$f_k(P) \leqslant f_k(Q), \text{ for each } 1 \leqslant k \leqslant d-1.$$

Joswig and Ziegler (2000) proved the special case $n = d + 1$. Initially, Kalai stated the conjecture for cubical $(d-1)$-spheres, but Joswig and Ziegler (2000) found a counterexample for $d = 4$ and $n = 6$.

A.7.5 (Problem 8.7.15; Perles and Shephard, 1967) Answer the following.

(i) Is the icosahedron nonfacet?

(ii) For each $d \geqslant 3$, what is the minimum number of vertices of a nonfacet d-polytope?

(iii) For each $d \geqslant 3$, does there exist a finite number $n(d)$ such that every simplicial d-polytope with at least $n(d)$ vertices is nonfacet? Is $n(d) = d + 4$?

(iv) For every $d \geqslant 5$, does there exist a simple d-polytope that is nonfacet?

A.7.6 (Problem 8.7.16) Let F be the pyramid over a $(d-2)$-cube, and let P be d-polytope whose facets are all combinatorially isomorphic to F. Bipyramids over cubical $(d-1)$-polytopes are examples of such polytopes. Suppose that P has n vertices.

Is it true that, for $d \geqslant 5$, the minimum number of k-faces is realised when P is a bipyramid over a capped cubical $(d-1)$-polytope?

For background on the problem, consult Blind and Blind (2003).

Appendix B

Topology

This appendix gathers the topological terms and results that we used but didn't define in the book.

B.1 Topological Spaces

A *topological space* is a pair (X, O) consisting of a set X and a set O of subsets of X, called *open sets*, such that the following hold.

(i) The sets $\varnothing$ and X are open.
(ii) Any union of open sets is open.
(iii) The intersection of every two open sets is open.

The set O is the *topology* of the space.

Open sets provide a neat characterisation of continuous functions between topological spaces (X, O) and (X', O'). A map $\varphi \colon X \to X'$ is *continuous* if the inverse image of opens sets is always open. A subset $Y \subseteq X$ is *closed* if and only if the subset $X \backslash Y$ is open. This gives rise to the next basic result.

Proposition B.1.1 *If (X, O) is a topological space, then the following hold.*

(i) *The sets $\varnothing$ and X are closed.*
(ii) *Any intersection of closed sets is closed.*
(iii) *The union of every two closed sets is closed.*

A bijective map $\sigma : X \to X'$ is a *homeomorphism* if both σ and σ^{-1} are continuous functions, which means that a set $Z \subseteq X$ is open if and only if $\sigma(Z) \subseteq X'$ is open. Homeomorphisms play the same role in topology as isomorphisms play in graph theory or linear algebra. We say that X and X'

are *homeomorphic spaces* if there is a homeomorphism between them. A basic pair of homeomorphic spaces is $\mathbb{R}^d$ and an open d-ball.

A *metric space* is a pair (X, d) consisting of a set X and a map $d\colon X \times X \to \mathbb{R}$, called a *metric*, such that the following hold.

(i) For all $x, y \in X$, we have that $d(x,y) \geqslant 0$, and that $d(x,y) = 0$ if and only if $x = y$.
(ii) For all $x, y \in X$, we have that $d(x,y) = d(y,x)$.
(iii) For all $x, y, z \in X$, we have that $d(x,z) \leqslant d(x,y) + d(y,z)$.

In this setting, a subset Y of $\mathbb{R}^d$ is open if, for every $\boldsymbol{x} \in Y$, there is an open d-ball centred at $\boldsymbol{x}$ and contained in Y. And the set $\mathbb{R}^d$ together with all its open sets is a topological space; this latter is refereed to as the *usual topology* of $\mathbb{R}^d$.

B.2 Jordan–Lebesgue–Brouwer's Separation Theorems

Lebesgue (1911) and Brouwer (1912) proved a natural generalisation of Jordan's curve theorem for simple closed curves (Jordan, 1882). We remark that Jordan's theorem is usually stated for simple closed curves in $\mathbb{R}^2$, but it could have been equally stated for simple closed curves in $\mathbb{S}^2$.

Theorem B.2.1 (Jordan–Lebesgue–Brouwer's separation theorem for spheres) [1] *Let $d \geqslant 2$. If C is a subset of $\mathbb{S}^d$ homeomorphic to the unit $(d-1)$-sphere, then $\mathbb{S}^d \backslash C$ consists of precisely two arcwise-connected components, each of which has C as its boundary.*

Lebesgue (1911) provided a proof sketch of the statement that $\mathbb{R}^d \backslash C$ consists of at least two arcwise-connected components; it is understood that the sketch of Lebesgue (1911) can be satisfactorily extended to a formal proof. Brouwer (1912) in turn proved that $\mathbb{R}^d \backslash C$ consists of at most two arcwise-connected components.

Since the unit sphere $\mathbb{S}^d$ with a point removed is homeomorphic to $\mathbb{R}^d$ (Proposition 3.1.7), Jordan–Lebesgue–Brouwer's separation theorem for spheres implies the following.

Theorem B.2.2 (Jordan–Lebesgue–Brouwer's separation theorem) [2] *Let $d \geqslant 2$. If C is a subset of $\mathbb{R}^d$ homeomorphic to the unit $(d-1)$-sphere, then $\mathbb{R}^d \backslash C$*

[1] A proof is available in Deo (2018, thm. 6.10.5).
[2] A proof is available in Deo (2018, thm. 6.10.6).

consists of precisely two arcwise-connected components, each of which has C as its boundary.

Remark B.2.3 Jordan's curve theorem for simple closed curves is recovered when $d = 2$ since a simple closed curve is a space homeomorphic to the unit one-dimensional sphere.

There is also a neat generalisation of Jordan's curve theorem for polygonal arcs (3.1.2), due to Perles et al. (2009). As in the case of Jordan's curve theorem for polygonal arcs, this generalisation has a very accessible proof. Recall that the underlying set of a polytopal complex $\mathcal{C}$, denoted by $\operatorname{set}\mathcal{C}$, is the set of points in $\mathbb{R}^d$ that belong to at least one polytope in $\mathcal{C}$.

Theorem B.2.4 (Jordan–Lebesgue–Brouwer's separation theorem for strongly connected complexes) *Let $d \geqslant 2$. Suppose that $\mathcal{C}$ is a strongly connected $(d-1)$-complex in $\mathbb{R}^d$ in which every $(d-2)$-face is contained in exactly two $(d-1)$-faces. Then $\mathbb{R}^d \setminus \operatorname{set}\mathcal{C}$ consists of precisely two polygonally connected components, each of which has $\mathcal{C}$ as its boundary.*

Remark B.2.5 Jordan's curve theorem for polygonal arcs is recovered when $d = 2$.

B.3 Compact Sets and Continuous Functions

A standard result in analysis is that a continuous function on a compact set is bounded and it attains its minimum and its maximum on the set.

Proposition B.3.1[3] *Let X be a compact set in $\mathbb{R}^d$ and let $\varphi\colon X \to \mathbb{R}$ be a continuous function. Then φ is bounded in X and it attains its minimum and maximum at points $\boldsymbol{a}, \boldsymbol{b}$ of X:*

$$\varphi(\boldsymbol{a}) = \min\{\varphi(\boldsymbol{x}) \mid \boldsymbol{x} \in X\},$$
$$\varphi(\boldsymbol{b}) = \max\{\varphi(\boldsymbol{x}) \mid \boldsymbol{x} \in X\}.$$

Furthermore, $\varphi(X)$ is compact.

B.4 Postscript

For further information on the material in this appendix, we recommend Webster (1994), Morris (2020), and Jänich (1984).

[3] A proof is available in Webster (1994, thm. 1.9.2, cor. 1.9.3).

Appendix C

Graphs

This appendix introduces the definitions and results of graph theory that are used but not defined in the body of the book. It aims to provide the graph-theoretical background for the book.

We discuss different notions of graphs, subgraphs, and walks on graphs in Sections C.1 to C.3. We then study connected graphs in Section C.4, including trees and complete graphs. Higher connectivity of graphs is the theme of Section C.5; this continues in Section C.6, where we look at a theorem of Menger (1927) on the connectivity and edge connectivity of graphs.

Graphs of 3-polytopes are planar graphs. Sections C.7 to C.9 cover these graphs. Section C.7 studies basic properties of planar graphs, while Section C.8 presents characterisations due to Kuratowski (1930) and Whitney (1932). Section C.9 reviews the result that the cycle space of a plane graph is isomorphic to the bond space of its dual graph.

C.1 Basic Definitions and Results

A *graph* is a pair $G = (V, E)$ consisting of a set V of *vertices* and a set $E \subseteq [V]^2$, disjoint from V, of *edges*; we denote by $[V]^2$ the set of all 2-element subsets of V. We always assume that V is finite. The vertex set of a graph G is denoted by $\mathcal{V}(G)$ and the number of elements in $\mathcal{V}(G)$ by $v(G)$. Similarly, the edge set of G is denoted by $\mathcal{E}(G)$ and the number of elements in $\mathcal{E}(G)$ by $e(G)$. A graph with precisely one vertex is *trivial*; otherwise it is *nontrivial*.

In some instances, for a vertex or an edge of a graph G, we write $x \in G$ instead of $x \in \mathcal{V}(G)$ or $x \in \mathcal{E}(G)$. Also, for an edge $e = \{x, y\}$, we often write $e = xy$, or simply xy, and we say that the vertices x and y are the *endvertices* or *ends* of e, that x and y are *adjacent* or *neighbours*, and that x and y are *incident* with e. We also say that e is *incident* with x and y. The set $\mathcal{N}_G(x)$

of neighbours of a vertex x in G is the *neighbourhood* of x. Additionally, two edges e, f are *adjacent* if they share an endvertex. A set of vertices or a set of edges is *independent* if no two of its elements are adjacent, and an independent set of edges in a graph is a *matching* of the graph.

Let $G = (V, E)$ be a graph and let $X, Y \subseteq V$. We denote by $\mathcal{E}_G(X, Y)$ the set of all edges from a vertex $x \in X$ to a vertex $y \in Y$, and we write $\mathcal{E}(x, Y)$ and $\mathcal{E}(X, y)$ rather than $\mathcal{E}(\{x\}, Y)$ and $\mathcal{E}(X, \{y\})$. And in the case $Y = V \backslash \{x\}$, we write $\mathcal{E}_G(x)$ instead of $\mathcal{E}(x, Y)$, as this is the set of all edges in G incident with x. The cardinality of $\mathcal{E}_G(x)$ is the *degree* of x, and it is denoted by $\deg_G(x)$. A vertex of degree zero is an *isolated vertex*. It is clear that $\deg_G(x) = \#\mathcal{N}_G(x) = \#\mathcal{E}_G(x)$. We denote by $\delta(G)$ and $\Delta(G)$ the minimum and maximum degrees of a vertex in G, respectively. If all vertices have the same degree r, the graph is *regular*, or r*-regular* if we want to be specific.

Euler (1736) showed a simple relation between vertex degrees and the number of edges.

Lemma C.1.1 (Handshaking lemma; Euler, 1736) *For any graph G, we have that*

$$\sum_{x \in \mathcal{V}(G)} \deg x = 2e(G).$$

Two graphs are *equal* if they have the same sets of vertices and edges; otherwise, the graphs are *unequal*. But two graphs may have essentially the same structure and yet be unequal, as in the case when we relabel the vertices of one graph and obtain the second graph. This similarity is captured by graph isomorphisms. An *isomorphism* from a graph G to a graph G' is a bijection $\sigma : \mathcal{V}(G) \to \mathcal{V}(G')$ that preserves adjacency between vertices: $\{x, y\} \in \mathcal{E}(G)$ if and only if $\{\sigma(x), \sigma(y)\} \in \mathcal{E}(G')$. In this case, we say that G and G' are *isomorphic*. We do not distinguish between isomorphic graphs, and thus write $G = G'$.

Graph operations produce new graphs from old graphs $G := (V, E)$ and $G' := (V', E')$. The *union* of G and G' is the graph $G \cup G' := (V \cup V', E \cup E')$, and the *intersection* of G and G' is the graph $G \cap G' := (V \cap V', E \cap E')$. If $G \cap G' = (\emptyset, \emptyset)$, we say that G and G' are *disjoint* and refer to their union as a *disjoint union*. The *Cartesian product* of G and G' is the graph $G \times G'$ defined on $V \times V'$ and with two pairs (u, u') and (v, v') being adjacent if either $uv \in E$ and $u' = v'$, or $u'v' \in E'$ and $u = v$. The *complement* of G, denoted $\overline{G}$, is a graph on $\mathcal{V}(G)$ where two vertices are adjacent if and only if they are nonadjacent in G.

We often study graphs by looking at smaller graphs contained in them. Let $G = (V, E)$ and $G' = (V', E')$ be graphs. We say that G' is a *subgraph* of G

if $V' \subseteq V$ and $E' \subseteq E$, and we say that G' is a *spanning subgraph* if $V' = V$ and $E' \subseteq E$. In the case that G' is a subgraph of G and $G' \neq G$, then G' is a *proper subgraph* of G. Moreover, if G' is a subgraph of G that contains all the edges in G with both ends in V', then we say that G' is an *induced subgraph* or that G' is induced by V', and we write $G[V']$. We can derive subgraphs from a graph $G = (V, E)$. If $D \subseteq [V]^2$, then $G - X$ denotes the spanning subgraph of G whose edges are in $E \backslash D$, and $G + D$ denotes the graph on the vertex set V whose edges are in $E \cup D$. In the case $X \subseteq V \cup E$ and $X = \{x\}$, we prefer $G - x$ to $G - \{x\}$.

We end this subsection with a remark on maximality and minimality. A subgraph G' of a graph G is *maximal* with respect to some particular property if no subgraph of G with the property has G' as a proper subgraph; and G' is *minimal* if no subgraph of G with the property is a proper subgraph of G'.

C.2 Walks

A *path* is a graph whose vertices can be arranged in a sequence in such a way that two vertices are adjacent if and only if they are consecutive in the sequence. We adopt the notation and terminology from Diestel (2017, sec. 1.3). For a path $L := x_1 \dots x_n$, we say that the vertices x_1 and x_n are the *endvertices* or *ends* of L and the vertices $x_2, \dots, x_{n-1}$ are the *inner* vertices of L. Two paths are *independent* if they share no inner vertex. For $1 \leqslant i \leqslant j \leqslant n$, we write

$$x_i L = x_i \dots x_n,$$
$$L x_i = x_1 \dots x_i,$$
$$x_i L x_j = x_i \dots x_j.$$

This notation extends to the concatenation of paths: if L and M are paths in a graph that share a vertex y, then $xLyMz$ is a path that goes from x to y on L and then from y to z on M. A *cycle* on three or more vertices is a graph whose vertices can be arranged in a cyclic sequence in such a way that two vertices are adjacent if and only if they are consecutive in the sequence. The *length* of a path or a cycle is the number of its edges. A path of length n is an *n-path*, while a cycle of length n is an *n-cycle*.

For a subgraph G' of a graph G, we call a nontrivial path $L := x_1 \dots x_n$ in G a *G'-path* if $\mathcal{V}(L) \cap \mathcal{V}(G') = \{x_1, x_n\}$. A path from a vertex x to a vertex y in the graph G is referred to as an $x - y$ *path*, and the *distance* between x and y, denoted $\text{dist}_G(x, y)$, is the length of a shortest $x - y$ path in G. The maximum distance between any two vertices in G, denoted diam G, is the *diameter* of G.

Walks generalise the notions of paths and cycles. A *walk* of length n in a graph G is a nonempty, alternating sequence $x_1e_1x_2e_2 \dots e_nx_{n+1}$ of vertices and edges in G such that x_i and x_{i+1} are the ends of the edge e_i, for all $i < n + 1$. The vertices x_1 and x_{n+1} are the *endvertices* or *ends* of the walk and the other vertices are its *inner* vertices. If x_1 and x_{n+1} coincide, the walk is *closed*. A *trail* is a walk with no repeated edges, while a path is a walk with no repeated vertices. And a cycle is a closed walk with no repeated vertices except the first and last vertex.

C.3 Directed Graphs and Multigraphs

In some instances, we need to consider graphs where the edges have directions, namely digraphs, or graphs with multiple edges between two vertices, namely multigraphs.

A *digraph* G is a pair (V, E) of disjoint sets satisfying $E \subseteq V \times V$, where V is nonempty and finite. The elements of V and E are the *vertices* and *arcs* of the digraph G, respectively.

For an arc $e = (x, y)$, the first vertex x is its *tail*, and the second vertex y is its *head*. The head and tail of an arc are its *endvertices* or *ends*. In this case, we say that y is an *outneighbour* of x and that x is an *inneighbour* of y. The *indegree* of a vertex x, denoted $\deg_G^-(x)$, in a digraph G is the number of arcs having x as a head, while the *outdegree* of x, denoted $\deg_G^+(x)$, is the number of arcs having x as a tail. We denote by $\mathcal{N}_G^+(x)$ the set of outneighbours of a vertex x in G and by $\mathcal{N}_G^-(x)$ its set of inneighbours.

A *multigraph* is a triple $G = (V, E, I)$ consisting of a set V of *vertices*, a set E, disjoint from V, of *edges*, and an incidence relation I that assigns to each edge an unordered pair of two, not necessarily distinct, vertices. We always assume that V is finite.

Two or more edges in a multigraph G having the same endvertices are said to be *multiple edges*: edges e and e' such that $I(e) = I(e')$. A *loop* is an edge whose ends are equal. The *degree* of a vertex in the multigraph G, denoted $\deg_G(x)$, is the number of edges in G incident with x, counting each loop twice. In a drawing of G, the degree of a vertex is geometrically the minimum number of crossings that occur while one traverses the small circle around the dot representing the vertex.

Unless we explicitly make clear that we are considering digraphs or multigraphs, the reader should assume that we are working with graphs.

C.4 Connected Graphs

A nonempty graph is *connected* if there is a path between every two of its vertices; otherwise it is *disconnected*.

Proposition C.4.1 *The vertices of a connected graph can be linearly ordered as $v_1, \ldots, v_n$ so that the subgraph induced by $\{v_1, \ldots, v_i\}$ is connected for every $i \in [1 \ldots n]$.*

Proof If the vertices $v_1, \ldots, v_i$ have already been chosen, then let v_{i+1} be a vertex that is adjacent to a vertex in $G[v_1, \ldots, v_i]$ but is not itself in $G[v_1, \ldots, v_i]$. That $G[v_1, \ldots, v_i]$ is connected follows from induction on i. □

A path $L := x_1 \ldots x_n$ in a graph is an *$X - Y$ path* if $\mathcal{V}(L) \cap X = \{x_1\}$ and $\mathcal{V}(L) \cap Y = \{x_n\}$. A maximal connected subgraph of a graph is a *component* of the graph. In a graph $G = (V, E)$, we say that a set $Z \subseteq V \cup E$ *separates* two sets $X, Y \subseteq V$ in G if every $X - Y$ path in G contains an element of Z. And we say that the set Z *separates* two vertices u and v of G if Z separates the sets $\{u\}, \{v\}$, *and* $u, v \notin Z$; in this case, we must have that u and v are nonadjacent. Additionally, we say that the set Z *separates* G if it separates two vertices of G. A separating set of vertices is a *separator*, and a separator of cardinality r is an *r-separator.* A *trivial* separator is one whose vertices are incident with a single vertex; otherwise the separator is *nontrivial.* The vertex in a 1-separator is a *cut vertex*, and an edge that separates its ends is a *bridge.* Bridges have an easy characterisation.

Proposition C.4.2[1] *An edge is a bridge if and only if it lies in no cycle.*

Some separating sets of edges receive special names. A separating set D of edges in a graph $G = (V, E)$ is an *edge cut* in G if there exists a nonempty proper subset $X \subseteq V$ such that $D = \mathcal{E}(X, V \backslash X)$. We write $\mathcal{E}(X, \overline{X})$ instead of $\mathcal{E}(X, V \backslash X)$ when there is no ambiguity about the universal set V. An *r-edge cut* is an edge cut with r edges. A *trivial* edge cut is one whose edges are incident with a single vertex; otherwise the edge cut is *nontrivial.*

Every minimal separating set D of edges in a nontrivial graph G is an edge cut. To see this, note that some component S of $G - D$ arose by deleting all edges in the edge cut $\mathcal{E}(\mathcal{V}(S), \overline{\mathcal{V}(S)})$. We give a special name to such edge cuts: a minimal nonempty edge cut is a *bond.*

[1] A proof is available in Bondy and Murty (2008, prop. 3.2).

Proposition C.4.3[2] *Let G be a connected graph and let X be a nonempty proper subset of $\mathcal{V}(G)$. A nonempty edge cut $\mathcal{E}(X, \overline{X})$ is a bond if and only if both $G[X]$ and $G[\overline{X}]$ are connected.*

Trees

A graph is *acyclic* if it has no cycles, and a connected acyclic graph is a *tree*. We call an acyclic graph a *forest*, as its components are trees. Following Diestel (2017, sec. 1.5), we write xTy for the unique path between vertices x and y in a tree T. The basic properties of trees are gathered next.

Proposition C.4.4 (Basic properties of trees)[3] *The following assertions are equivalent for a graph G.*

(i) *G is a tree.*
(ii) *Every two vertices of G are linked by a unique path.*
(iii) *G is connected but the removal of each edge disconnects it.*
(iv) *G is connected with n vertices and $n-1$ edges.*
(v) *G contains no cycles but adding an edge between any two nonadjacent vertices create one.*

A vertex of degree one in a tree T is a *leaf*, and the endvertices of a longest path in T are vertices of degree one. Thus, every tree has at least two leaves. In fact, a tree T has at least $\Delta(T)$ leaves (Problem C.10.1).

A *spanning tree* is a spanning subgraph that is also a tree. Every connected graph has a spanning tree.

When we single out a vertex r in a tree T, the *root* of the tree, every other vertex on the tree can be characterised by its position relative to the root. If two vertices x and y are adjacent in T, then we say that x is the *parent* of y if x lies in the unique path rTy from r to y in T; in this case, we say that y is a *child* of x. More generally, we say that a vertex x is an *ancestor* of a vertex y provided x is on the path rTy. Then we call y a *descendent* of x. The vertices at distance ℓ from the root form the ℓth *level* of the tree. The root is considered to be at level zero.

Some types of rooted spanning trees deserve special names. A *depth-first search tree*, or simply *DFS-tree*, of a nonempty connected graph G is a spanning tree T of G with a designated root. The process begins at the root, marking it as visited, the other vertices of G are added via the following

[2] A proof is available in Bondy and Murty (2008, prop. 2.15).
[3] A proof is available in Diestel (2017, prop. 1.5.1).

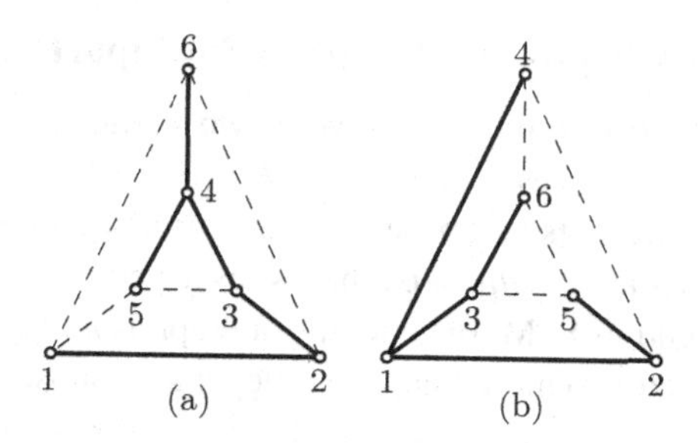

Figure C.1 A DFS-tree and a BFS-tree of a graph G; the labels represent the order in which the vertices are added to the tree. (a) A DFS-tree of G highlighted in bold. (b) A BFS-tree of G highlighted in bold.

recursive step: if every vertex of G is in T, stop and return T. Otherwise, add to T a vertex x not yet visited that is a neighbour of the current vertex. Mark x as visited, designate it as the current vertex, and repeat the recursive step starting from x. If there is no such neighbour, backtrack to the most recent vertex x with unvisited neighbors, make x the current vertex, and rerun the recursive step from x. See Fig. C.1(a). DFS-trees provide a characterisation of cut vertices in a connected graph.

Proposition C.4.5[4] *Let G be a connected graph and let T be a DFS-tree of G. The root of T is a cut vertex of G if and only if it has at least two children.*

Similarly, we define a *breadth-first search tree*, or simply *BFS-tree*, of a nonempty connected graph G as a spanning tree T of G with a root r at level $\ell := 0$ and whose other vertices are added via the following recursive step: if every vertex of G is in T, stop and return T. Otherwise, let $\ell := \ell + 1$, and add to T all vertices of G at distance ℓ from r, in the same order as the parents were added in level $\ell - 1$; that is, for vertices x and y at distance $\ell - 1$ from r, if x was added to T before y, then on level ℓ the children of x are added before the children of y. Rerun the recursive step. See Fig. C.1(b).

The construction of a BFS-tree yields the following.

Proposition C.4.6[5] *Let G be a connected graph and let T be a BFS-tree of G. Then the following hold.*

(i) *Every edge of G is incident with vertices on the same or consecutive levels of T.*
(ii) *The vertices at level ℓ in T coincide with the vertices at distance ℓ from the root.*

[4] A proof is available in Bondy and Murty (2008, thm. 6.6).
[5] A proof is available in Bondy and Murty (2008, thm. 6.2).

Complete Graphs and Complete Multipartite Graphs

A *complete graph* is a graph where every two vertices are adjacent, and a complete graph on n vertices is denoted by K^n (Fig. 3.9.2(a),(c)).

A graph is *bipartite* if its vertex set can be partitioned into two independent subsets X and Y, called *partite sets*; that is, every edge in the graph has one end in X and one end in Y. More generally, a graph is *multipartite* if its vertex set can be partitioned into more than one independent subset; if the partite sets have cardinalities $n_1, \ldots, n_r$, respectively, we say that the multipartite graph is on $n_1, \ldots, n_r$ vertices. A multipartite graph where every two vertices from different partite sets are adjacent is *complete*. A complete multipartite graph on $n_1, \ldots, n_r$ vertices is denoted by $K^{n_1,\ldots,n_r}$.

A tree is bipartite. A cycle is bipartite if and only if it has even length. Moreover, in a connected bipartite graph G, the partite sets of each spanning tree are the partite sets of G. These three facts provide a simple characterisation of bipartite graphs due to König (1936).

Proposition C.4.7 (König, 1936)[6] *A graph is bipartite if and only if every cycle is bipartite.*

C.5 Higher Connectivity

A nontrivial graph G is *r-connected*, for $r \geqslant 0$, if it has more than r vertices and no two vertices of G are separated by fewer than r other vertices. It follows that every nonempty graph is 0-connected. The greatest integer r such that G is r-connected is its *connectivity* $\kappa(G)$. It is clear that $\kappa(G) \leqslant v(G) - 1$, with equality if and only if G is a complete graph. It is often convenient to allow the trivial graph to be both 0-connected and 1-connected. Under this convention, a graph is 1-connected if and only if it is connected. The following is plain.

Proposition C.5.1 *A (nontrivial) r-connected graph has minimum degree at least r.*

In the same way that a connected graph can be decomposed into components, a 2-connected graph can be decomposed into *blocks*, maximal connected subgraphs without a cut vertex. The main properties of blocks are gathered next.

Proposition C.5.2[7] *Let G be a graph. Then the following assertions hold.*

[6] A proof is available in Bondy and Murty (2008, thm. 4.7).
[7] A proof is available in Bondy and Murty (2008, prop. 5.3).

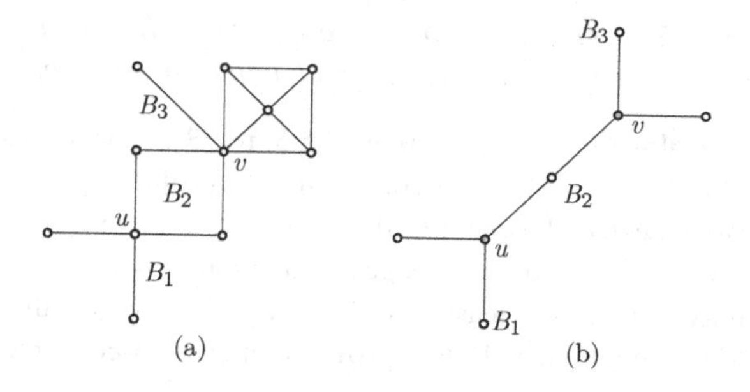

Figure C.1 Blocks and block graph of a graph G. (a) The five blocks and two cut vertices of G(b) The block graph of G.

(i) *Every block is either a maximal 2-connected subgraph, a bridge, or an isolated vertex.*
(ii) *Every two blocks of G have at most one vertex in common, which is a cut vertex of G.*
(iii) *Each cycle of G lies in precisely one block.*

Because every two nondisjoint blocks of a graph G intersect in one cut vertex, we can associate a bipartite graph Z to G in which one partite set consists of the cut vertices of G and the other partite set consists of the blocks of G, and in which a cut vertex of Z is adjacent to each a block containing it. We call Z the *block graph* of G (Fig. C.1).

Proposition C.5.3 *The block graph of a connected graph is a tree.*

A consequence of Proposition C.5.2 is the following.

Proposition C.5.4 (2-connected graphs)[8] *The following assertions are equivalent for a graph G with at least three vertices.*

(i) *G is 2-connected.*
(ii) *G has only one block.*
(iii) *Every two vertices of G lie in a common cycle.*
(iv) *Every two edges of G lie in a common cycle.*
(v) *G has no cut vertices.*
(vi) *For every vertex $v \in G$, $G - v$ is connected.*

Adding a G-path to a 2-connected graph G results in a 2-connected graph. Starting from a cycle, this operation produces all 2-connected graphs.

[8] A proof is available in Diestel (2017, sec. 3.1).

Proposition C.5.5[9] *A graph is 2-connected if and only if it can be constructed from a cycle by successively adding a G-path to the current graph G.*

There are analogues to Proposition C.5.5 for 3-connected graphs. If a graph G is r-connected, then the contraction G/e of G by an edge e is at least $(r-1)$-connected (Problem C.10.4), and for some edges e, G/e is not r-connected (Fig. 3.9.1(b),(c)). An edge e in a graph G is called *contractible* if the connectivity of G/e is at least that of G; otherwise we say that the edge is *noncontractible*. Thomassen (1980a) proved that every 3-connected graph has a contractible edge.

Lemma C.5.6 (Thomassen, 1980a) *Every 3-connected graph G other than K^4 has an edge e for which G/e remains 3-connected.*

The proof of Lemma C.5.6 in Thomassen (1980a) gives an algorithm, which is polynomial in the number of vertices of a 3-connected graph, for finding a contractible edge in the graph (Problem C.10.2). A different algorithm can be devised from a result of Elmasry et al. (2013) stating that every DFS-tree of a 3-connected graph contains a contractible edge (Problem C.10.2). Moreover, we mention that Lemma C.5.6 has counterparts for connected and 2-connected graphs (Problems C.10.5 and C.10.6), but it does not extend to higher connected graphs: for every integer $r \geqslant 4$, there are r-connected graphs other than K^{r+1} with no contractible edges (Problem C.10.3).

By Lemma C.5.6, a 3-connected graph can be reduced to K^4 by a sequence of edge contractions. We gather the details below.

Theorem C.5.7[10] *A graph G is 3-connected if and only if there is a sequence $G_1, \ldots, G_n$ of graphs satisfying the following two properties:*

(i) *$G_1 = K^4$ and $G_n = G$;*
(ii) *G_i has an edge e such that both ends of e have degree at least three and $G_{i-1} = G_i/e$, for every $i \in [2 \ldots n]$;*
Additionally, each graph G_i in the sequence is 3-connected.

Graph operations that produce r-connected graphs from existing r-connected graphs are often useful; one such example is given below.

Lemma C.5.8 *Let G be an r-connected graph, and let G' be a graph obtained from G by adding a new vertex x and adding r edges incident with x and r other vertices in G. Then G' is also r-connected.*

[9] A proof is available in Diestel (2017, prop. 3.1.1).
[10] A proof is available in Diestel (2017, thm. 3.2.5).

Proof Consider a separator X of G'; we show that $\#X \geqslant r$. If $x \in X$, then $X \setminus \{x\}$ separates G; thus $\#X \geqslant r+1$. Suppose instead that $x \notin X$. If $\mathcal{N}_{G'}(x) \subseteq X$ then $\#X \geqslant r$, as $\#\mathcal{N}_{G'}(x) \geqslant r$. Hence, assume there is a neighbour y of x not in X. This implies that y and x are in a component of $G' - X$, which in turn implies that X separates G. As a result, $\#X \geqslant r$, as desired. □

If removing any edge of an r-connected graph results in a graph that is no longer r-connected, then we say that the graph is *minimally r-connected*. This minimality implies the existence of a vertex of degree r. It is clear that every r-connected graph can be reduced to a minimally r-connected one by deleting a finite number of suitable edges.

Theorem C.5.9 (Halin, 1969) *A minimally r-connected graph has a vertex of degree r.*

A similar notion is that of critically connected graphs. We say that a graph is *critically r-connected* if removing any vertex of the graph results in a graph that is no longer r-connected. Unlike minimally r-connected graphs, critically r-connected graphs need not have a vertex of degree r (Chartrand et al., 1972).

r-Edge-Connected Graphs

A nontrivial graph G is *r-edge-connected*, for $r \geqslant 0$, if no two vertices of G are separated by fewer than r edges. The greatest integer r such that G is r-edge-connected is its *edge connectivity* $\kappa'(G)$. It follows that $\kappa'(G) = 0$ if G is disconnected. The edge connectivity of a graph can also be seen as the minimum integer r for which there is an r-edge cut. We adopt the convention that the trivial graph is both 0-edge-connected and 1-edge-connected, but is not r-edge-connected for any $r > 1$. As in the case of r-connected graphs, the following is also plain.

Proposition C.5.10 *A (nontrivial) r-edge-connected graph has minimum degree at least r.*

Subdividing an edge preserves 2-edge connectivity.

Lemma C.5.11 *If an edge is subdivided in a 2-edge-connected graph, then the resulting graph is also 2-edge-connected.*

Proof Let G be a 2-edge-connected, and let G' be obtained from G by subdividing an edge e of G by a new vertex v. Proposition C.4.2 yields that a graph is 2-edge-connected if and only if every edge lies in a cycle; see also

Proposition C.5.12. It follows that every edge of G lies in a cycle. It is now clear that every edge of G' belongs to a cycle: every cycle in G containing e gives rise to a cycle in G' containing v and the two new edges incident with it. Hence G' is 2-edge-connected. □

Lemma C.5.11 facilitates characterisations of 2-edge-connected graphs similar to those in Proposition C.5.4 for 2-connected graphs.

Proposition C.5.12 (2-edge connectivity) [11] *The following assertions are equivalent for a graph G with at least three vertices.*

(i) *G is 2-edge-connected.*
(ii) *Every edge lies in a cycle.*
(iii) *Every two vertices of G lie in a common closed trail.*
(iv) *Every two edges of G lie in a common closed trail.*
(v) *G has no bridges.*
(vi) *For every edge $e \in G$, $G - e$ is connected.*

If we delete an end of each edge in an edge cut, all the edges of the cut are deleted. This suggests, but does not prove, that $\kappa(G) \leqslant \kappa'(G)$. Whitney (1932) showed that the result is, however, true.

Proposition C.5.13 (Whitney, 1932)[12] *If G is a nontrivial graph, then $\kappa(G) \leqslant \kappa'(G) \leqslant \delta(G)$.*

The *line graph* of a graph G, written $L(G)$, is the graph whose vertices are the edges of G with two vertices being adjacent in $L(G)$ if and only if they are adjacent as edges in G. Clearly, isomorphic graphs have isomorphic line graphs; Whitney (1932) showed that the converse holds except in two exceptions (Problem C.10.10). Line graphs facilitate results related to the edge connectivity of the original graphs.

Chartrand and Stewart (1969) proved a number of results on the connectivity of line graphs, including the next two.

Proposition C.5.14 (Chartrand and Stewart, 1969) *For $r \geqslant 2$, a graph is r-edge-connected if and only if it has minimum degree at least r and an r-connected line graph.*

Proposition C.5.15 (Chartrand and Stewart, 1969) *If a graph is r-edge-connected, then its line graph is $(2r - 2)$-edge-connected.*

[11] A proof is available in West (2001, ex. 4.2.11).
[12] A proof is available in Diestel (2017, prop. 1.4.2).

As in the case of r-connected graphs, if removing any edge of an r-edge-connected graph results in a graph that is no longer r-edge-connected, we say that the graph is *minimally r-edge-connected*. This minimality implies the existence of a vertex of degree r.

Theorem C.5.16 (Mader, 1971)[13] *A minimally r-edge-connected graph have a vertex of degree r.*

Mader (1971) actually proved that every minimally r-edge-connected graph has at least $r+1$ vertices of degree r. After this work, estimating the number of vertices of degree r in r-edge-connected graphs has been considered by many authors, including Cai (1993) and Mader (1974, 1995).

In complete analogy with critically r-connected graphs, we say that a graph is *critically r-edge-connected* if removing any vertex of the graph results in a graph that is no longer r-edge-connected. Like minimally r-edge-connected graphs, critically r-edge-connected graphs have a vertex of degree r (Mader, 1986).

C.6 Menger's Theorem

This section is devoted to the famous theorem of Menger (1927) and some of its consequences.

Theorem C.6.1 (Menger, 1927) *Let G be a graph, and let X and Y be two subsets of its vertices. Then the minimum number of vertices separating X from Y in G equals the maximum number of disjoint $X-Y$ paths in G.*

Diestel (2017, sec. 3.3) offers three very accessible proofs of this theorem.

Following Diestel (2017, sec. 3.3), we denote by $\kappa_G(X,Y)$ the minimum number of vertices separating X from Y in G. We write $\kappa_G(x,Y)$ instead of $\kappa_G(\{x\},Y)$, and for two distinct vertices x,y write $\kappa_G(x,y)$ instead of $\kappa_G(\{x\},\{y\})$. We make several remarks.

Remark C.6.2 For subsets X,Y,Z of vertices in a graph G, the number $\kappa_G(X,Y)$ obeys some simple rules.

(i) Each of the sets X and Y separates X from Y, and so $\kappa_G(X,Y) \leqslant \min(\#X,\#Y)$.
(ii) If G is edgeless, then $\#(X \cap Y) = \kappa_G(X,Y)$.

[13] A proof is available in Lovász (2007, prob. 6.49).

(iii) If $X \subseteq Y$, then $\#X = \kappa_G(X,Y)$.
(iv) If $Z \subseteq Y$, then $\kappa_G(X,Z) \leqslant \kappa_G(X,Y)$.

From Menger's theorem (C.6.1), for subsets X, Y of vertices in a graph G it follows that the number $\kappa_G(X,Y)$ coincides with the maximum number $\lambda_G(X,Y)$ of disjoint $X - Y$ paths in G. Below, we list other consequences.

Corollary C.6.3 *Let G be an r-connected graph, and let X and Y be two subsets of its vertices, each of cardinality at least r. Then there are r disjoint $X - Y$ paths in G.*

When we apply Menger's theorem to the line graph of a graph G, we get an edge version of the theorem for the graph. Let X and Y be two disjoint subsets of $\mathcal{V}(G)$, let $\kappa'_G(X,Y)$ denote the minimum number of edges separating X from Y in G, and let $\lambda'_G(X,Y)$ denote the maximum number of edge-disjoint $X - Y$ paths in G.

Theorem C.6.4 (Menger's theorem, Edge version) *Let G be a graph, and let X and Y be two disjoint subsets of its vertices. Then the minimum number of edges separating X from Y in G equals the maximum number of edge-disjoint $X - Y$ paths in G.*

Proof We form a new graph G' from G by adding two new vertices x and y so that x is adjacent to every vertex in X and y is adjacent to every vertex in Y. We then apply Menger's theorem (C.6.1) to the line graph of G' and the sets $X' := \mathcal{E}_{G'}(x)$ and $Y' := \mathcal{E}_{G'}(y)$. It follows that the minimum number $\kappa_{L(G')}(X',Y')$ of vertices in $L(G')$ separating X' from Y' coincides with the maximum number $\lambda_{L(G')}(X',Y')$ of disjoint $X' - Y'$ paths in $L(G')$. Since we have the equalities

$$\begin{aligned}\kappa'_G(X,Y) = \kappa'_{G'}(x,y) = \kappa_{L(G')}(X',Y') &= \lambda_{L(G')}(X',Y') \\ &= \lambda'_{G'}(x,y) = \lambda'_G(X,Y),\end{aligned}$$

the theorem is established. □

From the edge version of Menger's theorem (C.6.4), for *disjoint* subsets X, Y of vertices in a graph G it follows that the number $\lambda'_G(X,Y)$ coincides with the minimum number $\kappa'_G(X,Y)$ of edges separating X from Y in G. Both versions of Menger's theorem can be applied to two distinct vertices.

Corollary C.6.5[14] *Let x and y be two distinct vertices of a graph G.*

[14] A proof is available in Diestel (2017, cor. 3.3.5).

(i) *If x and y are nonadjacent, then the minimum number of vertices separating x from y in G equals the maximum number of independent $x - y$ paths in G.*

(ii) *The minimum number of edges separating x from y in G equals the maximum number of edge-disjoint $x - y$ paths in G.*

Both versions of Menger's theorem yield characterisations of r-connected graphs and r-edge-connected graphs due to Whitney (1932).

Corollary C.6.6[15] *Let G be a graph.*

(i) *The graph G is r-connected if and only if every pair x, y of distinct vertices is joined by at least r independent $x - y$ paths in G.*

(ii) *The graph G is r-edge-connected if and only if every pair x, y of distinct vertices is joined by at least r edge-disjoint $x - y$ paths in G.*

We remark that Corollary C.6.6(i) cannot be improved. Consider the complete bipartite graph $G := K^{r,r+1}$. Then G is r-connected and it is not possible to find $r + 1$ independent paths between two vertices from the partite set with $r + 1$ vertices. In the same vein, Corollary C.6.6(ii) is best possible (find an optimal example).

Also due to Whitney (1932) is the corollary that, in a graph with nonadjacent vertices, there is a minimum separator disconnecting two nonadjacent vertices.

Corollary C.6.7[16] *Let $G = (V, E)$ be a graph with at least two nonadjacent vertices. Then*

$$\kappa(G) = \min \{\kappa_G(x, y) \mid x, y \in V, xy \notin E\}.$$

In Proposition C.5.4(iii), we saw that every two vertices in a 2-connected graph lie in a common cycle. A theorem of Dirac (1960) extends this result to r-connected graphs. For a vertex x and a subset Y of vertices in a graph, we call a set of $x - Y$ paths an *$x - Y$ fan* if every two such paths intersect only at x. An $x - Y$ fan with exactly r paths is an *r-fan*.

Theorem C.6.8 (Dirac, 1960) *Let G be an r-connected graph and let X be a set of r vertices in G, with $r \geqslant 2$. Then G has a cycle that includes all the vertices of X.*

Proof The case $r = 2$ was presented in Proposition C.5.4(iii), and so an induction on r can start. Assume that $r \geqslant 3$. The graph G is $(r - 1)$-connected,

[15] A proof is available in Diestel (2017, thm. 3.3.6).
[16] A proof is available in Bondy and Murty (2008, thm. 9.2).

and thus the induction hypothesis gives a cycle C that includes all the vertices of $X \setminus \{x\}$, for some vertex $x \in X$. If $x \in C$ we are done, so assume otherwise. Let $X = \{x_1, \ldots, x_{r-1}, x\}$.

Suppose that $v(C) \geqslant r \geqslant 3$. From Menger's theorem (C.6.1) follows the existence of an r-fan S from x to $\mathcal{V}(C)$: apply Corollary C.6.3 to the sets $\mathcal{N}_G(x)$ and $\mathcal{V}(C)$. The set $X \setminus \{x\}$ divides C into $r-1$ edge-disjoint segments, say into $[x_1, x_2], \ldots, [x_{r-1}, x_1]$. By the pigeonhole principle, two of the paths in S, say L_u and L_v, touch C in one of these segments, say $[x_1, x_2]$. Let $\{u\} := \mathcal{V}(L_u) \cap \mathcal{V}(C)$, let $\{v\} := \mathcal{V}(L_v) \cap \mathcal{V}(C)$, and let C' be the $u-v$ path on C not contained in $[x_1, x_2]$. It follows that the cycle $xL_u uC'vL_v x$ contains all the vertices in X.

Suppose that $v(C) = r - 1$ and $r \geqslant 4$. Again, we have an $(r-1)$-fan S from x to $\mathcal{V}(C)$. And there are paths in S, say L_u and L_v, to two consecutive vertices u and v of C. Let C' be the subpath uCv in C with more vertices. The cycle $xL_u uC'vL_v x$ contains all the vertices in X. □

C.7 Properties of Planar Graphs

In Chapter 2, we proved Euler–Poincaré–Schläfli's equation for polytopes (Theorem 2.12.17), which relates the numbers of faces of a d-polytope. For the case of 3-polytopes, this equation looks like

$$f_0 - f_1 + f_2 = 2,$$

where f_i is the number of i-faces of the 3-polytope. This is the famous relation of Euler (1758a,b). From a realisation of a 3-polytope in $\mathbb{R}^3$ we can obtain a realisation of its graph in the unit 2-sphere $\mathbb{S}^2$: assume the polytope contains the origin and consider a radial projection of the boundary of the polytope to $\mathbb{S}^2$ (see Proposition 3.2.11 and Section 3.9). Hence Euler–Poincaré–Schläfli's equation holds for 3-connected planar graphs. It turns out that this equation holds for all connected planar graphs.

Theorem C.7.1 (Euler's equation) *For a connected plane graph with f_0 vertices, f_1 edges, and f_2 faces, we have that*

$$f_0 - f_1 + f_2 = 2.$$

There are more than twenty proofs of Euler's equation (Eppstein, n.d.). Rather than a proof of the theorem, we give four simple applications of it. A plane graph is *maximal* if it is no longer plane after adding any edge or, equivalently, if it is not a spanning subgraph of another plane graph; this definition is about graphs, and so multiple edges or loops are not allowed.

Proposition C.7.2 *Let G be a plane graph. Then the following hold.*

(i) *If G has at least three vertices, then G has at most $3v(G) - 6$ edges.*
(ii) *The graph G has exactly $3v(G) - 6$ edges if and only if every one of its faces is a triangle if and only if G is a maximal plane graph.*
(iii) *If G is a maximal plane graph with at least four vertices, then its minimum degree is at least three.*
(iv) *If G has at least four vertices, then G has at least four vertices of degree at most five.*

Proof (i)–(iii) The proofs of these statements are standard; see, for instance, Diestel (2017, sec. 4.4).

We give a proof of (iv). Suppose that G has at most three vertices of degree five or less. If G is not maximal, then we can add edges to G and still obtain a plane graph with at most three vertices of degree five or less. Thus we can assume that G is maximal. The minimum degree of G is at least three by (iii), in which case the handshaking lemma (C.1.1) gives that

$$2e(G) = \sum_{v \in \mathcal{V}(G)} \deg_G v \geqslant 6(v(G)-3)+3\times 3 = 6v(G)-18+9 = 6v(G)-9,$$

which contradicts (i) and so proves (iv). □

As with maximal plane graphs, we say that a planar graph is *maximal* if it is not a spanning subgraph of another planar graph. If we want to be precise, we say that a maximal planar graph is *edge-maximal*. Proposition C.7.2 can be used to prove the equivalence of planar realisations of maximal planar graphs and maximal plane graphs.

Proposition C.7.3 [17] *A planar realisation of a maximal planar graph is a maximal plane graph, and vice versa.*

A maximal plane graph with at least three vertices triangulates the 2-sphere, and by Proposition C.7.3 a realisation of a maximal planar graph, namely an edge-maximal planar graph, with at least three vertices also triangulates the 2-sphere. For surfaces other than the 2-sphere this may, however, not be true. While every graph that triangulates the Klein bottle (and any surface) is edge-maximal, the graph K^7 minus an edge is noncomplete, edge-maximal in

[17] A proof is available in Diestel (2017, prop. 4.2.8).

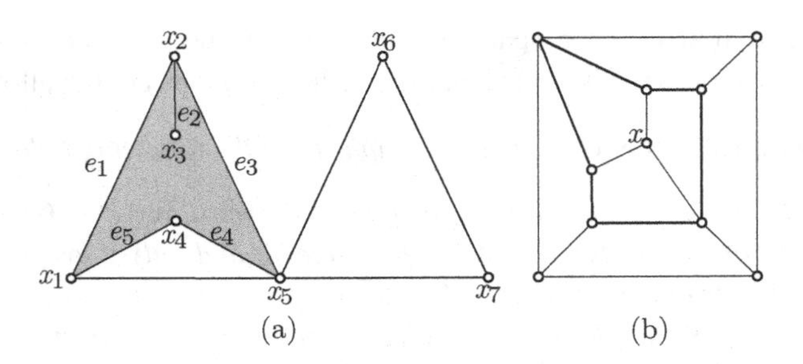

Figure C.7.1 Boundary walks in connected plane graphs. (a) Boundary of faces that are not cycles. The boundary $x_1e_1x_2e_2x_3e_2x_2e_3x_5e_4x_4e_5x_1$ of the face highlighted in grey is not a cycle. The boundary of the unbounded face is not a cycle either. (b) A 3-connected plane graph where every boundary of a face is a cycle. The link of the vertex x is depicted in heavier lines.

the Klein bottle, and does not triangulate the surface (Franklin, 1934). Other examples are presented in Davies and Pfender (2021).

In a connected plane graph, the boundary of a face needs not be a cycle but can always be regarded as a closed walk; see the boundary of the face in Fig. C.7.1(a) highlighted in grey. However, in 2-connected plane graphs, the boundary of face is always a cycle (Fig. C.7.1(b)).

Proposition C.7.4[18] *Each face in a 2-connected plane graph is bounded by a cycle.*

We adopt a terminology from polytope theory. For a 3-connected plane graph G and a vertex x of G, let $\mathcal{F}_x$ be the faces of G that contain x. The *link* of x in G, denoted $\text{lk}(x, G)$, is the subgraph of G induced by the vertices and edges in $\mathcal{F}_x$ that are disjoint from x; for an example, see Fig. C.7.1(b).

Proposition C.7.5 *The link of a vertex in a 3-connected plane graph is a cycle, which contains the neighbours of the vertex.*

Proof Let x be a vertex of a 3-connected plane graph G. The graph $G-x$ is 2-connected, and so all its faces are bounded by cycles (Proposition C.7.4). If F is the face of $G-x$ that contained x (before we removed it), then the neighbours of x all lie in the boundary C of F. It is now plain that C is the link of x. □

The following proof is inspired by the second proof of Balinki's theorem (4.1.1) on the d-connectivity of graphs of d-polytopes.

[18] A proof is available in Diestel (2017, prop. 4.2.6).

Theorem C.7.6 *Every maximal plane graph with at least four vertices is 3-connected. Furthermore, a minimum separator contains the vertices of a nonfacial cycle of the graph.*

Proof In a maximal plane graph, the link of a vertex is the subgraph induced by the neighbours of the vertex. The theorem follows from a proof of the following statement.

$$\text{Every maximal plane graph } G \text{ with at least four vertices is 3-connected. Additionally, for each vertex } x \text{ in a minimum separator } X \text{ of } G\text{, the set } X\backslash\{x\} \text{ is a separator of the link of } x \text{ in } G. \tag{C.7.6.1}$$

A maximal plane graph is connected: in a disconnected plane graph we can always add an edge between two components without disrupting planarity. If G is a complete graph, then it is K^4 and the theorem holds; if $G = K^n$ for $n \geqslant 5$, then it would contain K^5 as a subgraph and would be nonplanar (Theorem 3.1.4). Let X be a separator in G of minimum cardinality and let y and z be two nonadjacent vertices separated by X (Corollary C.6.7). Then $y, z \notin X$. By Menger's theorem (C.6.1), we may assume that there are $\#X$ independent $y - z$ paths in G, each containing precisely one vertex from X. Let L be one such $y - z$ path and let x be the vertex in $X \cap \mathcal{V}(L)$; say that $L = u_1 \ldots u_m$ such that $y = u_1$, $u_j = x$, and $u_m = z$.

The graph C of the link of x in G is a cycle (Proposition C.7.5). Additionally, the neighbours of x are all part of C, and so $u_{j-1}, u_{j+1} \in C\backslash X$. There are two independent $u_{j-1} - u_{j+1}$ paths in C. We must have that $X\backslash\{x\}$ separates u_{j-1} from u_{j+1} in C, since X separates y from z in G. Hence $\#(X\backslash\{x\}) \geqslant 2$, which establishes that G is 3-connected.

The aforementioned path L was arbitrary among the $\#X$ $y - z$ paths separated by X, and each such path contains a unique vertex of X. Hence, for each $x \in X$, the set $X\backslash\{x\}$ separates the link of x. As a consequence, every vertex $x \in X$ has degree at least two in the subgraph of G induced by X, which means that X contains a cycle. This cycle cannot be a face of the graph. □

A minor adaptation of the proof of Theorem C.7.6 shows the following.

Corollary C.7.7 *Every graph in which the link of every vertex is $(r - 1)$-connected is r-connected.*

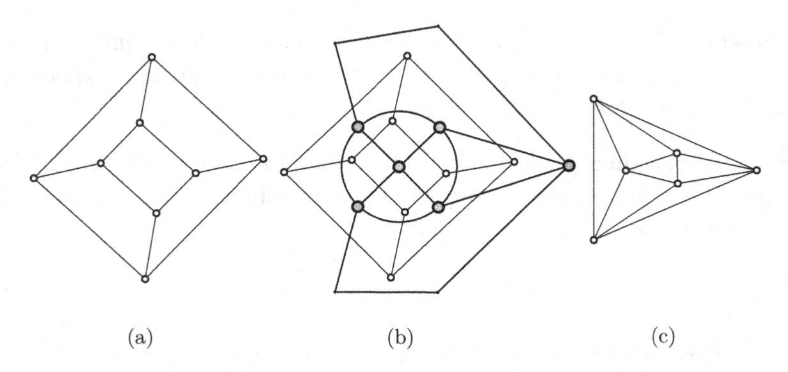

Figure C.7.2 A plane graph and its dual graph. (a) A connected plane graph G. (b) The dual graph G^* of G is depicted in heavier lines. (c) A planar realisation of the dual graph G^* with edges as line segments.

Dual Graphs

Consider a connected plane graph G. The *dual graph* $G^*(P)$ of G is a graph on the faces of G where two vertices f^* and g^* of G^* are incident with an edge e^* if and only if the corresponding faces f and g of G share an edge e; we say that the edges e and e^* are *dual*. The dual graph of a plane multigraph is defined analogously. We remark that when G has no bridges, G^* has no loops.

The dual graph G^* of a connected plane graph G is a plane multigraph. To get a planar realisation of G^*, we place each vertex f^* of G^* inside the corresponding face f of G, and then we draw an edge e^* between the vertices f^* and g^* so that it crosses its dual edge e, the edge shared by the faces f and g of G. This is the *standard planar realisation* of G^*, which is illustrated in Fig. C.7.2(b)–(c).

Proposition C.7.8 [19] *The dual graph G^* of a connected plane graph G is connected. Furthermore, the dual graph of G^* is isomorphic to G.*

Let G represent a plane graph. In a planar realisation of the dual graph G^* of G, to each edge e of G there corresponds the edge e^* of G^* dual to e. We extend this correspondence to sets of edges:

$$\text{For a subset } S \in \mathcal{E}(G), \text{ let } S^* := \{e^* \in \mathcal{E}(G^*) \mid e \in S\}. \qquad \text{(C.7.9)}$$

From the definitions of S^* and dual graphs, we obtain a relation between cycles of a graph and bonds of its dual graph.

[19] A proof is available in Bondy and Murty (2008, prop. 10.9).

Proposition C.7.10[20] *Let G be a connected plane multigraph, and let G^* be the dual graph of G. An edge set S of G is the edge set of a cycle in G if and only if S^* is a bond of G^*.*

C.8 Planarity Criteria

This section is devoted to two characterisations of planar graphs, one due to Kuratowski (1930) and one due to Whitney (1932).

Theorem C.8.1 (Kuratowski, 1930)[21] *A graph is planar if and only if it contains no subdivision of K^5 or $K^{3,3}$.*

Wagner (1937) noted that in Kuratowski's theorem it suffices to consider minors rather than topological minors.

Lemma C.8.2[22] *A graph contains K^5 or $K^{3,3}$ as a minor if and only if it contains K^5 or $K^{3,3}$ as topological minor.*

According to Lemma C.8.2, a graph with K^5 as a minor has K^5 or $K^{3,3}$ as a topological minor. It is easy to come across graphs that contain K^5 as a minor but not as a topological minor. The Petersen graph has K^5 as a minor (Fig. 3.9.2(b)) and $K^{3,3}$ as a topological minor, but it has no subdivision of K^5 as it is a 3-regular graph.

A consequence of Lemma C.8.2 is that Kuratowski's theorem (C.8.1) on topological minors is equivalent to Wagner's theorem (C.8.3) on minors.

Theorem C.8.3 (Wagner, 1937) *A graph is planar if and only if it contains neither K^5 nor $K^{3,3}$ as a minor.*

A common approach to prove Kuratowski's theorem (C.8.1) is to prove it for 3-connected graphs and then reduce the general case to this case via a result like Lemma C.8.4.

Lemma C.8.4[23] *Let G be a graph containing no subdivision of K^5 or $K^{3,3}$. If the addition of any edge to G creates such a subdivision, then G is a maximal planar graph.*

[20] A proof is available in Diestel (2017, prop. 4.6.1).
[21] A proof is available in Thomassen (1980b).
[22] A proof is available in Diestel (2017, lem. 4.4.2).
[23] A proof is available in Diestel (2017, thm. 4.4.5).

Whitney's Theorem on Abstract Duals

While Kuratowski's theorem (C.8.1) is perhaps the most well-known planarity criterion, Whitney (1932) gave another such criteria.

A multigraph G^* is an *abstract dual* of a multigraph G if there is a bijection between $\mathcal{E}(G^*)$ and $\mathcal{E}(G)$ and a bijection between the bonds of G^* and the edge sets of cycles in G. The definition of a dual graph and Proposition C.7.10 give that the dual graph of a connected plane graph G is an abstract dual graph of G, but not every abstract dual of G is a dual graph of G. Whitney (1932) characterised planar multigraphs in terms of abstract duals.

Theorem C.8.5 (Whitney, 1932)[24] *A multigraph is planar if and only if it has an abstract dual.*

C.9 Duality, Cycle Spaces, and Bond Spaces

If you recall, a separating set D of edges in a nontrivial graph $G = (V, E)$ is an edge cut if $D = \mathcal{E}(X, V \backslash X)$, and a bond in G is a minimal nonempty edge cut of G. We can also think of an edge cut B in G as a spanning subgraph of G whose edges are those in B. With this description of edge cuts, the following statements follow easily.

Proposition C.9.1[25] *The following hold.*

(i) *The symmetric difference of two edge cuts of a graph is an edge cut of the graph.*
(ii) *A subgraph of a graph is an edge cut if and only if it is an edge-disjoint union of bonds.*

A consequence of Proposition C.9.1 is that the set of all edge cuts of a graph G forms a linear space $\mathcal{Y}(G)$ over the field $GF(2)$ with respect to the symmetric difference of edge cuts and the bonds of G span $\mathcal{Y}(G)$. The linear space $\mathcal{Y}(G)$ is called the *bond space* of G. By combining this with Proposition C.7.10, we can relate the cycle and bond spaces via the duality of graphs.

[24] A proof is available in Diestel (2017, thm. 4.6.3).
[25] A proof is available in Diestel (2017, prop. 1.9.3).

Theorem C.9.2[26] *The cycle space of a plane graph is isomorphic to the bond space of its dual graph.*

Let G be a graph. If to each subset X of $\mathcal{E}(G)$ we associate the spanning subgraph of G whose edge is X, then the subsets of $\mathcal{E}(G)$ form a linear space over $GF(2)$ under the symmetric difference of spanning subgraphs of G. This linear space is the *edge space* of G and is denoted by $\mathcal{X}(G)$. It follows that the bond space and cycle space of a graph are subspaces of its edge space; in fact, they are orthogonal complements (Diestel, 2017, thm. 1.9.4).

C.10 Problems

C.10.1 Prove that a tree T has at least $\Delta(T)$ leaves.

C.10.2 This problem explores algorithms for finding contractible edges in a graph.

(i) Write an algorithm, polynomial in the number of vertices, for finding a contractible edge in a 3-connected graph.
(ii) Prove that every DFS-tree of a 3-connected graph other than K^4 contains a contractible edge (Elmasry et al., 2013).
(iii) Using (ii), write an algorithm for finding a contractible edge in a 3-connected graph.
(iv) Find 3-connected graphs that have a DFS-tree containing exactly one contractible edge and 3-connected graphs with a spanning tree containing no contractible edge (Elmasry et al., 2013).
(v) Prove that every spanning tree in a 3-connected, 3-regular graph other than K^4 contains a contractible edge (Elmasry et al., 2013).

C.10.3 For every integer $r \geqslant 4$, find an r-connected graph other than K^{r+1} with no contractible edges.

C.10.4 Let $r \geqslant 2$. Prove that, for every r-connected graph G and every edge e of G, the graph G/e is at least $(r-1)$-connected.

C.10.5 Prove that, in a connected graph G, there is an edge e of G such that the graph G/e is connected.

C.10.6 Prove that, in a 2-connected graph G with at least four vertices, there is an edge e of G such that the graph G/e is 2-connected.

[26] A proof is available in Diestel (2017, prop. 4.6.2).

C.10.7 Prove that, for every graph G and every $X \subset \mathcal{V}(G)$, it holds that

$$\#\mathcal{E}(X, \overline{X}) = \sum_{v \in X} \deg v - 2e(G[X]).$$

C.10.8 Prove that, for every 3-regular graph G, it holds that $\kappa(G) = \kappa'(G)$.

C.10.9 Prove that removing n sets of r edges each from an r-edge-connected graph G leaves at most $2n$ components.

C.10.10 (Whitney, 1932) Prove that two connected graphs with isomorphic line graphs are themselves isomorphic unless one is K^3 and the other is $K^{1,3}$.

C.10.11 (Beineke, 1968, 1970) This problem explores a characterisation of line graphs based on forbidden induced subgraphs given by Beineke (1968, 1970).

(i) Find the nine graphs with at most six vertices that are not line graphs but every induced subgraph of them is.
(ii) A connected graph G is the line graph of some graph if and only if G does not contain any of the nine graphs identified in Part (i) as an induced subgraph.

C.10.12 (Liu's criterion; Liu, 1990) Let G be a connected graph with at least $r + 1$ vertices. If, for any two vertices x, y of G at distance two, there are r independent $x - y$ paths in G, then G is r-connected.

C.10.13 Let G be a connected graph with at least two vertices. If, for every two adjacent vertices x, y of G there are at least r edge-disjoint $x - y$ paths in G, then G is r-edge-connected.

C.10.14 Let G be an r-edge-connected graph. Prove that the line graph of G is r-connected and $(2r - 2)$-edge-connected.

C.10.15 (Mader, 1985) Let $r \geqslant 4$. Prove that, for every pair of vertices s and t in an r-edge-connected loopless multigraph, there exists a $s - t$ path such that the removal of its edges results in an $(r - 2)$-edge-connected loopless multigraph.

C.10.16 (Thomassen, 1981) Prove that every $(r + 3)$-connected graph G contains an induced cycle C for which $G - \mathcal{V}(C)$ is r-connected.

C.10.17 Prove that the faces of a connected plane G are all bounded by cycles if and only if G is 2-connected.

C.10.18 Prove that a graph is planar if and only if each of its blocks is planar.

C.10.19 Show that the dual graph of a 3-connected plane graph is 3-connected.

C.10.20 Let $Z \subseteq \mathbb{R}^d$ be a convex open set, and let $\varphi \colon Z \to \mathbb{R}$ be a differentiable function with continuous second-order partial derivatives. Further, let $\boldsymbol{x} = (x_1, \ldots, x_d)^t$ and define the *Hessian matrix* $\nabla^2\varphi$ of φ as

$$\nabla^2\varphi(\boldsymbol{x}) := \begin{pmatrix} \frac{\partial^2\varphi}{\partial x_1 \partial x_1}(\boldsymbol{x}) & \cdots & \frac{\partial^2\varphi}{\partial x_1 \partial x_n}(\boldsymbol{x}) \\ \vdots & \ddots & \vdots \\ \frac{\partial^2\varphi}{\partial x_n \partial x_1}(\boldsymbol{x}) & \cdots & \frac{\partial^2\varphi}{\partial x_n \partial x_n}(\boldsymbol{x}) \end{pmatrix} \tag{C.10.0.1}$$

Prove the following.

(i) The function φ is convex if and only if its Hessian matrix is positive semidefinite for every $\boldsymbol{x} \in Z$.
(ii) If the Hessian matrix of φ is positive definite for every $\boldsymbol{x} \in Z$, then the function is strictly convex.

C.10.21 Let $\varphi \colon \mathbb{R} \to \mathbb{R}$ be a twice differentiable function. Prove that if $\varphi''(x) > 0$ for every x in the domain of φ, then φ is strictly convex.

C.10.22 Prove that, if a strictly convex function $\varphi \colon \mathbb{R}^d \to \mathbb{R}$ has a minimum on a convex set of $\mathbb{R}^d$, then the minimum is unique.

C.11 Postscript

The result on the existence of contractible edges in 3-connected graphs (Lemma C.5.6) is from Thomassen (1980a), and the proof in Thomassen (1980a) is hard to beat in simplicity. Many authors have considered the distribution of contractible edges; many such results can be found in Kriesell (2002).

The proof of Theorem C.6.8 is inspired by the presentations in Bondy and Murty (2008, thm. 9.6) and West (2001, thm. 4.2.24).

The faces of a connected plane graph are bounded by cycles if and only if the graph is 2-connected, and so the sufficient condition of Proposition C.7.4 is also necessary (Problem C.10.17). The proof of Theorem C.7.6 on the 3-connectivity of maximal plane graphs is adapted from a proof of Balinki's theorem (4.1.1) in Pineda-Villavicencio (2021).

References

Abbott, T. G. 2008. *Generalizations of Kempe's universality theorem.* M.Phil. thesis, Massachusetts Institute of Technology. (353)

Adiprasito, K. 2019. FAQ on the g-theorem and the hard Lefschetz theorem for face rings. *Rend. Mat. Appl., VII. Ser.*, **40**(2), 97–111. (393)

Adiprasito, K., Papadakis, S. A., and Petrotou, V. 2021. *Anisotropy, biased pairings, and the Lefschetz property for pseudomanifolds and cycles.* arXiv:2101.07245. (393)

Adiprasito, K. A., and Benedetti, B. 2014. The Hirsch conjecture holds for normal flag complexes. *Math. Oper. Res.*, **39**(4), 1340–1348. (335, 338)

Adiprasito, K. A., and Sanyal, R. 2016. Relative Stanley-Reisner theory and upper bound theorems for Minkowski sums. *Publ. Math., Inst. Hautes Étud. Sci.*, **124**, 99–163. (277, 304)

Adler, I., and Dantzig, G. 1974. Maximum diameter of abstract polytopes. *Math. Program. Study*, **1**, 20–40. (335, 336)

Alon, N., and Kalai, G. 1985. A simple proof of the upper bound theorem. *Eur. J. Comb.*, **6**(3), 211–214. (393)

Altshuler, A., and Shemer, I. 1984. Construction theorems for polytopes. *Isr. J. Math.*, **47**(2–3), 99–110. (105, 153)

Andreev, E. M. 1970a. On convex polyhedra in Lobachevskij spaces. *Math. USSR, Sb.*, **10**, 413–440. (212)

Andreev, E. M. 1970b. On convex polyhedra of finite volume in Lobachevskii space. *Math. USSR, Sb.*, **12**, 255–259. (212)

Ardila, F., and Develin, M. 2009. Tropical hyperplane arrangements and oriented matroids. *Math. Z.*, **262**(4), 795–816. (110)

Asimow, L., and Roth, B. 1978. The rigidity of graphs. *Trans. Amer. Math. Soc.*, **245**, 279–289. (352)

Asimow, L., and Roth, B. 1979. The rigidity of graphs. II. *J. Math. Anal. Appl.*, **68**(1), 171–190. (360)

Athanasiadis, C. A. 2009. On the graph connectivity of skeleta of convex polytopes. *Discrete Comput. Geom.*, **42**(2), 155–165. (231, 252)

Athanasiadis, C. A. 2011. Some combinatorial properties of flag simplicial pseudomanifolds and spheres. *Ark. Mat.*, **49**(1), 17–29. (221)

Avis, D., and Moriyama, S. 2009. On combinatorial properties of linear program digraphs. Pages 1–13 of: *Polyhedral computation*. CRM Proc. Lecture Notes, vol. 48. Providence, RI: AMS. (176)

Babson, E., Finschi, L., and Fukuda, K. 2001. Cocircuit graphs and efficient orientation reconstruction in oriented matroids. *Eur. J. Comb.*, **22**(5), 587–600. (271)

Babson, E. K., Billera, L. J., and Chan, C. S. 1997. Neighborly cubical spheres and a cubical lower bound conjecture. *Isr. J. Math.*, **102**, 297–315. (263, 391, 399)

Bajmóczy, E. G., and Bárány, I. 1979. On a common generalization of Borsuk's and Radon's theorem. *Acta Math. Acad. Sci. Hung.*, **34**(3), 347–350. (43)

Balinski, M. L. 1961. On the graph structure of convex polyhedra in n-space. *Pac. J. Math.*, **11**, 431–434. (161, 231, 232, 254)

Bárány, I. 1982. A generalization of Carathéodory's theorem. *Discrete Math.*, **40**(2–3), 141–152. (42)

Bárány, I. 2021. *Combinatorial convexity*. Providence, RI: AMS. (42, 43)

Bárány, I., and Rote, G. 2006. Strictly convex drawings of planar graphs. *Doc. Math.*, **11**, 369–391. (198)

Barnette, D. W. 1969. A simple 4-dimensional nonfacet. *Isr. J. Math.*, **7**, 16–20. (339, 389)

Barnette, D. W. 1971. The minimum number of vertices of a simple polytope. *Isr. J. Math.*, **10**, 121–125. (236, 350, 371, 372, 374, 378, 392)

Barnette, D. W. 1973a. Generating planar 4-connected graphs. *Isr. J. Math.*, **14**, 1–13. (221)

Barnette, D. W. 1973b. Graph theorems for manifolds. *Isr. J. Math.*, **16**, 62–72. (254)

Barnette, D. W. 1973c. A proof of the lower bound conjecture for convex polytopes. *Pac. J. Math.*, **46**, 349–354. (236, 340, 350, 371, 372, 374, 378, 392, 393)

Barnette, D. W. 1974a. The projection of the f-vectors of 4-polytopes onto the (E, S)-plane. *Discrete Math.*, **10**, 201–216. (341, 343)

Barnette, D. W. 1974b. An upper bound for the diameter of a polytope. *Discrete Math.*, **10**, 9–13. (319, 323, 327, 339)

Barnette, D. W. 1982. Decompositions of homology manifolds and their graphs. *Isr. J. Math.*, **41**, 203–212. (254)

Barnette, D. W. 1987a. 5-connected 3-polytopes are refinements of octahedra. *J. Comb. Theory, Ser. B*, **42**(2), 250–254. (221, 228, 395)

Barnette, D. W. 1987b. Two "simple" 3-spheres. *Discrete Math.*, **67**, 97–99. (212, 272, 397)

Barnette, D. W., and Grünbaum, B. 1969. On Steinitz's theorem concerning convex 3-polytopes and on some properties of planar graphs. Pages 27–40 of: *The many facets of graph theory (Proc. Conf., Western Mich. Univ., Kalamazoo, Mich., 1968)*. Berlin: Springer-Verlag. (212)

Barnette, D. W., and Grünbaum, B. 1970. Preassigning the shape of a face. *Pac. J. Math.*, **32**, 299–306. (212)

Barnette, D. W., and Reay, J. R. 1973. Projections of f-vectors of four-polytopes. *J. Comb. Theory, Ser. A*, **15**, 200–209. (341, 342)

Batagelj, V. 1989. An improved inductive definition of two restricted classes of triangulations of the plane. Pages 11–18 of: *Combinatorics and graph theory*, vol. 25. Warsaw: Banach Cent. Publ. (221, 228, 395)

Bayer, M. 1987. The extended f-vectors of 4-polytopes. *J. Comb. Theory, Ser. A*, **44**, 141–151. (386, 387, 393)

Bayer, M. M. 2018. Graphs, skeleta and reconstruction of polytopes. *Acta Math. Hung.*, **155**, 61–73. (272)

Bayer, M. M., and Billera, L. J. 1984. Counting faces and chains in polytopes and posets. Pages 207–252 of: Green, C. (ed), *Combinatorics and algebra (Proc. Conf., Boulder/Colo. 1983), Contemp. Math.*, vol. 34. Providence, RI: Amer. Math. Soc. (393)

Bayer, M. M., and Billera, L. J. 1985. Generalized Dehn–Sommerville relations for polytopes, spheres and Eulerian partially ordered sets. *Invent. Math.*, **79**, 143–157. (340, 381, 390)

Beineke, L. W. 1968. *On derived graphs and digraphs*. Beitr. Graphentheorie, Int. Kolloquium Manebach (DDR) 1967, 17–23 (1968). (426)

Beineke, L. W. 1970. Characterizations of derived graphs. *J. Comb. Theory*, **9**, 129–135. (426)

Berger, M. 2009. *Geometry I*. Universitext. Berlin: Springer-Verlag. Translated from the 1977 French original by M. Cole and S. Levy. (16, 42, 153)

Billera, L. J., and Lee, C. W. 1981. A proof of the sufficiency of McMullen's conditions for f-vectors of simplicial convex polytopes. *J. Comb. Theory, Ser. B*, **31**(3), 237–255. (372, 378)

Björner, A., and Vorwerk, K. 2015. On the connectivity of manifold graphs. *Proc. Am. Math. Soc.*, **143**(10), 4123–4132. (254)

Björner, A., Edelman, P. H., and Ziegler, G. M. 1990. Hyperplane arrangements with a lattice of regions. *Discrete Comput. Geom.*, **5**(3), 263–288. (270)

Blind, R., and Blind, G. 1994. Gaps in the numbers of vertices of cubical polytopes, I. *Discrete Comput. Geom.*, **11**(3), 351–356. (184, 229)

Blind, G., and Blind, R. 1998. The almost simple cubical polytopes. *Discrete Math.*, **184**(1–3), 25–48. (93)

Blind, G., and Blind, R. 1999. Shellings and the lower bound theorem. *Discrete Comput. Geom.*, **21**(4), 519–526. (374, 393)

Blind, G., and Blind, R. 2003. On a class of equifacetted polytopes. Pages 69–78 of: *Discrete geometry*. Monogr. Textbooks Pure Appl. Math., vol. 253. New York: Dekker. (389, 392, 399)

Blind, R., and Mani-Levitska, P. 1987. Puzzles and polytope isomorphisms. *Aequationes Math.*, **34**(2–3), 287–297. (256)

Bollobás, B., and Thomason, A. 1996. Highly linked graphs. *Combinatorica*, **16**(3), 313–320. (255)

Bondy, J. A., and Murty, U. S. R. 2008. *Graph theory*. Graduate Texts in Mathematics, vol. 244. New York: Springer. (xii, 164, 407, 408, 409, 410, 417, 422, 427)

Bonichon, N., Felsner, S., and Mosbah, M. 2007. Convex drawings of 3-connected plane graphs. *Algorithmica*, **47**(4), 399–420. (198)

Bonifas, N., Di Summa, M., Eisenbrand, F., Hähnle, N., and Niemeier, M. 2014. On sub-determinants and the diameter of polyhedra. *Discrete Comput. Geom.*, **52**(1), 102–115. (335)

Brehm, U., and Sarkaria, K. S. 1992. *Linear vs piecewise linear embeddability of simplicial complexes*. 92/52. Max-Planck-Institut für Mathematik. (225)

Bremner, D., and Schewe, L. 2011. Edge-graph diameter bounds for convex polytopes with few facets. *Exp. Math.*, **20**(3), 229–237. (335)

Bricard, R. 1897. Mémoire sur la théorie de l'octaèdre articulé. *J. Math. Pures Appl.*, **3**, 113–148. (366)

Brightwell, G. R, and Scheinerman, E. R, 1993. Representations of planar graphs. *SIAM J. Discrete Math.*, **6**(2), 214–229. (214)

Brinkmann, G., Greenberg, S., Greenhill, C., McKay, B. D, Thomas, R., and Wollan, P. 2005. Generation of simple quadrangulations of the sphere. *Discrete Math.*, **305**(1), 33–54. (221, 228, 230, 395)

Brøndsted, A. 1982. A dual proof of the upper bound theorem. Pages 39–43 of: *Convexity and related combinatorial geometry (Norman, Okla., 1980)*. Lecture Notes in Pure and Appl. Math., vol. 76. New York: Dekker. (393)

Brøndsted, A. 1983. *An introduction to convex polytopes*. Graduate Texts in Mathematics, vol. 90. New York: Springer-Verlag. (xi, 32, 42, 70, 87, 88, 152, 153, 229, 393)

Brouwer, L. E. J. 1912. Beweis des Jordan'schen Satzes für den n-dimensionalen Raum. *Math. Ann.*, **71**, 314–319. (401)

Bruggesser, H., and Mani, P. 1971. Shellable decompositions of cells and spheres. *Math. Scand.*, **29**, 197–205. (123, 128)

Bui, H. T., Pineda-Villavicencio, G., and Ugon, J. 2024. The linkedness of cubical polytopes: Beyond the cube. *Discrete Math.*, **347**(3), 113801. (245)

Bui, H. T., Pineda-Villavicencio, G., and Ugon, J. 2021. The linkedness of cubical polytopes: The cube. *Electron. J. Comb.*, **28**, P3.45. (249, 253)

Bunt, L. N. H. 1934. *Bijdrage tot de theorie der convexe puntverzamelingen*. Thesis, University of Groningen, Amsterdam. (43)

Cai, M.C. 1993. The number of vertices of degree k in a minimally k-edge-connected graph. *J. Comb. Theory, Ser. B*, **58**(2), 225–239. (415)

Carathéodory, C. 1907. Über den Variabilitätsbereich der Koeffizienten von Potenzreihen, die gegebene Werte nicht annehmen. *Math. Ann.*, **64**, 95–115. (20, 42)

Chartrand, G., and Stewart, M. J. 1969. The connectivity of line-graphs. *Math. Ann.*, **182**(3), 170–174. (414)

Chartrand, G., Kaugars, A., and Lick, D. R. 1972. Critically n-connected graphs. *Proc. Am. Math. Soc.*, **32**(1), 63–68. (413)

Conforti, M., Cornuéjols, G., and Zambelli, G. 2014. *Integer programming*. Vol. 271. Cham: Springer. (58)

Connelly, R., and Guest, S. D. 2022. *Frameworks, tensegrities, and symmetry*. Cambridge: Cambridge University Press. (351, 371, 392)

Cordovil, R., and Duchet, P. 2000. Cyclic polytopes and oriented matroids. *Eur. J. Comb.*, **21**(1), 49–64. (101)

Courdurier, M. 2006. On stars and links of shellable polytopal complexes. *J. Comb. Theory, Ser. A*, **113**(4), 692–697. (154)

Cremona, L. 1890. *Graphical statics. Two treatises on the graphical calculus and reciprocal figures in graphical statics*. Oxford: Clarendon Press. (198)

Criado, F., and Santos, F. 2017. The maximum diameter of pure simplicial complexes and pseudo-manifolds. *Discrete Comput. Geom.*, **58**(3), 643–649. (338)

Danaraj, G., and Klee, V. 1974. Shellings of spheres and polytopes. *Duke Math. J.*, **41**, 443–451. (173, 229)

Dancis, J. 1984. Triangulated n-manifolds are determined by their [n/2]+ 1-skeletons. *Topol. Appl.*, **18**(1), 17–26. (263)

Dantzig, G. B. 1963. *Linear programming and extensions*. Princeton, NJ: Princeton University Press. (306)

Davies, J., and Pfender, F. 2021. Edge-maximal graphs on orientable and some nonorientable surfaces. *J. Graph Theory*, **98**(3), 405–425. (420)

Davis, C. 1954. Theory of positive linear dependence. *Am. J. Math.*, **76**(4), 733–746. (34)

De Loera, J. A., Kim, E. D., Onn, S., and Santos, F. 2009. Graphs of transportation polytopes. *J. Comb. Theory, Ser. A*, **116**(8), 1306–1325. (110)

De Loera, J. A., Rambau, J., and Santos, F. 2010. *Triangulations: Structures for algorithms and applications*. Algorithms and Computation in Mathematics, vol. 25. Berlin: Springer-Verlag. (154)

de Verdière, Y. C. 1991. Un principe variationnel pour les empilements de cercles. *Invent. Math.*, **104**(1), 655–669. (212)

Dehn, M. 1905. Die *Euler*sche Formel im Zusammenhang mit dem Inhalt in der nichteuklidischen Geometrie. *Math. Ann.*, **61**, 561–586. (131)

Dehn, M. 1916. Über die Starrheit konvexer Polyeder. *Math. Ann.*, **77**, 466–473. (364, 366, 367)

Del Pia, A., and Michini, C. 2016. On the diameter of lattice polytopes. *Discrete Comput. Geom.*, **55**(3), 681–687. (333, 335)

Deo, S. 2018. *Algebraic topology. A primer*. 2nd ed. Vol. 27. New Delhi: Hindustan Book Agency; Singapore: Springer. (401)

Develin, M. 2004. LP-orientations of cubes and crosspolytopes. *Adv. Geom.*, **4**(4), 459–468. (175, 226)

Deza, A., and Pournin, L. 2018. Improved bounds on the diameter of lattice polytopes. *Acta Math. Hung.*, **154**(2), 457–469. (335)

Deza, A., and Pournin, L. 2019. Diameter, decomposability, and Minkowski sums of polytopes. *Can. Math. Bull.*, **62**(4), 741–755. (299, 302)

Diestel, R. 2017. *Graph theory*. 5th ed. Graduate Texts in Mathematics, vol. 173. Berlin: Springer-Verlag. (xii, 165, 181, 182, 215, 255, 405, 408, 411, 412, 414, 415, 416, 417, 419, 420, 423, 424, 425)

Dirac, G. A. 1960. In abstrakten Graphen vorhandene vollständige 4-Graphen und ihre Unterteilungen. *Math. Nachr.*, **22**, 61–85. (417)

Doolittle, J. 2018. *A minimal counterexample to strengthening of Perles's conjecture*. arXiv:809.00662. (271)

Doolittle, J., Nevo, E., Pineda-Villavicencio, G., Ugon, J., and Yost, D. 2018. On the reconstruction of polytopes. *Discrete Comput. Geom.*, **61**(2), 285–302. (272, 397)

Edelsbrunner, H. 2012. *Algorithms in combinatorial geometry*. 1st ed. Berlin: Springer. (153)

Eggleston, H. G., Grünbaum, Branko, and Klee, Victor. 1964. Some semicontinuity theorems for convex polytopes and cell-complexes. *Comment. Math. Helv.*, **39**, 165–188. (105, 153)

Ehrenborg, R., and Hetyei, G. 1995. Generalizations of Baxter's theorem and cubical homology. *J. Comb. Theory, Ser. A*, **69**(2), 233–287. (227)

Eisenbrand, F., Hähnle, N., Razborov, A., and Rothvoß, T. 2010. Diameter of polyhedra: Limits of abstraction. *Math. Oper. Res.*, **35**(4), 786–794. (335, 336, 338, 339)

Elmasry, A., Mehlhorn, K., and Schmidt, J. M. 2013. Every DFS tree of a 3-connected graph contains a contractible edge. *J. Graph Theory*, **72**(1–2), 112–121. (412, 425)

Eppstein, D. n.d. *Twenty-one proofs of Euler's formula: $V - E + F = 2$*. Accessed 22 September 2021. www.ics.uci.edu/~eppstein/junkyard/euler/. (418)

Erickson, J. 2020. *Tutte's spring embedding*. http://jeffe.cs.illinois.edu/teaching/topology20/notes/12-spring-embedding.pdf. (230)

Euler, L. 1736. Solutio problematis ad geometriam situs pertinentis. *Comment. Acad. Sci. U. Petrop.*, **8** (Reprinted in Opera Omnia Series Prima, Vol. 7. pp. 1-10, 1766.), 128–140. (404)

Euler, L. 1758a. Demonstratio nonnullarum insignium proprietatum quibas solida hedris planis inclusa sunt praedita. *Novi Comm. Acad. Sci. Imp. Petropol.*, **4**, 140–160. (130, 418)

Euler, L. 1758b. Elementa doctrinae solidorum. *Novi Comm. Acad. Sci. Imp. Petropol.*, **4**, 109–140. (128, 130, 418)

Ewald, G. 1996. *Combinatorial convexity and algebraic geometry*. Graduate Texts in Mathematics, vol. 168. New York: Springer-Verlag. (393)

Ewald, G., and Shephard, G. C. 1974. Stellar subdivisions of boundary complexes of convex polytopes. *Math. Ann.*, **210**, 7–16. (150)

Farkas, J. 1898. Die algebraischen Grundlagen der Anwendungen des Fourier'schen Principes in der Mechanik. *Ungar. Ber.*, **15**, 25–40. (152)

Farkas, J. 1901. Theorie der einfachen Ungleichungen. *J. Reine Angew. Math.*, **124**, 1–27. (152)

Fáry, I. 1948. On straight line representation of planar graphs. *Acta Sci. Math.*, **11**, 229–233. (158)

Felsner, S., and Rote, G. 2019. On primal-dual circle representations. Pages 8:1–8:18 of: Fineman, Jeremy, and Mitzenmacher, Michael (eds), *Simplicity in Algorithms (Proc. 2nd Symposium, San Diego, 2019)*. OpenAccess Series in Informatics (OASIcs), vol. 69. Schloss Dagstuhl–Leibniz-Zentrum für Informatik. (230)

Firsching, M. 2020. The complete enumeration of 4-polytopes and 3-spheres with nine vertices. *Isr. J. Math.*, **240**, 417–441. (228, 307, 395)

Fleming, B., and Karu, K. 2010. Hard Lefschetz theorem for simple polytopes. *J. Algebr. Comb.*, **32**(2), 227–239. (393)

Flores, A. 1934. Über n-dimensionale Komplexe, die im R_{2n+1} absolut selbstverschlungen sind. *Ergeb. Math. Kolloq.*, **6**, 4–7. (155, 225)

Fortuna, E., Frigerio, R., and Pardini, R. 2016. *Projective geometry: Solved problems and theory review*. Vol. 104. Switzerland: Springer. (42)

Fourier, J. B. J. 1827. Analyse des travaux de l'Académie Royale des Sciences pendant l'année 1824. *Partie mathématique*. (46)

Francese, C., and Richeson, D. 2007. The flaw in Euler's proof of his polyhedral formula. *Am. Math. Mon.*, **114**(4), 286–296. (128)

Franklin, P. 1934. A six color problem. *J. Math. Phys.*, **13**, 363–369. (420)

Friedman, E. J. 2009. Finding a simple polytope from its graph in polynomial time. *Discrete Comput. Geom.*, **41**(2), 249–256. (256, 264, 265, 269)

Friedmann, O. 2011. A subexponential lower bound for Zadeh's pivoting rule for solving linear programs and games. Pages 192–206 of: *Integer programming and combinatoral optimization (Proc. 15th international conference, IPCO 2011, New York, 2011)*. Berlin: Springer. (306)

Fukuda, K. 2004. From the zonotope construction to the Minkowski addition of convex polytopes. *J. Symbolic Comput.*, **38**(4), 1261–1272. (303, 304)

Fukuda, K. 2022. *Frequently asked questions in polyhedral computation*. https://people.inf.ethz.ch/fukudak/. (52)

Fusy, Éric. 2006. Counting d-polytopes with $d + 3$ vertices. *Electron. J. Comb.*, **13**(1), R23. (144)

Gale, D. 1954. Irreducible convex sets. Pages 217–218 of: *Proceedings of the International Congress of Mathematics*, vol. II. (278)

Gale, D. 1963. Neighborly and cyclic polytopes. Pages 225–232 of: *Proc. Sympos. Pure Math*, vol. 7. (95)

Gallet, M., Grasegger, G., Legerský, J., and Schicho, J. 2021. Combinatorics of Bricard's octahedra. *C. R., Math., Acad. Sci. Paris*, **359**(1), 7–38. (367)

Gallier, J. 2011. *Geometric methods and applications*. 2nd ed. Texts in Applied Mathematics, vol. 38. New York: Springer. (8, 42, 153)

Gallivan, S. 1985. Disjoint edge paths between given vertices of a convex polytope. *J. Comb. Theory, Ser. A*, **39**(1), 112–115. (231, 247, 254, 255, 396)

Gawrilow, E., and Joswig, M. 2000. Polymake: A framework for analyzing convex polytopes. Pages 43–73 of: Kalai, G., Ziegler, G.M. (eds) *Polytopes—Combinatorics and computation*. DMV Sem., vol. 29. Basel: Birkhäuser. (262)

Geelen, J. 2012. *On how to draw a graph*. www.math.uwaterloo.ca/~jfgeelen/Publications/tutte.pdf. (230)

Gluck, H. 1975. Almost all simply connected closed surfaces are rigid. Pages 225–239 of: *Geometric topology (Proc. Conf. Park City 1974)*, Lect. Notes Math. vol. 438 Berlin: Springer. (361)

Goodman, J. E., O'Rourke, J., and Tóth, C. D. (eds). 2017. *Handbook of discrete and computational geometry*. 3rd ed. Boca Raton, FL: Chapman & Hall/CRC. (110, 272, 353, 393)

Graham, R. L., Knuth, D. E., and Patashnik, O. 1994. *Concrete mathematics: A foundation for computer science*. 2nd ed. New York: Addison-Wesley. (375)

Graver, J. E. 2001. *Counting on frameworks: Mathematics to aid the design of rigid structures*. Vol. 25 of The Dolciani Mathematical Expositions, Washington, DC: Mathematical Association of America. (392)

Graver, J., Servatius, B., and Servatius, H. 1993. *Combinatorial rigidity*. Providence, RI: American Mathematical Society. (392)

Gruber, P. M. 2007. *Convex and discrete geometry*. Grundlehren der Mathematischen Wissenschaften [Fundamental Principles of Mathematical Sciences], vol. 336. Berlin: Springer. (xi, 128, 154)

Grünbaum, B. 1963. Unambiguous polyhedral graphs. *Isr. J. Math.*, **1**, 235–238. (101, 102, 104, 153)

Grünbaum, B. 1965. On the facial structure of convex polytopes. *Bull. Am. Math. Soc.*, **71**, 559–560. (220, 230)

Grünbaum, B. 1969. Imbeddings of simplicial complexes. *Comment. Math. Helv.*, **44**, 502–513. (225)

Grünbaum, B. 2003. *Convex polytopes*. 2nd ed. Graduate Texts in Mathematics, vol. 221. New York: Springer-Verlag. (xi, 101, 102, 131, 133, 135, 136, 142, 151, 153, 154, 212, 218, 220, 228, 230, 256, 257, 258, 272, 303, 320, 340, 341, 344, 378, 395)

Grünbaum, B., and Shephard, G. C. 1987. Some problems on polyhedra. *J. Geom.*, **29**, 182–189. (214)

Haase, C., and Ziegler, G. M. 2002. Examples and counterexamples for the Perles conjecture. *Discrete Comput. Geom.*, **28**(1), 29–44. (271)

Halin, R. 1966. Zu einem Problem von B. Grünbaum. *Arch. Math.*, **17**, 566–568. (220, 228, 395)

Halin, R. 1969. A theorem on n-connected graphs. *J. Comb. Theory*, **7**, 150–154. (413)

Halmos, P. R. 1974. *Finite-dimensional vector spaces*. 2nd ed. New York: Springer-Verlag. (42)

Helly, Ed. 1923. Über Mengen konvexer Körper mit gemeinschaftlichen Punkten. *Jahresber. Dtsch. Math.-Ver.*, **32**, 175–176. (42)

Hirata, T., Kubota, K., and Saito, O. 1984. A sufficient condition for a graph to be weakly k-linked. *J. Comb. Theory, Ser. B*, **36**(1), 85–94. (255)

Hodge, W. V. D., and Pedoe, D. 1994. *Methods of algebraic geometry* (Cambridge Mathematical Library). Cambridge: Cambridge University Press. (10)

Hoffman, A. J., and Kruskal, J. G. 1956. Integral boundary points of convex polyhedra. *Ann. Math. Stud.*, **38**, 223–246. (58)

Holmes, B. 2018. On the diameter of dual graphs of Stanley-Reisner rings and Hirsch type bounds on abstractions of polytopes. *Electron. J. Comb.*, **25**(1), P1.60. (339)

Holt, F., and Klee, V. 1999. A proof of the strict monotone 4-step conjecture. Pages 201–216 of: *Advances in discrete and computational geometry (Prof. Conf. 1996 AMS-IMS-SIAM South Hadley, MA, 1996). Contemp. Math.*, vol. 223. Providence, RI: Amer. Math. Soc. (253)

Holt, F. B., and Klee, V. 1998. Many polytopes meeting the conjectured Hirsch bound. *Discrete Comput. Geom.*, **20**(1), 1–17. (332, 339)

Hopcroft, J. E., and Kahn, P. J. 1992. A paradigm for robust geometric algorithms. *Algorithmica*, **7**(4), 339–380. (211, 230)

Höppner, A., and Ziegler, G. M. 2000. A census of flag-vectors of 4-polytopes. Pages 105–110 of: Kalai, G., Ziegler, G.M. (eds) *Polytopes—Combinatorics and computation*. DMV Sem., vol. 29. Basel: Birkhäuser. (343)

Huck, A. 1991. A sufficient condition for graphs to be weakly k-linked. *Graphs Comb.*, **7**(4), 323–351. (250, 251)

Ishizeki, T., and Takeuchi, F. 1999. Geometric shellings of 3-polytopes. Pages 132–135 of: *Proc. 11th Canadian Conference on Computational Geometry*. (175)

Izmestiev, I. 2009. Projective background of the infinitesimal rigidity of frameworks. *Geom. Dedicata*, **140**, 183–203. (363)

Jänich, K. 1984. *Topology*. Undergraduate Texts in Mathematics. New York: Springer-Verlag. (402)

Jockusch, W. 1993. The lower and upper bound problems for cubical polytopes. *Discrete Comput. Geom.*, **9**(2), 159–163. (391, 399)

Jordan, C. 1882. *Cours d'analyse de l'École Polytechnique*. 3 vol. Paris: Gauthier-Villars. (157, 229, 257, 401)

Joswig, M. 2000. Reconstructing a non-simple polytope from its graph. Pages 167–176 of: Kalai, G., Ziegler, G.M. (eds) *Polytopes—Combinatorics and computation*. DMV Sem., vol. 29. Basel: Birkhäuser. (269, 271)

Joswig, M. 2002. Projectivities in simplicial complexes and colorings of simple polytopes. *Math. Z.*, **240**(2), 243–259. (230)

Joswig, M., and Ziegler, G. M. 2000. Neighborly cubical polytopes. *Discrete Comput. Geom.*, **24**(2–3), 325–344. (391, 399)

Joswig, M., Kaibel, V., Pfetsch, M. E., and Ziegler, G. M. 2001. Vertex-facet incidences of unbounded polyhedra. *Adv. Geom.*, **1**(1), 23–36. (310)

Joswig, M., Kaibel, V., and Körner, F. 2002. On the k-systems of a simple polytope. *Isr. J. Math.*, **129**, 109–117. (226, 269)

Kaibel, V. 2003. Reconstructing a simple polytope from its graph. Pages 105–118 of: Jünger, M., Reinelt, G., Rinaldi, G. (eds) *Combinatorial optimization—Eureka, you shrink!* Lecture Notes in Comput. Sci., vol. 2570. Berlin: Springer. (264)

Kalai, G. 1987. Rigidity and the lower bound theorem. I. *Invent. Math.*, **88**(1), 125–151. (340, 350, 373, 374, 385, 389, 392)

Kalai, G. 1988a. A new basis of polytopes. *J. Comb. Theory, Ser. B*, **49**(2), 191–209. (390)

Kalai, G. 1988b. A simple way to tell a simple polytope from its graph. *J. Comb. Theory, Ser. A*, **49**(2), 381–383. (171, 256, 272)

Kalai, G. 1990. On low-dimensional faces that high-dimensional polytopes must have. *Combinatorica*, **10**(3), 271–280. (179, 181, 229, 389)

Kalai, G. 1992. Upper bounds for the diameter and height of graphs of convex polyhedra. *Discrete Comput. Geom.*, **8**(4), 363–372. (335, 336)

Kalai, G. 1997. Linear programming, the simplex algorithm and simple polytopes. *Math. Programming*, **79**(1–3), 217–233. (306)

Kalai, G., and Kleitman, D. J. 1992. A quasi-polynomial bound for the diameter of graphs of polyhedra. *Bull. Am. Math. Soc., New Ser.*, **26**(2), 315–316. (320, 336)

Kalai, G. (coordinator). 2010. *Polymath 3: Polynomial Hirsch conjecture*. http://gilkalai.wordpress.com/2010/09/29/polymath-3-polynomial-hirsch-conjecture. (338)

Kallay, M. 1979. *Decomposability of convex polytopes*. Ph.D. thesis, The Hebrew University of Jerusalem, Jerusalem. (295)

Kallay, M. 1982. Indecomposable polytopes. *Isr. J. Math.*, **41**(3), 235–243. (286, 291, 297, 300, 303, 304)

Kallay, M. 1984. Decomposability of polytopes is a projective invariant. Pages 191–196 of: *Convexity and graph theory (Jerusalem, 1981)*. North-Holland Math. Stud., vol. 87. Amsterdam: North-Holland. (302, 304)

Karavelas, M. I., and Tzanaki, E. 2016a. A geometric approach for the upper bound theorem for Minkowski sums of convex polytopes. *Discrete Comput. Geom.*, **56**(4), 966–1017. (277, 304)

Karavelas, M. I., and Tzanaki, E. 2016b. The maximum number of faces of the Minkowski sum of two convex polytopes. *Discrete Comput. Geom.*, **55**, 748–785. (277, 304)

Karmarkar, N. 1984. A new polynomial-time algorithm for linear programming. *Combinatorica*, **4**, 373–395. (306)

Kawarabayashi, K., Kostochka, A., and Yu, G. 2006. On sufficient degree conditions for a graph to be k-linked. *Comb. Probab. Comput.*, **15**(5), 685–694. (255)

Kelmans, A. K. 1980. Concept of a vertex in a matroid and 3-connected graphs. *J. Graph Theory*, **4**, 13–19. (182)

Khachiyan, L. G. 1979. A polynomial algorithm in linear programming. *Dokl. Akad. Nauk SSSR*, **244**, 1093–1096. (52, 306)

Kim, E. D., and Santos, F. 2010a. *Companion to "An update on the Hirsch conjecture"*. arXiv:0912.4235. (339)

Kim, E. D., and Santos, F. 2010b. An update on the Hirsch conjecture. *Jahresber. Dtsch. Math.-Ver.*, **112**(2), 73–98. (339)

Klee, V. 1964. A combinatorial analogue of Poincaré's duality theorem. *Can. J. Math.*, **16**, 517–531. (131)

Klee, V. 1964a. Diameters of polyhedral graphs. *Can. J. Math.*, **16**, 602–614. (318, 335)

Klee, V. 1964b. On the number of vertices of a convex polytope. *Can. J. Math.*, **16**, 701–720. (106, 153, 392)

Klee, V. 1964c. A property of d-polyhedral graphs. *J. Math. Mech.*, **13**, 1039–1042. (253)

Klee, V., and Minty, G. J. 1972. *How good is the simplex algorithm?* Pages 159–175 of: *Inequalities III (Proc. 3rd Symp., Los Angeles 1969)*. (306)

Klee, V., and Walkup, D. W. 1967. The d-step conjecture for polyhedra of dimension $d < 6$. *Acta Math.*, **117**, 53–78. (305, 306, 307, 308, 310, 311, 312, 317, 320, 335, 338, 339)

Kleinschmidt, P., and Onn, S. 1992. On the diameter of convex polytopes. *Discrete Math.*, **102**(1), 75–77. (333)

Koebe, P. 1936. *Kontaktprobleme der konformen Abbildung*. Ber. Sächs. Akad. Wiss. Leipzig, Math.-phys. Kl. **88**, 141–164. (212)

König, D. 1936. *Theorie der endlichen und unendlichen Graphen. Kombinatorische Topologie der Streckenkomplexe.* XI + 258 S. Leipzig, Akademische Verlagsgesellschaft (Mathematik in Monographien, Bd. 16). (410)

Krein, M., and Milman, D. 1940. On extreme points of regular convex sets. *Stud. Math.*, **9**, 133–138. (43)

Kriesell, M. 2002. A survey on contractible edges in graphs of a prescribed vertex connectivity. *Graphs Comb.*, **18**(1), 1–30. (427)

Kuratowski, K. 1930. Sur le probleme des courbes gauches en topologie. *Fundam. Math.*, **15**, 271–283. (403, 423)

Kusunoki, T., and Murai, S. 2019. The numbers of edges of 5-polytopes with a given number of vertices. *Ann. Comb.*, **23**(1), 89–101. (343)

Larman, D. G. 1970. Paths on polytopes. *Proc. Lond. Math. Soc.*, **20**, 161–178. (327, 336)

Larman, D. G., and Mani, P. 1970. On the existence of certain configurations within graphs and the 1-skeletons of polytopes. *Proc. London Math. Soc.*, **20**, 144–160. (220, 226, 228, 229, 231, 245, 247, 249, 254, 255, 395)

Lauritzen, N. 2013. *Undergraduate convexity: From Fourier and Motzkin to Kuhn and Tucker*. Hackensack, NJ: World Scientific Publishing. (38, 42, 152)

Lebesgue, H. 1911. Sur l'invariance du nombre de dimensions d'un espace et sur le théorème de *M. Jordan* relatif aux variétés formées. *C. R. Acad. Sci., Paris*, **152**, 841–843. (401)

Lee, C. W. 1991. Regular triangulations of convex polytopes. Pages 443–456 of: P. Gritzmann and B. Sturmfels (eds) *Applied geometry and discrete mathematics: The Victor Klee Festschrift*. DIMACS Ser. Discrete Math. Theoret. Comput. Sci., vol. 4. (111, 153)

Lee, C. W. 2013. *Polytopes: Course notes*. www.ms.uky.edu/~lee/ma714fa13/notes.pdf. (392)

Ling, J. M. 2007. New Non-Linear Inequalities for Flag-Vectors of 4-Polytopes. *Discrete Comput. Geom.*, **37**(3), 455–469. (388, 390)

Liu, G. Z. 1990. Proof of a conjecture on matroid base graphs. *Sci. China Ser. A*, **33**, 1329–1337. (426)

Lockeberg, E. R. 1977. *Refinements in boundary complexes of polytopes*. Ph.D. thesis, University College London. (227, 228, 230, 253, 395, 396)

Lovász, L. 2007. *Combinatorial problems and exercises*. 2nd ed. Providence, RI: AMS Chelsea Publishing. (415)

Lovász, L. 2019. *Graphs and geometry*. Providence, RI: Colloquium Publications. (194, 211, 212, 214, 230, 361, 362, 371, 392)

Lutz, F. H. 2004. Small examples of nonconstructible simplicial balls and spheres. *SIAM J. Discrete Math.*, **18**(1), 103–109. (393)

Lutz, F. H. 2008. Combinatorial 3-manifolds with 10 vertices. *Beitr. Algebra Geom.*, **49**(1), 97–106. (393)

MacLane, S. 1937. A combinatorial condition for planar graphs. *Fundam. Math.*, **28**, 22–32. (181, 182)

Mader, W. 1971. Minimale n-fach kantenzusammenhängende Graphen. *Math. Ann.*, **191**, 21–28. (415)

Mader, W. 1974. Kantendisjunkte Wege in Graphen. *Monatsh. Math.*, **78**, 395–404. (415)

Mader, W. 1985. Paths in graphs, reducing the edge-connectivity only by two. *Graphs Comb.*, **1**(1), 81–89. (426)

Mader, W. 1986. Kritisch n-fach kantenzusammenhängende Graphen. (Critically n-edge- connected graphs). *J. Comb. Theory, Ser. B*, **40**, 152–158. (415)

Mader, W. 1995. On vertices of degree n in minimally n-edge-connected graphs. *Comb. Probab. Comput.*, **4**(1), 81–95. (415)

Maharry, J. 1999. An excluded minor theorem for the octahedron. *J. Graph Theory*, **31**(2), 95–100. (221, 228, 230, 395)

Maharry, J. 2000. A characterization of graphs with no cube minor. *J. Comb. Theory, Ser. B*, **80**(2), 179 – 201. (221, 230)

Maksimenko, A. M. 2009. The diameter of the ridge-graph of a cyclic polytope. *Discrete Math. Appl.*, **19**(1), 47–53. (335)

Marden, A., and Rodin, B. 1990. On Thurston's formulation and proof of Andreev's theorem. Pages 103–115 of: Ruscheweyh, S., Saff, E. B., Salinas, L. C., and Varga, R. S (eds), *Computational methods and function theory (Proc. Conf., Valparaíso/Chile 1989)*. Lect. Notes Math., vol. 1435. Berlin: Springer. (212)

Matoušek, J. 2002. *Lectures on discrete geometry*. Graduate Texts in Mathematics, vol. 212. New York: Springer-Verlag. (153)

Matoušek, J. 2003. *Using the Borsuk-Ulam theorem*. Berlin: Springer-Verlag. (225)

Matschke, B., Santos, F., and Weibel, C. 2015. The width of five-dimensional prismatoids. *Proc. Lond. Math. Soc.*, **110**(3), 647–672. (331, 332)

Maxwell, J. C. 1864. XLV. On reciprocal figures and diagrams of forces. *Philos. Mag*, **27**(182), 250–261. (198)

Maxwell, J. C. 1869. On reciprocal figures, frams, and diagrams of forces. *Trans. R. Soc. Edinb.*, **26**, 1–40. (198)

McMullen, P. 1970. The maximum numbers of faces of a convex polytope. *Mathematika*, **17**, 179–184. (94, 153, 302, 340, 374, 375, 378, 393, 397)

McMullen, P. 1971a. The minimum number of facets of a convex polytope. *J. Lond. Math. Soc.*, **3**, 350–354. (347)

McMullen, P. 1971b. The numbers of faces of simplicial polytopes. *Isr. J. Math.*, **9**, 559–570. (340, 377, 378)

McMullen, P. 1976. Constructions for projectively unique polytopes. *Discrete Math.*, **14**(4), 347–358. (150, 152)

McMullen, P. 1987. Indecomposable convex polytopes. *Isr. J. Math.*, **58**(3), 321–323. (289, 292, 304)

McMullen, P. 1993. On simple polytopes. *Invent. Math.*, **113**(2), 419–444. (393)

McMullen, P., and Shephard, G. C. 1971. *Convex polytopes and the upper bound conjecture*. London: Cambridge University Press. (141, 142, 153, 154)

McMullen, P., and Walkup, D. W. 1971. A generalized lower-bound conjecture for simplicial polytopes. *Mathematika*, **18**, 264–273. (372, 374)

Menger, K. 1927. Zur allgemeinen Kurventheorie. *Fundam. Math.*, **10**(1), 96–115. (403, 415)

Mészáros, G. 2015. *Linkedness and path-pairability in the Cartesian product of graphs*. Ph.D. thesis, Central European University. (253)

Mészáros, G. 2016. On linkedness in the Cartesian product of graphs. *Period. Math. Hung.*, **72**(2), 130–138. (253)

Meyer, W., and Kay, D. C. 1973. A convexity structure admits but one real linearization of dimension greater than one. *J. Lond. Math. Soc., II. Ser.*, **7**, 124–130. (22)

Meyer, W. J. 1969. *Minkowski addition of convex sets*. Ph.D. thesis, The University of Wisconsin. (297, 300, 301, 304)

Mihalisin, J., and Klee, V. 2000. Convex and linear orientations of polytopal graphs. *Discrete Comput. Geom.*, **24**(2–3), 421–435. (174)

Minkowski, H. 1896. *Geometrie der zahlen*. Leipzig: Teubner. (43, 152)

Minkowski, H. 1911. *Gesammelte Abhandlungen von Hermann Minkowski. Unter Mitwirkung von Andreas Speiser und Hermann Weyl, herausgegeben von David Hilbert. Band I, II*. Leipzig: Teubner. (43)

Mohar, B. 1993. A polynomial time circle packing algorithm. *Discrete Math.*, **117**(1), 257–263. (214)

Mohar, B., and Thomassen, C. 2001. *Graphs on surfaces*. Baltimore, MD: Johns Hopkins University Press. (157, 158, 182, 212, 214, 229, 230)

Morris, S. A. 2020. *Topology without tears*. www.topologywithouttears.net/topbook.pdf. (402)

Motzkin, T. 1935. Sur quelques propriétés caractéristiques des ensembles convexes. *Atti Accad. Naz. Lincei, Rend., VI. Ser.*, **21**, 562–567. (43)

Motzkin, T. 1936. *Beiträge zur Theorie der linearen Ungleichungen*. Thesis, University of Basel. (46, 152)

Motzkin, T. S. 1957. Comonotone curves and polyhedra. *Bull. Am. Math. Soc.*, **63**. (374)

Mulmuley, K. 1993. *Computational geometry: An introduction through randomized algorithms*. Englewood Cliffs, NJ: Prentice Hall. (393)

Murai, S., and Nevo, E. 2013. On the generalized lower bound conjecture for polytopes and spheres. *Acta Math.*, **210**(1), 185–202. (374)

Naddef, D. 1989. The Hirsch conjecture is true for (0,1)-polytopes. *Math. Program.*, **45**, 109–110. (333)

Nevo, E. 2022. *Private Communication.* (258)

Nevo, E., Pineda-Villavicencio, G., Ugon, J., and Yost, D. 2019. Almost simplicial polytopes I. The lower and upper bound theorems. *Can. J. Math.*, **72**(2). (393)

Okamura, H. 1984. Multicommodity flows in graphs. II. *Jpn. J. Math., New Ser.*, **10**(1), 99–116. (255)

Pach, J., and Agarwal, P. K. 1995. *Combinatorial geometry.* New York: John Wiley & Sons. (212)

Pak, I. 2010. *Lectures on discrete and polyhedral geometry.* www.math.ucla.edu/~pak/book.htm. (392)

Papadakis, S. A., and Petrotou, V. 2020. *The characteristic 2 anisotropicity of simplicial spheres.* arXiv:2012.09815. (393)

Perles, M. A., and Prabhu, N. 1993. A property of graphs of convex polytopes. *J. Comb. Theory, Ser. A*, **62**(1), 155–157. (235, 252)

Perles, M. A., and Shephard, G. C. 1967. Facets and nonfacets of convex polytopes. *Acta Math.*, **119**, 113–145. (389, 391, 399)

Perles, M. A., Martini, H., and Kupitz, Y. S. 2009. A Jordan–Brouwer separation theorem for polyhedral pseudomanifolds. *Discrete Comput. Geom.*, **42**(2), 277–304. (402)

Pestenjak, B. 1999. An algorithm for drawing planar graphs. *Softw. Pract. Exper.*, **29**(11), 973–984. (207)

Pfeifle, J., Pilaud, V., and Santos, F. 2012. Polytopality and Cartesian products of graphs. *Isr. J. Math.*, **192**(1), 121–141. (272)

Pilaud, V., Pineda-Villavicencio, G., and Ugon, J. 2023. Edge connectivity of simplicial polytopes. *Eur. J. Comb.*, **113**, 103752. (231, 236, 237)

Pineda-Villavicencio, G. 2021. A new proof of Balinski's theorem on the connectivity of polytopes. *Discrete Math.*, **344**, 112408. (254, 427)

Pineda-Villavicencio, G. 2022. *Cycle space of graphs of polytopes.* arXiv: 2208.02579. (229)

Pineda-Villavicencio, G., and Schröter, B. 2022. Reconstructibility of matroid polytopes. *SIAM J. Discrete Math.*, **36**(1), 490–508. (269)

Pineda-Villavicencio, G., and Ugon, J. 2018. *Polymake script to construct polytopes with isomorphic skeletons.* http://guillermo.com.au/pdfs/ConstructPolytope-A.txt. (263)

Pineda-Villavicencio, G., and Yost, D. 2022. A lower bound theorem for d-polytopes with $2d+1$ vertices. *SIAM J. Discrete Math.*, **36**, 2920–2941. (168, 350, 391, 399)

Pineda-Villavicencio, G., Ugon, J., and Yost, D. 2018. The excess degree of a polytope. *SIAM J. Discrete Math.*, **32**(3), 2011–2046. (168, 169, 229, 296, 304, 343)

Pineda-Villavicencio, G., Ugon, J., and Yost, D. 2019. Lower bound theorems for general polytopes. *Eur. J. Comb.*, **79**, 27–45. (344)

Pineda-Villavicencio, G., Ugon, J., and Bui, H. T. 2020a. Connectivity of cubical polytopes. *J. Combin. Theory Ser. A*, **169**, 105–126. (254)

Pineda-Villavicencio, G., Ugon, J., and Yost, D. 2020b. Polytopes close to being simple. *Discrete Comput. Geom.*, **64**, 200–215. (270)

Pineda-Villavicencio, G., Ugon, J., and Yost, D. 2022. Minimum number of edges of d-polytopes with $2d+2$ vertices. *Electron. J. Comb.*, **29**(3), P3.18. (168, 350, 391, 399)

Pisanski, T., and Žitnik, A. 2009. Representing graphs and maps. Pages 151–180 of: Gross, J., and Tucker, T. (authors) and Beineke, L., and Wilson, R. (eds), *Topics in topological graph theory*. Cambridge: Cambridge University Press. (207)

Poincaré, H. 1893. Sur la généralisation d'un théorème d'Euler relatif aux polyèdres. *C. R. Acad. Sci., Paris*, **117**, 144–145. (128)

Preparata, F. P., and Shamos, M. I. 1985. *Computational geometry: An introduction*. Berlin, Heidelberg: Springer-Verlag. (153)

Przesławski, K., and Yost, D. 2008. Decomposability of polytopes. *Discrete Comput. Geom.*, **39**(1–3), 460–468. (303, 304)

Przesławski, K., and Yost, D. 2016. More indecomposable polytopes. *Extr. Math.*, **31**, 169–188. (290, 295, 298, 303, 304)

Radon, J. 1921. Mengen konvexer Körper, die einen gemeinsamen Punkt enthalten. *Math. Ann.*, **83**, 113–115. (20, 42)

Reid, M., and Szendroi, B. 2005. *Geometry and topology*. Cambridge: Cambridge University Press. (42)

Ribó Mor, A., Rote, G., and Schulz, A. 2011. Small grid embeddings of 3-polytopes. *Discrete Comput. Geom.*, **45**(1), 65–87. (211)

Richter-Gebert, J. 2006. *Realization spaces of polytopes*. Berlin: Springer. (64, 197, 211, 228, 230, 395)

Robertson, N., and Seymour, P. D. 1995. Graph minors. XIII. The disjoint paths problem. *J. Comb. Theory, Ser. B*, **63**(1), 65–110. (240)

Roshchina, V., Sang, T., and Yost, D. 2018. Compact convex sets with prescribed facial dimensions. Pages 167–175 of: de Gier, J., Praeger, C. E., and Tao, T. (eds), *2016 MATRIX Annals*. Cham: Springer. (33)

Roth, B. 1981. Rigid and flexible frameworks. *Am. Math. Mon.*, **88**(1), 6–21. (364, 392)

Rowlands, R. 2022. Reconstructing d-manifold subcomplexes of cubes from their $(d/2+1)$-skeletons. *Discrete Comput. Geom.*, **67**, 492–502. (263)

Sachs, H. 1994. Coin graphs, polyhedra, and conformal mapping. *Discrete Math.*, **134**(1), 133–138. (212, 230)

Sallee, G. T. 1967. Incidence graphs of convex polytopes. *J. Combinatorial Theory*, **2**, 466–506. (161, 166, 229, 238, 251, 252)

Santos, F. 2012. A counterexample to the Hirsch conjecture. *Ann. Math.*, **176**(1), 383–412. (107, 108, 153, 305, 306, 328, 329, 331, 332, 339)

Santos, F. 2013. Recent progress on the combinatorial diameter of polytopes and simplicial complexes. *Top*, **21**(3), 426–460. (335, 336, 337, 338, 339)

Santos, F., Stephen, T., and Thomas, H. 2012. Embedding a pair of graphs in a surface, and the width of 4-dimensional prismatoids. *Discrete Comput. Geom.*, **47**(3), 569–576. (331)

Sanyal, R., and Ziegler, G. M. 2010. Construction and analysis of projected deformed products. *Discrete Comput. Geom.*, **43**(2), 412–435. (263)

Sarkaria, K. S. 1991. Kuratowski complexes. *Topology*, **30**(1), 67–76. (225)

Schild, G. 1993. Some minimal nonembeddable complexes. *Topology Appl.*, **53**(2), 177–185. (225)

Schläfli, L. (1901) 1850–52. *Theorie der vielfachen Kontinuität*. Vol. 38. Basel: Birkhäuser. (128)

Schlegel, V. 1883. *Theorie der homogen zusammengesetzten Raumgebilde* (On the theory of homogeneous composed space forms). Nova Acta, **44**, 343–459. (153)

Schneider, R. 2014. *Convex bodies: The Brunn-Minkowski theory*. 2nd ed. Vol. 151. Cambridge: Cambridge University Press. (278, 303)

Schramm, O. 1992. How to cage an egg. *Invent. Math.*, **107**(3), 543–560. (230)

Schrijver, A. 1986. *Theory of linear and integer programming*. New York: John Wiley & Sons. (152)

Seymour, P. D. 1980. Disjoint paths in graphs. *Discrete Math.*, **29**(3), 293–309. (240)

Shemer, I. 1982. Neighborly polytopes. *Isr. J. Math.*, **43**(4), 291–314. (263)

Shephard, G. C. 1963. Decomposable convex polyhedra. *Mathematika*, **10**, 89–95. (278, 293, 300, 304)

Shifrin, T., and Adams, M. 2011. *Linear algebra: A geometric approach*. 2nd ed. New York: W. H. Freeman. (42)

Silverman, R. 1973. Decomposition of plane convex sets. I. *Pac. J. Math.*, **47**, 521–530. (278, 289)

Smilansky, Z. 1986a. *Decomposability of polytopes and polyhedra*. Ph.D. thesis, Hebrew University of Jerusalem. (297, 301)

Smilansky, Z. 1986b. An indecomposable polytope all of whose facets are decomposable. *Mathematika*, **33**(2), 192–196. (301, 303, 397)

Smilansky, Z. 1987. Decomposability of polytopes and polyhedra. *Geom. Dedicata*, **24**(1), 29–49. (292, 300, 301, 302, 304, 397)

Smilansky, Z. 1990. A nongeometric shelling of a 3-polytope. *Isr. J. Math.*, **71**(1), 29–32. (175)

Soltan, V. 2015. *Lectures on convex sets*. Hackensack: World Scientific. (42)

Sommerville, D. M. Y. 1927. The relations connecting the angle-sums and volume of a polytope in space of n dimensions. *Proc. R. Soc. Lond., A*, **115**(770), 103–119. (131)

Sommerville, D. M. Y. 1958. *An introduction to the geometry of n dimensions*. New York: Dover Publications. (153)

Spielman, D. A. 2019. *Spectral and algebraic graph theory*. http://cs-www.cs.yale.edu/homes/spielman/sagt/sagt.pdf. (194, 230)

Stanley, R. P. 1975. The upper bound conjecture and Cohen-Macaulay rings. *Stud. Appl. Math.*, **54**, 135–142. (393)

Stanley, R. P. 1980. The number of faces of a simplicial convex polytope. *Adv. Math.*, **35**, 236–238. (378, 393)

Stanley, R. P. 1996. *Combinatorics and commutative algebra*. 2nd ed. Progress in Mathematics, vol. 41. Boston, MA: Birkhäuser. (393)

Steinitz, E. 1906. Über die *Euler*schen Polyederrelationen. *Arch. der Math. u. Phys.*, **11**, 86–88. (340, 341)

Steinitz, E. 1922. Polyeder und raumeinteilungen. *Encyclop. d. math. Wiss.*, **3**, 1–139. (198, 211)

Sturmfels, B. 1987. Cyclic polytopes and d-order curves. *Geom. Dedicata*, **24**(1), 103–107. (100)

Sturmfels, B. 1988. Some applications of affine Gale diagrams to polytopes with few vertices. *SIAM J. Discrete Math.*, **1**(1), 121–133. (145, 212)

Sturmfels, B. 1996. *Gröbner bases and convex polytopes*. Vol. 8. Providence, RI: American Mathematical Society. (110)

Sukegawa, N. 2019. An asymptotically improved upper bound on the diameter of polyhedra. *Discrete Comput. Geom.*, **62**(3), 690–699. (323)

Tay, T.-S. 1995. Lower-bound theorems for pseudomanifolds. *Discrete Comput. Geom.*, **13**(2), 203–216. (374)

Thomas, R., and Wollan, P. 2005. An improved linear edge bound for graph linkages. *Eur. J. Comb.*, **26**(3–4), 309–324. (249, 255)

Thomas, R., and Wollan, P. 2008. The extremal function for 3-linked graphs. *J. Comb. Theory, Ser. B*, **98**(5), 939–971. (254, 396)

Thomassen, C. 1980a. 2-linked graphs. *Eur. J. Comb.*, **1**(4), 371–378. (240, 249, 250, 254, 396, 412, 427)

Thomassen, C. 1980b. Planarity and duality of finite and infinite graphs. *J. Comb. Theory, Ser. B*, **29**(2), 244–271. (423)

Thomassen, C. 1981. Nonseparating cycles in k-connected graphs. *J. Graph Theory*, **5**, 351–354. (426)

Thurston, W. P. 1980. *The geometry and topology of three-manifolds*. Lecture notes at Princeton University, http://library.msri.org/books/gt3m/PDF/Thurston-gt3m.pdf. (212)

Timan, A. F. 1963. *Theory of approximation of functions of a real variable*. New York: Pergamon Press. (100)

Todd, M. J. 1980. The monotonic bounded Hirsch conjecture is false for dimension at least 4. *Math. Oper. Res.*, **5**, 599–601. (307, 309, 310)

Todd, M. J. 2002. The many facets of linear programming. *Math. Program.*, **91**(3), 417–436. (306)

Todd, M. J. 2014. An improved Kalai-Kleitman bound for the diameter of a polyhedron,. *SIAM J. Discrete Math.*, **28**, 1944–1947. (319, 320, 322, 339)

Tutte, W. T. 1963. How to draw a graph. *Proc. Lond. Math. Soc.*, **13**, 743–768. (155, 164, 165, 182, 184, 186, 214, 230)

Tverberg, H. 1966. A generalization of Radon's theorem. *J. Lond. Math. Soc.*, **41**, 123–128. (42)

Van Kampen, E. R. 1932. Komplexe in euklidischen Räumen. *Abh. Math. Semin. Univ. Hamb.*, **9**, 72–78. (155, 225)

Wagner, K. 1937. Über eine Eigenschaft der ebenen Komplexe. *Math. Ann.*, **114**, 570–590. (423)

Walkup, D. W. 1970. The lower bound conjecture for 3- and 4-manifolds. *Acta Math.*, **125**, 75–107. (372)

Webster, R. 1994. *Convexity*. New York: Oxford University Press. (xii, 3, 20, 21, 23, 24, 25, 27, 28, 32, 35, 37, 38, 42, 130, 140, 141, 144, 152, 153, 154, 229, 402)

Werner, A., and Wotzlaw, R. F. 2011. On linkages in polytope graphs. *Adv. Geom.*, **11**(3), 411–427. (231, 245, 249, 254, 255, 396)

West, D. B. 2001. *Introduction to graph theory*. 2nd ed. Upper Saddle River, NJ: Prentice Hall. (414, 427)

Weyl, H. 1935. Elementare theorie der konvexen polyeder. *Comment. Math. Helv.*, **7**, 290–306. (152)

Whiteley, W. 1984. Infinitesimally rigid polyhedra. I. Statics of frameworks. *Trans. Amer. Math. Soc.*, **285**(2), 431–465. (367, 370)

Whiteley, W. 1992. *Matroids and rigid structures*. Pages 1–53 of: White, N. (ed), Matroid Applications (Encyclopedia of Mathematics and its Applications). Cambridge: Cambridge University Press. (392)

Whiteley, W. 1996. Some matroids from discrete applied geometry. Pages 171–311 of: *Matroid theory (Seattle, WA, 1995)*. Contemp. Math., vol. 197. Providence, RI: Amer. Math. Soc. (388)

Whitney, H. 1932. Congruent graphs and the connectivity of graphs. *Am. J. Math.*, **54**(1), 150–168. (414, 417, 426)

Whitney, H. 1932. Non-separable and planar graphs. *Trans. Am. Math. Soc.*, **34**, 339–362. (403, 423, 424)

Whitney, H. 1933. 2-isomorphic graphs. *Am. J. Math.*, **55**, 245–254. (164)

Williamson Hoke, K. 1988. Completely unimodal numberings of a simple polytope. *Discrete Appl. Math.*, **20**(1), 69–81. (226)

Wotzlaw, R. F. 2009. *Incidence graphs and unneighborly polytopes*. Ph.D. thesis, Technical University of Berlin. (230, 252)

Xue, L. 2021. A proof of Grünbaum's lower bound conjecture for general polytopes. *Isr. J. Math.*, **245**, 991–1000. (340, 344, 392)

Xue, L. 2022. *A lower bound theorem for strongly regular CW spheres up to* $2d + 1$. arXiv:2207.13839. (350, 388)

Yaglom, I. M. 1973. *Geometric transformations III: Affine and projective transformations*. New York: Random House. (153)

Yang, Y. 2021. A note on the diameter of convex polytope. *Discrete Appl. Math.*, **289**, 534–538. (335)

Yost, D. 2007. Some indecomposable polyhedra. *Optimization*, **56**(5–6), 715–724. (303)

Ziegler, G. M. 1995. *Lectures on polytopes*. Graduate Texts in Mathematics, vol. 152. New York: Springer-Verlag. (xi, 46, 50, 74, 75, 102, 114, 123, 134, 136, 149, 152, 153, 154, 174, 212, 229, 303, 331, 338, 393, 394)

Ziegler, G. M. 2000. Lectures on 0/1-polytopes. Pages 1–41 of: Kalai, G., Ziegler, G.M. (eds) *Polytopes—Combinatorics and computation*. DMV Sem., vol. 29. Basel: Birkhäuser. (333)

Ziegler, G. M. 2007. Convex polytopes: extremal constructions and f-vector shapes. Pages 617–691 of: Miller, E., Reiner, V., and Sturmfels, B. (eds), *Geometric combinatorics*. IAS/Park City Math. Ser., vol. 13. Providence, RI: Amer. Math. Soc. (211)

Index of Symbols

Index

Printed in the United States
by Baker & Taylor Publisher Services